BIOMEDICAL PARTICLE ACCELERATORS

BIOMEDICAL PARTICLE ACCELERATORS

Waldemar H. Scharf
Institute of Radioelectronics
Warsaw Technical University
Warsaw, Poland

Translated from the Polish manuscript by:
Oskar A. Chomicki

English edition edited by:
Jeffrey V. Siebers

AIP PRESS
American Institute of Physics **New York**

In recognition of the importance of preserving what has been written, it is a policy of the American Institute of Physics, to have books of enduring value published in the United States printed on acid-free paper, and we exert our best efforts to that end.

AIP Press
American Institute of Physics
500 Sunnyside Boulevard
Woodbury, New York 11797-2999

Library of Congress Cataloging-in-Publication Data

Scharf, Waldemar
[Bio-medyczny Akcelerator Czastkowy. English]
Biomedical Particle accelerators/Waldemar H. Scharf.
p. cm.
"Translated from the Polish Manuscript by Oskar A. Chomicki; English ed. edited by Jeffrey V. Siebers."
Includes bibliographical references and index.
ISBN 1-56396-089-3
1. Particle accelerators. 2. Medical physics. 3. Nuclear medicine. I. Siebers, Jeffrey Vincent. II. Title.
[DNLM: 1. Particle Accelerators. QC 787.P3 S311b]. R920.S37 1993
615.8'4201'53973--dc20
DNLM/DLC
for Library of Congress 92-49010
CIP

This book is dedicated to my
youngest twin grandchildren -
Alexandra and Peter

Contents

CHAPTER 3 Electron Accelerators for Routine Radiotherapy

CHAPTER 4 Accelerator Laboratories for Routine Therapy

CHAPTER 5 Accelerators for Nonconventional Therapy

Preface

For the past 60 years the development of charged-particle accelerators has been rapid and uninterrupted.

At the early stage of accelerator development in the 1930s, the potential of using accelerators in radiation therapy was already noted, and the rapid increase in the number of cancer cases—in the USA alone estimated at that time at 100,000 yearly—constituted an additional and strong impetus to the use of new and improved machines in the treatment of tumors. As a result, over the past 50 years, about 3000 radiotherapy accelerators have been installed, treating millions of patients. Several hundred other machines have been used in radiobiological and medical research, opening up new prospects for subtle analytical and diagnostic techniques. Consequently, biomedical accelerators are now the largest group of charged-particle machines. They have become indispensable tools in the hands of medical physicists and physicians in many hospitals and medical facilities. It is to this group of accelerators that the present book, published by AIP Press, is devoted.

After presenting in Chapter 1 the fundamental problems of biomedical applications of accelerators, Chapter 2 provides the fundamental techniques of particle acceleration with emphasis placed on biomedical uses. Chapter 3 deals with electron accelerators, especially rf linacs, used in routine radiotherapy. Machines employed in intraoperational irradiations and in radiosurgery, as well as in total body irradiations are also discussed. Such problems as dosimetry, therapy planning, and management, together with the description of the relevant instrumentation, are presented in Chapter 4. Chapter 5 gives a summary of the state-of-the-art accelerators designed for prospective unconventional therapies using protons, heavy ions, mesons, and neutrons.

Chapter 6 is devoted to synchrotron radiation used in biomedical analyses and in angiography. Production of radionuclides, which is now increasingly more often a domain of accelerators, is described in Chapter 7. Accelerator processing techniques, of interest in sterilization and food preservation as well as in environmental protection, is discussed in Chapter 8. Finally, Chapter 9 provides a description of accelerator analytical techniques, especially the methods of microbeams, the particle-induced characteristic x-ray emission, and of elastic scattering, activation, and accelerator mass spectrometry. The bibliography has been brought to the end of 1991. The book also contains a list of firms (denoted [F1], [F2], ...) manufacturing biomedical accelerators and auxiliary equipment.

My first book on accelerators, entitled "Particle Accelerators and Their Uses" (Harwood Academic Publishers, New York, 1986) presented the history and development of accelerators. Specific applications of accelerators in technology and research have been elaborated in more detail in my second book "Particle Accelerators—Applications in Technology and Research" (Research Studies Press Ltd. and John Wiley and Sons, Tauton, New York, 1989).

The present book owes a great deal to Dr. F. T. Cole (Fermi National Accelerator Laboratory, USA) who has shown great interest and offered help in promoting my books on accelerators.

The author also wants to express his thanks to Professor L. H. Lanzl (Department of Medical Physics, Rush University, USA) for his valuable suggestions concerning Chapter 1. Dr. D. T. L. Jones (National Accelerator Centre, Faure, Rep. of South Africa) has kindly reviewed Chapter 5, and Professor S. V. Heselius (Accelerator Laboratory Åbo Akademi, Finland) offered his remarks on Chapter 7. Special thanks are due to my friend, Oskar A. Chomicki, who has put in a lot of work in rendering my Polish text into English and in preparing the whole text for publication. The book could not have been prepared so quickly and efficiently without the help of Mrs. Ewa Piliszek who put the text into the computer, and the great effort of Mr. A. J. Pruszyński's Publishing House "KOMPAZ" of Warsaw, Poland, which has given it its present form. Last, but not least, the Editor Dr. J. V. Siebers (Loma Linda University Medical Center, USA) and the Publishers, AIP Press, deserve the author's greatest appreciation for their assistance in publishing the book in its excellent form.

Warsaw, Poland *Waldemar H. Scharf*
August 1993

CHAPTER 1

Introduction

1.1 Accelerators in Radiotherapy

> *"If the trend of recent years were to continue, by the year 2000, one European in three would be affected by cancer at some time in his or her life, compared with one in four today" [1.1]*

1.1.1 Chronology of Development

Only a few months after the discovery by W.K.Roentgen in 1895 of a type of radiation, later called *roentgen rays* after its discoverer, or *X-radiation*, first attempts were made in Germany and the USA to use this radiation in the treatment for malignant tumors in patients. The earliest radiation sources were provided by *gas filled X-ray tubes*. It was W.D.Coolidge who, in 1913, did the groundwork for the present-day X-ray techniques by developing a *vacuum tube* with a hot tungsten cathode. The earliest tubes of this type, manufactured for general use, operated at a peak voltage of 140 kV with a 5 mA current [1.2]. By 1937, X-ray tubes used in radiation therapy operated at 400 kV with 5 mA of current and tubes operating at 200 kV had currents of up to 30 mA.

X-rays generated by these tubes were, unfortunately, fairly soft (Fig. 1.1.a) and from the medical point of view, the depth–dose curves were particularly disadvantageous since the maximum dose would be delivered at the skin surface and then would rapidly fall off with depth in tissue.

For this reason, during the early experience with vacuum X-ray tubes, first attempts were also made at searching for sources of a more penetrating radiation with attention focused

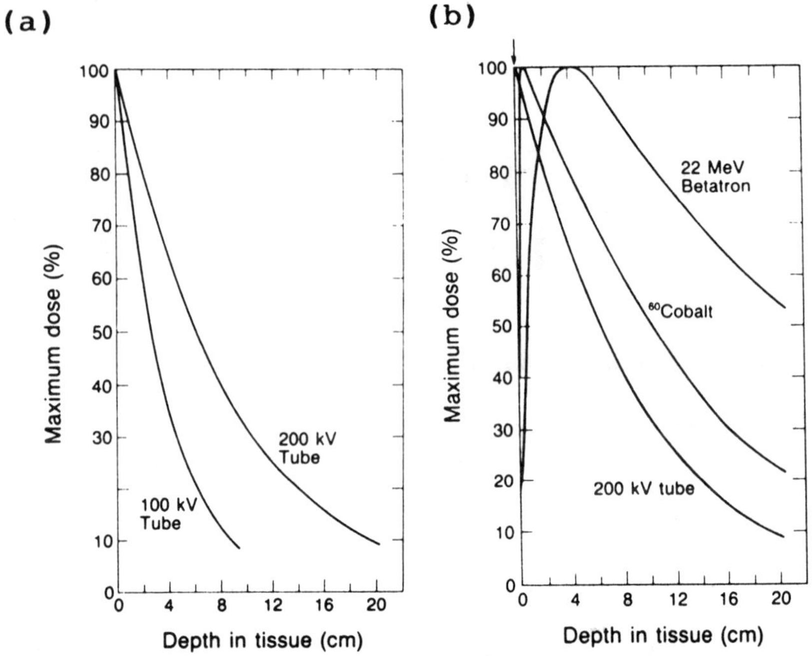

FIGURE 1.1 Depth-dose curves for beams from: (a) — 100 and 200 kV X-ray radiation, (b) — 200 kV X-ray and Co^{60} gamma radiation (1.17 and 1.33 MeV) and 22 MeV betatron X-rays (after W.M. Saunders [1.2]).

on *gamma emitting radionuclides.* The first device using this harder radiation was probably a *radium cannon* developed by Koenig at a Women's Hospital in Breslau (now Wroclaw) in 1912 for the treatment of pelvic cancer [1.3]. The first radium cannons, utilizing a few grams of radium-226, a radionuclide that emits gamma rays with energies ranging between 0.24 and 2.20 MeV, were installed at only some dozen or so places. The main obstacle to a wider use of these devices was their low availability and the high cost of radium. Therefore, as early as 1935, when it was generally accepted that the future developments in radiotherapy would involve megavoltage therapy, Dr K.Schmitz (USA) made the significant statement that

"Radium cannons containing four grams of radium ... require 30 hours to treat one patient. About 90 000 patients require some form of distant radiation (per year) ... Therefore, since about 200 patients can be treated with a 4 gram cannon per year, about 450 cannons of 4 grams each of radium would be required. This means that 1800 grams of radium at about $ 50 000 per gram would have to be procured to enable the medical profession to treat all cancer patients adequately with radium at a cost of 90 million dollars.

"Modern installations for roentgen therapy, containing 1000 kilovolt transformers can take care of one patient within 4 hours or about 750 patients per year. Hence, 120 roentgen stations would be necessary ... at an average investment of 60 thousand dollars per station or a total of 7 and a half million dollars.

"The amount of radium needed is not available and probably never will be. Transformers can be built in quantity and could soon be supplied..." [1.3].

Although future progress in megavolt therapy turned out not to be based on transformers, Dr. Schmitz's words were no doubt prophetic.

In the early half of the 1950s the era of cobalt sources for radiotherapy, containing several kilocuries of ^{60}Co, began. Cobalt-60, emitting gamma rays with energies of 1.17 MeV and 1.33 MeV, made it possible to obtain far better depth-dose curves than those for the X-radiation emitted by an X-ray tube (Fig.1.1b). The ^{60}Co depth dose curve had its maximum at about 5mm below the skin surface, thus markedly decreasing the dose to the skin when deeper layers had to be irradiated. The first three ^{60}Co sources were produced in 1952 and installed in Canada and the USA (Table 1.1).

In the early 1930s several attempts were made to use *resonance transformers* in order to attain peak voltages of 1 to 2 MV. However, accelerators equipped with transformers of this kind did not find wide application in radiotherapy as greater attention was being focused at that time on a high voltage accelerator developed in 1932 by R.Van de Graaff and later named the *Van de Graaff machine* after its inventor. The first hospital-based accelerator of this type was a 1 MeV air-insulated machine installed in Boston in 1937 (Table 1.1). The second machine, a pressure-insulated 1.25 MeV version, was installed in 1940 at the Massachusetts General Hospital also in Boston. The real era of Van de Graaff accelerators began in

TABLE 1.1 Chronology of the Development of Radiotherapy and Nuclear Medicine, with Special Consideration of Accelerator Techniques.

Year	Event
1895	K.Roentgen discovers X-rays
1913	W.E.Coolidge develops vacuum X-ray tube
1931	E.O.Lawrence develops a cyclotron
1937	1 MV Air-insulated Van de Graaff accelerator installed at Huntington Memorial Hospital (Boston)
1938-43	R.Stone treats 250 patients with fast neutrons at the University of California
1939	First medical cyclotron, Crocker (USA)
1940	1.25 MV pressure Van de Graaff accelerator at Massachusetts General Hospital
1946	Medical radioisotopes commercially available (USA)
1946	Wilson suggests medical use of protons and heavy ions
1946	20 MeV electron beam for cancer treatment developed using a betatron (Urbana, USA)
1949	First patient treated with 20 MV X-rays from betatron at the University of Illinois (Urbana, USA)
1951	First European radiotherapy betatron put in operation at the Cantonal Hospital in Zürich (Switzerland)
1951	Cassen develops a mechanical radioisotope scanner
1952	First three cobalt units produced and installed at Canadian facilities in Saskatoon and Ottawa (Canada), and at Oak Ridge (USA)
1952	First medical 8 MeV rf linear accelerator installed at Hammersmith Hospital in London
1955	Medical cyclotron MRC installed at Hammersmith Hospital in London
1956	First American 6 MeV linear accelerator to generate a 4 MV beam put in operation at Stanford (USA)
1957	First medical application of a proton beam in Sweden
1958	Anger demonstrates his scinticamera
1959	First scanning electron beam (5 to 50 MeV) from a linear accelerator developed for cancer therapy (Chicago, USA)
1962	Isocentric rf linear accelerator installed in the U.S.
1965	Resumption of neutron therapy studies at Hammersmith Hospital in London
1973	Positron camera (PET)
1976	First irradiation with a pion beam in an rf linear accelerator LAMPF (USA)
1982	First cyclotron used for neutron rotation therapy in the U.S.
1990	First superconducting cyclotron for neutron therapy (Harper, USA)
1990	First hospital based proton synchrotron used for radiotherapy (Loma Linda, USA)

1946, when the High Voltage Engineering Corporation was formed by R. Van de Graaff and began commercial production of 2 and 2.5 MeV machines. A total of 40 accelerators were built, but their production was discontinued in 1959. They were of quite great size and rather unmaneuverable, occupied a large hospital area, and involved high irradiation costs. In spite of that, they turned out to be very reliable and at least 12 of them were still in operation as late as 1983 [1.4].

The era of betatrons. Dramatic increase in the photon radiation energy was made possible by the development of a new accelerator, *the betatron*, by D.W.Kerst, who, in 1943, suggested its use as a radiation therapy tool [1.5]. The first patient was irradiated in 1949 with X-rays generated by 20 MeV electrons from a Kerst betatron installed in Urbana (USA). Later, the Allis-Chalmers company started commercial production of 20 - 25 MV betatrons for radiotherapy. In 1977, 45 betatrons were in operation in the USA alone [1.4].

In Europe, the earliest work on the medical application of betatrons was started in 1948 in Göttingen (6 MV Gund betatron [1.6]) and the first 31 MeV betatron, produced by the Brown-Boveri company and adapted for radiation therapy, was put in operation in 1951 at the Cantonal Hospital in Zürich (Switzerland) [1.6]. Through 1966, this company manufactured more than 40 betatrons (Fig. 1.2). The Siemens company (W.Germany) produced more than 70 betatrons in the same period, and in the early 1970s about 200 betatrons were in medical use the world over [1.7].

Betatrons have played a significant role in the development of radiation therapy since they deliver X-rays with far better properties, such as skin sparing, higher depth dose and less side scatter than those of the radiation generated by X-ray tubes and radioisotope sources. Fig. 1.3 shows a depth-dose distribution normalized at 12 cm for X-radiation from a 25 MV betatron and the beams generated by a ^{60}Co source and a 200 kV X-ray tube. It can easily be seen that, for the same absorbed dose at the depth of 12 cm, the dose to the skin and the subcutaneous layers is lowest with the betatron.

The betatron's major disadvantages, such as its weight which makes this machine unwieldy, the relatively low intensity of the X-ray beam, as well as the small treatment field area have contributed to the discontinuation of its production in the mid-

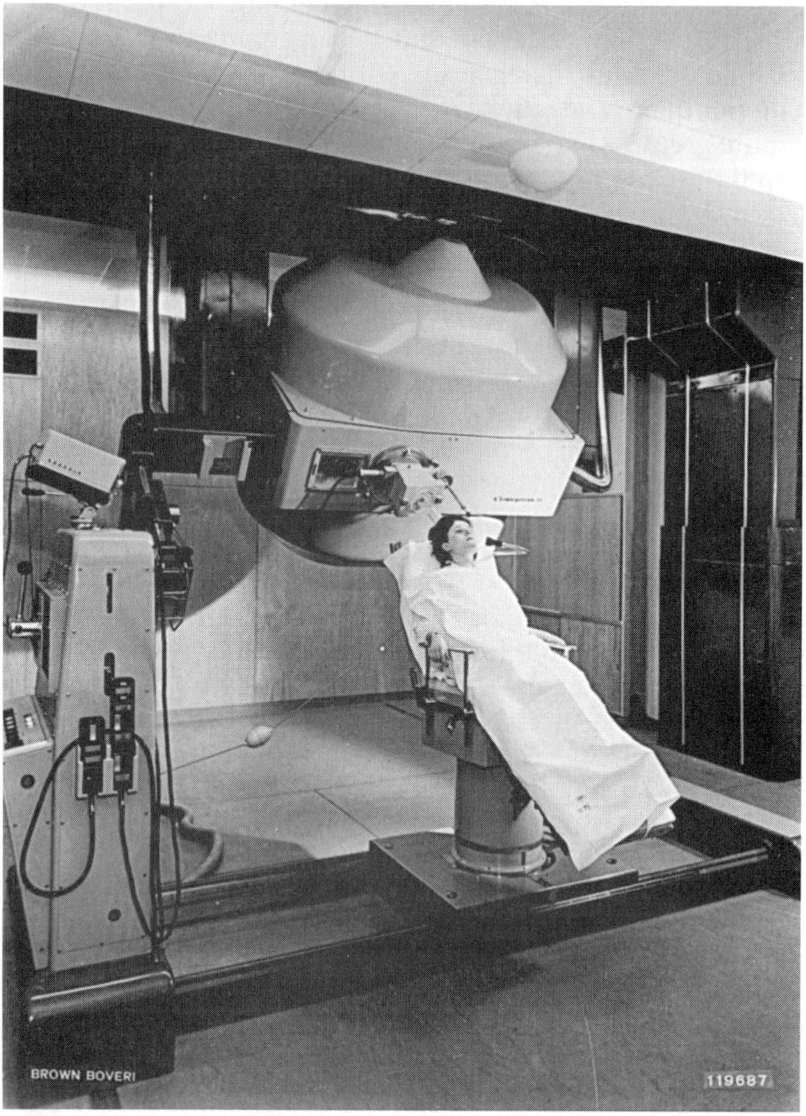

FIGURE 1.2 View of a medical betatron mfd. by Brown Boveri (Switzerland)

1970s. Since then, the use of betatrons has shown a gradual decline. In 1977, in the USA there were 45 betatrons in use, while in 1982 their number dropped to 32 [1.4].

The era of rf linear accelerators. The advances made during World War II in the construction of magnetrons for radar techniques have made it possible to use *microwave generators* for electron acceleration in radiation therapy [1.8, 1.9]. In England, the first *traveling-wave linear* accelerator was demonstrated by D.W.Fry in 1946 [1.10]. On this basis, C.W.Miller, in cooperation with Metropolitan Vickers Ltd. and the Atomic Energy Research Establishment at Harwell (England), designed the first British stationary *rf linear accelerator*, which was subsequently installed at Hammersmith Hospital in 1952 [1.11]. The first patient was treated on this machine in 1953, and the last one in 1969 [1.9].

In the same period, intensive work began in the USA, where H.S.Kaplan at the Stanford Medical Center, in cooperation with a team from the Stanford High Energy Physics Laboratory

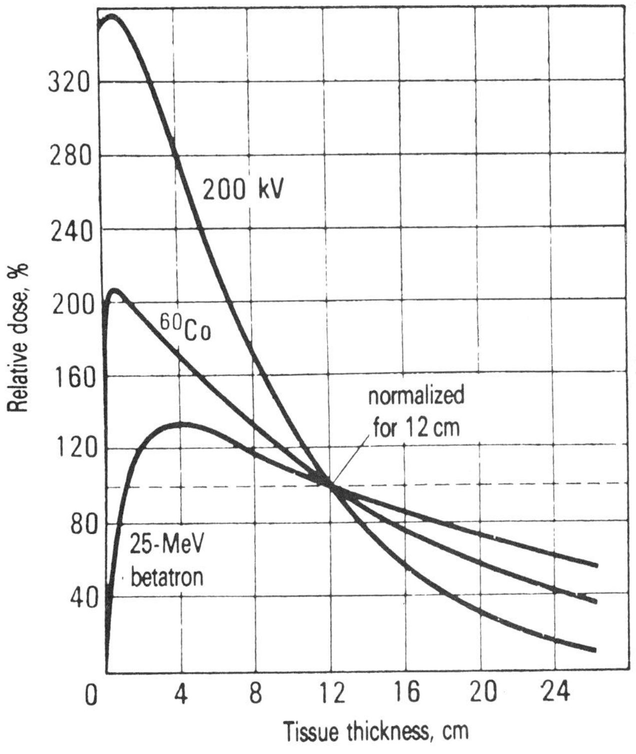

FIGURE 1.3 Depth-dose distributions in tissue for 250 kV X-ray, Co-60 gamma and 25 MV betatron X-ray radiations (doses are normalized at 12 cm depth point).

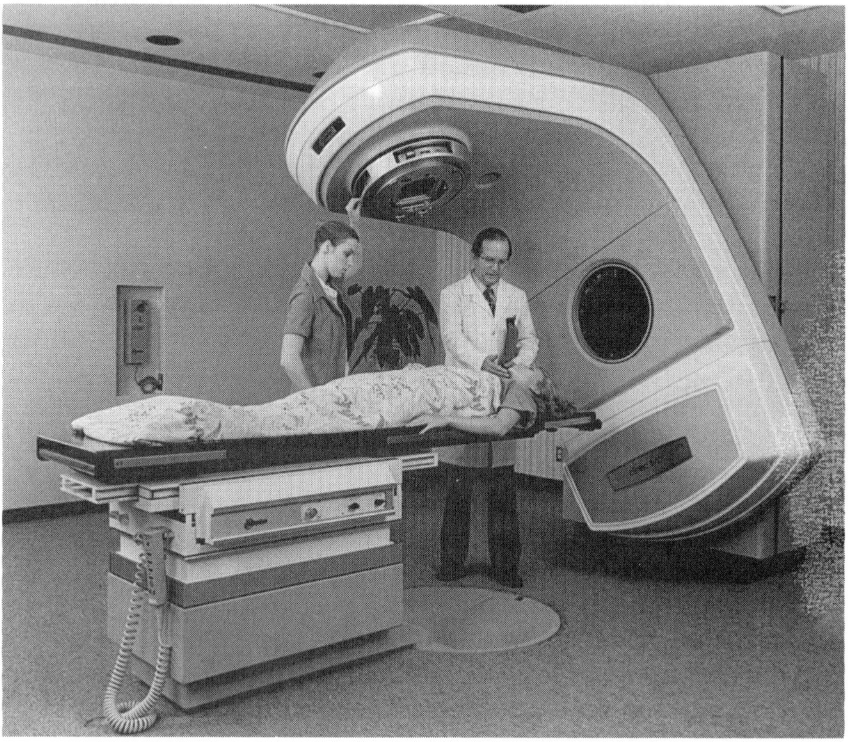

FIGURE 1.4 View of an rf linear accelerator mfd. by Varian Associates (USA).

(E.L.Ginzton, M.A.Bagshaw, W.W.Hansen and others [1.8, 1.12 and 1.13]), built a prototype linear accelerator machine for X-ray therapy at an energy of 5 MeV. It was installed at Stanford in 1955, and treatment was started in early 1956. Soon after, a clinical linear accelerator construction program was implemented [1.8] and Varian Associates designed and built a prototype isocentric accelerator which permitted a full rotation around the patient. This machine was installed at the UCLA Medical Center in 1962, and the first commercial machine was set up at the Stanford University School of Medicine (Palo Alto). Fig. 1.4 is a view of a modern medical accelerator manufactured by Varian Associates.

Thus, in radiation therapy the mid-1960s saw the beginning of the era of rf linear accelerators, which were soon to take up a dominant place on the world market of medical accelerators. At that time, the limitations of betatron technology became more evident. These were, above all, the low dose rate, 40 cGy/min (40 rad/min) at 100 cm treatment distance, and energy of 22 MeV as well as small treatment field areas which, at 100 cm, were 12.5x12.5 cm and 8 cm max. diameter for X-ray and electron therapy, respectively. The principal disadvantage of betatrons was that they could not be rotated a full 360° and thus were limited to arc therapy only. It is generally agreed that it was this fact, together with the availability of high dose rates, that contributed to the rapid growth in the number of rf linear accelerators and their dominant role among world's radiotherapy machines. In fact, linear accelerators can deliver dose rates of 200–400 cGy/min (200 to 400 rad/min), with field capabilities of up to 40x40 cm.

What is particularly worth noting is the extent to which rf linear accelerators have made inroads into the radiotherapy practice in the USA, which represents over 25% of the world capacity of megavoltage radiotherapy units [1.4]. In 1968, 79 linear accelerators had been installed in the USA, and in 1972 their number increased to 193. In 1968, linear accelerators were used in the treatment of some 300 thousand patients, who received about 3.5 million irradiations at the cost of 300 million dollars [1.14, 1.15 and 1.18].

More than one-half of the cancer patients in the United States receive curative or palliative radiation treatments. The number of new patients treated by radiation has been increasing at the rate of 2–4% per year during the past 10 years, and today, approximately 500,000 patients are being treated with radiation in the United States annually [1.60]. These patients receive their radiation treatment at various types of facilities with different administrative structures (government, hospitals, free-standing, Table 1.2). Table 1.2 shows that the number of all facilities increased by 11% during 1975–1986, whereas the number of patients treated annually increased by 44% in the same time period.

The availability of high energy treatment machines during 1978–1986 is also shown in Table 1.2. This shows a strong trend of increased use of accelerators compared with Co-60 units. In

TABLE 1.2 Type of Radiation Oncology Facilities and Availability of Radiotherapy Machines in the U.S.A. (after J.C.H. Chu et al. [1.60, 1.61]).

	Year				Change 1975–1986
	1975	1980	1983	1986	
	Type of facilities				
Government	43 (4%)	36 (3%)	40 (4%)	38 (3%)	–1%
Hospital	838 (81%)	868 (80%)	875 (81%)	895 (78%)	–3%
Freestanding	155 (15%)	142 (13%)	168 (15%)	217 (19%)	+4%
Unknown	0	37 (4%)	0	0	
Total	1036 (100%)	1083 (100%)	1083 (100%)	1150 (100%)	+11%
New Patients	311,942	377,837	400,494	445,397	+44%
	Availability of machines				
No. Treatment Machines	1345 (100%)	1625 (100%)	1765 (100%)	1980 (100%)	+46%
Linacs + Betatrons	416 (31%)	805 (49%)	1026 (58%)	1298 (66%)	+212%
Co-60 Machines	938 (69%)	820 (51%)	745 (42%)	672 (34%)	-28%
No. Machines withElectrons			385 (22%)	533 (27%)	

1975, the number of Co-60 units comprised nearly 70% of all machines, whereas in 1986 it fell to 34%. During that time, the number of linear accelerators (and betatrons) increased by more than 200%, whereas the number of Co-60 machines decreased by 28%. This change occurred while there was a 46% increase in the total number of radiation treatment machines in the U.S.A. Unfortunately, only 27% of the megavoltage machines are capable of providing electron beam treatments, a clear deficit in the availability of that modality. Between 1975 and 1986, the total number of new patients receiving radiation therapy also increased by 43% (Table 1.2), so that the average number of patients treated on each machine remained approximately the same [1.61].

The US 1986 Blue Book [1.62] "proposed standards of personnel, equipment, facilities and operations, which are necessary for the proper therapeutic use of ionizing radiation". It recommended one treatment machine for every 250 patients, and the national average for 1986 was 225 new patients for every machine, indicating, on average, that there was not any shortage of therapy machines in the U.S.A. Table 1.2 also shows that the number of new patients receiving radiation therapy increases at approximately 4% per year indicating the need for more than 70 new machines per year in addition to the replacement of the existing older units [1.61].

The future of radiation therapy in the U.S.A. remains firm on its own scientific merits. The long sought "magic cure" for cancer is still far in the future and, for the present, radiation therapy plays a crucial role in the treatment and cure of malignancies. The increased use of chemotherapeutic agents and their increased effectiveness have only improved the overall position of radiation therapy as an important link in the chain of cure for patients suffering from a wide variety of malignancies. Cancer, as we known it today, will not be 100% cured in our lifetime, but radiation therapy remains critical to the best cures available today and for the foreseeable future.

A similar trend in therapy is now evident in Japan [1.19]. In 1980, 170 rf linear accelerators were installed in hospitals and clinics, whereas in 1987 their number had grown to 370. Japan's medical establishments have a total of 416 accelerator units: 370 linear accelerators, 35 betatrons, 3 microtrons, and 8 medical cyclotrons. These 416 medical accelerators constituted 60% of all particle accelerators in operation in Japan in 1987 [1.19].

The world's total number of rf linear accelerators used in radiation therapy at present exceeds 3000. The commercial world market for these machines is estimated to have a value of about $ 250 million per year [1.20]. Cancer is a major health problem throughout the world and it is one of the major causes of death both in developing and developed countries. It was estimated in 1975 that about 6 000 000 people were suffering from cancer and 50% of them came from developing countries. In India alone, 1.5 million people were suffering from cancer (1989), and another 0.5 million are added every year [1.63]. These figures are going to double by the year 2000. There are only nine radiotherapy accelerators in operation in India. Now is the time to change the situation by making available more special sophisticated equipment and better planning for treatment of cancer patients in the developing countries. This will be facilitated if the scientific world can develop low-cost depend-

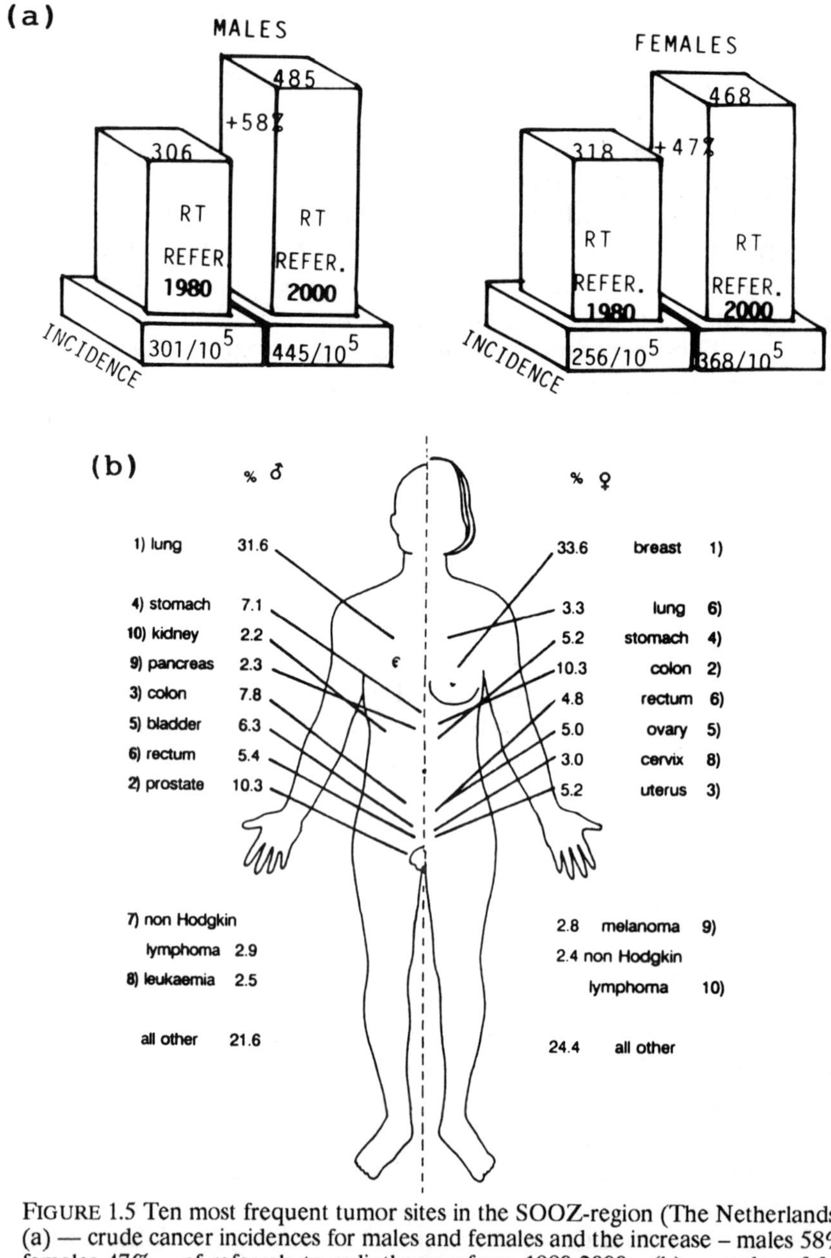

FIGURE 1.5 Ten most frequent tumor sites in the SOOZ-region (The Netherlands), (a) — crude cancer incidences for males and females and the increase – males 58%, females 47% – of referrals to radiotherapy from 1980-2000; (b) — order of frequency in the period 1978-1987 for males and females, excluding non-melanoma skin cancer and carcinoma in situ (after J.W.W. Coebergh and L.H. van der Heijden [1.66]).

able accelerators for such situations. In India, at present one out of every three cancer patients can be cured [1.63].

Cancer Registry. This work is being carried out on a global scale by the International Agency for Research on Cancer [1.64, 1.65]. Since not even all the developed countries can afford to have their whole territories covered by statistical data, areas are selected for complete and accurate cancer registration. In Europe, there are some ten-odd areas, listed in [1.64]. An example of registration of this kind is provided by the Dutch Einhoven Cancer Registry [1.66] (SOOZ area, SOOZ = Cooperative Association of Hospitals in Oncology).

The registry serves an area of about 2,500 km^2, comprising almost 1 million inhabitants or 7% of the Dutch population in 1985, with an average density of 400 per km^2, roughly the national average. Forty-seven percent of the population-at-risk lived in urban, 41% in sub-urbanized and 12% in rural municipalities. The data-base was computerized from 1974 onwards. Completeness could be assumed for most tumors from analyses of referral patterns and registration procedures as well as various comparisons with cancer mortality.

Fig. 1.5 shows, by way of example, some detailed analysis obtained from the above statistical data which are essential in the long-range planning process of radiation oncology services (Fig. 1.5a). From the incidence of the most frequent tumor sites (Fig. 1.5b) the total number and type of accelerators that have to be installed at cancer centers can be determined.

Cancer Programme. One out of every four Europeans will suffer some type of cancer during their lifetime. If nothing is done, this proportion will become one out of every three by the year 2000. The major goal of the *Europe against Cancer Programme* is to diminish the number of cancer casualties compared to the above-mentioned trend, by at least 15% by the year 2000 and by considerably more in the following decade [1.68].

Radiotherapy is the most widely used form of cancer treatment, being given to two out of every three patients. Approximately one quarter of cured cancer patients owe their five-year survival to radiotherapy alone, and another eight are cured by combined treatment with surgery and/or chemotherapy. Radiotherapy is, therefore, involved in almost half of curative cancer treatments. Moreover, radiotherapy is involved in the palliative treatment of many other patients.

Although, in principle, radiation is not selective for a cancer cell, the selective destruction of a tumor is achieved to some extent by the fact that (1) considerable efforts are made to concentrate the dose in the tumor tissue and so spare the surrounding normal tissue as much as possible, (2) some tumor cells are more radiosensitive than normal cells, (3) the reparability of malignant tissues is thought to be less efficient than that of the normal tissue [1.68].

1.1.2 Conventional Accelerator Therapy

The fundamental principles of radiation therapy can be presented briefly in the following set of approaches [1.21]:

Strategy: (1) Deliver a sufficiently high dose to eradicate the tumor, (2) Minimize the dose to the normal tissues, particularly those in the critical organs.

Tactics: (1) Make use of the advantages offered by the physical properties of various radiation beams or internal sources, (2) Make use of the biological advantages offered by the various radiation beams, such as the relative biological effectiveness (RBE), oxygen enhancement ratio (OER), redistribution, etc., (3) Increase the radiosensitivity of the tumor as compared with that of normal tissues by using a radiation sensitizer, (4) Decrease the radiosensitivity of normal tissues as compared with that of the tumor by using radiation protectors, and (5) Optimize the course of treatment by combining radiotherapy with other modalities, such as surgery, chemotherapy, immunotherapy, hyperthermia, etc.

Although the above principles seem to be quite straightforward, their implementation becomes a very complex problem. The curve representing *the probability of curing a tumor* (percent cure) with a dose of ionizing radiation absorbed by the treatment volume has a sigmoidal shape (Fig. 1.6). A similar curve typical for normal tissues may be called — *a complication curve.* The main difficulty is that these two curves are usually very close to each other along the dose scale. This calls for very careful optimization of the treatment planning (Sec.4.2).

On the basis of many years of experimentation and clinical observation, assuming the maximum permissible level of irradiation of normal tissues, empirical procedures have been established, referred to as *dosimetry protocols* (Sec. 4.2), according to

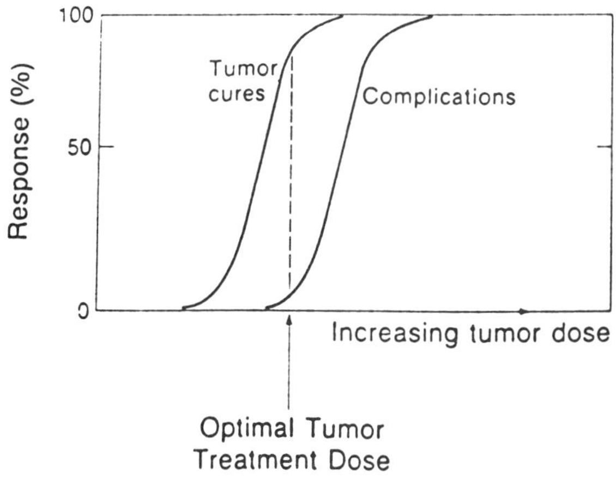

FIGURE 1.6 Dose-response curve for tumor cures (local control) and normal tissue as a function of dose.

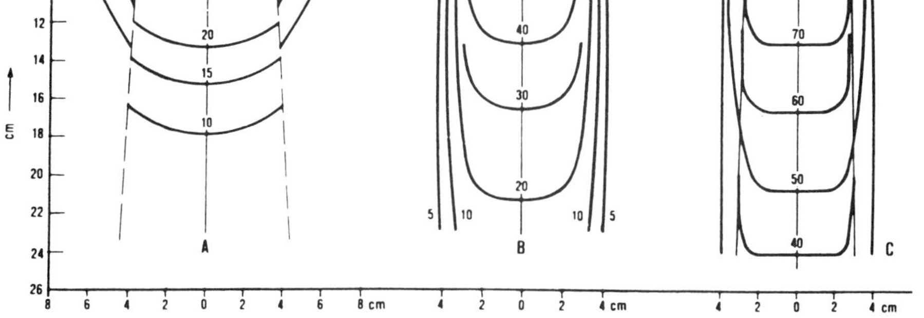

FIGURE 1.7 Isodose distributions: curve A — 200 kV X-rays (1.5 mm Cu filter, field size 15.6 cm^2, target surface distance TSD = 50 cm); curve B — ^{60}Co gamma rays (mean energy = 1.25 MeV, field size 5.5 cm dia., TSD = 80 cm); curve C — 25 MeV X-rays (field size 5 cm dia. TSD = 80 cm) [1.7].

which radiotherapy is administered in *daily fractions*. In a single treatment, the field area is irradiated with a relatively small dose of 1.7 to 2.0 Gy (170 to 200 rad) 5 days a week over 4 – 6 weeks [1.22]. The treatment can be performed in a stationary manner, i.e. the radiation head of the treatment unit is stationary relative

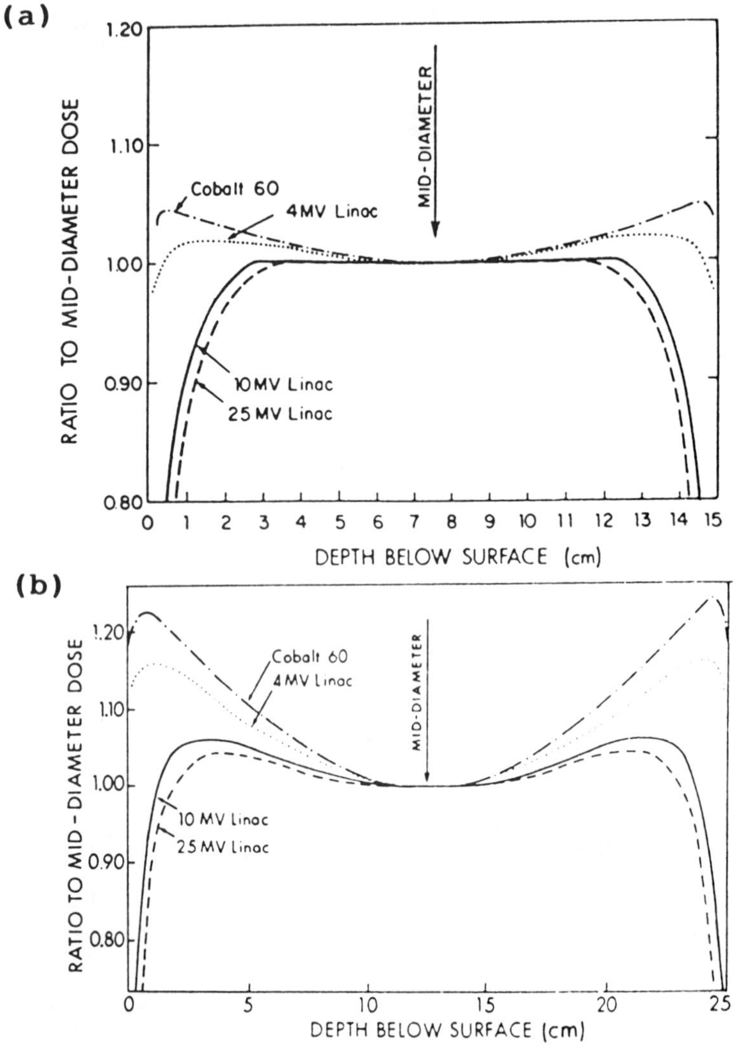

FIGURE 1.8 Comparison of the ratio of deep midline dose to lateral dose for Co-60 gamma rays and 4–25 MV X-ray beam for two treatment field diameters: (a) — 15 cm and (b) — 25 cm.

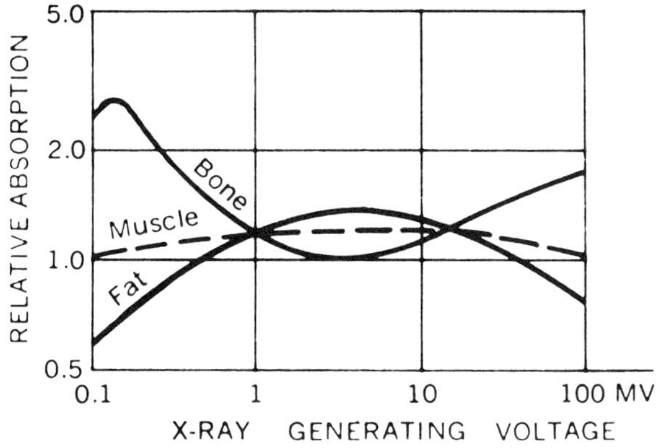

FIGURE 1.9 Relative absorption of ionizing radiation as a function of energy for muscle, bone and fat tissues.

to the patient. In this case single or multiple *treatment fields* are used, whereas, with present-day isocentric-mounted medical accelerators *rotational therapy* may be employed (Fig. 1.4), i.e. the accelerator head rotates through a given angle or a full 360° around the patient (Fig. 4.16). This latter method is derived from the so-called *arc therapy*, in which the radiation head follows a well-defined arc, either in a one-way or an oscillating manner (Fig. 1.14).

Photon therapy. Ionizing-radiation fields produced by radiation sources used in therapy can be represented by isodose curves, which most often are measured in water phantoms. Fig. 1.7 shows curves obtained for various types of radiation: 200 kV X-rays (A), Co-60 gamma-rays (B), and 25 MeV electrons from a betatron, converted to a 25 MV X-ray beam (C). The dose rate at the depth of 10 cm for Co-60 gamma rays is about 50%, whereas 25 MV X-rays produce about 80% of the maximum intensity at this depth.

Fig. 1.8 shows a comparison of the ratio of the deep midline dose to the lateral (subcutaneous) dose for Co-60 gamma rays and 4–25 MV X-ray beams. These curves represent parallel and opposed beams irradiating 15 cm and 25 cm diameter treatment fields: in the head, neck (Fig. 1.8.a) and thorax, and pelvis (Fig. 1.8.b), respectively. In both cases, the higher the energy, the better the uniformity of the dose distribution.

Another advantage is that with increasing X-ray energy, the relative (differential) absorption in various tissues, such as fat, bone, or muscle, is reduced. However, for low energies in the 0.1 to 1.0 MV range, bone receives a far greater dose than does the tumor tissue (Fig. 1.9).

(a)

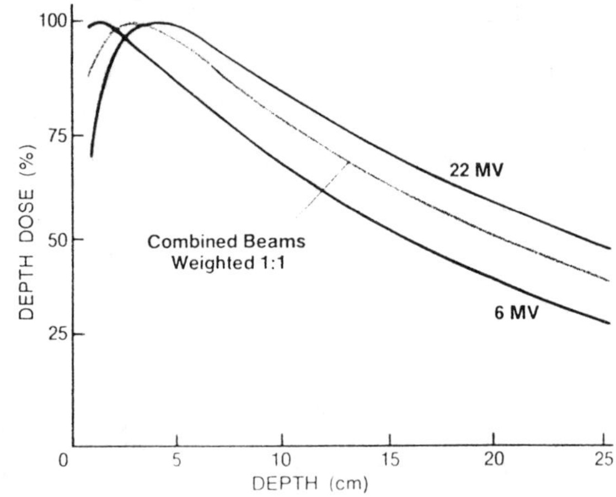

(b)

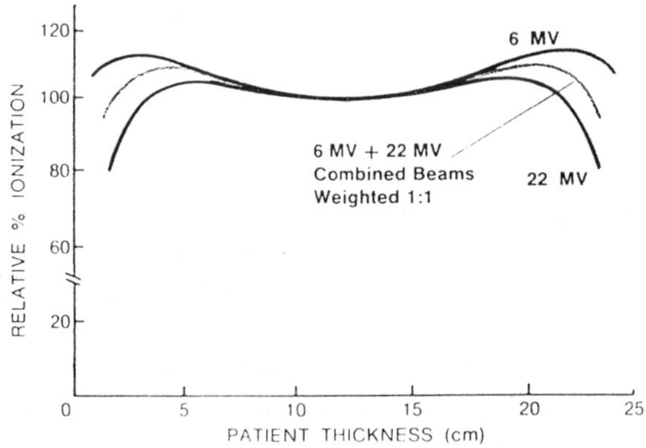

FIGURE 1.10 Dose distributions for a patient treated with 6 MV and 22 MV X-ray radiations and a mixed 6 MV + 22 MV radiation from a Clinac 2500 linear accelerator: (a) — depth-dose curves, (b) — lateral distribution for a 25 cm thick patient.

bearing on the acceleration structure since the particles accelerated in this case have nearly constant velocities. The structure may be homogenous, i.e. be of identical and constant composition.

It also follows from Fig. 2.1. that heavy particles, such as protons and especially heavy ions, behave differently from electrons in the 6-50 MeV energy range. They become relativistic only at energies several orders of magnitude higher than those for electrons. They attain high velocities much less quickly than very light particles, such as electrons, do.

Applications and energy ranges. Applications of various charged particles in biomedical techniques as well as the corresponding typical energy ranges are listed in Table 2.3.

TABLE 2.3 Biomedical Applications and Typical Energy Ranges for Various Charged Particles

Particle	Energy range	Application
Electrons	6 to 25 MeV	Routine electron and photon radiotherapy
	2 to 10 MeV	Radiosterilization, pasteurization, food preservation
	several hundred MeV to several GeV	Production of synchrotron radiation for angiography and analysis
Protons	2 to 10 MeV	Analytical methods
	several tens of MeV	Neutron production in experimental therapy
	several tens of MeV to 250 MeV	Proton radiotherapy
	500 to 700 MeV	Meson production in experimental therapy
	10 to 100 MeV	Medical radioisotope production
Deuterons	7 to 20 MeV	Medical radioisotope production
	5 to 20 MeV	Analytical methods
Heavy ions	70 to 700 MeV/nucleon	Experimental radiotherapy

2.2 High-voltage Linear Methods

2.2.1 Fundamentals

Electrostatic method (direct current). In accordance with the previous considerations charged-particle acceleration may in general be expressed as

$$E_k = \frac{m_0 c^2}{\sqrt{1 - \beta^2}} - m_0 c^2 = K\left[z\left(t\right)\right] dz \qquad (2.5)$$

where K is the electric field; z is the length of the acceleration path and the remaining notation is as in (2.3).

The simplest way to achieve acceleration is to use a system of two electrodes with an electrostatic potential difference between them. A charged particle in the space between the electrodes is then repelled by the electrode of the same sign and is attracted by the electrode of the opposite sign. For the particle to be accelerated efficiently, a vacuum must exist in the

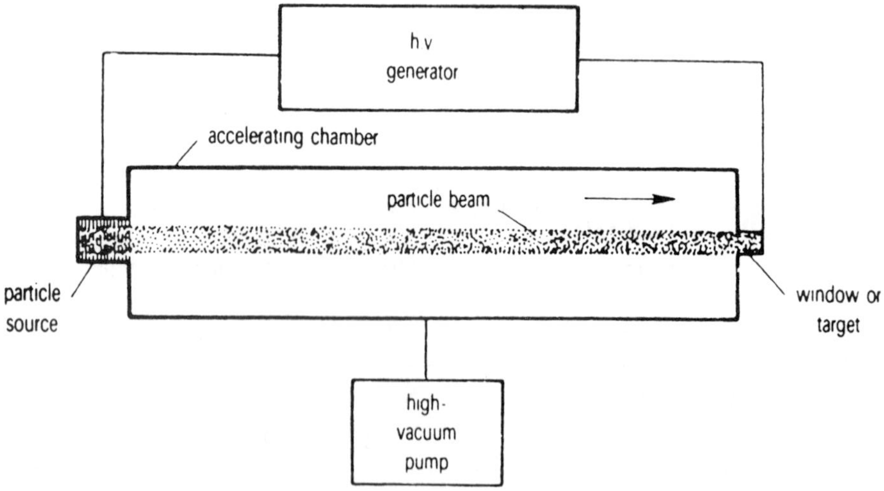

FIGURE 2.2 Principle of operation of a one-stage high voltage accelerator

space between the electrodes such that the mean path before the accelerated particle collides with gas molecules is much greater than the distance between the electrodes. Fig. 2.2 shows a basic schematic of a *high-voltage linear accelerator* which meets this requirement. It consists of an ion source, an accelerating chamber ending with a window or target, a high-vacuum pump and an hv generator. This accelerator is a *single-stage machine*. The maximum particle energy corresponds to the maximum voltage delivered by the generator. Currently, 100 kV to 30 MV generators exist with this design. Generators for 50 MV are under development. Since the accelerating chamber has the full voltage applied to it, its life-time is often limited. This is especially true of accelerators with comparatively high voltages. If the voltage delivered by a high-voltage generator to the accelerating chamber is V, a particle that has a charge q acquires the kinetic energy given by

$$E_k = q\,V \tag{2.6}$$

For electrons, $E_k = eV$. This is the reason why the energy of accelerated particles is commonly expressed in terms of the potential difference that should be applied to the particle if it had a unit charge. Energy of one *electron volt* is gained by a particle endowed with a charge equal to that of an electron when it is accelerated in an electric field whose potential difference is 1 V, thus 1 eV = $1.6 \cdot 10^{-19}$ J. Since 1 eV is a small unit, in charged-particle acceleration the following derived units are most often employed:

1 kiloelectron volt, 1 keV = 10^3 eV,

1 megaelectron volt, 1 MeV = 10^6 eV,

1 gigaelectron volt, 1 GeV = 10^9 eV, (1 GeV = $1.602 \cdot 10^{-10}$ J)

1 teraelectron volt, 1 TeV = 10^{12} eV.

It is obvious — as already mentioned — that the high voltage technique can only be used to obtain energies of the order of megaelectron volts (MeV). Higher energies, up to several TeV, can be reached with the use of other techniques (Sec. 2.7). In heavy ion acceleration, a single nucleon acquires an energy E_{kn}

when the chamber is supplied with the potential difference V. Therefore

$$E_{kn} = \frac{q}{A} V \tag{2.7}$$

where E_{kn} is the *kinetic energy per one nucleon*, most frequently expressed in keV/nucleon or MeV/nucleon, q is the ion charge and A is the ion mass number. The *ion charge*, q, is most often expressed in terms of electron unit charge, thus

$$q = ne \tag{2.8}$$

where n is the *degree of ion ionization*. For total ionization q = Ze, (n = Z), where Z is the ion atomic number. This condition is hardly ever attained in practice since highly ionized ions tend quickly to recombine.

It follows from Eq. (2.7) that by increasing the degree of ionization the ion can be accelerated with the same voltage V to a higher energy proportional to n. Thus, if a singly ionized ion traversing a path of 1 m in a field with a gradient of 1 MV/m acquires the energy of 1 MeV, a doubly ionized ion of the same element (q = 2e) will attain the energy of 2 MeV.

The efficiency of acceleration. The efficiency of a given method of charged-particle acceleration is dependent on the *energy gradient*, that is on the energy attained by a beam of accelerated particles after traversing a segment of its path in the accelerator. The energy gradient is most frequently expressed in kiloelectron volts or megaelectron volts per one meter path segment, i.e. keV m^{-1} or MeV m^{-1}. In accelerators of future generations the energy gradient will probably be of the order of GeV m^{-1}.

Depending on the type of accelerator, we make a distinction between a linear gradient and an effective gradient. The *linear gradient* finds use in linear accelerators where the path of the accelerated particle is rectilinear and the particle traverses a given segment of its trajectory only once, whereas the *effective gradient* is used in circular accelerators, where the acceleration is of a multiple character as the particles run many times in circular orbits. The effective gradient is in fact equal to the energy acquired during a full acceleration cycle along a 1 m long segment of the circumference. For example, if electrons are accelerated to an energy of 30 MeV in a betatron in a 1.5 m long circular orbit, the effective accelerating gradient is 20 MeV m^{-1}. In actual fact, acceleration requires about one million revolutions in the orbit when electrons travel a distance of 1500 km. The linear gradient, i.e. the acceleration attained during a single revolution is then 20 eV m^{-1}.

In the *electrostatic method*, the field along the path of the accelerated particle is constant with time and is independent of the path's coordinates. Thus, K(z) = const. This method was pioneered by I.D. Cockroft, E.T.S. Walton and R.I. Van de Graaff (1929-1933) who constructed the first high-voltage accelerators. The electrostatic method has been widely used up to the present day, but its main drawback lies in the considerable difficulties involved in obtaining voltages higher than about 10 MeV. The linear acceleration gradients achieved by this method are relatively small, ranging from 0.1 to 2 MeV m^{-1}.

2.2.2 High-voltage Generators

2.2.2.1 Van de Graaff Electrostatic Generators

Operating Principle. In the early 1930s, Van de Graaff patented the system of an *electrostatic generator*, now called after his name, the *Van de Graaff generator.* Its principle of operation is shown in Fig. 2.3.

A conveyor belt, made of an insulating material, is mounted between two cylinders, located up to several meters apart. The bottom cylinder coupled to a motor serves as a driving pulley so that the conveyor belt has a linear speed of a few to several tens of meters per second. A device for charging the conveyor belt is installed in the lower part of the generator. Near the belt is a needle point which acts as an emitter. Under the effect of a strong electric field positive and negative ions are produced in the gaseous medium round the point. If the latter has a positive potential it repels positive ions which are then deposited on the surface of the belt. These charges are distributed roughly uniformly on the belt surface, and are then carried upwards with the belt.

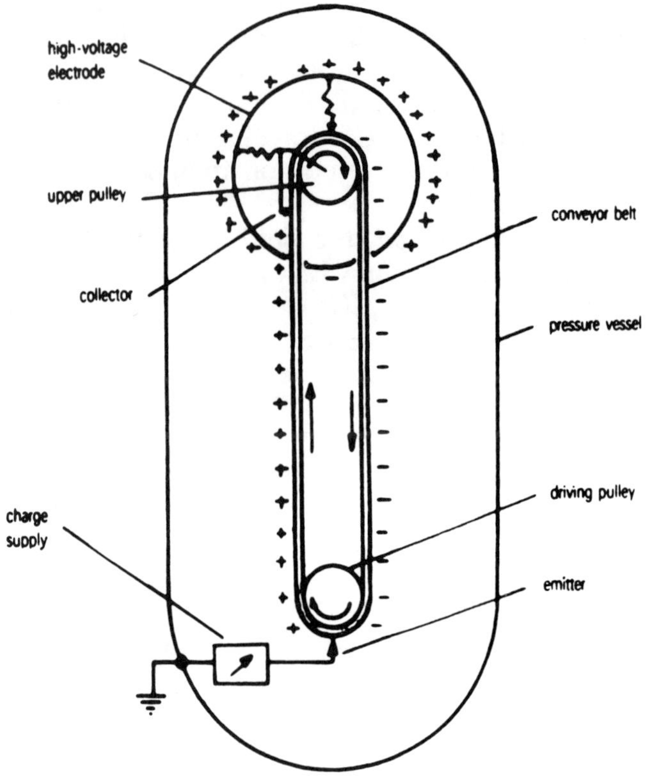

FIGURE 2.3 Principle of operation of a Van de Graaff accelerator.

Inside the spherical electrode, often referred to as the *dome*, called the *high voltage (hv) electrode* or *collecting electrode*, is another needle point referred to as a *collector*. The collector collects charges from the belt surface and then transfers them into the hv electrode, which thus increases the potential. In the example shown in Fig. 2.3 the potential is positive.

Inside the hv electrode (Fig. 2.3) we see another, additional needle point whose function is analogous to that of the charging system near the bottom pulley. In order that the belt should not return "uncharged" from the upper pulley, the needle point causes the belt to be charged this time with negative charges, which are then returned to the bottom pulley. A generator of this type is referred to as a generator with *two-fold or two-directional charging*.

Current and voltage ranges. The hv electrode of the Van de Graaff electrostatic generator obviously cannot be charged to any high potential. If such a generator operates in air, corona discharge between the electrode and the walls of the tank, in which it is housed, occurs at a voltage above a certain value. Discharges of this kind can then become transformed into spark discharges along the conveyor belt itself. The dielectric strength of atmospheric air cannot exceed a value of about 30 kV cm^{-1}. The hv electrode is usually made in the form of a sphere. If it has a radius of 1 m, the maximum voltage to which it can be charged is about 3 MV. If this value were exceeded the maximum dielectric strength allowable for air would also be exceeded.

In 1934, R.G.Herb and his co-workers were first to develop a *pressurized electrostatic accelerator*, one in which a compressed gas at high pressure is used as an insulator. Since then, this technique has been widely used in hv accelerators of various types, and the present-day accelerators with Van de Graaff generators are made exclusively as pressure-insulated machines. Typically, the voltages used range at present from several hundred kV to 25 MV. The largest accelerator of this type generates 30 MV (Daresbury Nuclear Structure Facility, England).

Nitrogen or methane derivatives were first employed as the insulating gas. Compressed air is not suitable for this purpose because of the fire hazard. Gases of large molecular mass, such as *freon* and *sulphur hexafluoride* (SF_6) (which is most frequently used as the insulator in accelerators) are now used since they have the greatest dielectric strength.

Current load capacity is dependent, among other things, on such parameters as the width, linear speed and the type of material of the belt. A 0.5 m wide conveyor belt travelling at 1.0 km min^{-1} has a current load capacity of about 1 mA. As a rule, the beam powers resulting from the above do not exceed a few kilowatts.

In the older types of Van de Graaff machines the charging belt was made from cotton fabrics, rubberized balloon silk or man-made fibers. Even those manufactured to high standards have a life time that does not exceed several thousand hours. Furthermore, belt replacement, particularly in accelerators with higher voltages, results in a loss of about two weeks in useful operation. In the Pelletron system developed by the American company, National Electrostatics Corp. (NEC), the conventional flat belt is replaced by a chain consisting of small metal cylinders with well rounded ends joined by links of solid nylon. The links swing on pins with self-lubricating sleeves therefore requiring no oil.

The chains have a life-time of the order of some ten thousand hours. Each single chain can carry up to 75 μA, giving a maximum of 150 μA charging current per chain for two-directional charging. If a larger number of parallel chains are employed the generator current load capacity may be proportionally increased. NEC is capable of updating older versions of accelerators equipped with charging belts by replacing them with Pelletron chains. The introduction of the Pelletron design was an essential step in the development of electrostatic accelerators, especially for attaining higher voltages, up to 20 or 30 MV.

Multistage accelerators (tandems) The impetus of developing this technique was to utilize a single Van de Graaff generator for two-fold acceleration of the same ion beam. Fig. 2.4 illustrates the operating principle of a *two-stage accelerator* commonly known as the *tandem*. A beam of positive ions is produced in a source outside the accelerator. This has the important advantage of enabling easy handling of the source, e.g. ion exchange, etc. The beam passes through a channel filled with gas at low pressure. Owing to the interaction between the gas and the ions, negative ions are produced by the addition of two electrons to the initially positive ions. The beam of negative ions is then directed to a magnetic analyzer which selects ions with a given q/m ratio. The ions are subsequently injected into the accelerating chamber whose initial electrode is grounded. The hv electrode coupled to the end of the first section of the accelerating chamber is thus at a positive potential of, for example, a few MV. The ions become accelerated to an energy corresponding to the above potential, then they pass through another chamber in the middle section of the accelerator, where the ions electrons are removed resulting in a beam of positive ions. This method is referred to as *stripping*, and the *stripper* can

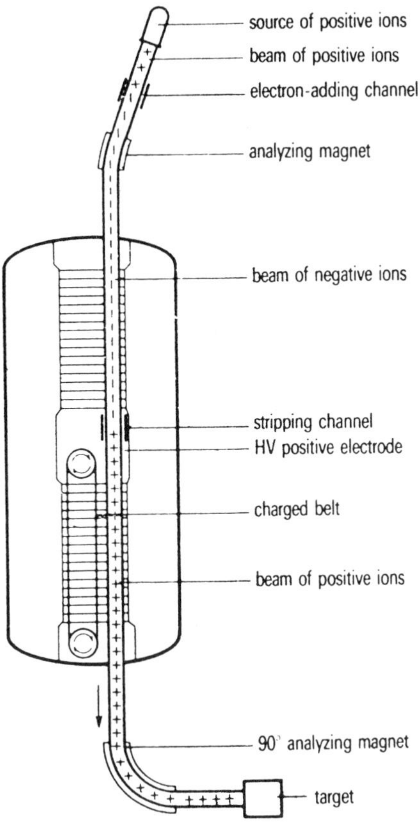

FIGURE 2.4 The operating principle of a two-stage (tandem) accelerator.

be a thin metal or carbon foil or a gas. Stripping is used not only in the high-voltage tandem accelerators, but also in cyclotrons (Sec. 2.6) and other types of ion machines.

After being converted to positive ions, the beam is again accelerated between the hv electrode and the end of the second accelerating section which is also grounded. Thus the voltage V_0 is utilized twice to impart double energy to the particles. Upon leaving the accelerator the beam is frequently directed again to the magnetic analyser before it strikes the target.

Instead of a two-stage system it is also possible to use a three-stage or four-stage system. The energies attainable in such systems are dependent on the degree of ionization of the accelerated ions, which can be as great as the atomic number Z, and on the number of stages of the accelerating system:

two-stage system	$E_{max} = V_0 (1 + q)$	(2.9a)
three-stage system	$E_{max} = V_0 (2 + q)$	(2.9b)
four-stage system	$E_{max} = V_0 (2 + 2q)$	(2.9c)

In accordance with the above equations, a *multi-stage system* with the maximum accelerating voltage V_0 has a possibility of augmenting the energy of the ions accelerated since it increases the charge q. This is why in present-day tandems, which are very often used for accelerating heavy ions, much attention is paid to the most intensive possible stripping. Current designs favor a multi-stage system of strippers, comprised of several foils, or a combination of foil and gas.

In the largest tandems in operation at present the energies per nucleon for uranium ions are close to 1 MeV, while those for less heavy ions, such as ^{16}O are as high as 10 MeV.

Present-day applications. The Van de Graaff accelerators played an important role in the development of acceleration techniques, particularly when the number of other techniques used to be relatively limited. At that time, they were also used in routine radiation therapy (Sec. 1.1.1). The number of electrostatic machines built so far is close to 700. In spite of the fact that 60 years have elapsed since the first machine was constructed, the interest taken in the Van de Graaff accelerators has not diminished. The main reason for this is that these accelerators provide beams of very high quality since they feature a comparatively small energy scatter and high terminal energy stability. Moreover, they fully meet the definition of a *constant voltage accelerator* as they produce no ripple voltage which is one of the main characteristics of other high voltage accelerators. An additional advantage of the Van de Graaff machines is that they are very versatile being able to operate with multistation systems (Fig. 2.5).

In addition to basic research in atomic and nuclear physics, present-day Van de Graaff accelerators are often used in analytical studies. The most common analytical methods, e.g. PIXE (Sec. 9.2.3) and RBS (Sec. 9.2.4), may be implemented with the use of ions with energies between 0.5 and 5 MeV [2.1], which can best be accelerated employing relatively inexpensive Van de Graaff tandems producing the microbeams (Sec. 9.1). Recent years have witnessed the development of dedicated laboratories which specialize in the application of accelerator techniques in biomedical microanalysis (Oxford University, Melbourne University) [2.2, 2.3].

Currently, Van de Graaff accelerators are being produced by two U.S. Companies: High Voltage Engineering Corp. (HVEC) [F1] and National Electrostatics Corp. (NEC) [F2].

Fig. 2.5 shows a lay-out of a research facility equipped with a tandem accelerator with a 2MV Pelletron-type Van de Graaff generator manufactured by NEC. The total area of approx. 260 m^2 is divided into three separate rooms. The accelerator room houses the accelerator proper, together with storage tanks filled with SF_6 which is used as an insulating gas, and compressors to pump the gas from the accelerator pressure tanks to the gas tanks when the accelerator is not in operation due to repairs,

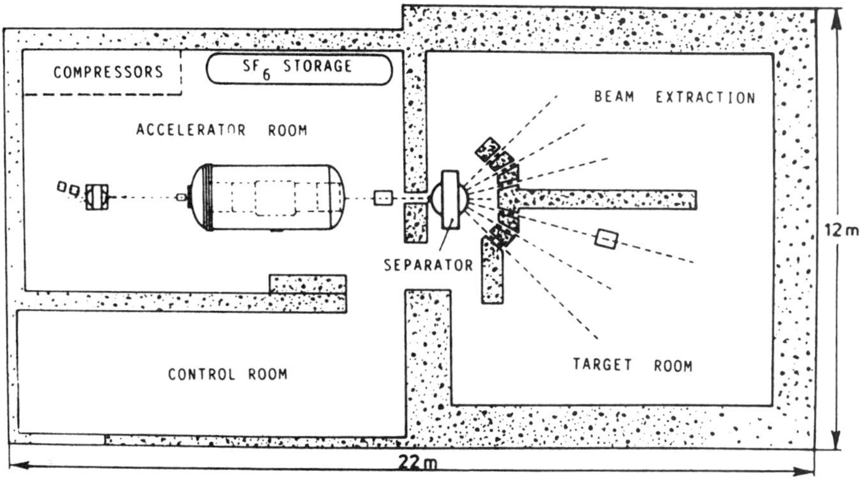

FIGURE 2.5 Lay-out of a research laboratory equipped with a Pelletron-type 2 UDM 2-MV tandem Van de Graaff accelerator (mfd by National Electrostatics Corp., U.S.A.)

replacement of parts, etc. In the control room, the control console enables the operator to control and monitor both the accelerator parameters and those of the equipment used in experiments.

Almost half of the total area of the facility is taken up by the laboratory room proper, called a *target room*, where the targets are bombarded with accelerated particles and experimental setups are located. The beam from a horizontally placed tandem accelerator is directed to the target room through a beam transport channel and then to a switching-analyzing magnet which distributes it to separate work-stations. Fig. 2.5 shows 6 such stations.

In addition to large tandem accelerators equipped with 25 MV Van de Graaff generators, the NEC company also manufactures a series of S-type tandems with generators of maximum voltages of 1 MV (3 SDH type) up to 3.0 MV. Fig. 2.6 is a view of a 5 SDM-type tandem equipped with a Van de Graaff generator with variable voltage of 0.30 to 1.6 MV. The accelerator proper together with the accelerating chamber is located in a pressure tank (middle part of Fig. 2.6) which is 0.76 m in diameter and 3.1 m long and is filled with 56 kg of the insulating gas SF_6. This tandem type accelerator can accelerate singly ionized ions in the energy range between 0.6 and 3.2 MeV; doubly ionized ions attain energies from 0.9 to 4.8 MeV.

Advances in the construction of the Van de Graaff machines, particularly those of tandems, are summarized at periodical conferences, such as those held in 1989 [2.4] and 1992 [2.5].

Pulsed operation of electrostatic accelerators. Electrostatic accelerators are in fact conventional accelerators operating in a continuous mode and, therefore, can fully be termed *direct voltage (direct current) accelerators*. In many research applications, especially in spectrometry the sample is irradiated in a pulsed fashion. This is why electrostatic accelerators are very often equipped by the manufacturer with a device that permits *pulsed operation*. The systems of this kind operate in two ways. The first method is based on switching particle sources by injecting the beam into the accelerator. In this case, the beam consists of single pulses, whose duration and repetition rate can usually be regulated. The second method involves systems that are installed at the accelerator's output and form the already ac-

FIGURE 2.6 View of a Pelletron-type 5 SDM tandem accelerator with 1.6 MV Van de Graaff generators (mfd by National Electrostatics Corp., USA)

celerated beam. The duration of the output pulses may in this case be of the order of nanoseconds or even picoseconds. NEC offers a system which makes it possible to produce proton or deuteron pulses with the amplitude of 1 mA and duration of 1-2 ns (FWHM) by controlling the ion source. The repetition rate can be regulated over a wide range from 62.5 Hz to 2 MHz, while the accelerated beam can be controlled by means of a so-called *drift system* which permits the production of heavy ion pulses with the duration time of 3 ns (FWHM).

2.2.2.2 Cascade Generators (Cockroft-Walton Type)

The asymmetric cascade circuit (Fig. 2.7) was developed in 1920 and is known in electrical engineering as the *Greinacher doubling voltage circuit*. In the early 1930s Cockroft and Walton were the first to use it in particle acceleration for the production of voltages of the order of several hundred kilovolts. This is the reason why accelerators equipped with *cascade generators* are often called *Cockroft-Walton accelerators*.

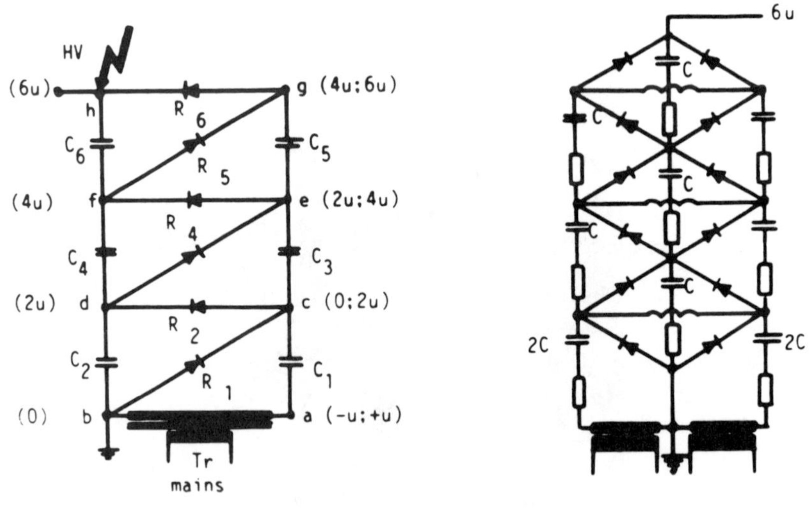

FIGURE 2.7 Cascade generator: (a) asymmetrical circuit, (b) symmetrical circuit.

The *asymmetric cascade generator* (Fig. 2.7a) includes n identical stages, called *cascades*, consisting of capacitors C and rectifiers R. The first stage, i.e. capacitor C_1 and rectifier R_1, is supplied from the secondary winding of transformer Tr that supplies in turn an alternating voltage with amplitude u. At point *a* of the circuit the transformer produces an alternating voltage oscillating between peak values $\pm u$. The first stage also includes rectifier R_1. In the positive halfperiod of the supply voltage the rectifier conducts and charges capacitor C_1 to the value u. In the next halfperiod the capacitor will not be able to discharge because the rectifier is then reverse-biased and does not conduct. As a consequence, the potential at point *c* is increased by the value u and now oscillates between 0 and u. This will cause the capacitor C_2 to be charged through rectifier R_2 to a voltage of 2u. Point *d* is then at a constant potential of 2u.

It follows that if the cascade generator circuit consists of n identical stages in operation without current load, a constant output voltage of $U_0=2n\cdot u$ is produced at the generator's output. The insulation of the elements making up each stage, the capacitors and rectifiers, must in this case withstand reverse voltages of only 2u.

The description of the cascade generator described above is true only for the state without load. If the generator is connected to accelerating chamber through which flows a current equal to that of the beam of accelerated particles, the true charging curves for individual cascades will be different from the ideal curves. This will be reflected in the voltage drop U and ripple ΔU in the output voltage.

In acceleration practice, an asymmetrical circuit (Fig. 2.7a) is most often replaced by a symmetrical circuit (Fig. 2.7b). However in both cases the voltage for an n-stage cascade generator is given by

$$U_n = \sqrt{2}\, n \cdot u_0 \tag{2.10}$$

where n is the number of cascades and u_0 is the true voltage of the secondary transformer Tr. The above circuits differ essentially in respect of the *voltage drop* ΔU and the *ripple voltage* δU.

Thus we have

asymmetrical circuit — symmetrical circuit

$$\Delta U = \frac{I}{fC}\frac{n}{3}\left(2n^2 + 1\right) \qquad \Delta U = \frac{I}{fC}\frac{n}{3}\left(\frac{n^2}{2} + 1\right) \qquad (2.11)$$

$$\delta U = \frac{I}{fC}\frac{n}{2}\left(n + 1\right) \qquad \delta U = \frac{I}{fC}\frac{n}{2} \qquad (2.12)$$

where I is the dc current, f is the supply voltage frequency, C is the capacitance of a given stage (cascade) and n is the number of stages. By comparing both equations it follows that as the number of stages is increased the voltage drop ΔU and the ripple voltage δU become considerably lower for the symmetrical circuit than those for the asymmetrical circuit.

Since the voltage drop and the ripple amplitude in the cascade circuits (Eqs. 2.11 and 2.12) vary inversely with ripple voltage frequency f, cascade generators often get their supply power from *frequency converters* which operate over the range between 0.5 and 19 kHz.

The construction of cascade accelerators was pioneered by the Swiss company, Haefely [F3], which developed and manufactured generators for typical voltages of 500-1000 kV and others for higher voltages of 2.5 to 4.0 MV. The currents available were as high as several hundred mA, thus the beam powers attained were up to several hundred kW. These generators were employed as injectors for higher energy accelerators as well as in radiation processing.

At present, cascade generators, apart from their application in radiation processing, are commonly employed in *accelerator electron microscopy*. The voltage range used varies from 0.5 MV to several MV. In addition, microscopy applications require (1) high stability of the supply voltage, (2) a high degree of precision in the manufacture of electromagnetic coils for beam focusing, (3) high stability of coil supply currents, and (4) a special design eliminating possible mechanical vibrations within the accelerator itself or from outside.

In order to obtain full utilization of the theoretically attainable resolution power of the *megavolt microscope*, the supply *voltage stability* must lie in the range between 10^{-5} and 10^{-6}, after making allowance for the ripple voltage. In other words, if the generator operates at 1 MV the acceptable instability should not exceed 1 to 10 V in 3 minutes, which is the maximum time for a single exposure. Another requirement that makes it difficult to achieve high stability is the need to have the output voltage controllable from 10 to 100% of the maximum voltage. In this case the current load capacity can, however, be neglected, because currents as low as a few microamperes are usually sufficient, and only a very small fraction of the beam (10^{-11} to 10^{-12} A) is used to produce a good picture.

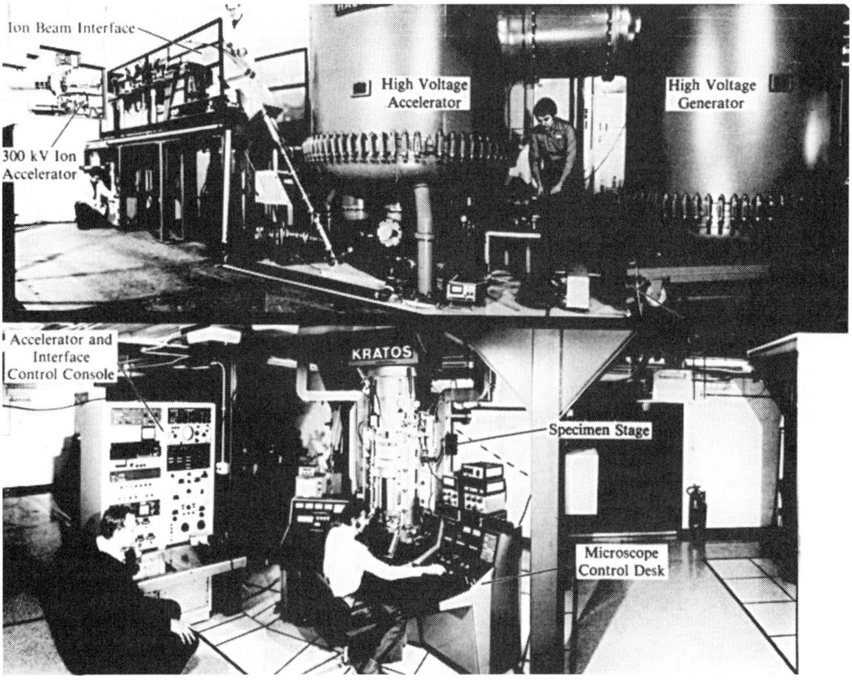

FIGURE 2.8 View of accelerator laboratory with Kratos 1.2 MV electron microscope, mfd. by Haefely (Tandem Facility, Materials Science Div., Argonne Ntl. Lab., USA)

The Haefely company has developed a series of *twin-type cascade accelerators* for the purposes of accelerator microscopy. Fig. 2.8 shows a facility at Argonne National Laboratory (USA) where one accelerator of this type has been installed. The cascade generator is housed in the right-hand pressure tank, shown in the upper part of the picture. The tank is coupled by a horizontal connection with another identical tank which contains the accelerator proper, the accelerator chamber and the electron source. The microscope itself is located on the lower platform (Fig. 2.8).

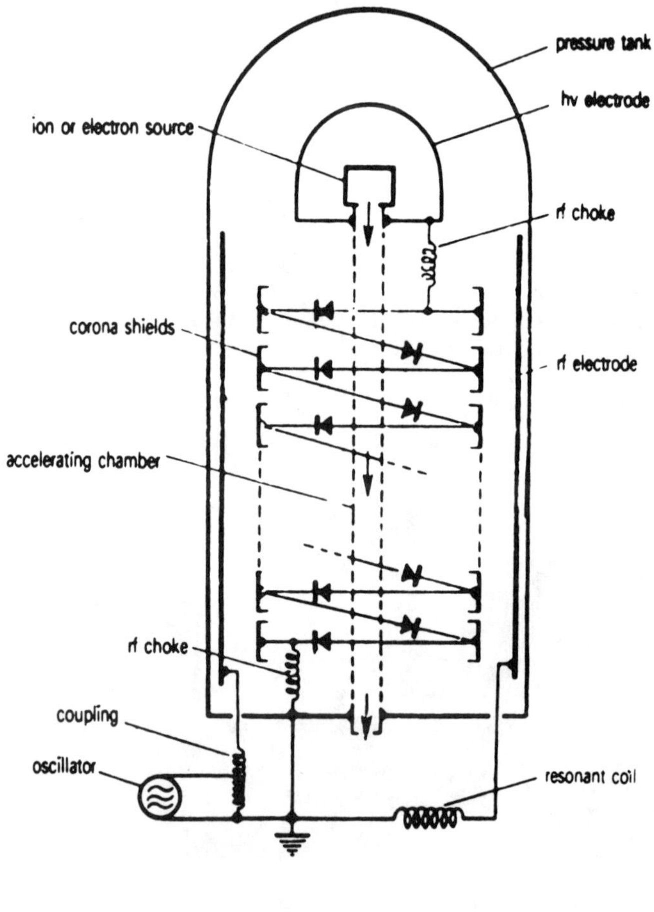

FIGURE 2.9 A schematic of Dynamitron-type accelerator (mfd. by Radiation Dynamics)

Typical voltages attained in the above model of twin-type Haefely machines are 1.0 MV, 1.2 MV, and 1.5 MV. Basic models feature stability of $\pm 1 \cdot 10^{-5}$, whereas those of the highest quality have a stability of $\pm 10^{-6}$, all the values being achieved in a 3-min period.

Accelerators equipped with cascade generators for use in megavolt microscopy are also manufactured by the Japanese company Nissin [F4], which has a comparatively large number of such accelerators for export to Europe and the USA.

Cascade generators with voltages of several hundred kV also find application in generating neutrons for radiation therapy (Sec. 5.3).

2.2.2.3 Dynamitrons

Dynamitron is a trade name used for accelerators manufactured by the US company Radiation Dynamics Inc. (RDI) [F5]. The dynamitron operating principle is shown in Fig. 2.9. Two large D-shaped electrodes supplied with power from an rf generator at 100 kHz are mounted inside a pressure tank. The voltage from these electrodes, which constitute a kind of antennae, induces the corresponding secondary voltages in separate segments of the receiving unit. The segments are coupled to rectifiers and the rectified voltages from each segment are then added up to supply an hv electrode. As seen from Fig. 2.9, the accelerating chamber is installed inside these segments. The tank is filled with an insulating gas, SF_6.

The direct voltage produced by a single segment is 50 kV. Among all other accelerators produced now in a standard series, Dynamitrons are characterized by the highest beam powers, amounting to 200 kW. As a result of the constant increase in power not only is the throughput rate increased in such typical radiation processing techniques as sterilization by radiation, but also the costs are greatly reduced. Fig. 2.10 illustrates the change in the cost of one kilowatt-hour of the electron beam in the years 1958 -1980. By increasing the beam power from 3 kW to 150 kW the cost of one kilowatt-hour was reduced by a factor of 20.

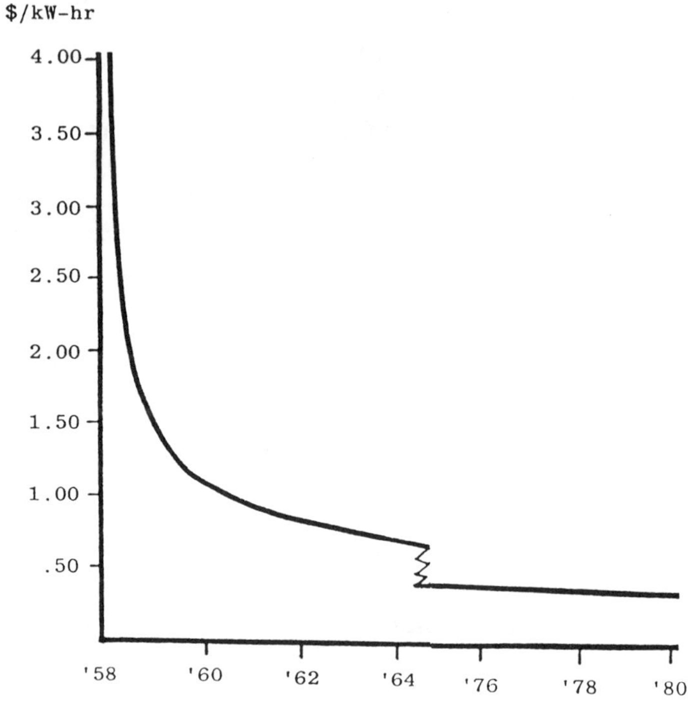

FIGURE 2.10 Radiation cost of 1 kW-hour of electron beam from Dynamitron-type accelerator vs. maximum beam power (time)

2.2.2.4 Insulating-core Transformers

Conventional high-voltage transformers with a grounded magnetic core are employed in particle acceleration, mainly because the grounded core technique involves some insulation difficulties. In order to overcome this drawback, R. Van de Graaff, who was first to develop an electrostatic generator (Sec. 2.2.2.1), also designed a so-called *insulating-core transformer*, known as an *ICT transformer*. Its design is schematically shown in Fig. 2.11.

The core is divided here into sections separated by thin insulating spacers. The core is excited through the primary winding in a three-phase system (only two phases, for simplicity, are indicated in Fig. 2.11) by a potential of 460 kV at 50 or 60 Hz. Each secondary section is coupled to a rectifier operating as a voltage divider and is an independent 50 kV unit provided with a characteristic equipotential end-ring. The transformer current load capacity is dependent on the core diameter and on the total output voltage. The transformer constitutes a compact unit, housed in a pressure tank filled with an insulating gas, SF_6.

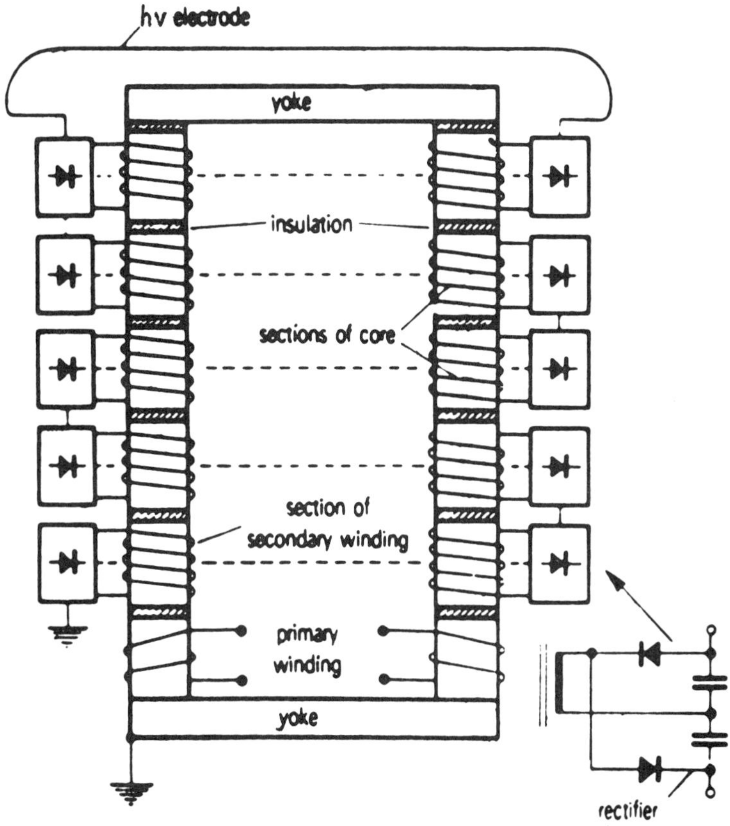

FIGURE 2.11 A schematic of an insulating-core transformer ICT (mfd. by HVEC)

Insulating-core transformers have long been in production by the US company, High Voltage Engineering Corp. (HVEC) [F1] and its European division, High Voltage Engineering Europa B.V. [F6]. They are utilized as electron accelerators for radiation processing with energies of 300 to 3000 keV and beam powers of the order of several tens of kilowatts. Their production is at present being continued by a Japanese branch of the company.

2.3 Linear rf Techniques

Spectacular advances in radioengineering, radar and television techniques, as well as a rapid progress in microwave generation technology have established solid foundations for charged particle methods based on the electrical component, E, of rf electromagnetic fields. These acceleration methods are suitable for both light particles, such as electrons, and heavy ions, since they make it possible to achieve energies between several MeV and several GeV.

2.3.1 Linear rf Structures for Ion Acceleration

The Wideröe method of rf linear acceleration. As far back as 1928 the Swiss scientist, R.Wideröe, published his fundamental paper, which contained, among other things, the theory of the betatron and his first results of linear acceleration of potassium ions to the energy of 50 keV. Fig. 2.12 shows schematically the system used for this purpose. It consists of three cylindrical drift electrodes. Both extreme electrodes are grounded and the middle electrode gets its power from an rf generator at 1 MHz and 25 kV. The distance, d, between the electrode gaps is chosen so that the following condition be fulfilled

$$\frac{d}{v} = \frac{1}{2f} \quad \text{or} \quad d = \frac{\beta_0}{2} \tag{2.13}$$

where f is the rf generator frequency; v is the ion velocity and β_0 is the electrode gap wavelength corresponding to the frequency f.

The time taken by the potassium ions to cross each electrode gap is equal to 1/2 of the half cycle of the rf voltage. Since at the time of Wideröe's first experiments generator frequencies much

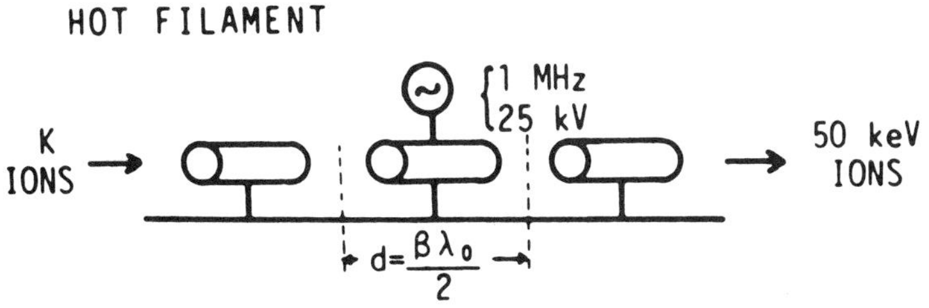

FIGURE 2.12 The Wideröe experimental system for rf linear acceleration of positive ions.

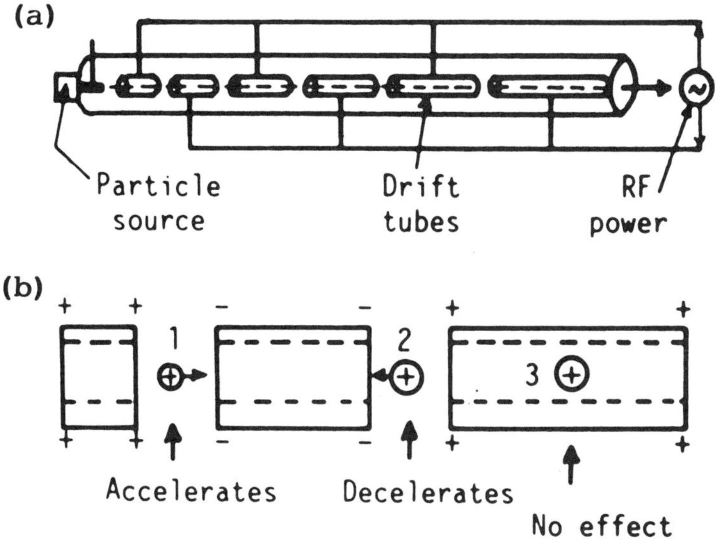

FIGURE 2.13 The Wideröe structure for linear acceleration of ions: (a) system of electrodes, (b) potential distribution; of 3 protons at a given time at different points, proton 1 is accelerated, proton 2 is decelerated, and proton 3 is screened against rf electric field.

higher than 1 MHz did not exist, acceleration of lighter particles that would be able to travel faster (Fig. 2.1) was not feasible.

Fig. 2.13 is a schematic diagram of an updated *Wideröe structure* to accelerate ions with an rf field. It consists of a system of cylindrical drift electrodes of increasing length (Fig.2.13a). The potential distribution between the adjacent electrodes is shown in Fig. 2.13b. At a given moment, acceleration takes place only at one gap. After crossing this gap, protons or other heavy

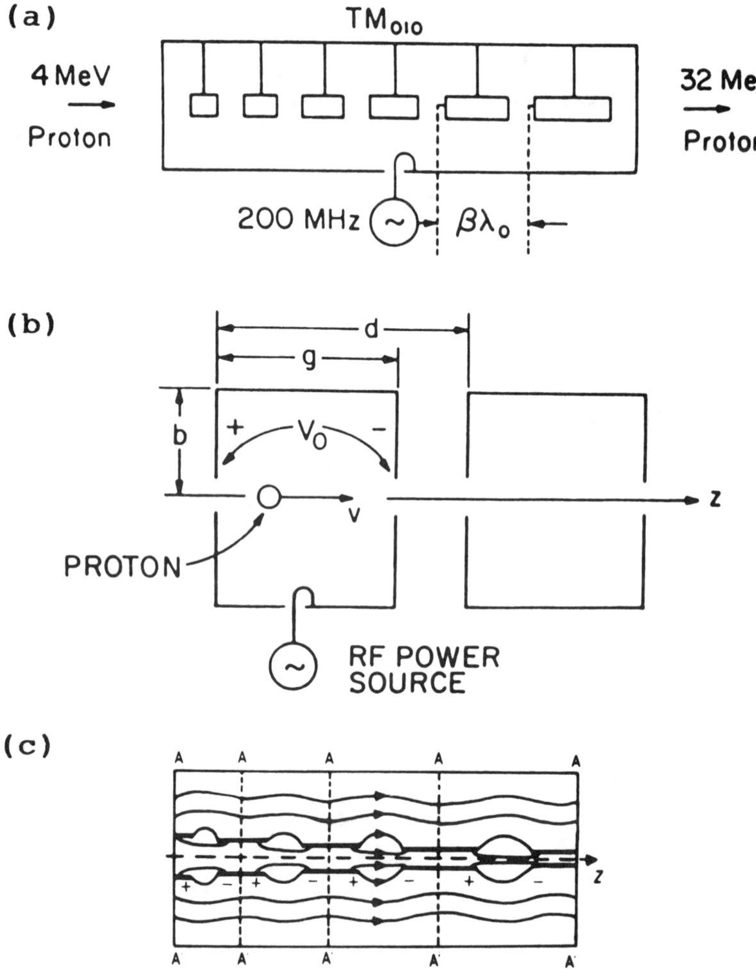

FIGURE 2.14 The Alvarez system for rf linear acceleration of ions: (a) schematic of experimental set-up, (b) potential distribution in resonator, (c) standing wave.

particles enter a cylindrical electrode which acts as a screen preventing particles from being decelerated. Since the acceleration voltage frequency, f, is constant with time and the velocity of heavy particles such as protons or heavy ions, increases relatively slowly with acceleration (small changes of β, Fig. 2.1), the screening electrodes become respectively longer (Fig. 1.13a), so that the particle beam arrives at the successive accelerating gaps when the rf voltage is in an appropriate phase.

The Wideröe system of electrodes, or the *Wideröe structure* in short, is suitable only for ion acceleration. It cannot be used for electron acceleration.

The Alvarez method of linear accelerator. In 1947, L.Alvarez and his team built a machine which accelerated protons to an energy of 32 MeV (Fig. 2.14a). Its accelerating structure, which differed significantly from the Wideröe structure, was supplied with power from a 200 MHz radar generator. The *Alvarez structure* consists of a set of resonators which have an rf voltage of the same phase applied to them. Inside each resonator exists a potential distribution as depicted in Fig. 2.14b. Therefore, the acceleration proper takes place in the resonator gaps. The edges of the adjacent resonators are equipped with cylindrical drift electrodes also of variable length, matched in such a way as to screen the particles against the decelerating effect of the field. In contrast to the Wideröe structure with electrodes which have the same potential applied to them along the whole length (Fig. 2.13b), the Alvarez structure has potentials of different sign applied to the electrode ends making up an integral part of the adjacent resonators. As a result, a type of a standing wave (Fig. 2.14c) is formed that ensures particle beam acceleration.

Typical linear gradients achieved in the Alvarez structures range from 1 to 2.5 MV m^{-1}, while the gradients at the corresponding accelerating gaps range from 6 to 10 MV m^{-1}. Like the Wideröe structures, the Alvarez structures are exclusively designed to effectively accelerate heavy particles in the relative velocity range of

$$0.04 < \beta = \frac{v}{c} < 0.5 \qquad (2.14)$$

The method of rf quadrupoles. The acceleration of ions with low energies, ($\beta < 0.04$), by employing the Alvarez structure is not effective. This is particularly true for high beam currents of particles such as protons, when the effect of the spatial charge causing the beam to defocus becomes stronger.

In 1970, I.M. Kapchinski and V.A. Teplyakov proposed an rf resonator with a symmetry corresponding to that of magnetic quadrupoles (see Fig. 2.24). This system was referred to as an *rf quadrupole* and denoted by the acronym RFQ (*Radio Frequency Quadrupole*). The schematic diagram of an RFQ resonator is shown in Fig. 2.15. It combines the action of focusing and bunching the beam, in addition to acceleration proper. Focusing is ensured by a transverse electrical gradient, which is of particular importance since magnetic focusing is least effective for small velocities of ions (Sec. 2.4.3).

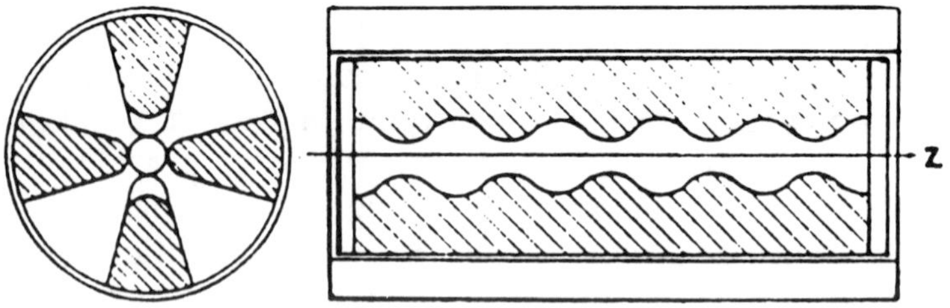

FIGURE 2.15 A schematic of a quadrupole resonator (RFQ)

In an RFQ resonator (Fig. 2.15) equipped with four wave-guides, longitudinal changes occur in the transverse gradient thanks to the wave-like modulation of the axial shape of the wave-guides. As a result, a longitudinal electrical component is formed, which ensures acceleration and bunching of particles. The acceleration is of continuous character and, therefore, it does not occur only in electrode gaps, or other structures. Equally strong is the effect of bunching, its efficiency being close to 100%. In practice, when the structure is 1-2 m long, protons (or heavy ions) can be accelerated from the initial energy of several tens of keV to the final energy of several MeV. RFQ resonators are now being increasingly often used in ion pre-acceleration. They constitute an injection system for subsequent stages where further acceleration is carried on.

2.3.2 Linear rf Structures for Electron Acceleration

The traveling and standing wave method. It follows from Fig. 2.1 illustrating the dependence of the particle relative velocity on energy, that the electrons, i.e. the lightest particles, become highly relativistic once they attain energies higher than 1 MeV. The velocity of a 5 MeV electron differs from that of light by less than 0.5%. Compared with protons, electrons become accelerated much faster, and subsequently acceleration goes on with very small changes in velocity. That is why the rf electric field can be much more easily utilized in electron acceleration than in ion acceleration. Instead of high frequency voltages between several tens and several hundred megacycles used in ion acceleration, still much higher frequencies, i.e. much shorter wave-lengths, can be employed in this method. In radiotherapeutical accelerators, a typical frequency of 3000 MHz (3 GHz) and the corresponding wave-length of 10 cm is used. Since the linear dimensions of resonators are proportional to wave-length, in rf electron accelerators resonators that are smaller and relatively simple in design can be employed. Typically, for the supply frequency of 3 GHz a single resonator is 10 cm in diameter and 2.5 – 5 cm in length.

Fig. 2.16a shows a set of simple cylindrical resonators forming a *wave-guide*. The field of the electromagnetic wave in the wave-guide is made up of electric field lines as indicated by solid lines with arrows, as well as magnetic lines, indicated by black dots and circles. The electrons move along the accelerator resonators (Fig. 2.16a) in the axial direction with a velocity equal to the traveling wave velocity. The traveling wave-length, λ, embraces 12 adjacent resonators. The wave-guide to produce a wave of this kind most frequently ends with a *characteristic wave resistance* which absorbs the energy of the propagated wave (Fig. 2.17a). In ordinary wave-guides with smooth walls, the *phase velocity* of the propagated wave is greater than the speed of light and is, therefore, always greater than the particle

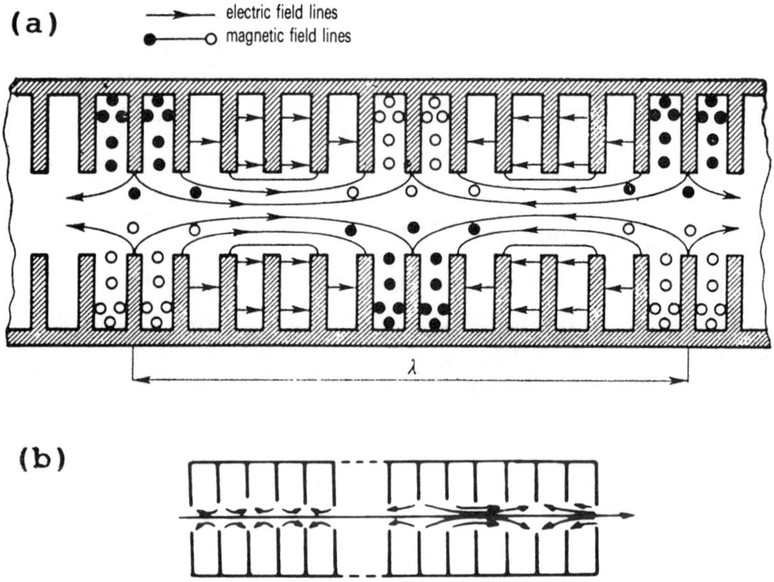

FIGURE 2.16 Formation of accelerating waves in rf electron accelerator: (a) travelling wave, (b) standing wave.

velocity. The phase velocity of the wave in the wave-guide can be regulated with a system of diaphragms which are chosen in such a way that the accelerated electrons move in step with the wave.

The system shown in Fig. 2.16a is suitable only for electron acceleration. Potential differences on the order of hundreds or even thousands of kilovolts can be achieved between the edges of resonators supplied with a very high frequency pulse voltage. These potential differences are thus much higher than those that would result from the dielectrical strength of the electrode gap. These high voltages are allowed since for a discharge to develop, a time longer than the very-short lasting action of voltage pulses is required. As a result, *electron acceleration linear gradients* attained are typically in the range between 10 and 20 MeV m^{-1}. Resonators have also been developed which make it possible to obtain much higher gradients, up to 100 - 150 MeV m^{-1}.

A schematic diagram of an electron rf linear accelerator equipped with a traveling wave is shown in Fig. 2.17a. The wave-guide is supplied at a point from a microwave generator with rf energy which travels, with some losses due to the absorption in the resonator walls, from the left to the right side of the accelerating structure. The energy excess is then absorbed without reflection in a *load* in the above mentioned wave resistance.

The field accelerating along the wave-guide axis can also be obtained with a so-called *standing wave*. A schematic diagram of an accelerator of this kind is shown in Fig. 2.17b. Here, the wave-guide is powered either from one end, as in the case illustrated in Fig. 2.17b, or at the middle part. The rf energy, after reaching the right end of the structure, undergoes multiple reflections. The returning wave becomes superimposed on the direct wave forming a stationary state, referred to as a standing wave. The way standing waves are formed in a system of cylindrical resonators is shown in Fig. 2.16b. The right-hand side of this figure shows a standing wave. Its half-wave length is equal to the distance at which 4 cavity resonators are located. A standing wave accelerator can be utilized both in electron and heavy ion acceleration. Since heavy ions, even those with large initial energies, would not be able to pass through the resonators

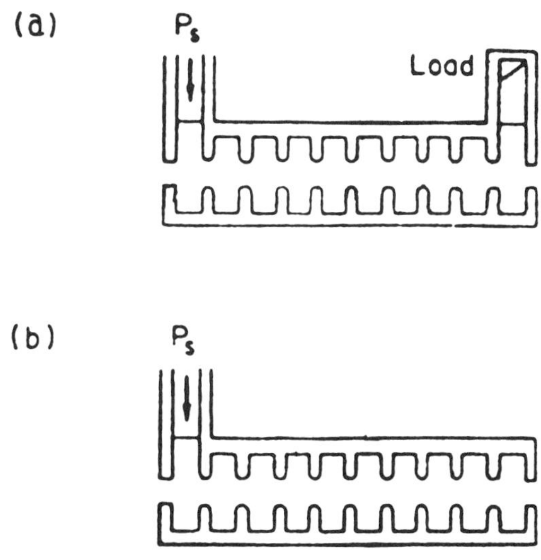

FIGURE 2.17 Systems of rf linear accelerators: (a) with a current wave, (b) with a standing wave.

during the half cycle of the accelerating voltage due to the ion relatively low velocities along the resonators' axis, it is essential that the resonators be screened with cylindrical electrodes of variable length (Fig. 2.14c).

Basic circuit of an rf electron accelerator. The schematic of the basic circuit of an accelerator intended for electron acceleration is shown in Fig. 2.18. The electron beam is produced in the accelerator source (gun) with a heated cathode operating in a triode circuit. The anode in the source is powered with a typical voltage of 80 kV which accelerates the electrons to a velocity of about 0.5 c (see Fig. 2.1). The grid to control the electron beam is usually negatively polarized at 50 V with respect to the cathode so that the electrons cannot traverse it. To "open" the source, the grid is supplied with positive pulses of a few hundred volts. The recently developed fast transistor pulse systems make it possible to produce pulses as short as 1 or 2 nanoseconds. The overwhelming majority of rf linear accelerators, however, operate with beam pulses in the 1-10 μs range at the repetition rates of up to 500 or even 1000 pps. The single-pulse current amplitude may be as high as 10 A.

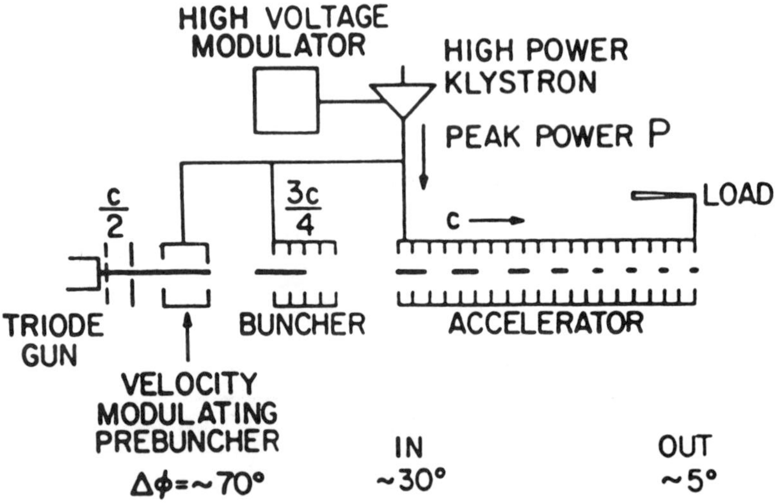

FIGURE 2.18 A schematic of an rf linear accelerator.

The rf supply voltage for separate accelerator stages (Fig. 2.18) is produced in generators equipped with magnetrons or klystrons operating normally at the *S-band* which encompasses a frequency range between 1.55 and 5.2 GHz, i.e. wave-lengths between 19.2 and 5.75 cm. It is only the frequency of 3 GHz (3000 MHz) or the corresponding wave-length of 10 cm that is utilized in radiotherapeutical accelerators.

The use of higher frequencies for powering resonators would be advantageous because, among other things, the accelerator could be reduced in size, as the dimensions of the rf resonators are proportional to the supply voltage wave-length. Frequencies exceeding 3 GHz have not, as yet, been used, since there are no rf generators that would have sufficient energy efficiency at such high frequencies. On the other hand, rf electron accelerators in research applications are frequently supplied with a voltage of a lower frequency of the order of several hundred MHz (*UHF-band*) or of about 1300 MHz (*L-band*).

The beam of electrons produced in the source (gun) first arrives at a *prebuncher* (Fig. 2.18) which is a single low-Q resonator, supplied with about 2 kW of rf power at the fundamental frequency. The prebuncher is intended for modulating the velocity of individual electrons with peak gap voltages of 10–15 V so that the relatively delayed electrons be accelerated more, and the electrons with velocities higher than the average velocity be accelerated less. As a result, a *longitudinal compression* or *electron bunching* with the smallest possible linear dimensions is obtained.

At a distance of 30 cm along the beam's linear direction (Fig. 2.18), about 70% of electrons are bunched within a phase interval $\sim \Delta\phi$ 70°. This is because the accelerating system proper in accelerators of this type is powered with an rf voltage (see Fig. 2.17), and, therefore, the phase of the bunch of electrons must be matched to the phase of this voltage. That is why in some rf electron accelerators an additional *bunching system* with a transverse electric field, often called a *rf chopper*, is used which can operate as an electric gate letting through single electron bunches every 12.5 ns at the frequency of 40 MHz. Another design is to combine the prebuncher and the chopper into one rf resonator at a subharmonic frequency with respect to the accelerator fundamental frequency. As a result, a sequence of more widely spaced and longer bunches are obtained at the output, and the intensity of a single pulse, i.e. the amount of charges in the bunch, increases. Combined with a fast pulse unit supplying power to the respective control grid, single high-current electron bunches can be produced.

The next component of the accelerator system is the buncher (Fig. 2.18) proper. The electron bunches at its output attain an energy of about 250 keV, which corresponds to a velocity of 0.75c. The bunches undergo further phase compression and become confined within a phase interval of 30°. Slower electrons become relatively more accelerated than the faster ones. The beam thus formed is then directed to successive accelerating sections (chambers). The number of such sections is very large, e.g. the length of the largest rf electron linear accelerator is approx. 3200 m. This accelerator has an energy of 11-40 GeV and is known as SLAC (Stanford University Accelerator Center, Stanford University, USA).

In the smaller rf linacs which are the main subject of our considerations, that is those with an energy of the order of several to several tens of MeV, the prebuncher and the buncher are usually combined with the first accelerating section (chamber). A magnetron with a pulse power of 2 MW is normally used as an rf source and acceleration is most frequently done with a standing wave (Fig. 2.16b). *Phase bunching* and *pre-acceleration* occur along a segment less than one metre long. After traversing this segment the accelerated electron beam becomes fully relativistic, i.e. its velocity differs from that of light by less than 1%. As follows from the curves in Fig. 2.1, 3 MeV electrons reach velocities of 99% c. It is this conversion of comparatively slow electrons generated by the source (Fig. 2.18) to fully relativistic electrons over the distance of a mere 1 metre that radically distinguishes rf electron accelerators from rf ion accelerators. As a consequence of a fast progress in electron bunching techniques achieved recently, the injection voltages in the range of 100-200 kV initially employed have been abandoned and replaced with those in the 1-30 kV range. Typically, approximately 30% of the injected electrons are captured, bunched and subsequently accelerated by the structure.

Linear accelerating gradients attained in rf electron accelerators usually range from 10 to 20 MeV m^{-1}. This is why terminal energies of the order of ten plus or several tens of MeV are obtained in the accelerating structures with lengths not higher than 2-3 m. Intensive research is now being carried out on the design of structures whose gradients range from 40 to 110 MeV m^{-1}. Commercial production of such structures would make it possible to reduce the accelerator length even further.

Evolution of accelerating structures. Fig. 2.19a is a schematic diagram of a typical $\pi/2$-type structure equipped with *circular diaphragms*, which was originally employed in accelerators with a standing wave. The efficiency of acceleration of this kind was low since there existed no accelerating gradient in every other resonator. The first step to improve the design was to shorten the lengths of *passive resonators* and to increase the lengths of *active resonators* (Fig. 2.19b). Regardless of their lengths, all the resonators were tuned up to the same frequency, preserving the stability characteristic of the $\pi/2$-type structures. Next, a new idea was introduced to the design of the accelerating wave-guides. It was to remove from the beam the passive *coupling resonators*, those that do not contribute to acceleration, by locating them at the side of the wave-guide. Accordingly, a wave-guide was constructed which had *side-coupled resonators* (Fig. 2.19c). These resonators act as elements coupling the accelerating resonators or cavities proper. Hence the name — *resonance coupling*. The design shown in Fig. 2.19d was developed at Los Alamos Scientific Laboratory (USA) and is commonly known as *LASL resonator*.

Comparison of traveling wave with standing wave structures. Typically, a structure with a traveling wave operates in a circuit with diaphragms (Fig. 2.16a) that are spaced at unequal distances in the buncher for the purpose of matching the phase velocity of the wave, v_p, to the that of the accelerated electrons, v_e (Fig. 2.20a). After attaining a sufficient degree of bunching and a sufficiently high velocity of the accelerated electron bunch the structure becomes uniform, i.e. the distances between diaphragms forming successive cavities are identical. In the earlier designs, $2\pi/4$-type structures were dominant. They were characterized by the fact that at a given instant electrons were accelerated in one out of any four successive resonators (the bottom part in Fig. 2.20a). In further design, three resonators with phase shift of $2\pi/3$ rad (120°) corresponded to a single wave-length, λ, and at a given moment the bunch was accelerated in one out of three successive resonators. Standing wave structures developed later made it possible to accelerate

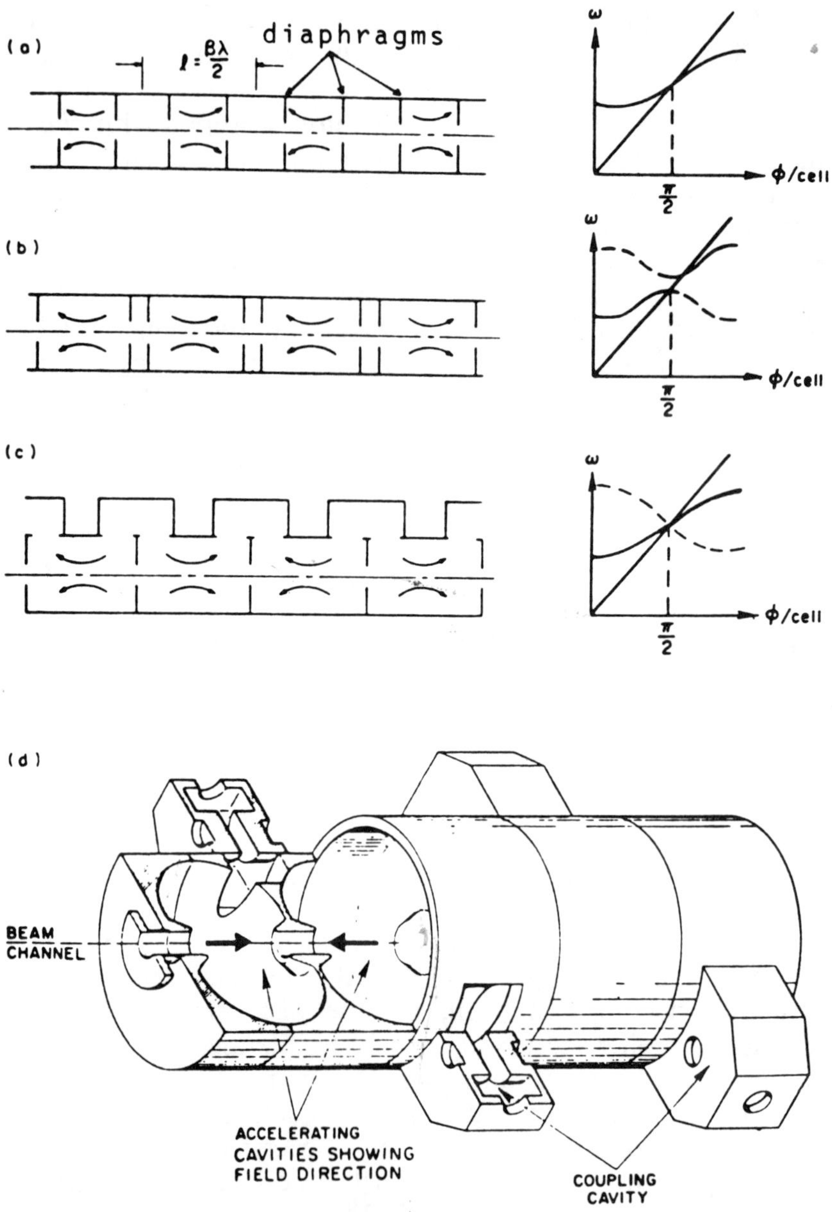

FIGURE 2.19 Evolution of $\pi/2$-type accelerating structures with circular diaphragms into side-coupled structures (after G.A. Loew and R. Talmen [2.7], description in the text).

(a)

(b)

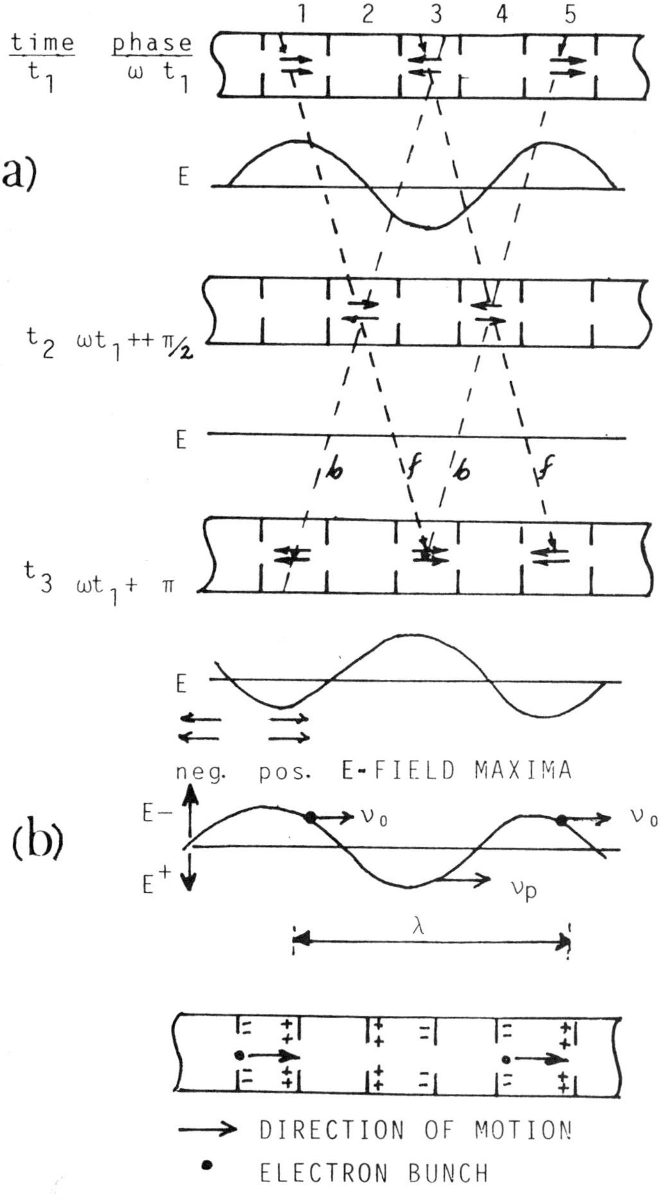

FIGURE 2.20 Comparison of operating principles of accelerating structures: (a) with a traveling wave, (b) with a standing wave

electron bunches at a given instant. Traveling wave structures have a wider frequency band and thus, for the same type of operation, are less susceptible to frequency variations as compared with the corresponding structures with a standing wave. Since the bunching resonator in the design featuring a traveling wave is adjacent to the rf energy supply which is installed close to the electron source (see Fig. 2.17a), the amplitude of the electrical component in the initial cavities is relatively independent of the beam current. The action of the bunching system is, therefore, resistant to the variations in the beam load structure. As a result, good overall efficiency of bunching is obtained for a wide energy range, the typical value being 35%.

Fig. 2.20b illustrates standing wave accelerating field patterns for three successive instants, t_1, t_2, t_3. The pattern at time t_1 recurs every one-half cycle after that at time t_3. The resulting field E, which is a sum of the forward component (denoted by f in Fig. 2.20b) and the reflected component (symbol b), displays an amplitude twice as large as that of the traveling wave structure, provided the I^2R losses in copper and beam load are negligible. As follows from Fig. 2.20b, the field is always equal to zero in every other cavity, e.g. nos. 2 and 4. For example, at instants t_1 and t_3 each component (the forward and the backward one) equals zero, and at instant t_3 both components are of equal value but opposite sign. As mentioned earlier this unfavourable situation can be rectified by using a design with a side coupling (Fig. 2.19c and d).

In traveling wave structures, the choice of resonator diaphragm aperture must be the result of a compromise. The aperture, on the one hand, should be large enough to ensure efficient rf energy propagation along the wave-quide, i.e. good coupling between the adjacent cavities. On the other hand, it should be small enough to maximize the half-axial electric component, E, of the beam accelerating field. This controversy has been eliminated in standing wave structures since the accelerating resonators are optimized with respect to the E component, and side coupling resonators are optimized in such a way as to obtain high efficiency of rf energy transport. Therefore, the accelerating resonators have small apertures and they can be equipped with endpieces which act as screening drift electrodes (see Fig. 2.19d). The gradient in such endpieces equals zero and the electron bunch crosses the resonator gaps when the accelerating electric field, E_2, is close to maximum.

In standing wave structures, all the resonators tend to exhibit identical values of the electric field since the rf energy undergoes multiple (some 100 times) oscillations, forward and backward in the wave-guide, to give shape to the resulting E and H components of the field. By loading the resonator with a beam, the component in all resonators, including the bunching cavity, can be reduced. Consequently, bunching in standing wave structures is efficient only in a narrow energy range which is typically equal to ±5%. To ensure proper bunching with beam current increase it is essential that the rf energy delivered to the wave-guide be correspondingly increased.

2.3.3 Parameters of rf Accelerating Structures

Microwave resonators ensure that acceleration is highly effective. In a 5 cm long resonator gap an accelerating potential of about 1 MV can be produced with pulse loss energy of some 0.2 MW in the resonator. The efficiency of this operation can be characterized by a *figure of merit or quality factor Q* defined as

$$Q = \frac{f_0}{\Delta f} = 2\pi f \,\frac{\text{energy stored in resonator}}{\text{energy loss per cycle}} \tag{2.15}$$

where f_0 is the resonant frequency, Δf is the frequency difference between the half power points. The energy accumulated in the resonator is, in fact, that of the E and H components of the field. Loss energy, on the other hand, is that released in the surface layers of the resonator walls due to the current transfer, leading to the increase in E and H fields. In the S-band (3 GHz), typical of radiotherapeutic accelerators, the unloaded resonator Q-factor is approx. $2 \cdot 10^4$ and higher. This value diminishes with the increase in resonator beam load and with the growth of coupling with adjacent cavities. If the Q-factor of a beam loaded resonator operating at 3 GHz is 10^4, its half-width frequency is $\Delta f = 300$ kHz. To ensure stable operation in such conditions, i.e. to have the energy decrease less than 1%, it is essential to use a system of *automatic frequency stabilization* to keep it within ±20 kHz.

To meet the above stabilization criterion, the *mechanical tolerances* in the manufacturing of the wave-guide, i.e. of a system of accelerating resonators, must be equal to 10^{-2} mm (10 μm). Very careful machining is thus important. The coefficient of thermal expansion for copper, from which resonators are made, is so high that the changes in the resonance frequency caused by expansion amount to 60 kHz °C^{-1}, and, therefore, the temperature of the resonator inside surface should be kept with an accuracy of better than 0.5°C. Greater changes in temperature can be compensated for by means of an *automatic frequency control*.

The energy V which is achieved by the electron traveling through the accelerating structure is determined as an integral of the first harmonic of the axial accelerating field, K_z, over the length of the structure, l, in metres multiplied by the electron charge e

$$E = e\int_0^l K_z \, dz \qquad [\,\mathrm{MeV}\,] \qquad (2.16)$$

The *unit shunt impedanace*, i.e. the impedance per unit length expressed in MΩ m^{-1}, is a useful parameter determining the quality of an accelerating structure. It enables the evaluation in a simple way of the *supply power*, P, that is necessary to produce the field E producing the desired energy V. The *shunt impedance*, r, is given by

$$r = \frac{K_z^2}{dP/dz} \qquad [\,\mathrm{M\Omega\, m^{-1}}] \qquad (2.17)$$

In other words, the impedance is a squared axial field K_z divided by the unit loss power, i.e. the *loss power* released in the resonator walls of unit length. Typically, in the S-band the r values are 50 - 120 MΩ m^{-1}, whereas the impedance in proton accelerating structures at 200 MHz is about 35 MΩ m^{-1}. The r-value should be as high as possible since in this way the supply power demand can be reduced. When r = 100 MΩ m^{-1} and the frequency is 3 GHz, the accelerating gradient of 70 MV m^{-1} can be achieved in a 1 m long structure when the pulse power is 50 MW. In radiotherapeutic accelerators smaller gradients are utilized, those between 10 and 20 MV m^{-1}, and, therefore, less costly magnetrons and klystrons can be employed featuring lower pulse power (Sec. 3.1.2)

2.3.4 Beam Temporal Structure

Magnetrons and klystrons employed at present as supply rf generators for linear accelerators, attain pulse powers from a few to some 40 MW. Such a high power can only be achieved in pulses that do not exceed several microseconds in duration. For example, if a 30 MW pulse power klystron operates at 3000 MHz it executes 3000 oscillations per one microsecond at full power, the pulse repetition rate ranging from a few hundred to a few thousand per second. Thus, for 1000 pps, the klystron mean power is 30 kW. This pulse mode of operation is described by the *filling coefficient* which is defined as the product of the duration of a single pulse and the pulse repetition rate. In the above example, the filling coefficient is equal to 10^{-3}(1 μs·1000 pps). That means that the klystron operates at full power at the mere 0.1% of the accelerator operating time. The filling coefficients for rf linear accelerators range from 10^{-3} to 10^{-1} (0.1% to 10%). Typically, for radiotherapeutical accelerators they are 10^{-3}.

Thus, in an rf linac, the accelerated beam is of a pulsed structure. In the above example, the electron beam at the accelerator's output consists of 1 μs pulses. At the repetition rate of 1000 pps, the interval that appears between the successive pulses is about 999 μs. This is referred to as *macrostructure*. In a single pulse of 1 μs duration the klystron executes 3000 oscillations operating at 3000 MHz. Since the electrons are bunched in narrow packets (bunches), acceleration of one packet (bunch) corresponds to a single oscillation. That is why in this case the output pulse consists of 3000 electron packets (bunches) and the beam structure of this kind is referred to as a *fine structure* or microstructure.

(a)

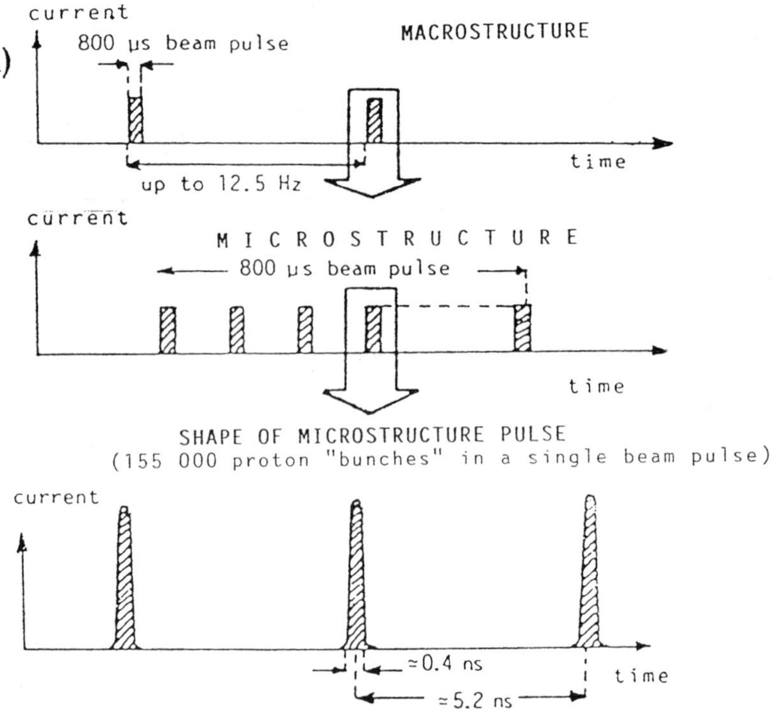

(b)

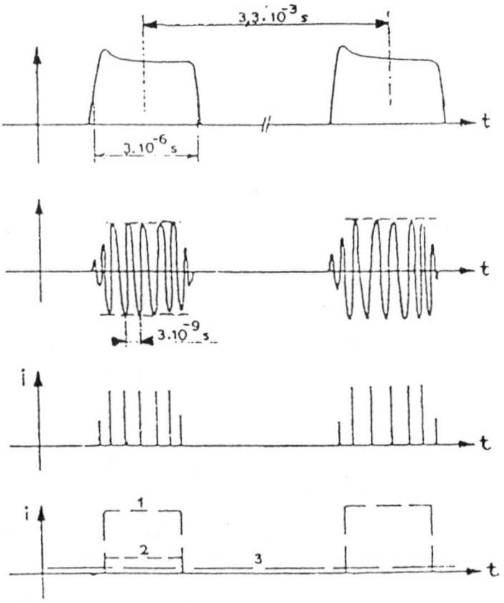

FIGURE 2.21 Temporal structure of accelerated beam in rf linear accelerator: (a) proton accelerator, (b) radiotherapeutic electron accelerator

Fig. 2.21a shows, by way of example, the micro- and macrostructure of the beam acceleration in a 10 MeV proton linear accelerator equipped with resonators operating at 193 MHz. The beam macrostructure includes 800 μs pulses whose maximum repetition rate is 12.5 Hz. Thus in one second an accelerated beam is on for a total of $12.5 \cdot 800 = 10.000$ μs or 10 ms. The *macrofilling coefficient* is then 10^{-2}, i.e. 1%. The beam microstructure is determined by the acceleration in the cavity resonators. A single macropulse, 800 μs in duration, consists, in fact, of 155 000 subpulses, each of 0.4 ns duration time and the intervals between the successive subpulses of about 5.2 ns (Fig. 2.21). The total duration time of the subpulses during a single macropulse is 62 μs. Thus the *microfilling coefficient* is 62/800 = 0.0775 or 7.75%.

The example shown in Fig. 2.21b refers to an electron accelerator employed in radiation therapy. Single 3 μs pulses (Fig. 2.21b-1) are repeated every 3.3 ms, the period of each oscillation (Fig. 2.21b-2) being 3 ns. As a result, the beam current consists, in fact, of very sharp peaks (Fig. 2.21b-3) with subnanosecond duration times. The whole situation is summed up in Fig. 2.21b-4. At the onset of bunching, the electron pulse current (1) is equal to 0.3 A in the phase angle of 120°, and it is increases to 3A at the angle of 12° at the end of bunching. The 3 μs pulse current is 0.1 A, while the mean current over several seconds is a mere 100 μA or 10^{-4}A.

What should be emphasized is that the peak beam current in rf linear accelerators exceed the mean values by a factor of 10^3 to 10^4. Hence, both in radiation therapy and in other types of irradiations (e.g. sterilization) the peak intensity is higher than the mean intensity by several orders of magnitude. This fact must be taken into account in accelerator beam dosimetry.

A detailed analysis of the theoretical fundamental of rf linear accelerator operation is presented in the monograph [2.6] as well as in the latest works [2.7 and 2.8]. The recent advances in rf linear accelerator development are summarized at periodical conferences, such as those held in 1988 and 1990 [2.9 and 2.10]. Rf linear accelerator applications in radiation therapy are dealt with in two small monographs [2.17 and 2.18].

2.4 Particles in a Magnetic Field. Magnets

2.4.1 Bending the Particle Orbit

If a particle with a total mass $m = Am_n$ and charge q travels with velocity v in a direction perpendicular to the lines of the magnetic field of induction B, it is subjected to the *Lorentz force* F_L (Fig. 2.22a) perpendicular to the directions v and B.

The force F_L is given by

$$F_L = v\,B\,q \tag{2.18}$$

The radius of curvature of the particle orbit, r, is determined by the equilibrium between the *centrifugal force* F_0 acting upon the particle and the force F_L, according to the equation

$$\frac{m\,v^2}{r} = v\,B\,q \tag{2.19}$$

centrifugal force = the Lorentz force

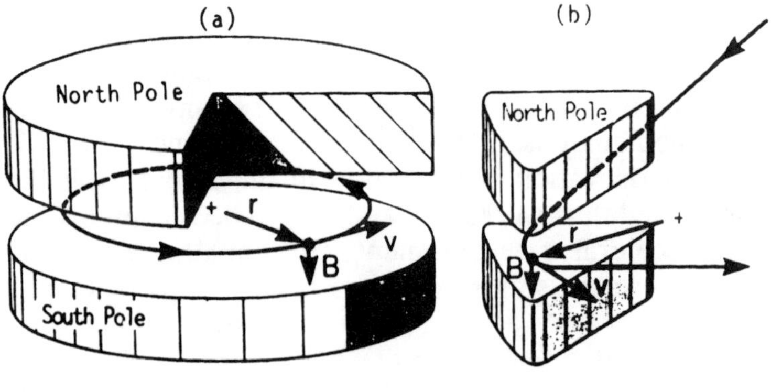

FIGURE 2.22 Motion of ions in magnetic field produced by (a) circular solid magnet, (b) sectional magnet.

The condition for the equilibrium of forces can also be presented as

$$B\,r = \frac{m_n\,v}{q/A} \tag{2.20}$$

where B is magnetic induction, r is the orbital radius, m_n is the nuclear mass, v is the particle linear velocity, q is the particle charge and A is the mass number. The product Br is called the *particle magnetic rigidity*. It follows from (2.19) that the higher the ratio of the charge to the mass number for a given particle the lower induction value B can be used for an orbit with the same radius r.

The particles, whose paths are curved by the magnetic induction B, travel in an orbit with the radius r with a frequency of revolutions f_r that can be expressed as

$$f_r = \frac{v}{2\pi\,r} = \frac{(q\,/\,A)\,B}{2\pi\,m_n} \tag{2.21}$$

Magnets of solid construction, such as those shown in Fig.2.22a, are employed in circular accelerators like cyclotrons, synchrotrons and microtrons. Magnets of this kind, whose diameters exceed 1-2 m are very expensive, and, therefore, the diameters of the largest solid electromagnets are never greater than 6-7 m.

The maximum energy for the particle beam in a circular accelerator is proportional to the product Br (the magnetic rigidity or *bending power*). With the present-day standard of magnetic materials technology, it is not possible to exceed the induction value of about 2.0 T using traditional materials at ambient temperatures. With these practical limitations imposed on the induction value, the particle energy, proportional to the product Br, can only be increased by increasing the orbital radius r. This requires an increase in the magnet diameters and, thus, a great increase in mass. This situation is depicted in Fig. 2.23. It follows from the main parameters of traditional magnets at ambient temperature, that the increase in magnetic rigidity calls for a corresponding increase in magnet mass from a few hundred tons to several thousand tons. By using the so-called *superconducting magnets* whose excitation windings operate at the temperature of liquid helium (the bottom line in Fig. 2.23) the mass of magnets can be radically reduced by about one order of magnitude. The reduction in the dimensions, weight, costs and

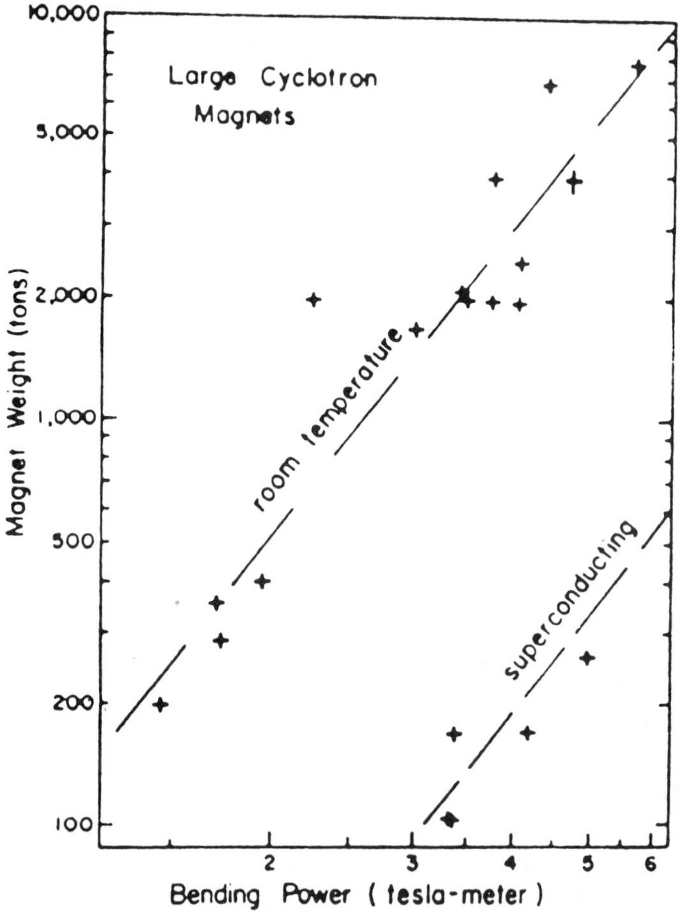

FIGURE 2.23 Dependence of magnet weight on magnetic rigidity Br (Tm) for large cyclotrons (after H. Blosser [2.11]).

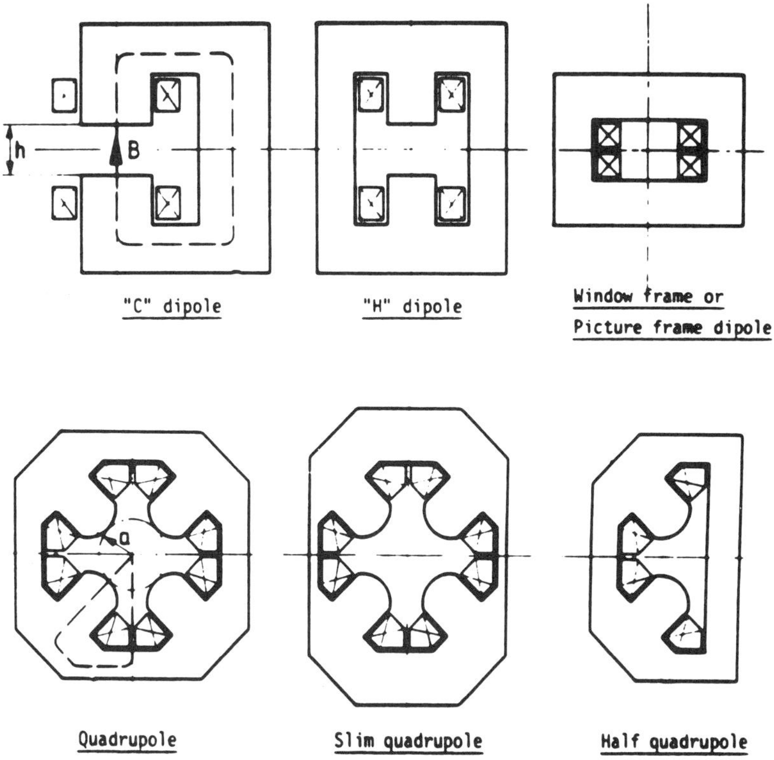

FIGURE 2.24 Construction of bipolar magnets used for bending particle trajectory and fourpolar magnets (quadrupoles) used for focusing particles.

supply power of superconducting accelerators is very attractive from the point of view of the deployment of such accelerators in hospital practice. This is why in recent years we have witnessed development of some new designs which will undoubtedly form a basis for future generations of medical accelerators, such as the MSU/Harper Hospital 50 MeV minicyclotron [2.19, 2.20], the Amersham/Oxford Instruments 12-17 MeV minicyclotron, the MSU synchrocyclotron [2.20] and the EULIMA cyclotron [2.21 and 2.22] (Sec. 5.1.2).

The reduction of magnet mass can be implemented with traditional methods by replacing solid construction (Fig.2.22a) with segmental structures (Fig. 2.22b). In the latter design, the paths of particles moving in an orbit are curved only in the portions of the orbit which include magnetic segments. Between these segments there are free areas where the particles move in an approximately linear path. The use of segmental magnets has made it possible to develop high-energy accelerators — electron and proton synchrotrons — where particles travel in orbits with diameters of the order of kilometers.

Typically, the systems used for bending particle orbits (Fig. 2.22) are bipolar magnetic structures and, therefore, are referred to as *magnetic dipoles*. Fig. 2.24 shows a schematic diagram of some of the dipoles used in charged-particle acceleration. The figure also shows magnetic structures consisting of four poles, the so-called *quadrupoles*, normally used in beam focusing techniques.

2.4.2 Beam Transport and Energy Analysis

After acceleration is completed, both in the linear and circular accelerators the beam is often extracted and is delivered to receiving units through a beamline *transport system*. During transport, the direction of the beam must often be changed. This is especially true for radiotherapeutic accelerators which permit rotational therapy (Sec.3.1.3). Fig. 2.25a shows the construction of the magnet that enables the transported beam to be bent by 90°.

Unfortunately, the particle beams in accelerators are not monoenergetic. Each beam has a *particle energy dispersion*, ΔE_k, around some central energy E_k. Energy dispersion is defined as an energy range that includes the overwhelming majority of particles, for example, 95% of the total number of particles.

The radius of curvature of the paths of particles traveling in the perpendicular magnetic field is, in turn, dependent on their momentum or energy. In a field with uniform induction, paths of particles with higher energies will be less curved than those with lower energies (Fig. 2.25b). This property is commonly taken advantage of in the analysis of particle energy. By placing a *diaphragm with a well-defined slit*, perpendicular to the beam direction, particles with accurately determined energy dispersion can be selected from the beam. Systems of this kind are often employed in acceleration techniques to diminish the energy dispersion of the accelerated beam. By regulating the slit's width, as shown in Fig. 2.25b, the dispersion of the beam

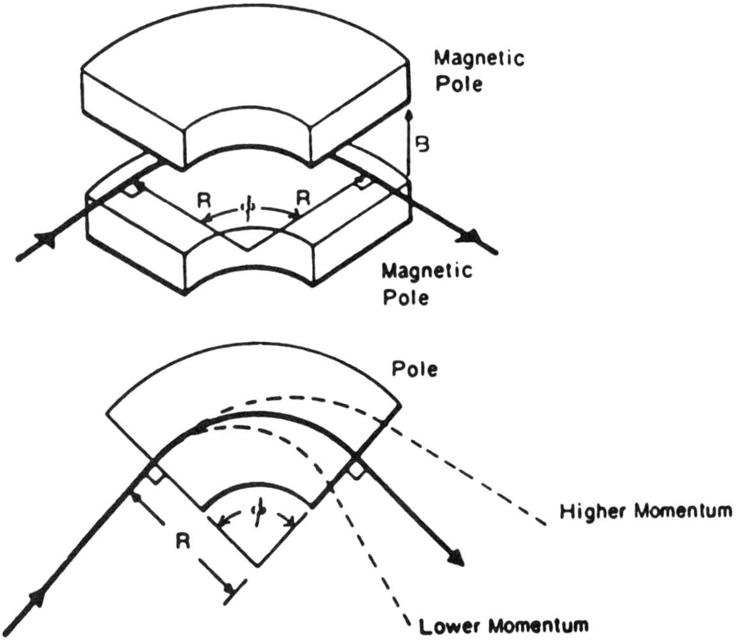

FIGURE 2.25 An analyzing magnet: (a) construction, (b) trajectories of particles with various momentum values. R — radius of trajectory curvature, ϕ — angle of beam deflection.

analyzed can be increased or decreased. These matters will be discussed in greater detail in Sec. 3.1.3.

2.4.3 Magnetic Focusing

In circular accelerators, the magnetic field, in addition to bending the particle's trajectory, plays another important role in the acceleration proper: it ensures both *horizontal and vertical focusing* of the beam. This is necessary because, due to the action of the Coulomb forces, the particle beam tends to increase its dimensions, i.e. to spread out. In circular accelerators provided with solid magnets (Fig. 2.22a) the tendency of the beam to spread out was originally counterbalanced by having the pole pieces shaped in such a way that the field from the center towards the edges of the pole pieces would decrease (Fig.2.26). As a result of the shape of the field, the particles that for some reason were located below or above the central plane became tightly bound to the plane. In this way vertical or axial focusing was carried out (Fig.2.26).

Thus the particle moving in an nonuniform magnetic field is subjected to a force that is proportional to the particle's distance from the symmetry plane. In practice, this force causes the particle to oscillate around the central plane which should hold the *"ideal" particle orbits*.

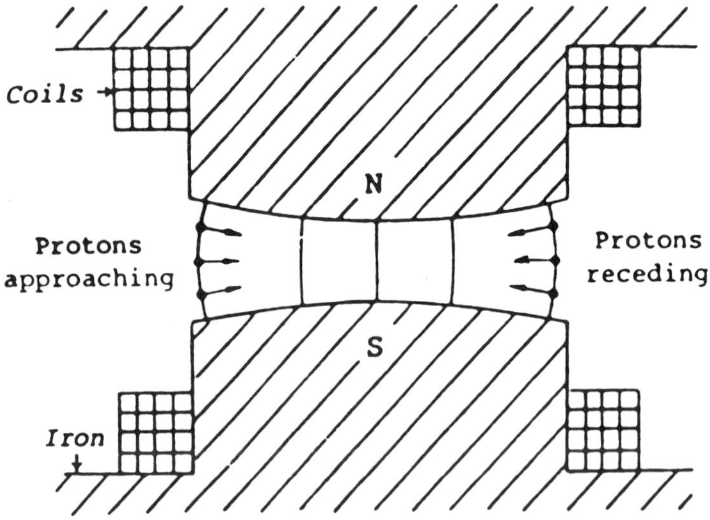

FIGURE 2.26 Magnetic field in a synchrocyclotron with weak focusing

These particle oscillations are referred to as *betatron vertical oscillations*. However, the above term is not quite accurate since particle oscillations are also found in traditional cyclotrons as well as in other circular accelerators. The betatron oscillations are essential in acceleration processes and determine the main parameters of circular accelerators.

In order to express the efficiency of magnetic focusing, a so-called *field index or field attenuation coefficient* denoted by the symbol n is introduced. It can be defined as

$$n = \frac{\text{relative changes in induction}}{\text{relative changes in radius}} = -\frac{dB/B}{dr/r} = -\frac{r\,dB}{B\,dr} \tag{2.22}$$

To ensure good vertical stability or proper axial focusing it is essential that the field distribution should lead to a decrease in induction with an increase in the radius, i.e. the following condition be fulfilled

$$0 < n < \infty \tag{2.23a}$$

In addition to effective vertical focusing, simultaneous focusing in the horizontal direction or radial focusing should also be ensured.

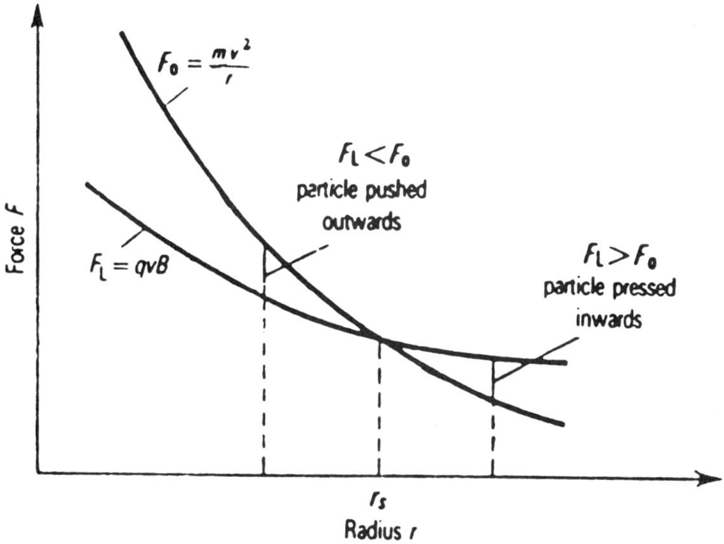

FIGURE 2.27 Radial focusing

Fig. 2.27 shows the Lorentz force and centrifugal force curves as a function of the orbital radius for a given particle velocity (v = const). According to Eq. (2.18), the centrifugal force becomes balanced with the Lorentz force for a stable orbit with a radius r_S. When the particle orbit deviates in any way from the above stable orbit, the particle will find itself subjected to a force F_L that is too small (the radius $r < r_S$) or too great ($r > r_S$). The particles deviating from the stable orbit r_S will thus be subjected to *quasi elastic forces* which will tend to turn them back to the stable orbit. It is the stabilizing forces of this kind that give rise to *radial betatron oscillations.*

The *condition for radial stabilization* is that with increasing radius the Lorentz force should decrease slower than the centrifugal force, or the force F_L should increase. To ensure simultaneous focusing in the vertical and horizontal (radial) directions in a circular accelerator equipped with magnets as shown in Fig. 2.26, the condition 2.23a is modified to the following form

$$0 < n < 1 \tag{2.23b}$$

The field index must always be less than one, i.e. it is small. Therefore, the efficiency of focusing is also low and the method is referred to as *weak focusing.* The accelerators that utilize the above method are called *constant gradient machines.*

The constancy of the gradient refers to the azimuth, a particle moving in a given orbit encounters identical values of the field at each point of the orbit. With increasing orbital radius the field decreases relative to the maximum value at the accelerator axis.

Method of strong focusing. In the early '50s a much more efficient method of focusing was developed, referred to as *strong focusing* or the *alternating gradient method.* At present it is most often implemented with fourpolar magnets, called *quadrupoles.* A schematic diagram of such a magnet has already been presented in Fig. 2.24. Now a distribution of forces acting on a beam is shown in Fig. 2.28. The four black dots represent four positively charged particles at various points relative to the axis of the system perpendicular to the plane of drawing. At the axial (central) point the resultant effect of the magnetic fields is zero, and therefore the particles moving exactly along the axis do not undergo any deflection in the quadrupole. Particles above or below the central point become "pushed" to this point by the magnetic field shown in Fig. 2.28b. A quadrupole depicted in Fig. 2.28b focuses the beam in the vertical direction, whereas in the horizontal direction the beam is, unfortunately, defocused.

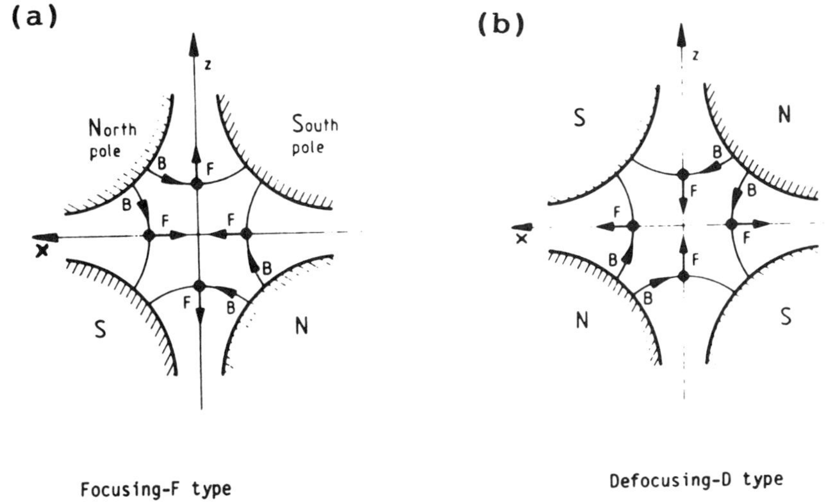

FIGURE 2.28 Fields and forces in quadrupoles: (a) focusing F-type magnet, (b) defocusing D-type magnet.

The electromagnet focusing the beam in the vertical direction as shown in Fig. 2.28b is referred to as a *type D magnet*, while that focusing in the horizontal direction (Fig. 2.28a) is called a *type F magnet*. The field indices, n, in magnets of this kind are as high as several tens or hundreds. A quadrupole system of magnets for strong focusing consists of a large number of successive magnets of type F and D.

In order to explain the idea of strong focusing, we shall make use of an optical analogue (Fig. 2.29). A system of two adjacent quadrupoles with their directions of action shifted (F + D or D + F) may be regarded as a system of concave and convex lenses. The resultant focal length, F, of two lenses whose focal lengths are f_1 and f_2 respectively, separated by a distance d, is given by

$$\frac{1}{F} = \frac{1}{f_1} + \frac{1}{f_2} - \frac{d}{f_1 f_2} \qquad (2.24)$$

If the lenses have focal lengths that have equal absolute values but opposite signs ($f_1 = -f_2$), Eq. (2.24) takes the form of $F = f^2/d$ which is always positive. In fact, the resultant focal length F remains positive when f_1 and f_2 have unequal absolute values as long as they are of opposite sign.

Although each quadrupole provides strong beam focusing only in one plane, while it defocuses the beam in the orthogonal plane, the resultant effect of the system of two such quadrupoles is always positive. Since the field index need not be limited to a value smaller than one, very strong focusing may be achieved by the above method.

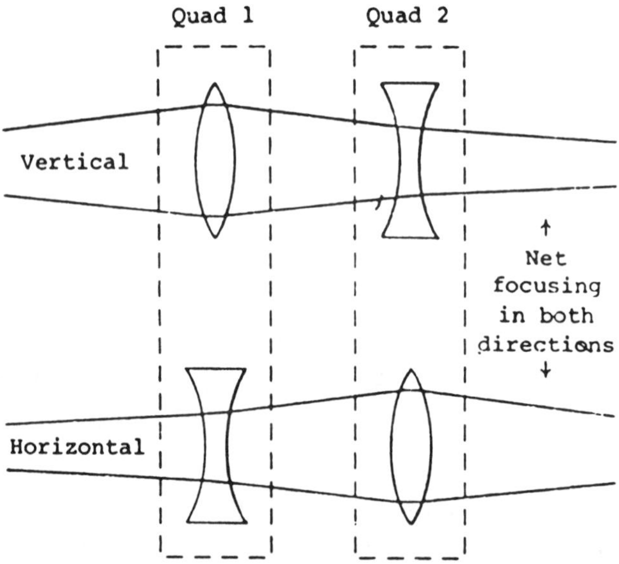

FIGURE 2.29 Optical analogy of strong focusing.

2.5 Betatron Methods

In the 1920s many researchers worked on the construction of a circular accelerator intended for accelerating electrons. This machine became later known as a betatron. The first such accelerator was built by D.W.Kerst in 1940. The electron energy attained in it was 2.35 MeV, while in the second machine of this type in 1942 it was already 20 MeV.

A betatron (Fig. 2.30a) consists of a ring-shaped accelerating chamber located between the poles of an electromagnet whose excitation windings are suppplied from a sinusoidal voltage (Fig. 2.30b). The electrons emitted by the source and injected into the so-called *stable orbit* with initial energies of the order of several tens keV move in a circular trajectory inside the chamber.

The magnetic field of the electromagnets plays a dual role in the betatron. First it causes the electron trajectory to curve and keeps the electrons in a circular orbit, and second, it induces a voltage that gives the electrons an acceleration tangential to the orbit. Fig. 2.31 is a schematic diagram of how a *betatron field* is generated. The system of electromagnets is supplied with an excitation current I. The electric rotational E component (Fig. 2.31b) is induced in the horizontal plane during the rise of the H component with closed field lines. The particles accelerated revolve in a circular orbit which lies in the plane of the induced *rotational field*. Typically, the gradients involved are, unfortunately, rather small as they most frequently are not higher than a few to several tens of electronvolts per revolution. Since it is the relativistic electrons with velocities close to the speed of light that are being accelerated, the number of revolutions they make during a single operation cycle is as high as several hundred thousand or million. As a result of acceleration of this kind, typical energies of the order of several or several tens of megaelectronvolts can be achieved.

The length of the betatron operation cycle is determined by the frequency of the power being applied. In Europe, conventional power is 50 Hz, while in the USA it is 60 Hz. For the example of a BBC betatron shown in Fig. 2.30 the operation cycle is completed in the rise phase of the magnetic flux shorter than 1/200 s (50 ms) for the frequency of 50 Hz. The electrons injected with an energy of 40 keV are accelerated and extracted when, in the above example, they attain a final energy of 31 MeV.

The end point of acceleration in a betatron (point b in Fig. 2.30b) occurs before the magnetic induction reaches its maximum value. For this reason the betatron electromagnet is

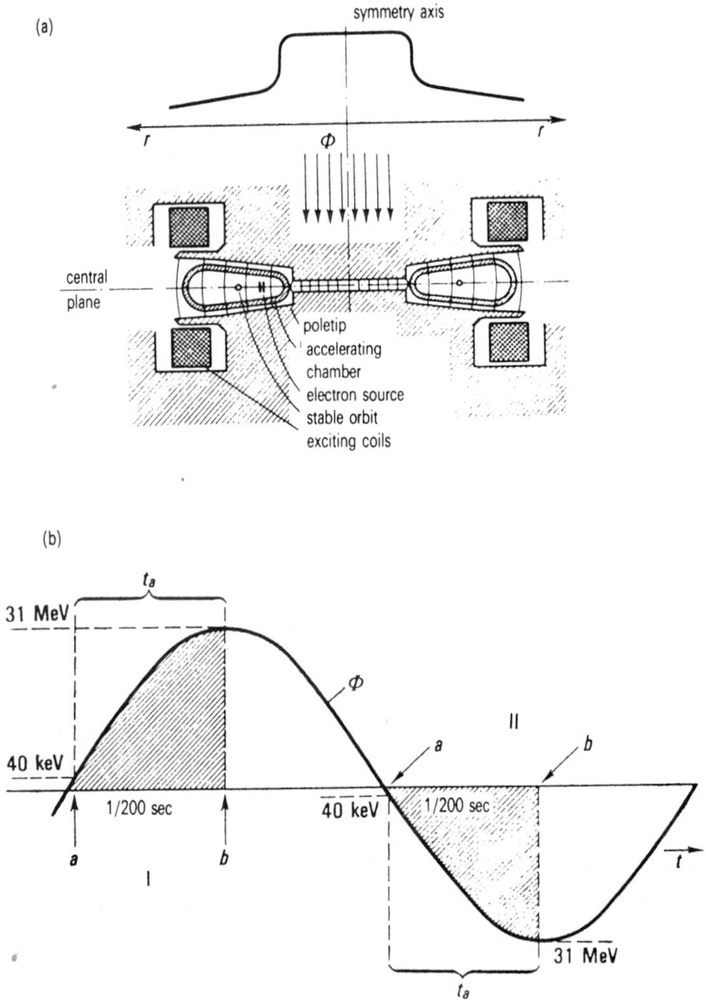

FIGURE 2.30 The betatron: (a) operating principle, (b) operating cycle.

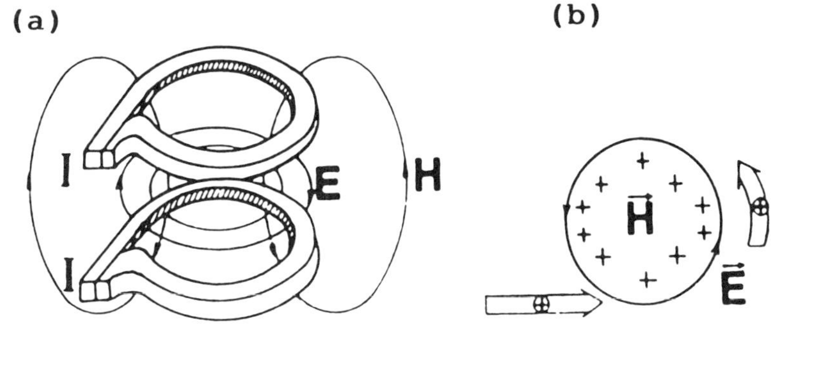

FIGURE 2.31 Betatron method of acceleration with a rotational electric field: (a) excitation system, (b) rotational electric field.

equipped with an additional winding, called an *expansion winding*. Shortly before the 50 ms acceleration cycle is completed, this winding is supplied with a strong current pulse of a few microsecond's duration, much shorter than the total operation cycle. The magnetic field thus induced by the expansion winding causes the orbital radius to increase, and the electrons start to move in an orbit with a radius larger than that of the stable orbit, so consequently, they gradually approach the chamber walls moving in a spiral trajectory. Finally, thanks to their high energies they penetrate the walls. In order to extract the electrons from the chamber a small *window* made, for example, from a thin metal foil is usually incorporated into the chamber wall. In a betatron intended for producing very hard X-rays (in photon therapy) the accelerated beam is deflected by means of the expansion winding in such a way that it strikes an X-ray producing *target* installed inside the accelerating chamber.

As already mentioned in Sec. 1.1.2., in the 1950s and 1960s betatrons played an important role in routine radiation therapy.

With the accelerated electrons converted to X-rays, the advantage of the betatrons was that the beam was very well focused, down to less than 1 mm^2. On the other hand, the disadvantages of the betatrons were their low maneuverability and the unattainability of high dose rates. For this reason, the production of betatrons was stopped in the mid-1970s. Since they feature high durability, betatrons can still be found at some places, operating as routine radiotherapy machines. Basic parameters and construction of betatrons are discussed in Sec. 3.2.

2.6 Cyclotron Methods

2.6.1 Conventional Cyclotrons

A *circular accelerator*, called a *cyclotron*, was designed by E.O.Lawrence and Livingstone in 1930. Lawrence was awarded the Nobel prize of 1939 "for the invention and development of a cyclotron, and especially for the results in artificial radioactivity of nuclides obtained with this device". The operating principle of a conventional cyclotron is shown in Fig. 2.32.

An ion source located in the central part of the machine emits a beam of ions which move inside two semicircular electrodes, called *dees*. The accelerated particles move in a nearly spiral trajectory which consists, strictly speaking, of two semicircles. The dees are installed inside a system of solid electromagnets (Fig. 2.22a). In a system of this kind, the *condition of the equilibrium of forces* is fulfilled between the Lorentz force and the centrifugal force according to Eqs. (2.18) and (2.19). The orbiting frequency of particles is then determined by Eq. (2.21).

Under the effect of an rf alternating voltage, an approximately uniform horizontal electric field or accelerating field is generated in the gap between the dees (Fig. 2.32). On the other hand, a constant magnetic field with induction of about 1.5 Wb m^{-2} acts throughout the whole chamber, bending the paths of the accelerated ions.

The frequency of the accelerated voltage is constant, usually within the range of 10 -30 MHz. The particle motion must be, naturally, fully synchronized with this frequency or its harmonic. Thus the single *orbiting period* T_0 must be also constant and independent of the particle energy

$$T_0 = \frac{1}{f_s} = \frac{2\pi\, m_n}{(q/A)\, B} = \text{const.} \qquad (2.25)$$

where f_s is the accelerating voltage frequency, synchronous with

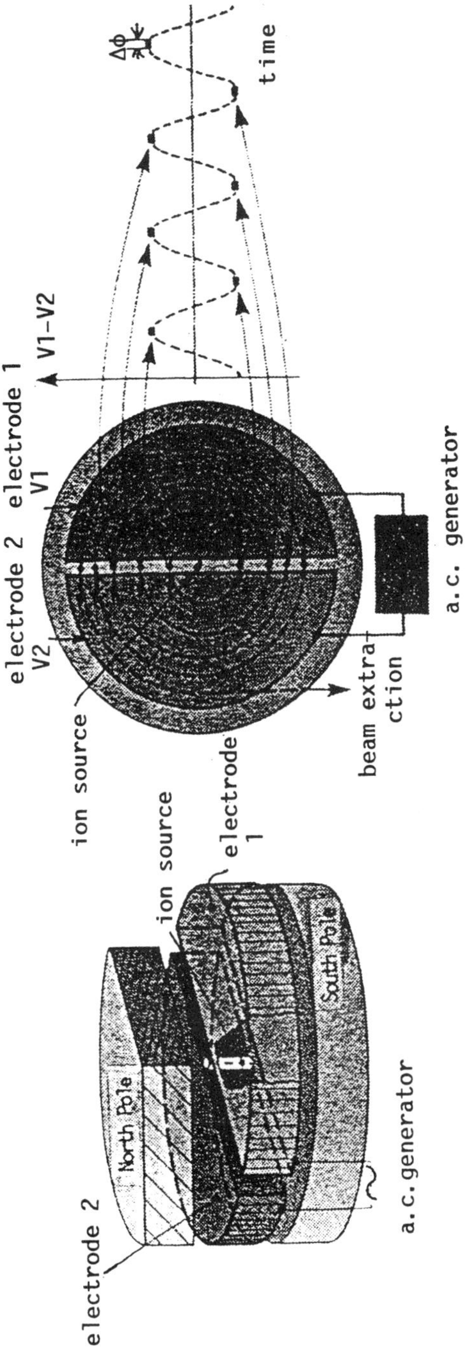

FIGURE 2.32 Operating principle of a conventional cyclotron.

the particle motion, m_n is the mass of the nucleon, q is the ion charge, A is the mass number and B is the magnetic field induction.

Unfortunately, with an increase in ion energy according to Eq. (2.4), a *relativistic increase of mass occurs*, which causes the process of acceleration to become desynchronized. The reason for this is that the accelerating voltage frequency in a conventional cyclotron is constant, or independent of time. As a result of the relativistic increase of mass, the particle becomes increasingly heavier and, consequently, cannot catch up with the changes in the accelerating voltage (see 2.25).

The practical *criterion of the desynchronization* of the acceleration process is based on the kinetic energy gain E_k equivalent to an increase of about 1% of the rest mass. Taking into account the values of rest masses of various particles we can obtain the following maximum energies to which ions may be accelerated in a conventional cyclotron:

protons (E_0 = 938 MeV)	$E_{k\,max}$	$\approx$	10 MeV
deuterons (E_0 = 1877 MeV)	$E_{k\,max}$	$\approx$	20 MeV
alpha particles (E_0 = 3733 MeV)	$E_{k\,max}$	$\approx$	40 MeV

Since the electron rest mass is 0.51 MeV and 1% of this value corresponds to an energy of a merely 5.1 keV, it is obvious that cyclotrons are not suitable for accelerating electrons.

Conventional cyclotrons have played an essential role in the development of charged-particle acceleration techniques. The fact that it was not possible to use them to achieve higher energies has gradually led to some limitation in their application. Nowadays, they are employed very seldom, if at all, only as small injectors for larger accelerator systems. Almost all the earlier conventional cyclotrons have been redesigned and turned into isochronous machines.

2.6.2 Isochronous Cyclotrons

Operating principle. In a conventional cyclotron, ions move in a uniform magnetic field. Therefore, at each point in an orbit of a radius r, exists the same induction B_r, independently of the value of the azimuth. In 1938, L.H. Thomas put forward an interesting modification of the cyclotron method, namely that of a totally different distribution of the magnetic field. In this new technique, the pole pieces are provided with segments forming *"magnetic ridges"* with gaps between sectors making up *"magnetic valleys"*. An ion, moving in an orbit of a given radius r, encounters sections whose magnetic induction is higher in the ridge than in the valley. Such a distribution of the magnetic field causes deformation of the particle trajectories compared with a spiral orbit in a conventional cyclotron. As a result, *isochronousness* is obtained over a much wider energy range. Even when a one percent increase of the relativistic mass is exceeded, condition (2.25) is still valid and the ions remain synchronous with the accelerating voltage. In addition, the particles are subjected to a magnetic field with an alternating gradient and, consequently, strong focusing also occurs (Sec. 2.4.3).

Since the particle orbit time is maintained constant, the Thomas cyclotron is often called an *isochronous cyclotron*. Cyclotrons of this type are denoted also by the acronym *FFAG (Fixed Field Alternating Gradient)*, as the magnetic field is constant with time, although areas of the alternating field gradient occur along the azimuth.

The *maximum unit energy* or the energy which can be imparted to a single nucleon is given by

$$E_{kn} = K\left(q/A\right)^2 \tag{2.26a}$$

$$K = \frac{R^2 B_{max}^2}{2\,m_n} \tag{2.26b}$$

where q is the ion charge, A is the ion mass number, R is the radius of the maximum orbit from which the beam is extracted, and B_{max} is the induction in the orbit R. The *K-coefficient*, most often expressed in terms of MeV/A, has values usually ranging between 100 and 600. In the new designs of isochronous cyclotrons with superconducting windings, the K-coefficient can be as high as 1000 – 1200.

Pole pieces with *magnetic segments or sectors*, as in Fig. 2.33a, are used in smaller isochronous cyclotrons with energies below 100 MeV. In larger cyclotrons, a segmental design (Fig. 2.33b)

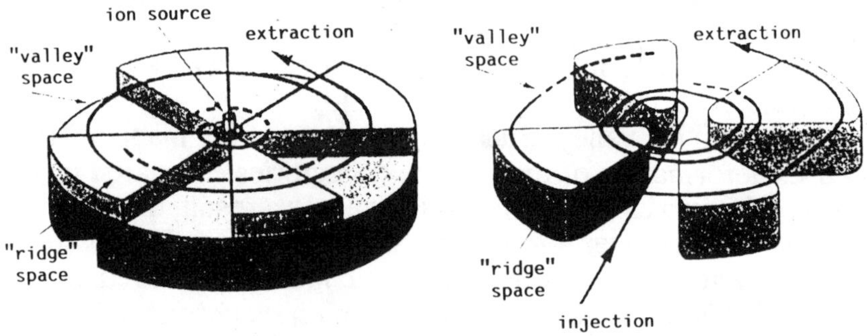

FIGURE 2.33 Isochronous cyclotron: (a) pole pieces with magnetic "valleys" and "ridges", solid magnet, (b) cyclotron with sectional magnets and external particle injection.

is used in which spaces between the adjacent segments are void of the magnetic field, correspond to magnetic valleys. Instead of using an ion source located in the central part of the accelerator (Fig. 2.33a), in the segmental design, the beam, after being preaccelerated, is injected from the outside. The beam injection energy is normally of the order of several tens of MeV and is often obtained in another smaller cyclotron cooperating with the main cyclotron. A segmental isochronous cyclotron, i.e. one with separated sectors (Fig. 2.33b) is denoted in the English language specialized literature by the acronym *SSC (Separated Sector Cyclotron)*.

It follows from Eq. (2.26a) that the unit energy of the accelerated ions depends on the square of the q/A ratio. Fig. 2.34 illustrates this dependence for an isochronous SPC 120 Scanditronix cyclotron. An accelerator of this type, of segmental or separated sector design (Fig. 2.33b), was developed for the Hahn-Meitner Institut (West Berlin). As an injector, a CN (HVEC)-type 6 MV tandem was employed. This system, known as *VICKSI* (Van de Graaff Isochron-Cyclotron Kombination für Schwere Ionen), is an example of a so-called *combined accelerator*, intended for heavy ion acceleration. The acceleration proper is implemented by two (or more) accelerators and the beam stripping between these accelerators is often carried out with the aim of increasing the charge-to-mass ratio, and, consequently, the total energy of the accelerated ion.

Cyclotron construction. Technique of H^- acceleration. The construction of a typical cyclotron for accelerating positive ions is shown schematically in Fig. 2.35a. A vacuum chamber (6) with dees (3) supplied from an rf generator (5) is located between two pole pieces of a solid electromagnets (see Figs. 2.32 and 2.33). The particles travel in a spiral orbit (Fig. 2.32) and are accelerated in a step-wise fashion attaining energies of 30 - 40 keV in a single orbit. The acceleration itself takes place in the gaps between the dees. The radii of the successive orbits increase in proportion to the square root of the total energy.

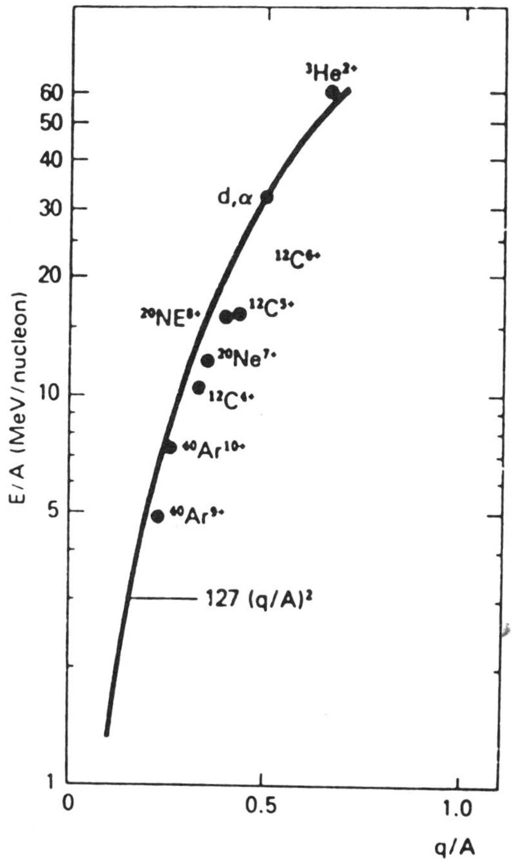

FIGURE 2.34 Unit energy E_{kn} vs. charge-to-mass ratio for an isochronous SPC 120 cyclotron (mfd. by Scanditronix)

The maximum attainable final energy is thus dependent on the magnet diameter and induction.

Towards the end of the acceleration process, the ions approach the *extraction system*, a system intended for extracting the beam. In Fig. 2.35a, which shows a system where positive ions are accelerated, a semicircular condenser, called a *deflector*, and an auxilliary magnet, called a *focalizer*, serve to extract the beam. The deflector electrodes are powered by a pulsed voltage which produces a gradient of 50-100 $kVcm^{-1}$ in the vicinity of the deflector. As a result, the beam is deflected out of the spiral trajectory and can leave the accelerator structure. The focalizer permits the beam path to be elongated and slightly focused.

Recently, instead of accelerating positive ions, *negative ions*, such as H^- or D^-, are becoming widely utilized. Negative ions are accelerated in the cyclotron (Fig. 2.35b) in the first stages of acceleration in a way similar to that used for positive ions. In the final stages of acceleration, however, they are made to strike a thin metal foil (4), where *stripping*, (a change of polarity) takes place. This process resembles a process in a tandem accelerator (Sec. 2.2.2.1). This means that the accelerated beam which initially consisted of negative ions of H^-, after stripping, is now made up of protons which leave the acceleration space. The field bending the particles in the system shown in Fig. 2.35b (at right angles to the plane of the figure) has a direction, of course, opposite to that of the system for accelerating positive ions, presented in Fig. 2.35a).

The above design helps to simplify the cyclotron construction by eliminating the electrostatic system of beam extraction. At the same time, the energy can be changed without the need to return the rf components of the cyclotron and change the magnetic field. For this purpose, foil (4) (Fig. 2.35b) is shifted onto a respective orbit. When the foil is shifted from outside the system, i.e. when the orbit radius is decreased, protons with respectively lower energies are obtained. Compared with the cyclotron which accelerates positive ions, in the above design the induced radioactivity is much lower. This is because in the process of extracting the beam (Fig. 2.35a) some positive ions strike the comparatively large surface of the electrodeflector, and thus cause in it high radioactivity. By totally eliminating the deflector the intrinsic radioactivity is considerably reduced.

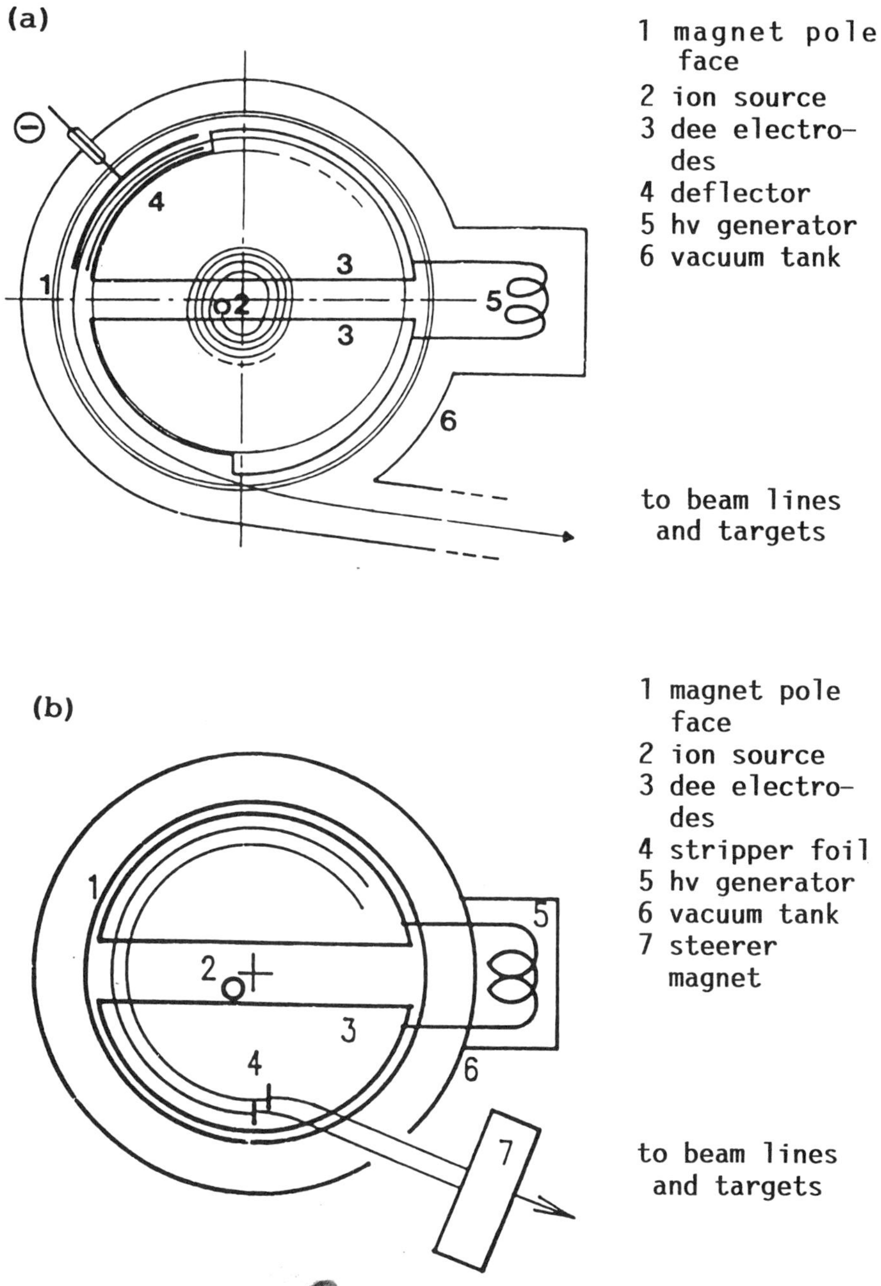

FIGURE 2.35 Cyclotrons for acceleration of (a) positive ions, (b) negative ions (H^-).

Energy classification of biomedical cyclotrons. Due to the reasons expounded in Sec. 1.4, cyclotrons have found wide application in the production of radionuclides and radiation therapy. The chronology of cyclotron installation is depicted in Fig. 2.36. Biomedical cyclotrons can be classified in five groups according

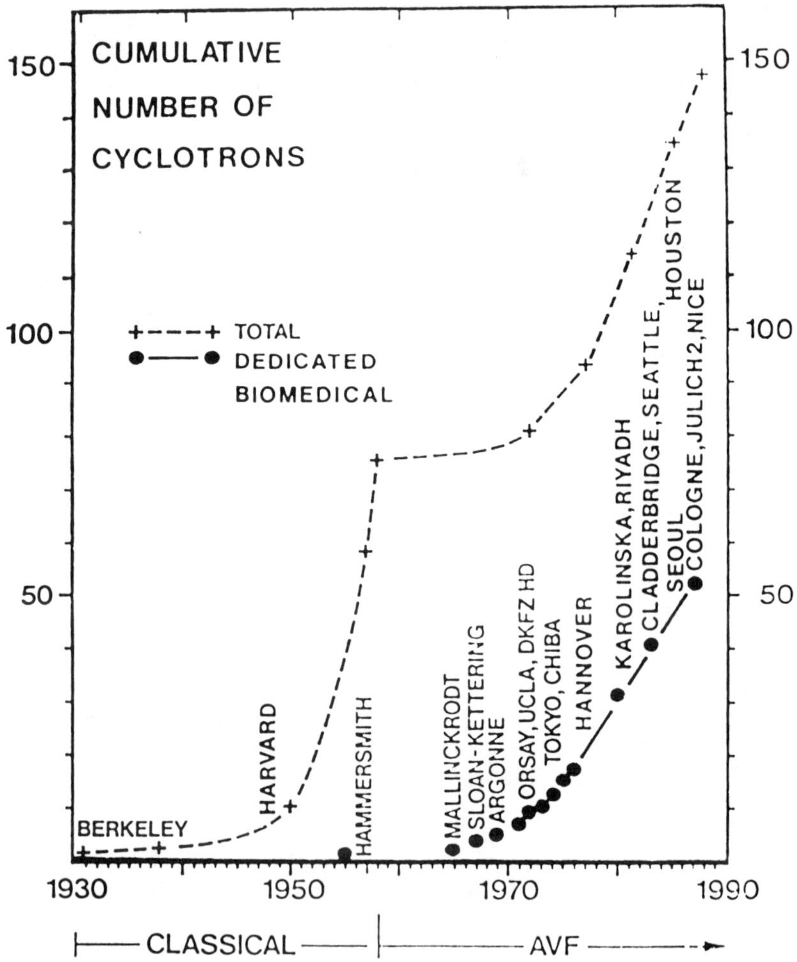

FIGURE 2.36 Chronology of cyclotron prevalence between 1930 and 1990, especially those of the isochronous type used in medicine and biology (after G. Wolber [1.58]).

to the type of their application. The first group comprises *"baby"* or *microcyclotrons* accelerating protons to 10–11 MeV or deuterons to about 7 MeV for the production of short-lived positron emitters. The second group are the so-called *"mini-cyclotrons"* intended for accelerating protons to energies of about 17 MeV or deuterons to 8–10 MeV. Like the cyclotrons of the first group, minicyclotrons are mainly used in the production of larger amounts of positron emitting nuclides, and, to some extent, long-lived radioisotopes. The third group includes larger *compact cyclotrons* with energies around 30 MeV for protons and 15 MeV for deuterons. These cyclotrons are suitable for the production of all the radionuclides used in medicine and biology. Due to their high yield they can operate in commercial factories producing radioisotopes (Secs. 7.1 and 7.4).

Finally, the fourth group comprises cyclotrons with energies higher than 40 MeV (H^+), intended for the production of radionuclides and for generation of neutrons subsequently used in radiation therapy. The demand for obtaining a neutron beam characterized by a depth dose comparable with that of 4-6 MV photons, led to the design of several cyclotrons with energies of 60 to 70 MeV, suitable for both neutron generation and direct production of protons for the radiation therapy (Secs. 5.1 and 5.3).

The fifth group of cyclotrons is made up of very large isochronous cyclotrons with sectional construction (Fig. 2.33b), which makes it possible to accelerate protons to the radiation therapy range of 200 to 250 MeV, as well as to produce mesons for experimental therapy.

Due to the highly differentiated construction of the above machines and to the classification of accelerators adopted in the present book, the following cyclotrons will be dealt with in more detail separately: neutron and meson generators (Sec. 5.2 and 5.3), proton accelerators for radiation therapy (Sec. 5.1.2) and systems of accelerators for the production of medical radionuclides (Sec. 7.1).

Fig. 2.37 shows the relation between the maximum proton energy and the cost of biomedical cyclotrons and systems for neutron therapy built between 1968 and 1988 [1.58]. Table 2.4 lists parameters of some cyclotrons representing various groups of minicyclotrons and other installations for medium energies.

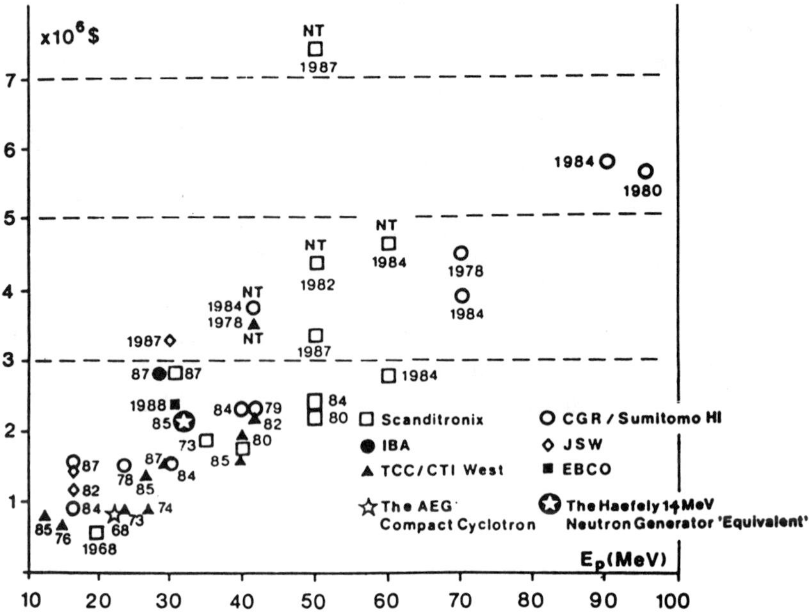

FIGURE 2.37 Dependence of maximum proton energy on cost of biomedical cyclotrons between 1968 and 1988 (after G. Wolber [1.58]).

Cyclotrons are dealt with at periodical International Conferences on Cyclotrons and Their Applications. Conference materials also include detailed lists of facilities installed. The latest conference (12th) was held in Berlin in 1989 [2.25], and the next one is planned for 1992 in Vancouver.

2.6.3 Synchrocyclotrons

Both in the conventional cyclotrons and in the isochronous cyclotrons, the accelerating voltage frequency is constant. The effect of the relativistic increase of mass can be compensated for, to some extent, by varying the frequency of the accelerating voltage synchronously with the increase in mass. The cyclotron of this type, in which the accelerating voltage frequency is modulated, is referred to as a *synchrocyclotron*, a *frequency-modulated cyclotron* or, in Russian terminology, a *phasotron*.

The fundamental equation which describes the particle motion in a magnetic field (2.19) can also be presented in the following form

TABLE 2.4 *Commercially Available Cyclotrons* for Biomedical Applications with End-energies of 17 and 30 MeV, status 1987/88 (after G. Wolber [1.58]).

Energy class[1)]	Model Manufact.[2)]	Energies (MeV) p	d	Neg. Ion 3)	Weight (tons)	El.Power required (kVA) dissipated (kW)	Cooling required (10^3 kcal/h)	Price[4)] MDM
	MC16F Scanditronix	17	8,5		18	120 100		2,6
17 MeV	370 CGR-SHI	17	10		20	200		2,8
	BC1710A JSW	17	10		32	200	150	2,8[5)]
	MC30H$^-$ Scanditronix	20-32	15	(+)	35	240 160	120	5,0[5)]
	CYCLONE II IBA	15-30	8-15	+	50	120 100	50	5,0[5)]
	4302 CTI-West	30	17	(+)	44	250	175	2,7
30 MeV	BC3015A JSW	13, 22, 30	15		65	300 260	250	5,8[5,6)]
	480 CGR-SHI	4-30	7-15		32	260		
	EBCO	15-32	-	(+)		120 100		4,2

1) lower energies (±11 MeV) and higher energies (40 MeV) included

2) CGR-SHI: CGR-design produced and sold by Sumitomo Heavy Industries/Tokyo, JSW: Japan Steel Works/Tokyo, CTI-West: Computer Tomographic Instr./Berkeley, IBA: Ion Beam Appl./Louvain-la-Neuve, EBCO Industries: Ltd., Richmond B.C., Canada

3) +: negative ion extraction realized and tested; (+): design only

4) rough prices excl. peripheral costs (transport, insurance, customs); 1$=1,80 DM

5) including costs for transport, installation and testing

6) inclusive $^3He^{++}$ and , inclusive spare parts

$$\frac{v}{r} = q\,c^2\,\frac{B}{E} = \omega_{1s} \tag{2.27}$$

It follows from this equation that ions can be accelerated when

$$B = \text{const.},\quad \omega_s = f(t),\quad \text{and}\quad E = f(t) \approx \frac{1}{\omega_s} \tag{2.28}$$

In Eqs. (2.26) and (2.27) ω_s denotes the particle angular velocity ($\omega = v/r$).

Eq. (2.27) describes the operation of a synchrocyclotron. Like a conventional cyclotron, the synchrocyclotron features a constant azimuth value of induction for each particle orbit.

The electromagnet is of a solid type shown in Fig. 2.22a. Since the accelerating voltage frequency is time dependent, the particle moves in step with the accelerating voltage when its relativistic mass increases. In this way, the main energy limit imposed on a conventional cyclotron is eliminated (Sec. 2.6.1).

Fig. 2.38 shows some parameters of a 600 MeV CERN synchrocyclotron whose dee diameter is 5.0 m and the diameter of the maximum orbit is 4.5 m. Induction vs. radius changes in a way similar to that of a conventional cyclotron: for the middle part of curve B its value is 1.94 T, whereas for the radius 2.27 m it drops to 1.81 T. The electromagnet weight is 2500 ton. The accelerating voltage amplitude varies from 5 to 30 kV, its frequency decreasing from 30.6 MHz to 16.6 MHz. The number of proton revolutions is 30000 and the energy increase per one revolution is 20 keV.

Fig. 2.38 also shows the variattions of the field index n as a function of the orbit radius. From the field index values it is clear that the synchrocyclotron is, in fact, a typical *circular accelerator with weak focusing* (Sec. 2.4.3).

Although the maximum particle energies theoretically attainable in the synchrocyclotron may be as high as several billion electron volts, those attained in practice are mostly limited by economic reasons. This is because the maximum energy, E_{max},

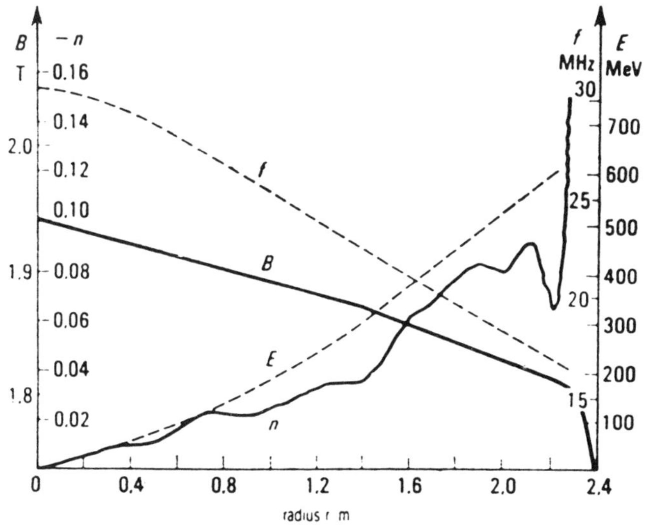

FIGURE 2.38 CERN 600 MeV synchrocyclotron parameters vs. orbital radius: B is magnetic induction; E is proton energy: f is the accelerating voltage frequency, and n is the field index.

in the synchrocyclotron is proportional to the square of the electromagnet radius and the large dimensions of the electromagnet involve increasing high costs. Therefore, the largest synchrocyclotron built to date (at the Leningrad Institute of Nuclear Physics, USSR) accelerates protons to a maximum energy of 1 GeV. In this respect synchrocyclotrons are no match for synchrotrons which make it possible to obtain protons and heavy ions with energies of tens and hundreds of GeV in a much more cost-effective way. An additional disadvantage of synchrocyclotrons as compared with isochronous cyclotrons is the relatively low intensity of the accelerated beams. This is because the isochronous cyclotron operates in a continuous mode whereas the synchrocyclotron, due to the frequency modulation of the accelerating voltage, operates only in a cyclic fashion.

In spite of the fact that synchrocyclotrons do not seem to have much future in biomedical applications, they have played an important part in the past and are still quite popular in biology and medicine. Of nine synchrocyclotrons in operation at present, eight machines are being used in such fields as proton therapy, radiobiological research and the production of radionuclides. For example, synchrocyclotrons in Berkeley (740 MeV) and at Harvard (160 MeV) are exclusively applied in biomedicine. However, the overall number of synchrocyclotrons has decreased: in 1975 there were 17 machines in operation, whereas in 1989 their number dropped to 9 [2.13].

Since the majority of the synchrocyclotrons in operation at present were built in the 1950s, or even earlier, a typical solution has been to modify or convert the older designs into the newer ones. A typical modified synchrocyclotron is that installed at the University of Uppsala (Sweden) [2.23 and 2.24]. It was constructed in 1946-51 for the proton energy of 185 MeV and was used between 1956 and 1977 for radiobiological research and clinical studies in proton therapy. At that time, a broad research program was being implemented at Uppsala to determine radiological and radiophysical aspects of proton therapy. The initial synchrocyclotron featuring 230 cm dia dees was converted into a laboratory depicted in Fig. 2.39.

This laboratory was again put in operation in 1986, and now is equipped with an accelerator which has a dual function. The initial system of electromagnets has been replaced by an isochronous cyclotron consisting of three segments. The accelerator can also operate as a typical frequency-modulated synchrocyclotron since frequency modulation is required for accelerating protons to an energy of more than 110 MeV and ^3H ions to even higher energies. The pole pieces are 280 cm in diameter and the magnetic unit weighs 600 t. The protons can be accelerated to 200 MeV, whereas other lighter ions to the correspondingly higher energies: for example, alpha particles to 50 – 200 MeV. The accelerated beam (Fig. 2.39) is directed to the neutron and radionuclide production unit located at the same level as the synchrocyclotron. It may also be sent to new biomedical units located about 5 m above the cyclotron floor and the beam is brought to this level by two 30° magnets. This area also incorporates, among other facilities, two treatment rooms, one for narrow beam (less than 5 cm in dia.) applications, and the other, for broad beams up to 30 cm in diameter. Separate facilities are intended for radiobiological and radiosurgery experiments. Analytical studies are made in

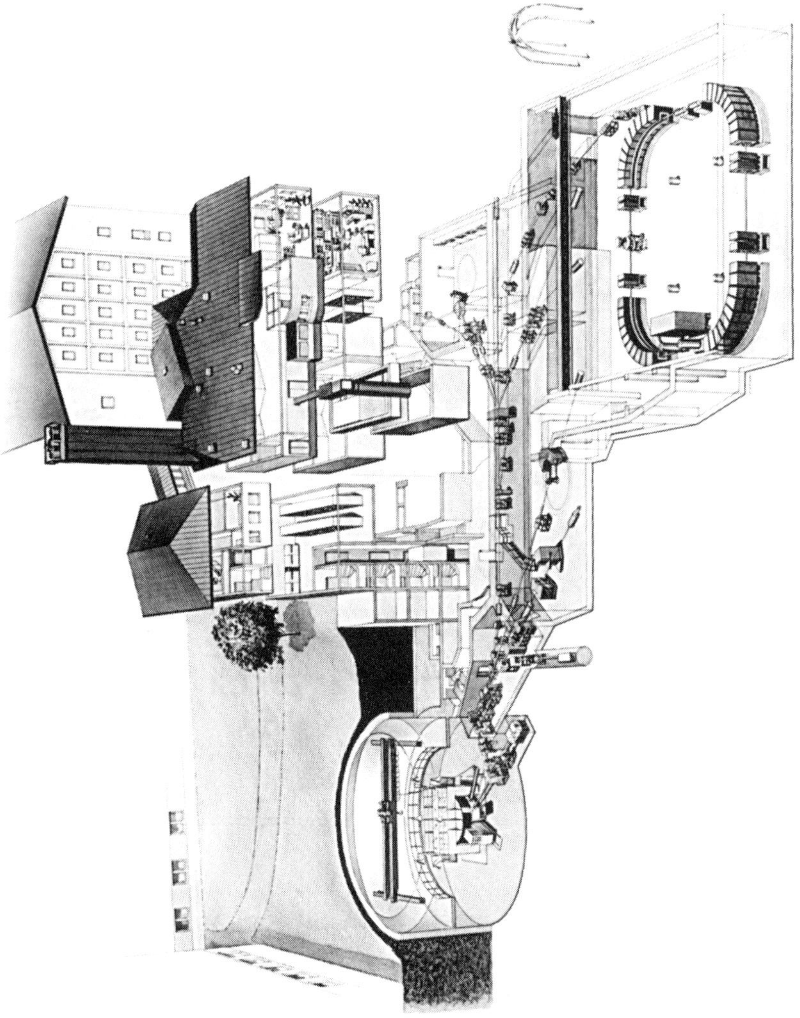

FIGURE 2.39 View of the facility for the converted 200 MeV synchrocyclotron at Gustaf Werner Institute, Uppsala, Sweden.

TABLE 2.5 Parameters on Some Synchrocyclotrons Used in Medicine and Radiobiology.

Location	Electromagnets dia. (cm)	Particle	Max. Energy	Current (pμA)[1]	Percentage of biomed. applications in total operating time
Buenos Aires, Argentina	180	d, alpha	28/56	14/3	20
Lyon, France	180	d, alpha	28/56	30/10	35[2]
Uppsala, Sweden	280	p d alpha	10-200 25-100 50-200		
Berkeley (LBL),[3] USA	461 (184")	p, d, alpha	740/460	1/1	100
Harvard, USA	241	p	160	1	95[4]
Dubna, USSR	600	p	680	7	
Leningrad, USSR	685	p	1000	0,64	

1) Equivalent internal beam current (pμA = particle μA)

2) Additionally, radionuclide production takes up 30% of the total operation time.

3) Shut down

4) Mainly proton radiotherapy; additionally radionuclide production takes up 2% of the total operating time.

another unit which employs a microbeam. The laboratory is equipped with a CELSIUS storage ring for physics studies.

After reconstruction, the accelerator at Uppsala can be operated both as a synchrotron proper and an isochronous cyclotron with K = 200. Parameters of other synchrocyclotrons used in radiotherapy and biomedical research are listed in Table 2.5.

2.7 Synchrotron Method

Principle of operation Fig. 2.40a shows a schematic of a *proton synchrotron*. Pre-accelerated particles are injected into an accelerating chamber equipped with a number of electromagnets (from several in small synchrotrons to several hundred in large machines) intended for bending the particle paths. The design of the bending magnets is shown in Fig. 2.24.

In synchrotrons, the acceleration proper is implemented with *resonators (gaps)* supplied with an rf voltage in a way similar to that in rf linear accelerators or cyclotrons. In an *electron synchrotron*, the orbital frequency of electrons is approximately constant during the total acceleration cycle, because the electrons achieve velocities close to the speed of light, c, at relatively low energies (Fig. 2.1). Thus the accelerating voltage frequency can also be kept constant. In proton synchrotrons, however, the velocity of the protons or heavy ions which slowly gain speed varies widely during the acceleration cycle. It is for this reason that an alternating frequency of the accelerating voltage, i.e. *frequency modulation* has to be applied in addition to the magnetic field alternating with time. In contrast to synchrocyclotrons (Fig. 2.38) in which the accelerating voltage frequency drops during acceleration, the proton synchrotrons feature an increase in frequency.

In a synchrotron (Fig. 2.40a), the particles are accelerated in an orbit with a constant radius (r_S = const). As the particle energy rises during the acceleration cycle, the magnetic field which keeps them in a stable orbit must also increase with time. Fig. 2.40b is a graph of this field featuring a characteristic truncated triangle. This example refers to a small medical synchrotron with a maximum proton energy of 250 MeV. The acceleration cycle is 2 sec, and the acceleration proper lasts 0.5 s. During that time, the pre-accelerated protons, injected at t = 0 into in a stable orbit, gain a maximum energy of 250 MeV. At the time of injection the induction in the orbit is 0.118 T, and 0.5 s later, i.e. when the protons attain the energy of 250 MeV, it rises to 1.52 T. After this, the magnetic field remains stable for 1 s (the flat section on the graph in Fig. 2.40b). This time, typical for synchrotrons, is referred to as the *flattop waiting period*, during which the beam of accelerated protons is gradually ex-

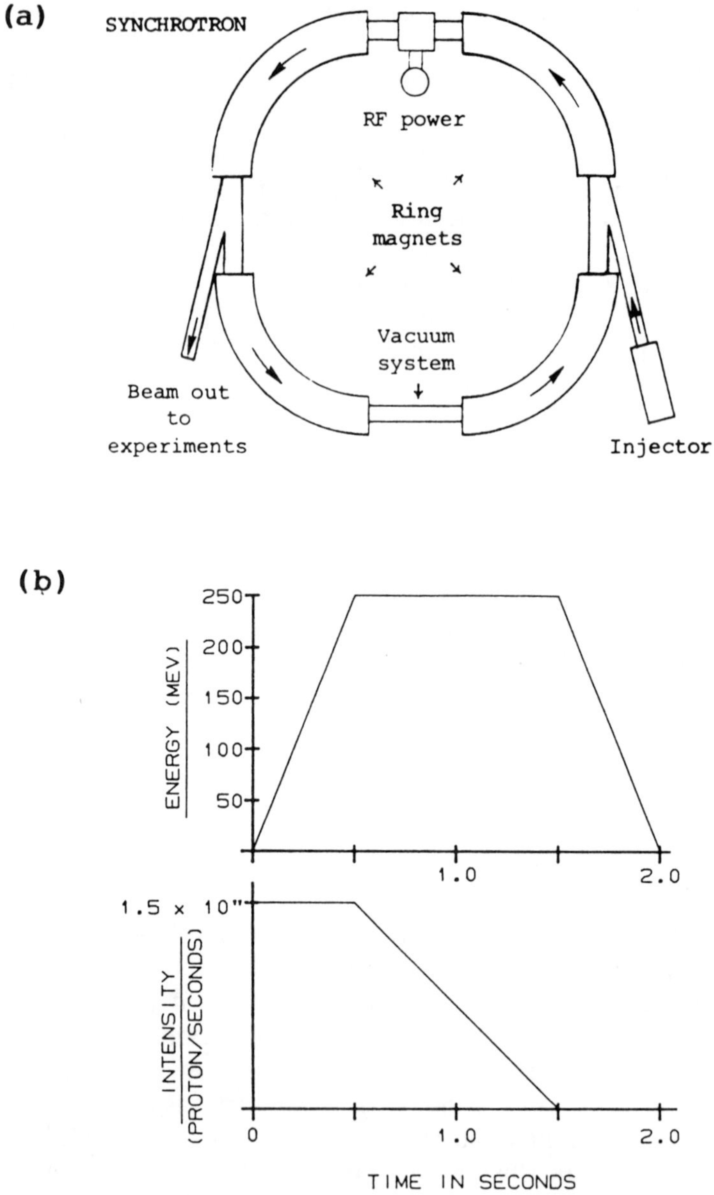

FIGURE 2.40 Proton synchrotron: (a) magnetic system circuit, (b) operation cycle of the 250 MeV medical proton synchrotron (Fermilab).

tracted (the bottom part of Fig. 2.40b). In the final stage of the cycle lasting again 0.5 s, the magnetic field falls from the maximum to its output value. Thus it is possible to obtain the repetition frequency of 30 cycles/min.

The conditions that have to be fulfilled for the *synchrotron mode* of accelerating particles can be presented in the following form

$$B = f_1(t), \qquad \omega = f_2(t), \qquad v = f_3(t),$$
$$E = f_4(t) \approx \frac{B}{\omega_s}, \qquad r_s = f_5(t) = \frac{v}{\omega_\sigma} \approx \text{const} \tag{2.29}$$

In large proton synchrotrons, the operation cycle ranges from several to several tens of seconds. Therefore, the frequency repetition of the subsequent acceleration cycles is rather small. However, the synchrotron method is at present the only one that affords proton acceleration energies exceeding 100 GeV.

Although both electron and proton synchrotrons belong mostly to the domain of basic research, particularly that of high energy physics, they have also played an essential role in biomedical applications, and it seems that this role will become more important in the near future.

Proton synchrotrons. The arguments presented in Sec. 1.1.3 and later discussed in detail in Sec. 5.1.1 and 5.1.2 make it quite clear why intensive work has been focused on the design and application of proton synchrotrons in radiotherapy. Fig. 2.41 shows unit energy ranges required for some therapeutic applications. For proton therapy the maximum energy required is 250 MeV. In this range accelerated protons can be produced by both isochronous cyclotrons and proton synchrotrons. Heavy ion therapy will probably be conducted exclusively with synchrotrons since isochronous cyclotrons cannot ensure that high enough unit energies will be achieved (Fig. 2.41).

In proton radiotherapy, synchrotrons can be successfully competitive against isochronous cyclotrons, mostly because of the lower investment costs. Table 2.6 lists machines used up to now in experiments using heavy particles in therapy. All the facilities operated to date utilize the existing proton synchrotrons, initial-

TABLE 2.6 Parameters of Some Biomedical Synchrotrons in Operation and on Drawing Board.

Location	Particle	Energy MeV/nucl	Ion current s^{-1}	Injection system
Facilities in Operation				
PARMS, Tsukuba, Japan	p	to 250[1)]	10^{11}	Linac
ITEP[2)], USSR	p	70-200		Linac
Bevatron, Brookhaven, USA	alpha, Ar	to 700	10^{9}	Linac
Fermilab-Loma Linda[3)], USA	p	70-250	10^{11}	RFQ
Facilities on Drawing Board				
Argonne, USA	p	2-250	10^{10}-10^{11}	Pelletron
Harvard, USA	p	50-250	10^{11}	RFQ
Bevatron[4)], Brookhaven, USA	alpha, Ar	to 700	10^{11}	Linac
NIRS[5)], Chiba, Japan	alpha, Ar	to 600	10^{9} -10^{10}	Linac
LBL, Berkeley, USA	alpha, Ar	to 800	10^{7} -10^{9}	Linac
PARMS[6)], Tsukuba, Japan	$p(H^-)$	120-230	10-20 nA	Pelletron

1) 500 MeV KEK Booster synchrotron; energy degraded to 250 MeV using carbon absorber (PARMS — Particle Radiation Medical Science Center)
2) 10 GeV proton synchrotron
3) Dedicated medical synchrotron
4) Updated Bevatron synchrotron
5) NIRS — National Institute of Radiological Sciences

ly intended for research in physics. They are, therefore, relatively complex in operation and designed for other than medical applications. Recently, this situation has dramatically changed since, as a result of previous work, a small medical synchrotron has been built at the Fermilab (Fig.2.42), for Loma Linda University Medical Center (LLUMC) [2.15 and 2.26].

This machine is now at the initial stage of operation. A source of 40 keV protons supplies a RFQ-accelerator assembly (Sec. 2.3) followed by a focusing system. The 425 MHz RFQ

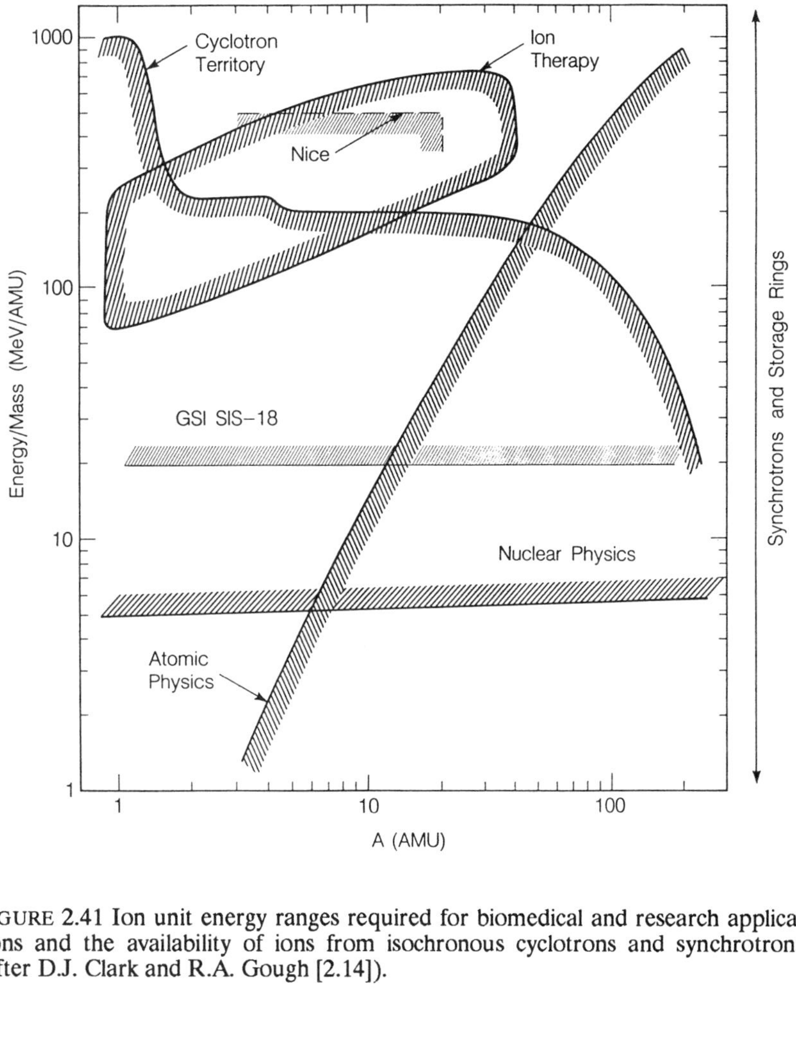

FIGURE 2.41 Ion unit energy ranges required for biomedical and research applications and the availability of ions from isochronous cyclotrons and synchrotrons (after D.J. Clark and R.A. Gough [2.14]).

pre-accelerates protons to an energy of 2.0 MeV. In Fig. 2.42, the proton source is located to the left of the synchrotron electromagnet ring, whereas the RFQ assembly is situated across the ring. The protons are injected into the synchrotron through a beam transport system which bends the beam's direction by about 180°.

The acceleration cycle has already been shown in Fig. 2.40b. After 0.5s, during which the magnetic field rises from 0.118 T to 1.52 T, the protons attain an energy of 250 MeV. In order to achieve this energy in 0.5s, a resonator, powered with rf energy whose frequency rises from 0.9 MHz at the beginning of the cycle to 10 MHz at its end, is used with an energy gain during a single revolution of 90 eV. Subsequently, the accelerated beam is injected into a 4-stand therapy facility, whose parameters are discussed in greater detail in Sec. 5.1.2.3.

The intensity of the accelerated beam of protons at the end of the acceleration process is about $1.5 \cdot 10^{11}$ and the beam can be extracted uniformly in 1 second (Fig. 2.40b). The efficiency of beam extraction is higher than 95%.

The diameter of the synchrotron magnetic system is less than 7.0 m, it means that the Fermilab accelerator is the smallest machine of all the existing proton synchrotrons.

At present, there are 16 proton synchrotrons with energies exceeding 1GeV. The largest is installed at the Fermilab (USA) and has a 2200 m diameter ring. It can accelerate protons to an energy of 1 TeV (10^{12} eV). Synchrotrons of this type feature low intensity of the beam, usually considerably less than 1 μA.

Electron synchrotrons. At present, 6 electron synchrotrons with energies higher than 1 GeV are in operation throughout the world. The upper limit of 12 GeV is currently regarded as a practical barrier above which it does not pay to use an electron synchrotron to accelerate electrons. This barrier is chiefly economic since, theoretically speaking, the electron energy can be increased at the expense of increasing the synchrotron radius, due to the fact that it is proportional to the product of $B \cdot r$. When the radius is increased, however, the cost of the accelerator also rises. From the physics point of view, the factor limiting the upper electron energy is the rapid radiation loss which increases with energy. The radiation loss is, in fact, equivalent to synchrotron radiation, which has found widespread application in radiobiological and biomedical re-

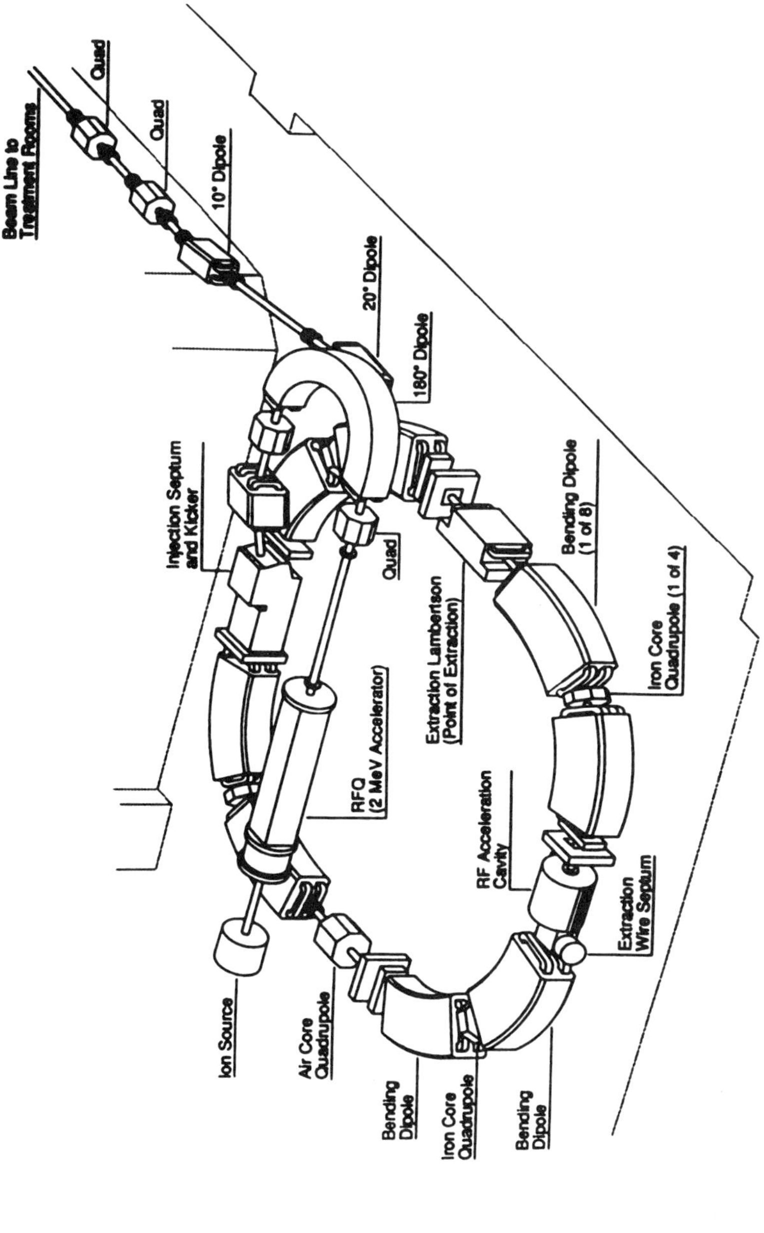

FIGURE 2.42 View of a 70-250 MeV medical proton synchrotron developed by Fermilab for Loma Linda University Medical Center (LLUMC)

search thanks to the development of electron synchrotrons. These problems will be dealt with in more detail in Chapter 6.

2.8 *Microtron Methods. Recirculatory Systems.*

Microtrons constitute a group of accelerators intended exclusively for accelerating electrons. The *microtron method of acceleration* (Fig. 2.43a) was first suggested by V.I. Veksler in 1944, but found its first applications in the early 1960s. In a microtron electrons move in circular orbits between the pole pieces of an electromagnet, similar to those shown in Fig 2.22a. The orbits, however, share one common point, at which an rf accelerating resonator is located (Fig. 2.43a).

Electrons are accelerated once during each revolution, and the resonator is powered from a generator operating in the *S-band*, with a frequency of 3 GHz and wave-length of 10 cm, which is similar to the resonators employed in rf linear accelerators (Sec. 2.3.2).

Typically, the energy the electrons gain when passing through the resonator (gap) is 0.5 MeV, although in some designs it may be as high as 1 MeV. Since microtrons usually operate in the 5 to 50 MeV range, the number of revolutions is relatively small, often not greater than 50. The electromagnet diameters usually range from 1 to 2 m. The beam intensities in microtrons are of the order of several tens of milliamperes per pulse, with mean intensities on the order of microamperes. Thus, as far as beam current, microtrons lie somewhere between betatrons and rf linear electron accelerators. The microtron's great advantage is that its energy dispersion is small (aprox. 10^{-3}). By using a movable beam extraction channel (Fig. 2.43) the output energy can easily be controlled in a step-wise fashion, e.g. every 0.5 MeV.

The number of microtrons in operation at present is less than 50. They are mostly employed as injection accelerators and in industrial radiography or other analytical applications. In the late 1970s the company Scanditronix started production of microtrons for therapeutic uses. For more details see Sec. 3.3.

Recirculating Systems. The basic design of a microtron has been developed into that of a *recirculating system (racetrack microtron).* The structure (Fig. 2.43b) which accelerates electrons to an energy of, for example, several MeV can also be used for multiple acceleration of the beam itself. An essential feature of this new unit is a system of electromagnets which bend the beam through 360° and re-introduce it to the accelerat-

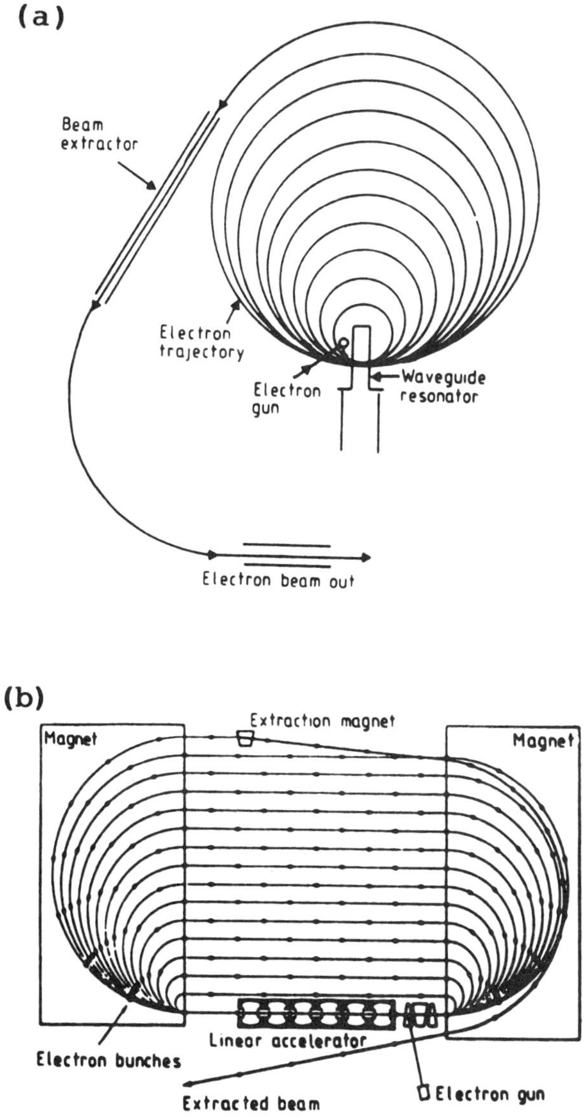

FIGURE 2.43 Microtron: (a) circular, (b) recirculating.

ing structure. Thus the efficiency of the accelerating system is improved, and a beam is obtained whose final energy is higher than the basic energy of the structure by a factor of 3 to 4. This design has also been employed in accelerators for radiotherapy (see Sec. 3.1.2.6).

2.9 Storage Ring Methods

In a synchrotron with a fixed target, the beam of accelerated particles usually bombards the target's atoms. However, the efficiency of this process quickly falls as the energy of the particles rises since an increasingly greater part of the energy of the particle colliding with a fixed atom becomes utilized on both particles' forward motion, i.e. in the direction of the motion of the primary particle, whereas the increasingly smaller part of that energy is turned into, what is essential from the experiment's point of view, the relative motion of both particles.

The phenomenon of this kind occurs only for high energies. For example, for 32 GeV protons, only 8 GeV, i.e. 25% of the initial energy, is converted into the relativic motion, whereas 24 GeV, i.e. 75%, is lost on the forward motion.

A radical increase in the efficiency of the energy utilization of the accelerated particles has been obtained using the method of colliding beams. Bunches of accelerated particles, made up by two independent beams traveling in opposite directions, are injected into a circular storage ring. Since the intensity of a single bunch is small, the process of filling the ring usually lasts from a few to several tens of minutes for one beam. As a result, high-current beams are obtained, with typical electron ring current of several tens of milliamperes and proton ring currents from a few to several tens of amperes. After the filling is finished, both beams are passed through an area referred to as an *interaction area*, where they undergo collisions similar to head-on collisions. Supposing one beam consists of 3 GeV electrons, and the other beam is made up of electron antiparticles, *positrons*, then the resultant energy, in the centre-of-mass system of coordinates, is 6 GeV. In the laboratory system of coordinates, in that of an electron accelerator with a fixed target, an equivalent energy effect would only be possible when the electrons have been accelerated to the as yet unattainable energy of 35 000 GeV.

The method of colliding beams belongs to the field of high-energy physics. In the context of the present book, storage rings are of interest only so far as the ideas used in their development can be employed in the design of intensive generators for the production of synchrotron radiation. For details, the reader is referred to Sec. 6.1 and 6.3.

2.10 Comparison of Various Methods

Table 2.7 lists the main parameters of various methods of particle acceleration with the view of the present or future applications in biology and medicine.

TABLE 2.7 Comparison of Parameters of Various Methods of Charged Particle Acceleration in Biomedical Applications (1992)

Method of Acceleration	Particle	Energies typical MeV	Attained max. MeV	Main Applications
High-voltage Linear:				
- Van de Graaff generator	electrons	0.25-20	35[1]	- analytical methods
- Cockroft-Walton cascade generator	protons and	0.1-2.0 (e)	4.0	- radiation processing
- Dynamitron	heavy ions	0.55-4.0 (e)	4.5	- sterilization, radiation processing
Rf Linear	electrons	5 - 50	40,000[2]	routine therapy, sterilization,
	heavy ions	20-200 (p)	800[3] (p)	experimental therapy, analysis
Betatron	only electrons	3-45	340[4]	routine therapy
Cyclotron:				
- conventional	only	a few	10(p)	
- isochronous	protons and	several tens(p)	590(p)[5]	- radionuclide prod., exp. therapy
- synchrocyclotron	heavy ions	several hundred	1000(p)[6]	- exp. therapy, radionuclide production

Method of Acceleration	Particle	Energies typical MeV	Attained max. MeV	Main Applications
Microtron	only electrons	5-25	50	- routine therapy
Synchrotron:				
- electrons	electrons	1-3 GeV	12 GeV[7]	- analysis (synchrotron radiation)
- protons, heavy ions	protons heavy ions	1-400 (p)	1000 GeV[8]	- exp. therapy

1) Daresbury Nuclear Structure Laboratory (England);
2) SLAC two mile accelerator, Stanford Linear Accelerator Center (USA);
3) LAMPF rf linear accelerator, Los Alamos Meson Physics Facility (USA);
4) Kerst Betatron (1952);
5) SIN isochronous cyclotron (Switzerland);
6) Leningrad synchrocyclotron, Gatchina (USSR);
7) Cornell electron synchrotron (USA);
8) Tevatron proton synchrotron (1 TeV = 1000 GeV), Fermi National Accelerator Laboratory, Batavia (USA)

CHAPTER 3

Electron Accelerators for Routine Radiotherapy

3.1 RF Linear Accelerators

3.1.1 Requirements Imposed on Accelerators for Radiotherapy

3.1.1.1 Clinical Requirements

Specific requirements. When treating neoplastic diseases the radiotherapist assumes great responsibility. Although radiotherapy is a useful method of treatment, being irreversible, it poses a threat similar to that of surgery. Moreover, it is a technique difficult to plan and carry out. The aim of a radiation treatment is to specify an optimum and well-determined dose of ionising radiation for an area or areas to be irradiated so that the dose is as uniform as possible over a target area and as low as possible outside it (Sec. 4.3.2, Fig. 4.16). A *safe dose* is that which is necessary to stop the cancerous process, to destroy the neoplasm and to prevent the relapse of the disease. The *efficiency of the treatment* is determined by the statistical *coefficient of percent survival*, which is, at the same time, the patient's first safety barrier.

Radiotherapy, unfortunately, affords only a relatively narrow margin between the doses which, after irradiating the target area, provide the longest possible survival time and those delivered to the normal tissue with serious pathological consequences (see Fig. 1.6). With the present-day development of radiotherapy it is possible to minimize considerably some of the more serious consequences, such as skin sclerosis or bone necrosis, thanks to the improved spatial radiation dose distribution. Hence, the notion of *survival quality* is a factor which both estimates the individual treatment progress and determines the patient's second safety barrier. In order to increase the percent coefficient of survival and the quality of survival, that is to ensure the highest possible degree of safety to the patient, it is essential that, with advance in technology, the *quality of treatment* is improved.

TABLE 3.1 Some Clinical Requirements of Megavoltage Radiotherapy Accelerators (after C.S. Nunan [1.16]).

Item	Major criteria
Precise delivered dose throughout target volume	Flatness of fields — all field sizes. Stability of field flatness versus angles of gantry and beam limiting device. Stability of penetrative quality.
Precise dimensions of target volume	Spatial precision of machine and radiation beam. Spatial precision of position indicators. Minimal dose to normal tissue Depth to maximum dose. Penetrative quality. Slope of fall-off of electron depth dose. Sharpness of dose profile shoulder. Width of penumbra at depth. Scatter from beam modifiers.
Wide variety of radiation modalities	Low and high X-ray energies. Low to high electron energies. Small to large field sizes. Rotational therapy.
Reliability	Unscheduled down time. Set-up time per field. Range and ease of equipment motions. Height of isocenter.
Moderate time to irradiate, for patient comfort, minimal motion	Dose rate (with beam modifiers).
Patient safety	Mechanical injury avoidance. Radiation injury avoidance.

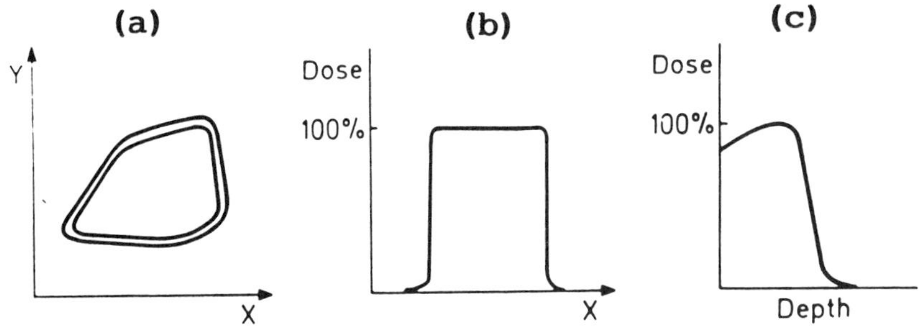

FIGURE 3.1 Basic parameters of an electron beam: (a) — profile, (b) — depth-dose distribution, (c) — nonuniformity of the treatment field.

It is, therefore, obvious that the success of the overall treatment and the safety of the patient depend to a large measure on the quality of radiation treatment at all its stages. Table 3.1 lists the fundamental clinical requirements of accelerator radiotherapy. To optimize treatments, the radiotherapist should have at his disposal a large variety of types and energies of radiation featuring high dose rates, large dimensions of the highly uniform irradiated fields, minimum scattering and contamination with other types of radiation. What is also of importance is the *stability and repeatability* of all the parameters, as well as the agreement between the beam's specifications with the requirements of radiation therapy.

The fundamental *beam characteristics* in radiotherapy are shown in Fig. 3.1. In what follows, some of the main clinical requirements will be dealt with in more detail.

Energy range. Based on the early experience with cancer therapy with the use of X-rays, in which the maximum dose in the target volume was limited by the skin dose, it soon became clear that increasing the energy of X-rays would be advantageous in patients with deep-seated tumors. With increasing energy, the X-rays deliver a lower *skin dose;* the *beam lateral (transverse) scattering* is lower because of lower angles of the scattered photons, and the radiation absorption in the bone due to the Compton effect also is lower (Fig. 1.9). All these factors have led to the preference of ever greater energies and to the design and operation of machines reaching energies as high as

50 MeV (e.g. Mevatron 50). Some of these advantages have become rather illusory, since, with increase in energy, the *beam penumbra* becomes wider as a result of a larger number of high-energy secondary electrons, and the unwanted *neutron radiation* appears, which in effect contaminates the beam. Moreover, the therapeutic quality of the beam deteriorates at very high energies due to poor dose distributions, and the dose to bone is elevated as a result of pair production. At present, in photon radiotherapy, the *upper energy range of 15 to 20 MeV is considered optimal.*

This has led the manufacturers to the maximum accelerator energy 22-25 MeV (Tables 3.5 and 3.12). The lower limit in photon therapy is determined by clinical requirements of the treatment of head and neck tumours or some lymphatic vessels for which the energies of 4 to 6 MeV are optimal. It follows from Table 1.4 that routine clinical applications of the electron therapy encompass the same energy range from 4 to about 20 MeV. When the electron energy is increased above 18 MeV the uniformity of the field is difficult to maintain.

Precision of the dose delivered. The aim of radiotherapy is to destroy the cells of the primary tumor with as little as possible damage to the surrounding tissues. The difficulty of this task is that the sensitivity of the tumor and that of normal tissues is a steep function of the dose (Fig. 1.6) and, therefore, small changes in the dose delivered, e.g. ±5%, can lead to large local changes in the tissue, of the order of ±20% [3.1]. According to the analysis quoted in [3.2], the therapist should be able to deliver a *dose distribution* to the patient so that the dose at the tumor or at another critical organ is known to within ±5%, since the changes in radiation dose of ±10% can cause marked changes in the probability of normal tissue necrosis or the recurrence of the primary tumor.

It follows from the theory of errors that the total error committed during radiotherapeutic procedures in a given patient can be expressed as a square root of partial errors, i.e.

$$\text{Total error} = \sqrt{\begin{aligned}&(\text{diagn. error})^2 + (\text{plan. error})^2 + \\ &+ (\text{treatm. prep. error})^2 + (\text{treatm. error})^2 + \\ &+ (\text{human error})^2 + (\text{X error})^2 + \ldots\end{aligned}} \qquad (3.1)$$

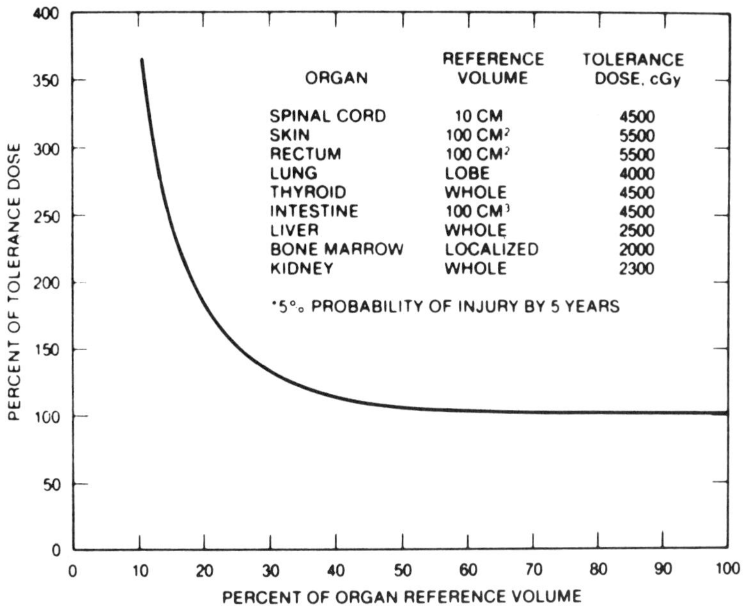

FIGURE 3.2 Dependence of injury dose on organ volume (after C.S. Nunan [1.16]).

As early as 1976, the *International Commission on Radiation Units and Measurements (ICRU)* in its Report no. 24 (ICRU-24) formulated the following recommendation: *"If the eradication of the primary tumor is sought, the accuracy in the absorbed dose to the target volume should lie within* $\pm 5\%$". The practical implementation of this recommendation in therapy, in spite of the fact that it was stated 16 years ago, is very difficult and in many cases — quite impossible. In very careful dose calibration in a phantom, the error is $\pm 2.5\%$ (95% confidence limit of 2σ). This value does not include other errors involved in the accelerator operation. All the errors are listed in Table 3.2 which makes use of the recommendations of the *International Electrotechnical Commission (IEC)* for medical accelerators and those of the ICRU [3.4].

It follows from Table 3.2 that the uncertainty in the delivered dose to a selected point at a given distance from the beam axis is $\pm 4.2\%$ (95% confidence limit). Since the treatment planning computational error may also lie within the identical range of

TABLE 3.2 Uncertainty in Dose at Off-axis Point in Target Volume (95% confidence limit, 2 standard deviations) (after C.S. Nunan [1.16]).

Calibration devices (ICRU-24):	Uncertainty
Physical constants	1.1%
Standardization of beam at Nat'l Stds. Lab.	0.5%
Calibration of secondary instrument at Reg. Cal. Lab.	0.4%
Calibration of field instrument (optimal model)	1.0%
Calibration of treatment beam (optimal model)	1.7%
Delivery of dose to tissue phantom (optimal model)	0.7%
Calibration root mean square	2.5%
Linac (IEC suggested tolerances):	
Dose monitoring system:	
Reproducibility	±0.5%
Proportionality (linearly)	±2.0%
Dependence on equipment position	±1.5%
Stability throughout the day	±1.0%
Stability throughout the week	±1.0%
Dose monitoring system root mean square	2.9%
Stability of dose due to 5 mm tolerance in SSD indicator	±1.0%
Stability of dose at depth due to ±1% instability in penetrative quality	±0.4%
Stability of flatness with angular position	±1.5%
Beam stability root mean square	1.8%
Total linac root mean square	3.1%
Total of calibration and linac root mean square:	4.2%
Other (AAPM)	
Imprecision in treatment planning computation	4.2%
Contributions from organ motion and changes in patient anatomy	?
Total of all contributions	>6.0%

±4% [3.5], the total error may increase to ±6% (95% confidence limit). In other words, for some 9% of the total number of patients, the uncertainty in the delivered dose to the tumor volume may exceed the assumed level of ±5%.

Summing up, even under optimal conditions, the ±5% accuracy is difficult to achieve due to the various factors causing partial uncertainties. That is why the requirements imposed on the accelerator itself and on all the quality assurance procedures (Sec. 3.1.6.6) should be as high as possible [1.16]

A more detailed discussion of the present-day techniques of achieving high degree of accuracy in photon and neutron therapies is given in [3.6].

Precise patient and treatment field positioning. Some organs, such as kidneys, spinal cord, lung and rectum are characterized by relatively high radiosensitivity. In some radiotherapeutic cases it is not possible to ensure an adequate dose to the tumor volume without some serious irradiation of the critical organs or fragments thereof. Radiation sensitive organs and normal tissues can tolerate higher doses without injury if the irradiated volume is not large or reduced (Fig. 3.2). The organs and tumors in the patient's body regions such as abdomen and pelvis can move as much as 1 to 3 cm relative to some landmarks due to patient's breathing and his individual anatomy. Therefore, the *precise patient positioning* during routine sequential therapies encounters practical limitations. These spatial variations in the position of the tumor volume require that the treatment fields planned should be larger than the true tumor volume. In this situation, fields with sharp edges must be used so that the injury to the neighbouring critical organs is limited. The spatial tolerances, recommended at present by the IEC, are ±2 mm for each individual motion and position indicator. On this assumption, the total cumulative maximum error in the position of the photon field axis and its edges relative to the true position of the rotation axis (isocenter) is ±4 mm (cf Table 3.7).

Errors in the determination of adjacent fields can also become a source of localized *overdoses or underdoses*, which may result either in injury to normal tissues or recurrence of the primary tumor. Hence, there is a need for accurate positioning of the edges of the radiation, not only at the patient's skin but also at the depth corresponding to the critical organs. As the probability of injury rapidly increases with dose, while it only slowly increases with the volume irradiated (Fig. 3.2), this problem can be relieved by tapering or successive shifting the adjacent edges of the radiation fields.

The range of therapeutic methods. Fig. 3.3 shows depth-dose distributions for parallel opposed 6 and 18 MV X-ray fields at the depths of 10 and 30 cm, which corresponds approximately to head and neck tumors and abdominal and pelvic region tumors. For example, when an 8 cm thick treatment volume at mid-depth in a patient 30 cm thick is irradiated with a given dose, the

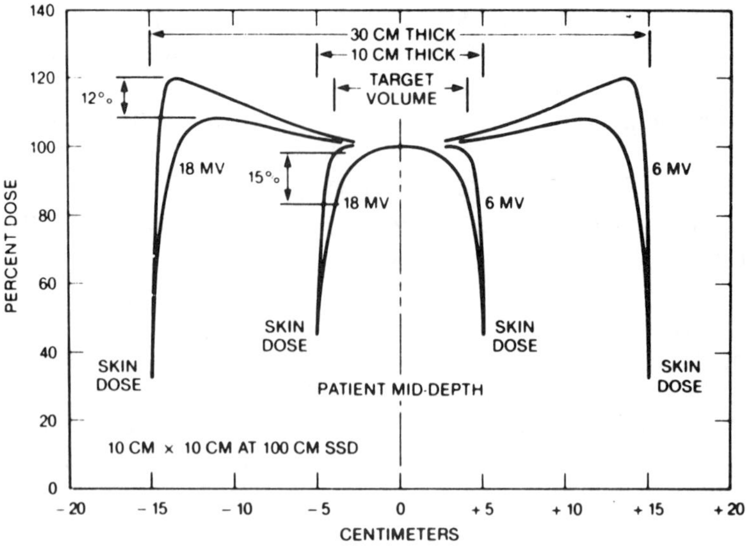

FIGURE 3.3 Depth-dose distributions for X-ray therapy with 6 and 18 MV parallel opposed beams (after C.S.Nunan [1.16]).

normal tissues receive a dose which for a 18 MV beam is significantly lower (by 12%) than that for a 6MV beam. For a patient who is 10 cm thick, however, the same treatment volume irradiated with a 6 MV field receives a considerably more uniform dose (by 15%) due to the more rapid *build-up of the dose* with depth. Thus, by using doses with various ratios of low and high energy X-rays (compare Fig. 1.10) therapy planning is much more flexible. Hence, the need for both low and high energy beam generation in the same accelerator becomes ever more essential (Sec. 3.1.2.5).

Another requirement in treatment planning is the need for a large range of *field sizes with uniform (flat) dose distributions* featuring sharp edges and small penumbra at the depth of the treatment volume. In some applications, fields are required whose sizes are up to 40x40 cm^2 at isocenter, with diagonals of approx. 50 cm.

What is most essential is the *reliability* of the accelerator operation. Any unforeseen break-downs may greatly interfere with treatment schedules. Also essential is the ease of patient and beam positioning, since the time of patient's preparation for treatment considerably exceeds the treatment time proper.

Patients exposed to radiation treatment are often in serious condition and have to be placed in a *supine position*. Correct positioning and immobilizing of a patient during treatment is of paramount importance. It should be as close as possible to that used during simulation. Isocentric irradiation is most favorable when the patient lies on his back since then the position of the various body organs can be best reproduced. For this, the accelerator gantry must be made to rotate through a full 360° angle.

Adequate space must be allocated to allow free rotation of the gantry about the patient while not interfering with patient positioning and immobilization (Figs. 3.4 and 3.5).

Beam spatial stability. Isodose contours are typically measured for a full set of conditions, including gantry angle, modifying devices (wedges, etc.) and field sizes. In order to achieve a uniform dose over the target volume, the *unflatness of the treatment fields* should be accounted for. In the planning of the total uncertainty in the determination of the delivered dose distribution a significant contribution may be due to the non-uniformity of the field contour as a function of gantry rotation and beam modifying accessories. This non-uniformity results from the instability of the position and direction of the electron beam striking the X-ray conversion target.

When generating X-rays the head must be equipped with a *flattening filter* (Sec. 3.1.4 and Fig. 3.38). A case in point is shown in Fig. 3.4 for a 18 MV X-ray beam. The point P, which is being irradiated, is located within the patient's body at a distance of 10 cm from the isocenter. This point is intercepted both by rays emitted from point B at the point source on the X-ray conversion target and by rays from point A emitted 10 mm from the target focal point. The flattening filter attenuates ray B in the ratio of 3:1 (attenuation from 300% to 100% intensity) while it attenuates ray A in the ratio of 6:1 (attenuation from 300% to 50% intensity). Thus, the displacement of the electron beam on the X-ray target by 10 mm causes 100% field asymmetry, since the dose at one side of the field has 50% intensity and on the other side it has 150% intensity. If, moreover, the angle of the electron beam generating ray A tilts by 100 mrad, the lobe of the X-ray emission curve (Fig. 3.26) is displaced in the flattening filter so that the output intensity changes from 600% to 100%. Hence, the asymmetry at absorption point P is still more aggravated. Thus, in order to ensure 1% field symmetry in the slightly exaggerated case discussed above, the mean position and the mean angle of the electron beam striking the X-ray conversion target should be maintained within ±0.1 mm and ±1 mrad for 18 MV X-rays.

This is an instructive illustration of the technical consequences of the radiological requirements imposed on accelerators for radiation treatment.

3.1.1.2 Standards

For over 20 years, intensive efforts have been made to establish international recommendations on standards for radiotherapy devices, including medical accelerators and their application. Activity of this kind has been pursued, first of all, by the *International Electrotechnical Commission (IEC)* and the *International Commission on Radiation Units and Measurements (ICRU)*. The basic standards set up by IEC and issued through 1991 are listed in Table 3.3.

The IEC was set up in 1906 and includes now 45 member nations representing 80% of the world's population. It has 200 technical committees and 500 working groups. *Group One of Subcommittee 62C* is responsible for developing standards for high energy equipment used in radiotherapy. The main advantage of establishing international recommendations is that both the recommendations and definitions of concepts used in radiotherapy are standardized so that differences which might appear when these standards are settled independently by each country are mostly eliminated. Therefore, the manufacturers and users as well as government agencies can make use of uniform terminology (Table 3.4) and standards when producing and using accelerators. Such recommendations also concern the accelerator auxiliary equipment and, most of all, dosimetry apparatus. Due to the application of standards, a realistic level for safety protection of patients and staff can be ensured. The intent is not to limit or restrict the variety of the machines produced to some standardized designs but to introduce a certain standard of quality in all accelerators independent of the different technical solutions used in their manufacture. Such a

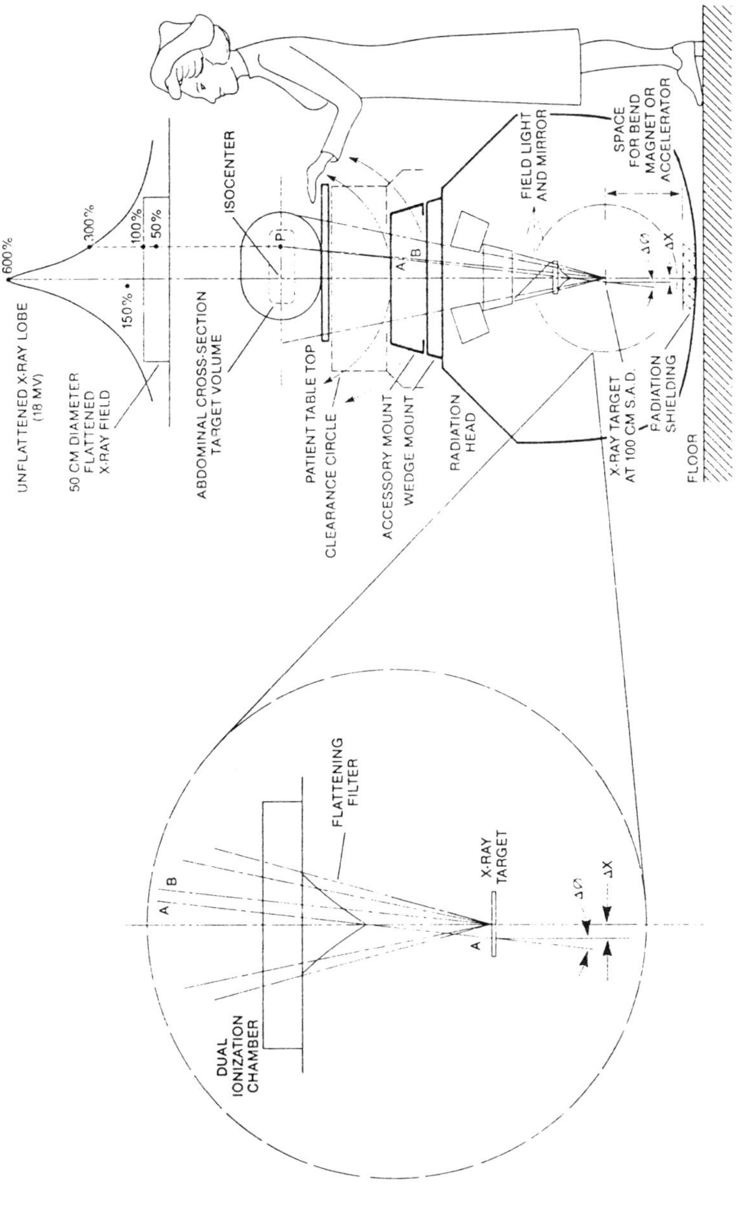

FIGURE 3.4 X-ray field symmetry (after C.S. Nunan [1.16]).

TABLE 3.3 List of Basic IEC (International Electrotechnical Commission) Documents Related to Radiotherapy Accelerators.

No.	Document	Title	Contents
1a	601-2-1(1981), 65pp	Safety of medical electrical equipment: Part 2 — Particular requirements for medical electron accelerators (1-50MeV), Section 1 — General information, Section 2 — Radiation safety for equipment.	Requirements for satisfactory radiation safety for electron medical accelerators operated routinely by qualified personnel. Include X-ray and electron beams of energies between 1 and 50 MeV and dose rates at 1 m of 0.001 - 1 Gys^{-1} and for source-patient treatment distances between 0.5 and 2 m
1b	601-2-1 Modification no. 1 (1984), 21pp	Suppl. 1, Section 3 — Electrical and mechanical safety requirements for equipment	
1c	601-2-1 Modification (1990)	Suppl. 2, Radiation safety	
2	-976 (first edition 1989), 147pp	Medical electron accelerators Functional performance characteristics	This standard describes a set of measurements requiring two to three months' work to complete. It specifies test procedures to be performed by the manufacturer at the design and construction stage, but does not specify acceptance test to be performed after installation at the purchaser's site.
3	-977 (first edition 1989), 109pp	Medical electron accelerators in the range 1 MeV to 50 MeV Guidelines for functional performance characteristics	Recommendations both to manufacturers and to users. They provide guidance to manufacturers on the needs of radiotherapists in respect of the performance of electron accelerators and to users wishing to check the manufacturer's declared performance characteristics, to carry out acceptance tests and to check periodically the performance throughout the life of the equipment.
4	601-2-8 (1987) 72pp	Part 2 — Particular requirements for the safety of therapeutic X-ray generators	Particular requirements for X-ray radiotherapy generators operating in the range of 10 kV to 400 kV

standard would eliminate any ambiguity in defining and presenting the basic machine characteristics.

Table 3.4 lists some selected definitions used in the terminology recommended by the IEC (the sequence of items as in the original document). Beginners or less advanced readers are advised to become acquainted with this table.

There are IEC standards published on two different kinds of subjects:

- safety standards for radiation, mechanical, and electrical safety,
- performance standards for quality characteristics (Table 3.3).

A disclosure standard for performance purposes seems to be the adequate form to allow for competition, new technical developments, and adaptability to local conditions of the user. Thus, the items of performance standards of IEC are:

(1) Criteria and definition of performance characteristics by which the quality of radiotherapy equipment can be described (see Sec. 3.1.6).

(2) Test procedures and test conditions by which the compliance with values specified by the manufacturer can be demonstrated.

(3) Obligation of the manufacturer to declare the actual performance characteristics for this equipment in a standardized format.

The procedure of developing the standards by the IEC SC-62C Working Group One takes considerable time since it involves circulation of drafts to National Committees. The IEC standards are then used by some countries in developing their own national standards. This is especially true of European Community countries. Some parts of the IEC radiation safety recommendations have also been applied in preparing the Suggested State Regulations for Control of Radiation (SSRCR) which are now in force in several US States. In the U.S.A., the regulation of accelerators and their users has been left in the domain of the individual states. Uniformity between states' regulations is encouraged by the Conference of Radiation Control Program Directors, Inc. in the publication of the SSRCR [3.109].

TABLE 3.4 Selected Definitions After IEC Standard no. 601-2-1 (first edition 1981), Appendix A.

Terminology and definitions

Stray radiation
For ionizing radiation, all radiation except that of the radiation beam under consideration specified according to radiation type, radiation quality and extent but including its residual radiation.

Scattered radiation
Radiation which during passage through material has been deviated from the original direction and has possibly been modified in radiation energy.

Leakage radiation
Ionizing radiation emerging after transmission through the protective radiation shielding of the radiation source.

Filtration
Modification of characteristics of the ionizing radiation on passing through matter, for example by: (1) preferential absorption of certain components of heterogeneous X or gamma radiation accompanying its attenuation; (2) modification of the distribution of Kerma in air over a cross-section of the radiation beam.

Build-up
Phenomenon of the increase with depth of the absorbed dose due to the release of secondary charged particles and to scattered radiation in matter beyond the entrance surface.

Nominal energy
As a characteristic of medical electron accelerators, radiation energy describing: (1) for X-radiation the energy of the electrons striking the target and (2) for electron radiation the energy of the electrons in the useful beam.

Target
The part of an particle accelerator onto which is directed a beam of accelerated particles to produce X or neutron radiation.

Medical electron accelerator
In radiotherapy, electron accelerator where the useful beam of radiation consists of the beam of accelerated electrons or is generated by the beam of accelerated electrons.

Radiation head
Structure from which the useful beam emerges.

Gantry
In equipment for radiotherapy, part of the equipment supporting and allowing possible movements of the radiation head.

Primary dose monitoring system
Dose monitoring system which is intended to terminate irradiation when a preselected value of the measured radiation quantity is reached.

Secondary dose monitoring system
Dose monitoring system which is intended to terminate irradiation in the event of failure of the primary dose monitoring system.

Added filter
Filter that can be positioned in, or taken out of the useful beam. This includes removable added filters (removable without tools) and fixed added filters (removable only with tools).

Field-flattening filter
Filter designed to make the energy fluence rate or absorbed dose rate reasonably constant across a radiation beam.

Beam scattering filter
Filter used in order to scatter a beam of electrons.

Wedge filter
Added filter effecting progressive attenuation across all or part of the useful beam.

Termination of irradiation
Stopping of irradiation without the possibility of restarting without re-selection of all operation conditions (that means return to the preparatory state): (1) when the preselected value of dose monitor units is reached, or (2) when the preselected value of elapsed time is reached, or (3) by deliberate manual act, or (4) operation of an interlock.

Interruption of irradiation
Stopping of irradiation and movements with the possibility of continuing without reselecting operating conditions (that means return to the ready state).

Virtual source
Point from which the radiation appears to originate.

Useful beam
All ionizing radiation from the radiation source which passes through the aperture of the diaphragm or the beam limiting device.

Axis of the beam
Straight line defining the direction of the useful beam. In radiotherapy, the line passing through the virtual source and the centre of the plane figure formed by the edge of the final beam limiting devices.

Radiation field
Plane section of the useful beam perpendicular to the axis of the beam.

Field size
The dimensions of an area in a plane perpendicular to the axis of the beam at a specified depth in a phantom and defined by specified isodose lines.

Treatment field
Area at the patient's surface which is to be irradiated.

Treatment volume
Region in the patient, to which is delivered the radiant energy.

Beam-limiting device
Device to restrict ionizing radiation to the required useful beam.

Diaphragm
Beam limiting device with either a fixed or an adjustable single aperture.

Isocentre
In radiology, centre of the smallest sphere through which the axis of the beam passes in all conditions.

Beam applicator
Device performing at least one of the following functions: (1) indicates the axis of the beam; (2) indicates the radiation field; (3) ensures a minimum distance from the radiation source to entrance surface; (4) serves as compression device.

Geometrical field size
Geometrical projection as seen from the centre of the front surface of the radiation source on a plane perpendicular to the axis of the beam of the distal end of the beam limiting device. The field is thus of the same shape as the aperture of the beam limiting device. The geometrical field size may be defined at any distance from the virtual source.

Normal treatment distance
(1) For electron irradiation, specified distance measured along the axis of the beam from the virtual source of electrons to the entrance surface. (2) For X-radiation, specified distance measured along the axis of the beam from the virtual source of X-radiation to the isocentre or, in equipment without isocentre, to a specified plane.

Dose monitoring unit (MU)
In a dose monitoring system unit in which a measured quantity is displayed and from which absorbed dose can be calculated.

Phantom
In medical radiology, volume of a material behaving in essentially the same manner as tissue, with respect to absorption or scattering of the ionizing radiation in question.

Controlled area
Area subject to special regulations for the purpose of protection against ionizing radiation and to which access is controlled.

Interlock
The means of preventing or terminating an operation unless certain predeterminned conditions are fulfilled.

Stand-by state
State of an equipment which can be maintained for long periods and from which it is possible to move rapidly into operation. For medical accelerator, state in which working levels of vacuum, temperature and others are maintained, but without the possibility to select the essential operating conditions.

Preparatory state
State of an equipment in which it is possible to select essential operating conditions. For some equipment this may be a special case of the stand-by state.

Ready state
State of equipment in which all operating conditions have been selected and all interlocks satisfied.

As long as 20 years ago, the International Commission on Units and Measurements (ICRU) issued general recommendations on dosimetry procedures for photons [3.3] and electrons [3.8]. The hope was to have these recommendations supplemented with concepts or ideas from a given country or region, which would then be used for routine dosimetric procedures, requirements or improvements. Such recommendations, referred to as local recommendations, have been subsequently set up in the U.K. [3.9 through 3.11], USA [3.12 through 3.14], W. Germany [3.15 through 3.17] and the Scandinavian countries [3.18]. The latter ones, having been prepared by the *Scandinavian Association of Clinical Practice (NACP)*, were also adopted by other countries and are now, for example, in force in India. The purpose of setting up the above recommendations, international and national alike, is to provide the medical physicists with a code of practical and uniform dosimetric procedures, called *Quality Assurance*, in radiation therapy establishments. A Code of this kind lists all the necessary tests and measurements aimed at observing the equipment performance at installation in order to determine that it is working properly, or that it is currently working as well as it did at the time of installation, or that repairs or modifications have improved recent performance, all which is necessary to ensure high quality of radiation therapy. These problems will be discussed in more detail in Sec. 3.1.6.

3.1.2 Fundamental Systems

3.1.2.1 Chronology of Development

The main chronology of the development of accelerators used in radiation therapy has already been presented in Sec. 1.1 (Table 1.1). Now, the fundamental features of the technical development of linear accelerators will be discussed from the time of the construction of the first machines in the 1950s up till now, that is the early 1990s.

The first *isocentric linac system* was developed in the UK in the early 1950s [3.19]. It produced a 4 MV photon beam with an accelerating wave-guide whose length was 100 cm. Copper cups forming a complete accelerating stack were contained in a tank evacuated by an oil diffusion pump which had to remain upright

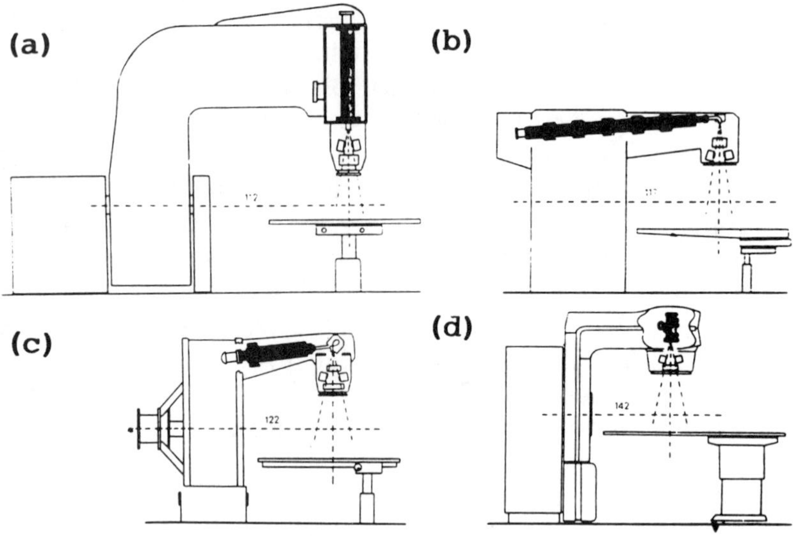

FIGURE 3.5 Development of radiation therapy linear accelerators (shaded parts represent accelerating structures drawn to the same scale; the distance of the isocenter from floor is given in cm): (a) — 4 MeV AEI linac with an in-line treatment head, mfd. in 1955; (b) — the SL75 Philips linac with a horizontal accelerating structure and a 90° bending magnet, suitable for full rotation therapy, mfd. in 1964; (c) — the Dynaray 4 linac with a 270° achromatic magnet, mfd. in 1971; (d) — a present-day system with an in-line treatment head, without beam bending, e.g. 4 MeV LMR-4 Toshiba linac with 35 cm long accelerating structure (after W.H.Sutherland [3.19]).

during the gantry's rotation. Since the whole structure was quite long, the gantry could only be rotated over an arc of 120°. In view of the limited microwave power the typical dose rates never exceeded 100 cGy min^{-1}. The maximum uniform field sizes were as low as 20x20 cm^2 at the source-to-rotation axis distance of 100 cm, and the daily throughput was 30 patients. The life time of individual parts ranged from 10 months (magnetron) to 3 months (electron source).

The subsequent development of radiation therapy linacs is shown in Fig. 3.5. The design of the first generation of linacs in the mid-1950s was based on the so-called *in-line position of the accelerating wave-guide* (Fig. 3.5a). The linear energy gradient was 4 MeV m^{-1}, which made it impossible to be rotated round the table, with partial rotation axis placed rather low, at 112 cm above the floor (Fig. 3.5a). In the 1960s, machines appeared featuring a horizontally mounted fairly long accelerating waveguide that generated high energy beams. This structure made use of a *magnet bending* the beam through a 90° angle and thus making the vertical dimension of the waveguide considerably shorter.

In this design, the treatment head became small enough to pass under the table during a full 360° rotation of the gantry with the isocenter placed relatively low (Fig. 3.5b).

In the early 1970s, a new generation of *compact accelerators* was developed, which used *achromatic magnets*, emitting a beam through an angle of about 270° (Fig. 3.5c). On substituting achromatic magnets for non-achromatic magnets the beam parameters and the stability of the treatment fields were greatly improved.

What was important in the above design was that a *waveguide with a traveling wave* was replaced by a *structure using a standing wave*. This made it possible to increase the linear gradients up to the currently obtainable 12 to 18 MeV m^{-1}, which, in turn, led to considerably shorter accelerating structures with comparable energies. In this way, the accelerator became smaller, and the rectilinear waveguide could again be employed (Fig. 3.5d), and therefore bending magnets could be eliminated. Full rotation of the treatment head was ascertained, though the position of the isocenter had to be increased. Such a design is being currently employed in accelerators with lower energies of 4 to 6 MV.

In consequence of the above development, the maximum treatment fields were considerably increased, up to the presently available 40x40 cm^2 or more, at the rotation axis plane (Fig. 3.6)

The *photon dose* rates have been increased from the initial values of 100-200 cGy min^{-1} to 400-500 cGy min^{-1}. In this way, the exposure times became much shorter even for wedge filters. In order to achieve optimum treatment of both thin and thick layers, the initial energy range of 4-5 MV has been extended to 4-25 MV. At the same time, multiple-energy mode of operation has been implemented, that is the possibility of using two photon energies and a wide electron energy range (Sec. 3.1.2.5) has been made use of.

The *reliability of accelerator* operation has been greatly improved. For some critical parts such as microwave tubes or

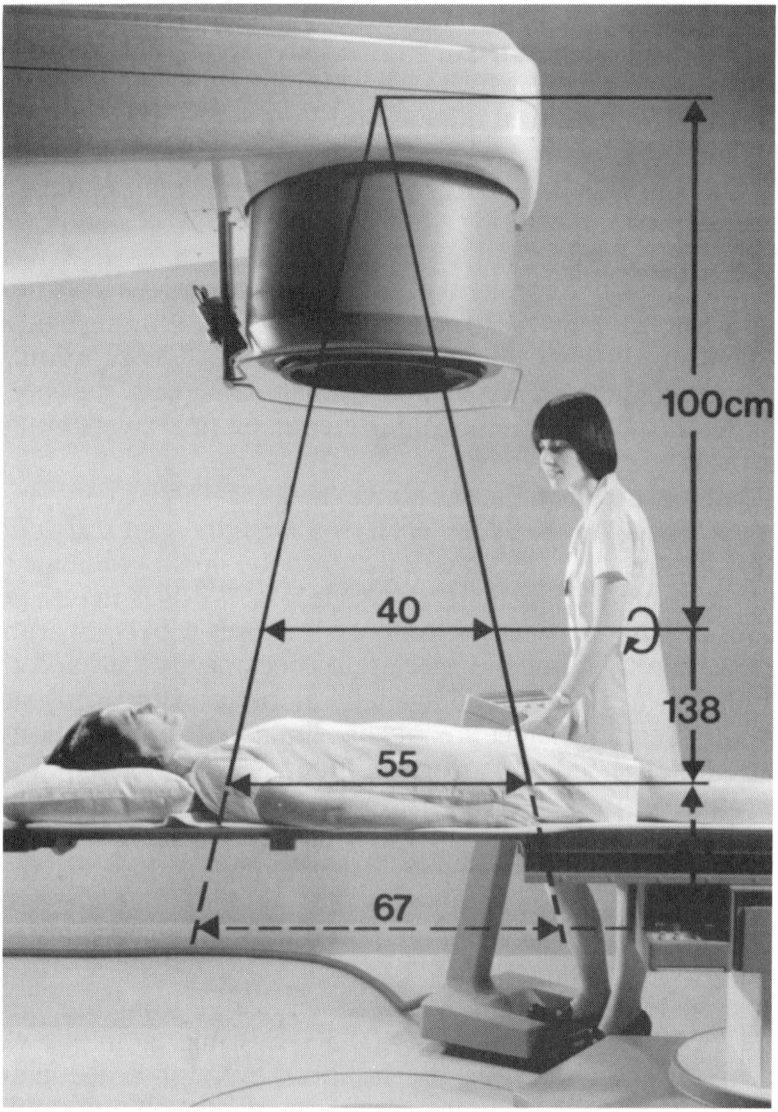

FIGURE 3.6 Linear dimensions of treatment fields (SL75-20 Philips accelerator).

electron guns, it is up to several years. The guaranteed life-time of electron guns is, at present, as long as 5 years. The previously used vacuum oil pumps have been replaced with ion pumps. This, together with the development in the design and construction of the accelerating structure, has led to the improvement in the quality of the vacuum, which, in turn, eliminated sparking and breakdowns, previously common especially at higher energy gradients.

3.1.2.2 Generation of rf Power

As already mentioned in Sec. 2.3.2, accelerators used in radiotherapy are powered with an rf energy from microwave tubes, such as *magnetrons and klystrons* operating at the *S-band* of the frequency of 3 GHz (3000 MHz), equivalent to the wavelength of 10 cm.

The operating principle of an rf klystron is depicted in Fig. 3.7a. A continuous electron beam produced in the source (cathode) and accelerated by a D.C. voltage, is passed through a buncher cavity which acts like a cavity described in Sec. 2.3.2 (Fig. 2.18). As a result of bunching in step with the modulating frequency, electron bunches arrive at the drift tube. Finally, in the catcher cavity the electron's kinetic energy is transferred to the oscillating electromagnetic field of the microwave. Thus, the low-power microwave which is fed into the input leaves the klystron much amplified, and the klystron is, in fact, an rf power amplifier.

Instead of a linear arrangement, like the one in the klystron (Fig. 3.7a), the magnetron features a circular set of cavities (Fig. 3.7b) which are grouped around the cathode in the center. Typically, six cavities drilled in a solid copper block are used. The anode is located in the external circuit. The cavities operate in a strong constant magnetic field with induction B parallel to the axis of the magnetron. A pulsed electric field applied between the cathode and the anode causes electrons produced by thermal emission to move in the direction of the anode. The magnetic and the pulse electric fields combined turn this motion into a spiral, and the electrons in the cavities convert their energy into rf energy.

A typical magnetron operating in the S-band can produce rf pulses with peak power of 2-5 MW. With a *filling coefficient* for typical accelerators used in radiotherapy of (10^{-3}) (Sec. 2.3.4), the above power in the pulse has the mean value of 2 kW. It is because the magnetron which operates at full power and generates 2 μs pulses subsequently repeated at the frequency of

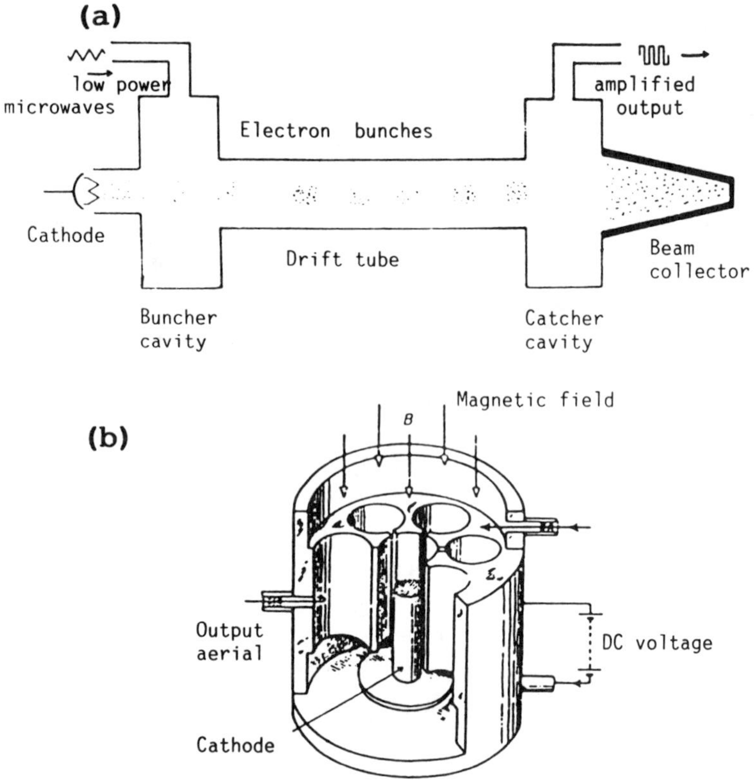

FIGURE 3.7 Principle of operation of microwave tubes: (a) — klystron, (b) — magnetron.

500 pulses per second is, in fact, equivalent to one that, during 17 hours, operates at full power for only 1 min! Klystrons, with respect to the generated power, surpass magnetrons since the klystron peak power ranges from 5 to 40 MW (mean power 5 to 40 kW).

Currently, radiation therapy accelerators are powered from magnetrons or klystrons. An advantage of magnetrons is that they are small enough to be installed directly in the accelerator gantry. Due to its bigger size, the klystron can only be mounted as a stationary device next to the gantry, which requires an additional rotational rf connection. Since magnetrons are self-excited tubes and thus do not require any additional exciting rf circuitry, they are much less expensive, but their life-times are also shorter and, therefore, they must be more frequently replaced. They are usually employed to energize accelerators with low and medium energies of 4 to 20 MeV. Higher energy machines require higher power supplies and, therefore, use klystrons. The life-times of the present-day klystrons are of the order of several tens of thousand hours.

Both magnetrons and klystrons are equipped with special *fine tuning devices* for the range of about 250 MHz relative to the fundamental frequency of 3 GHz. The tuning devices make it possible to compensate for the variations in frequency due to, for example, variations in temperature. Powered by a system of automatic frequency control (cf. Fig. 3.10), they ensure stable and optimal frequency of operation.

3.1.2.3 Accelerating Structures

Operating conditions. A typical load curve for an accelerator employed in radiotherapy is shown in Fig. 3.8. The two top curves show the dependence of electron kinetic energy on the beam intensity, the solid line representing the traveling wave accelerating structures and the dashed line the standing wave structures. In addition, the *X-ray output* as a function of beam intensity for accelerators using a beam of high energy electrons converted to bremsstrahlung is also indicated. Fig. 3.8 also shows the *efficiency of conversion from rf power to electron beam power.* To avoid any misunderstanding, the reader should note that the *overall efficiency* of the accelerator depends also on the *conversion of ac power to rf.*

The electron kinetic energy decreases linearly with beam current, i.e. with the accelerator load. The maximum load of the structure, or maximum current, is used during X-ray emission, that is when electrons are converted to X-rays. Due to the reasons discussed in more detail in Sec. 3.1.4, in order to achieve an identical dose rate for electrons and photons, the primary electron beam currents in the X-ray treatment mode must be greater than those used in the therapy with electrons by a factor of about 100. The X-ray output curve has a maximum when plotted as a function of beam current (Fig. 3.8).

The accelerating structure usually operates when its impedance is matched to that of the rf power source. The match between the structure and the source is specially recommended

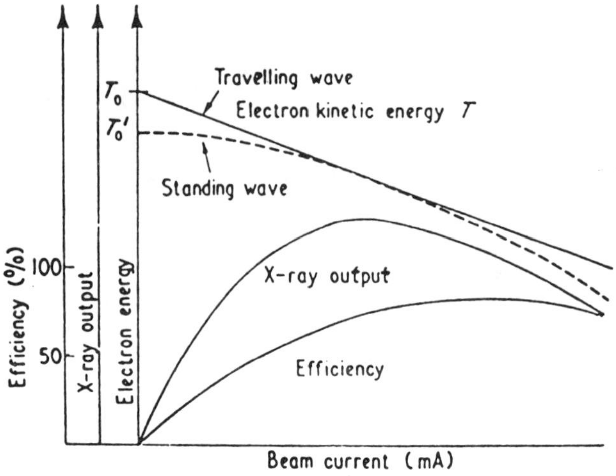

FIGURE 3.8 Dependence of electron beam energy, efficiency and X-ray dose rate on beam current.

because the *energy reflections* from the structure towards the power source must be restricted or, if possible, eliminated. The operating conditions are chosen in such a way that the energy reflected will be zero for an assumed optimum value of the electron beam current. The optimum operating range corresponds to the region where the load lines for traveling wave and standing wave structures have a point of tangency (Fig. 3.8). Strong reflections are also present at the beginning and end of the pulse (cf. Fig. 2.21). In order to suppress the reflections, the rf power source is coupled with the accelerating structure by means of a device referred to as a *circulator* (in British terminology — *isolator*). It absorbs almost all of the power reflected from the structure, which would otherwise be returned to the power source causing its detuning. The system of this kind is of special importance when magnetrons which are very sensitive to power reflections from the load are used.

Both the *Q-factor of the structure and the loss power*, dP/dz, the power released in the structure's walls (Sec. 2.3.3), are dependent on the precision of the manufacturing of the structure and the method of power supply. The fundamental parameter of the structure is the *zero energy*, E_0, the energy attained by the accelerated electrons at zero current load (Fig. 3.8). The actual energy for a load beam current of I can be expressed as

$$E_k = E_0 - k_1 I \qquad [\mathrm{MeV}] \qquad (3.2)$$

where E_k is the electron kinetic energy at the actual load; E_0 is the electron kinetic energy at zero current load; k_1 is the coefficient of energy attenuation and the characteristic parameter of the particular structure. Eq. (3.2) defines the load line for a traveling wave structure which — as depicted in Fig. 3.8 — is rectilinear. When the impedance is matched such a structure behaves like a *non-resonant, pure resistive load* to the power source; hence the straight load line.

For the *standing wave structures, at present most commonly used*, the load line (the dotted curve in Fig. 3.8) ceases to be rectilinear. Except for the tangential portion at the traveling wave line, which corresponds to the correct match, this curve is equivalent to that of a resonant circuit with a reactive components of the load zeroing in the region of the match. Therefore, the energy which corresponds to the zero load may be written as

$$E_0 = k_2 \sqrt{l \cdot r \cdot P_{Cu}} \qquad [\mathrm{MeV}] \qquad (3.3)$$

where k_2 is a constant usually less than unity, l is the structure length [m]; r is the shunt resistivity [$\mathrm{M\Omega\ m^{-1}}$] and P_{Cu} is the loss power in the structure copper [MW].

Hence the structure rf output power, P_0, can be expanded as

$$P_0 = P_{Cu} + P_L + P_e \qquad [\mathrm{MW}] \qquad (3.4)$$

where P_{Cu} is the loss power in the structure walls; P_L is the residual or reflected power dissipated in a load, and P_e is the power transferred to the accelerated electron beam, i.e.

$$P_e = E_b \cdot I_b \qquad [\mathrm{MW}] \qquad (3.5)$$

In Eq. (3.5) E_b is the beam energy and I_b is the beam current.

Parameter E_0 permits intercomparing various structures at zero load. For example, if $l = 1$ m, $P_0 = 2$ MW and $r = 50$ MΩ m^{-1}, the energy E_0 at zero load is 10 MeV. This energy is, in standing wave structures, a square root function of the parameters in Eq. (3.3). If the initial energy, E_0, is to be retained and the structure length halved, the peak power, P_{Cu}, must be doubled.

Design of accelerating structures. The vast majority of the present-day radiation therapy accelerators are equipped with standing wave structures with *side-coupled waveguides* (Fig. 2.19c), developed at *Los Alamos Scientific Laboratory and known as LASL resonators*. Details of the design modifications and parameters for various types of structures are given in [3.20]. Typically, the structure made from high-quality copper or brass, is a series set of resonant cavities, singly tuned for the operating frequency of approx. 3000 MHz. The cavities are fabricated by welding rolled disks machined to micrometer accuracy with diamond tools in an airconditioned room. Thus the Q-factor as high as 15000 may be obtained. The thermal energy released in the walls of the cavities is absorbed by the water running in a system of small tubes welded directly to the structure.

Accelerating structures most frequently form an integral part with the *electron gun and the beam deflection chamber*. Together they resemble a microwave power tube and they undergo similar technological processes, such as baking-out, degassing, evacuation to very high vacuum and sealing. The pressures of 10^{-9} Tr in the structure are considered low enough for an accelerator in a stand-by position. The vacuum is usually maintained by means of an ion-absorption pump, permanently coupled to the structure. In order to obtain sufficiently low pressure the pump must be switched on 30 min prior to the accelerator operation if the break in the pumping has not exceeded 15 days.

Fig. 3.9 represents an accelerating structure used in radiation therapy machines manufactured by the Siemens company. The electron gun included in the structure carries a 5-year warranty.

Electron guns intended for cooperation with traveling-wave structures are usually dismountable.

If the *gun heater* burns out it can be replaced on the spot. In the simplest form, the heater is a tungsten spiral filament mounted inside a cylindrical electrode. The whole system is keyed with pulses from a modulator (Fig. 3.10), whose amplitude is of the order of several tens of kilovolts. This causes the electron beam to be pre-accelerated to velocities of 25-50% of the velocity of light. Electrons thus accelerated enter a *bunching section* of the structure. In standing wave accelerators, the electron gun is most frequently nondismountable and is welded to the structure (Fig. 3.9). In this case, the cathode life-time must be much longer, and a conventional tungsten heater is replaced by a corrugated tungsten or an oxide cathode heated indirectly and featuring high electron emission. The cathode of this kind can operate at lower voltages and, there-

FIGURE 3.9 View of an accelerating structure of a radiation therapy accelerator as seen from the electron gun (mfd. Siemens).

fore, have a longer life-time. The electron gun is sometimes equipped with an additional electrode which corresponds to the grid in a triode and makes it possible to precisely modulate the source, i.e. accurately define the pulse length.

3.1.2.4 Fundamental Block Circuits

Accelerator system. The *general system of an accelerator*, as a block diagram, is depicted in Fig. 3.10. The operating cycle is initiated by a *modulator* which generates HV pulses delivered to an rf generator equipped with a magnetron or a klystron. The modulator also supplies pulses to the electron gun. The gun pulse may be delayed slightly so as to defer electron injection until the accelerating structure is nearly filled with rf energy. The length of the modulator pulses is of the order of few microseconds.

In some designs the rf generator is coupled to the accelerating structure with the *circulator (isolator)*. The circulator is used to prevent energy reflected from the structure from reaching the rf generator. Otherwise, for example, the magnetron frequency could be "pulled", which would result in loss of phase stability and reduce electron final energy, and, sometimes, in electrical breakdown. The accelerating structure is also coupled to the rf generator by means of an *automatic frequency control (AFC)* servo. The detection for the AFC is provided by two additional internal resonant cavities tuned to the frequency for example, of 1.5 MHz below or above the resonant frequency. When the instantaneous operating frequency changes in either direction the AFC system causes the magnetron to be tuned back to the postulated operating frequency.

To minimize the possibility of *breakdowns* caused by intense electric fields, the structure is placed in an envelope (shield) pressurized to one or two atmospheres of a gas of high dielectric strength such as freon. Structure elements, which have a marked effect on frequency variations and those which dissipate heat are, in addition, supplied with water circulating at a well-determined temperature to carry off heat and to stabilize the operating temperature.

The accelerator may include a linear or straight-through *treatment head*: the electron beam (Fig. 3.10) arrives at the head from the accelerating structure without changing direction (cf

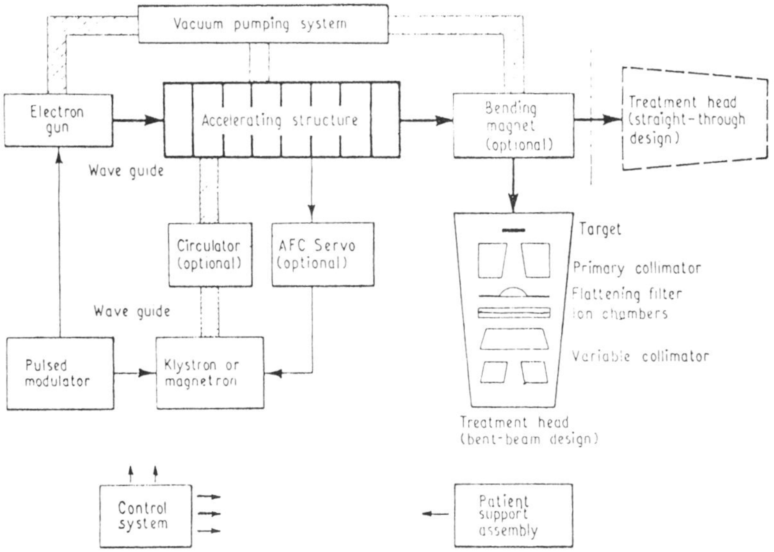

FIGURE 3.10 Block diagram of a radiation therapy linac.

Fig. 3.5d). The design of this kind is used at present only in low-energy accelerators (4-6 MV) since it requires a short structure several tens of cm in length, which permits a full isocentric rotation (Fig. 3.4). A more common design is that of the *bent-beam*. The accelerator is then equipped with a magnet typically bending the beam through 90° or 270° (Sec. 3.1.3).

The treatment head depicted schematically in Fig. 3.10 is one used in the X-ray therapy. The electron beam strikes a *target* and becomes converted to X-rays. The X-rays are then passed through a *primary collimator* and enter a *conical flattening filter*. *Ion chambers* of the dosimetry system and *variable collimators* to define precisely the treatment field (Sec. 3.1.4) are located adjacent to the primary collimator. The treatment head also contains other elements which *modify the beam shape* and an optical system which determines the field size and the skin-target distance. If the accelerator is intended only for X-ray therapy, the target is often welded to the accelerating structure near the last cavity in straight-through design to improve heat dissipation and simplify construction. If, on the other hand, the accelerator is intended for both X-ray and electron beam therapy, a retractable X-ray target is mounted either inside or outside the vacuum; during electron beam generation the target is removed from the beam's path. The electron beam is then extracted outside through a vacuum small window of the smallest possible thickness in order to minimize energy losses and electron scattering.

In accelerators for routine radiation therapy the design currently used is that shown in Fig. 3.11. If the magnetron is employed as a source of rf energy, the whole electronic system, together with the accelerating structure and the treatment head is contained in a *gantry* which can rotate through a full 360° angle around the rotation axis which passes through the rotation point X called isocenter (in the right and center part of Fig. 3.11). In some designs, the head can rotate through an angle of 365° to 370°; the rotation speed is controlled typically in the range between 0.1 and 1.0 rev min^{-1}.

The possible axes of rotation and directions of linear movement are identified in Fig. 3.11 (after IEC Documents 601-2-1 and 976, Table 3.3). The movements shall be designated as follows:

axis (1)	-	rotation of the gantry,
axis (2)	-	roll of the radiation head,
axis (3)	-	pitch of the radiation head,
axis (4)	-	rotation of the diaphragm system (beam limiting devices),
axis (5)	-	isocentric rotation of the table,
axis (6)	-	rotation of the table top,
axis (7)	-	pitch of the table (treatment coach),
axis (8)	-	roll of the table,
direction (9)	-	height of the table,
direction (10)	-	lateral displacement of the table,
direction (11)	-	longitudinal displacement of the table,
direction (12)	-	axis 1 to radiation source.

In addition, the following geometric parameters (not indicated in Fig. 3.11) are also specified:

direction (13)	-	height of the radiation source, from the isocenter,
dimension (14)	-	dimension X of the radiation field at a specified distance,
dimension (15)	-	dimension Y of the radiation field at a specified distance.

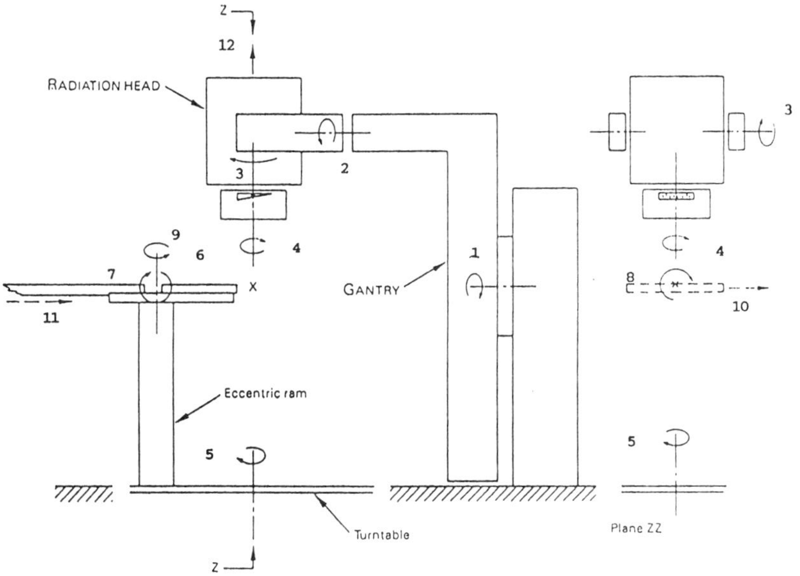

FIGURE 3.11 The rotary gantry (schematic) after IEC Document 976 (Table 3.3): (1) rotation of the gantry, axis 1; (2) roll of the radiation head, axis 2; (3) pitch of the radiation head, axis 3; (4) rotation of the diaphragm system, axis 4; (5) isocentric rotation of the table, axis 5; (6) rotation of the table top, axis 6; (7) pitch of the table, axis 7; (8) roll of the table, axis 8; (9) height of the table, direction9; (10) lateral displacement of the table, direction 10; (11) longitudinal displacement of the table, direction 11; (12) distance axis 1 to radiation source.

Axes (1) to (8), directions (9), (12) and (13) and dimensions (14) and (15) shall be provided with scales. Rotation scales shall be graduated in units of degrees, using positive numbers, for example ...358°, 359°, 0°, 1°, 2°... irrespective of movement available. Linear scales shall be graduated in units of centimetres with subdivision at intervals of 5 mm or less.

In the system shown in Fig. 3.11, the patient lies on the *treatment coach* which has a wide range of linear and rotational movements. For most treatment techniques, the patient lies prone or supine and is immobilized. The tumor center must be located at the *isocenter*, a point defined by the orthogonal intersection of the axis of rotation of the head and the therapy beam axis. Patient positioning is usually performed using three orthogonal light (laser) beams projected vertically from the ceiling of the therapy room and from its two sides. Due to these very

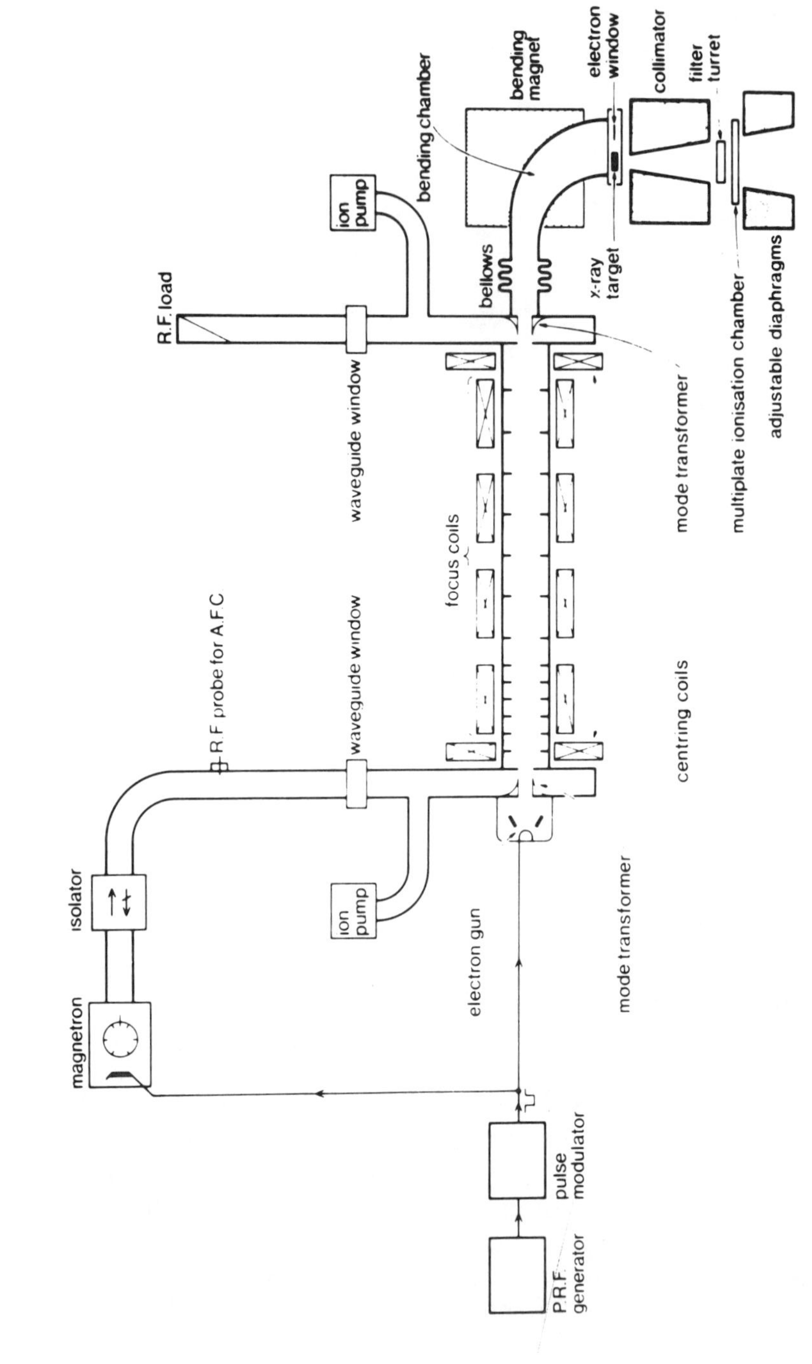

FIGURE 3.12 A schematic of a SL75 Philips traveling-wave radiation therapy accelerator.

small aperture beams and skin markers the irradiated field can be very well localized. The patient is subsequently positioned and the gantry can be fixed in a predetermined position (*field therapy*) or can be rotated (*rotational or arc therapy*).

Thanks to their compact construction, the present-day radiation therapy accelerators can be installed in rooms whose height is a mere 2.5m.

Traveling-wave accelerator system. This system is shown in Fig. 3.12. The magnetron and the electron gun are modulated, like in the system presented in Fig. 3.10, by periodic pulses from a modulator. The rf energy generated in the magnetron passes from the left hand side of the accelerating structure through an isolator and the rf waveguide window. The isolator lets the rf energy pass from the magnetron to the structure, but it prevents the possible return of the energy reflected from the magnetron. This arrangement, together with the automatic frequency control, ensures stability of the magnetron operation. The rf energy then passes through the structure, where it is mostly transferred to the beam of accelerated electrons. The residual energy is transferred to the load resistance coupled to the right hand end of the accelerating structure through an rf window.

The accelerating structure accommodates coils of two types of electromagnets. *Focusing multipole electromagnets (quadrupoles)* in the central part have the function of reducing the beam diameter and maintaining it at a predetermined and the lowest possible value. The electromagnets of the second type control, through a coupling to a position sensitive dosimetric system, center the beam and maintain it in a proper position. The importance of beam positioning stability for radiological parameters has been discussed in Sec. 3.1.1.1 (Fig. 3.4).

At its output, the accelerating structure is connected *through vacuum-tight "bellows"*, to a *magnetic bending chamber* which bends the beam through 90°. The bent beam may be extracted directly through a vacuum tight window, or subjected to conversion to X-rays.

Among the advantages of a traveling-wave system is that of a small *energy spread* in the accelerated beam, usually in the ±3% range (FWHM), and the relative ease with which the load can be changed. This change is of particular importance when the X-ray modality is changed to the electron modality, which requires large current reductions, down factors of up to 1000. In spite of the fact that traveling-wave systems are still being produced by one of the manufacturers, standing-wave accelerators have become far more prevalent.

A standing wave accelerator. A system of this kind can best be discussed with the example of Clinac 18 (Varian) accelerator whose major subsystems are shown in Fig. 3.13. The modulator, together with the Klystron generator and other auxiliary circuits, is housed in a separate stationary cabinet. The accelerating structure proper 1.4 m long with side-coupling cavities is installed horizontally in the rotating gantry. The structure is adopted to accelerating electrons to the energy of 10 MeV in the X-ray modality. These electrons are delivered to the magnet system which provides achromatic beam bending of 270°, and then are converted to X-rays in a target located in the gantry. In electron modality, the structure can generate electrons with energies of 6, 9, 12, 15 and 18 MeV. The structure is powered with rf energy via a rotating rf joint located between the movable frame and the stationary cabinet.

The distance between the X-ray target and the axis of rotation is 100 cm and that above the floor is 129.5 cm. A *beam stopper*, as an optional accessory, can be mounted in the lower gantry area opposite the treatment head (Fig. 3.13). It can be retracted mechanically from the beam region.

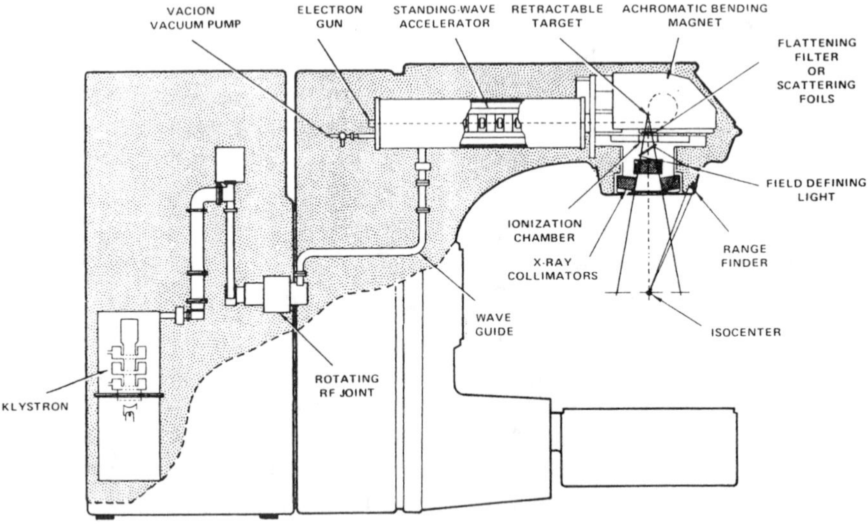

FIGURE 3.13 A schematic of a Varian Clinac 18 standing-wave radiation therapy accelerator.

The treatment head, in addition to a beam modifying assembly (collimators, flattening filters), includes an *optical system (range finder and field defining light)* to project the treatment field and localize the position at the isocenter.

Fundamental control and dosimetry systems. In order to meet the basic clinical requirements presented already in Sec. 3.1.1.1, electron and X-ray beams produced by the radiation therapy accelerator must display high stability of position and direction as well as energy of the particles accelerated. Stability of intensity within the treatment field and in the penumbra is also required. All these parameters are interdependent and, therefore, any change in any one of them causes corresponding changes in others.

For example, if one discusses the accelerating energy instability of the beam in the 8 MV X-ray modality, ±4% variations in energy are reflected in ±1% changes in the dose absorbed in the tissue at 10 cm along the beam axis.

Electron beam energy instability within ± 4% is at present regarded as acceptable. In electron therapy, the following empirical relationship is usually made use of: the depth of the 80% dose in cm is roughly equal to one third of the beam energy in MeV. Hence, an ±0.5 MeV change in the beam energy causes a shift of the point corresponding to the 80% depth dose by about 2 mm, which is tolerable. If the energy instability for 20 MeV electrons were 10%, the 80% depth dose shift would be as large as 7 mm, which is unacceptable. That is why a permissible change in energy is taken to be 0.5 MeV for the energy range of 5 to 20 MeV. However, recent recommendations by the IEC increase stability requirements so that the depth-dose stability in electron therapy should be such that energy variations are lower than ±1% for energies exceeding 10 MeV.

An example of the control and stabilization of the main beam parameters in the SL accelerators manufactured by the Philips company is shown in Fig. 3.14. In this system, an 8-section *ionization chamber* is used to monitor the radiation fields. Two sections monitor the front and rear edges, i.e. the opposite edges of the beam and are intended to control both the symmetry and beam energy. These two parameters are interdependent because the accelerator is equipped with a 90° bending magnet in which electron energy changes give rise to changes in field edges. In order to monitor these energy-sensitive positions of the field, the above chamber sections are provided with D-shaped electrodes. The output signal, after processing, is used to control the electron gun emission. By modifying the current through the structure the final beam energy can be controlled (Fig. 3.8). By coupling to a 90° bending magnet, which is also an energy spectrometer, the output beam energy stability is ±3% relative to the nominal value.

The four subsequent ion chamber sections are arranged in pairs and they monitor the radiation field: one pair controls the radiation field on the gun-target axis at points approximately midway between the field centre and the periphery, while the other pair controls the radiation field in the transverse axis. The signals from both pairs of sections make it possible to determine the beam position at the entry to the bending magnet. The shifts in this position can be corrected by two steering coils on the accelerating structure: one displaces the beam in the vertical direction, and the other in the horizontal direction. Thus, the stabilization of the field uniformity can be maintained.

The above discussion was concerned with the conversion of electrons to X-rays. When using the electron therapy modality, the *electron beam current is reduced* by a factor of 100 to 1000. At such small beam load, the final energy is not sensitive to changes in electron gun emission. The energy spread in this modality is practically dependent only on the bending spectrometer energy resolution. Consequently, energy control by controlling the electron gun emission is not effective and, therefore, unnecessary. The electron gun emission, however can be made use of in stabilizing the dose rate. The cathode current (gun emission), and thus the dose rate are very sensitive to the cathode heater temperature. The dose rate, therefore, may vary considerably depending on the instability of the gun heater current and other temperature-dependent variations.

The system depicted in Fig. 3.14 also shows two other ionization chamber sections: one plays the role of the *main dosimetry channel*, and the other as the *secondary channel.* A voltage signal from the chamber supplying the main channel is compared with a reference voltage which corresponds to the preset dose-rate. If any deviation occurs, the gun heater current is adequately adjusted. To stabilize the field symmetry, which ensures the field uniformity (flatness) in the electron therapy mode, the

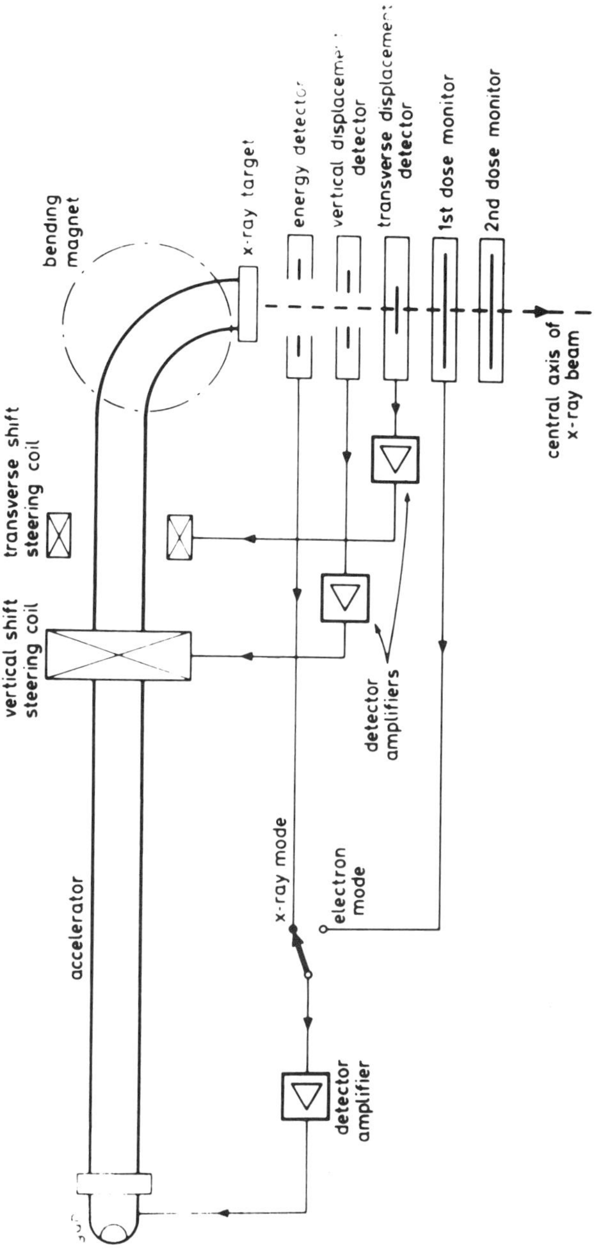

FIGURE 3.14 Beam parameter control system (Philips SL75).

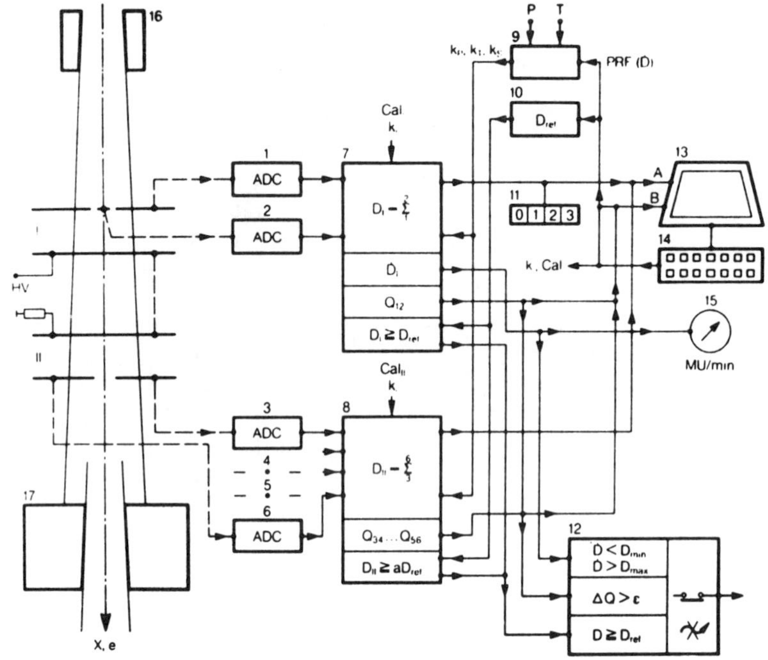

FIGURE 3.15 Block diagram of a dosimetry system (BBC Dynaray); description in the text.

system which is responsible for X-ray field uniformity is also used.

For the measurement of the dose delivered to the patient, i.e. of the fundamental parameter in radiation therapy, dual-channel dosimetry systems are used. A schematic of such a radiation monitoring system, BBC Dynaray systems, is shown in Fig. 3.15.

Thin-walled ionization chambers I and II are located between the pre-collimator (16) and main collimator (17); these two chambers have a total of six compartments (3 sections per chamber). The charges, which are generated in burst in the chambers by radiation pulses, are transformed into digital signals in charge balance converters (1) to (6) with the resolution corresponding to 0.01 cGy dose. Further signal processing is implemented digitally by means of self-monitoring microprocessors (7) and (8), installed in dosimetric channels I and II.

The signals from the chambers are first of all corrected in order to compensate for the recombination losses unavoidable at high dose rates (coefficient k_S). Subsequent corrections take into account fluctuations in pressure, p, and temperature, T, of the air in these chambers (coefficients k_p and k_T).

Using the keyboard (14), the user can enter into the calculation service program the weighing factors k_i and the calibration factors "Cal", which are dependent on the type of radiation and its energy. The four-digit monitor values of both channels, as well as the preset values and the instantaneous dose rates and irradiation times are shown on a common display screen (13). Signals A and B correspond to normal and calibration modes, respectively.

The two independent channels I and II operate in a continuous mode. If the maximum allowed dose rate in either of them is exceeded, the beam is switched off. The same happens if any malfunctioning in the operation of the channels occurs. The radiation monitoring system is provided with a mechanical dose counter (11) for the storage of actual values in case of a power failure. Field flatness is continuously checked for symmetrical and asymmetrical uniformity by seven quotient signals produced by integration of partial signals from the six chamber compartments. The results are displayed in the service program and may be used, by means of the beam interlocking circuit (12), to switch off the irradiation when the preset dose is reached or in the case of a fault. In addition to the dose, the dose rate is also monitored to make sure that it falls within pre-determined limits. For this purpose, special circuits with very short time constants are included in the analogue part of the input-converters (1) to (6). If the maximum allowed dose rate is exceeded they switch off the radiation beam. The chamber voltage is also monitored by dual channels — a general principle adopted by BBC.

Internal system of accelerator control. Present-day *fractionation radiotherapy* requires correct and repeated (reproducible) *positioning of the patient* some 20 to 30 times over four to six weeks. The treatment positioning must maximally approximate that during simulation. The method most often used for this purpose is the *isocentric technique*. Correct patient set up is aided by light projection systems (lateral and longitudinal). The patient is most frequently required to lie on his back since in this position he is most stable and his anatomy can best be reproduced at each treatment session. When the correct set up is achieved the patient, if possible, is also *immobilized*. During treatment, high dose rates are recommended because in this way long-lasting patient immobilization which is often difficult to achieve and uncomfortable for the patient can be avoided. This is specially essential when moving organs or tissues adjacent to them are to be treated.

The next stage consists in proper setting up of the accelerator parameters. The diagram shown in Fig. 3.16 shows the accelerator head which is equipped with a manual control systems which make it possible to set the settings corresponding to such parameters as filter and diaphragm positions, etc. manually by the operator in the treatment room. In general, all the movable elements are set by servo mechanisms. This is true both for the treatment table top, controlled by a separator manual control, and the treatment head. These movable elements include special *position sensors*, and other critical elements which incorporate *function sensors*. The signals from the sensors are subsequently fed to an *internal micro- or mini-computer system* which assists the operator.

When all the systems and controls are set, what is most important is to verify their settings by comparing them with those required by the treatment plan. Patient irradiation cannot be started unless the above comparison is made and agreement established. When agreement cannot be achieved all the deviations from the pre-programmed procedure must be accepted by the operator. *This means that regardless the degree of automation of the control system, it is the operator who has the final say in the treatment.*

The view of a modern *control console* (Varian Clinac 4) is presented in Fig. 3.17; Clinac 6 has a similar console. A monitor panel is located in the left-hand upper corner. It provides a means for monitoring the main accelerator parameters and for adjusting the dose rate. A set of indicators allow the operator to identify constantly the machine subsystems and their parameters. They also display the status of the interlock systems, which stop the accelerator operation if any irregularities or malfunctions appear. The following parameters are usually monitored: the level and temperature of the cooling water (excess temperature is indicated), flow, vacuum, terminal switch, which

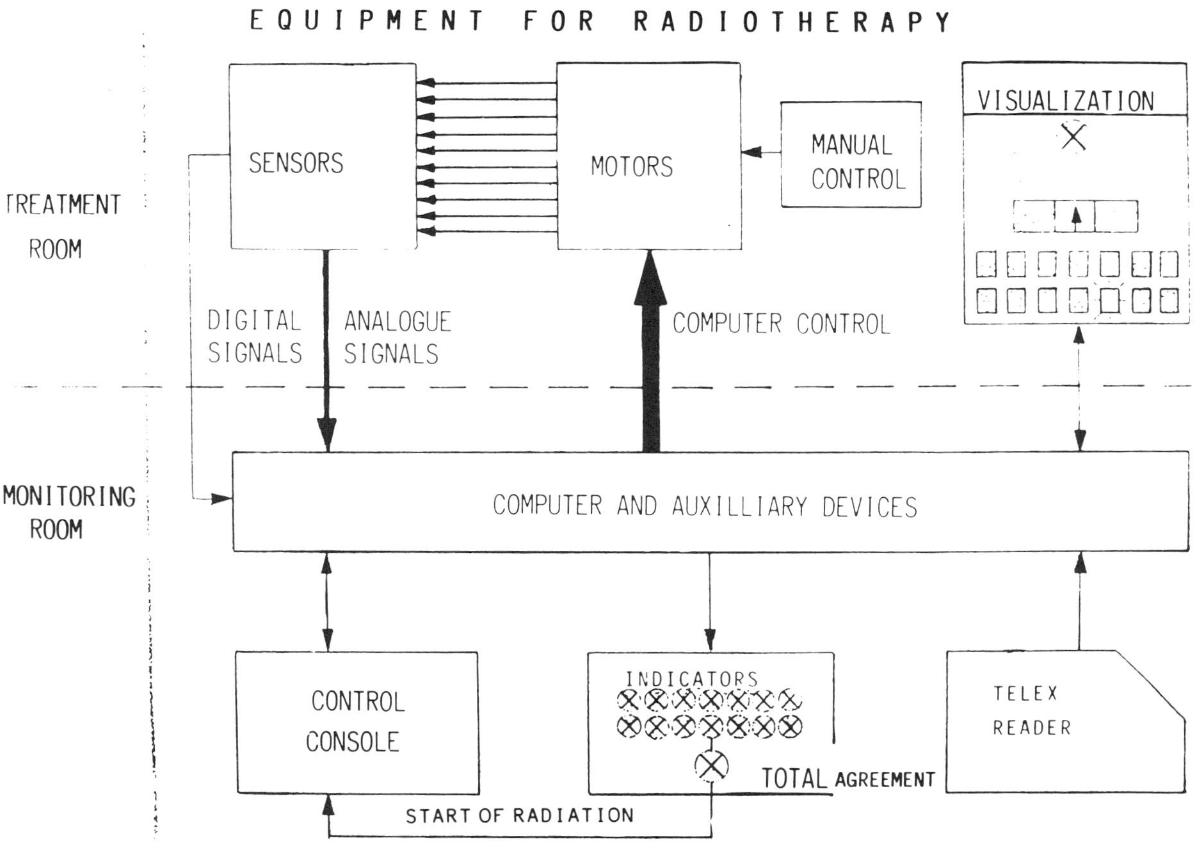

FIGURE 3.16 Block diagram of internal control system from operator console.

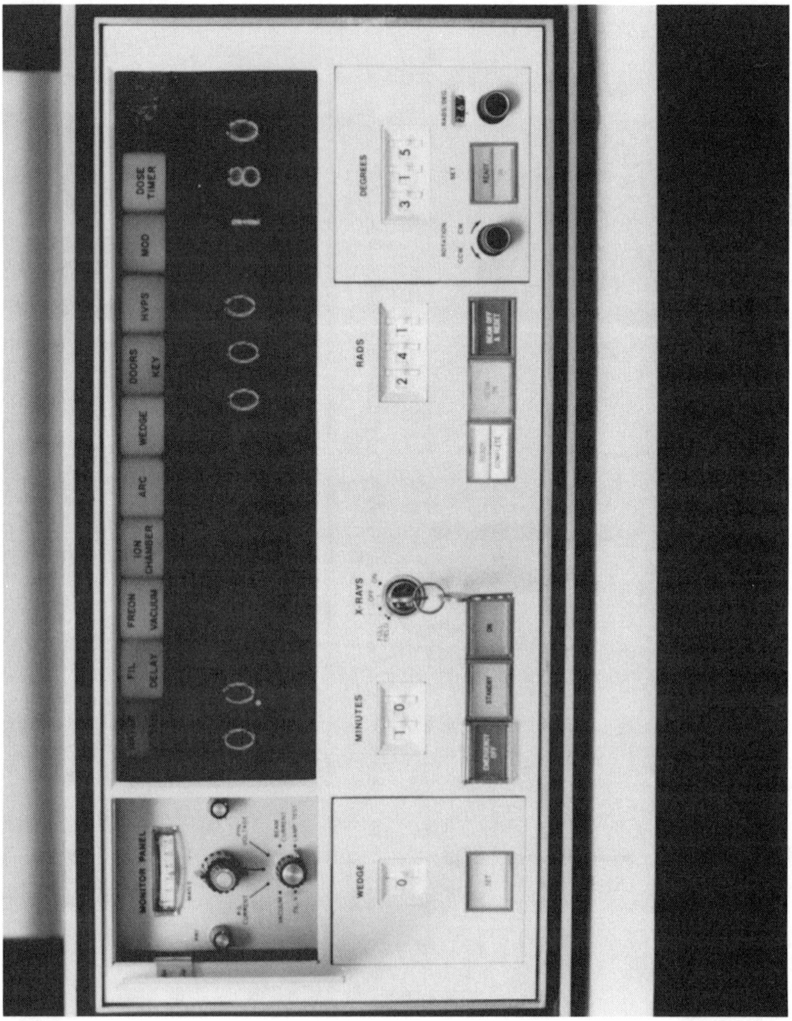

FIGURE 3.17 View of an operator control console (Varian Clinac 4).

indicates the gantry's end position, modulator current overload, freon pressure in the rf system, time delay necessary to properly pre-heat the machine, treatment room door interlock, etc. Wedge select interlock prevents treatment unless the wedge selected by the operator and the actual wedge used are identical. Dosimetry key D (Cliniac 6) is intended for resetting logic related to the dosimetry system. If the dosimetry system stops the irradiation, key D makes it possible to check if the system operates correctly before the next treatment is initiated. The indicators placed at the top of the console include (from the left): a timer, a dose rate indicator displaying monitor units per minute, two digital indicators for dosimetry channel 1 (MU1) and 2 (MU2) incorporating right indicator for gantry angle control; this indicator permits the operator to control gantry rotation during the therapy.

In the Cliniac 6 system three modes of operation are possible: (1) *"full field"*, (2) *"fixed field"* for single or multi-port therapy, and (3) *"arc clockwise" or "arc counterclockwise"* for rotational therapy.

From the console shown in Fig. 3.17, before starting treatment, the operator can set the prescribed dose and dose rate, maximum treatment time and the maximum dose, the gantry angle and dose rate per 1° in arc therapy.

External auxiliary systems. For the *patient throughput* of the order of several tens of patients per one shift, the operator has to handle a large number of data, between 2000 and 3000. This operation, if performed manually, is very time-consuming and can involve errors. As early as 1960s it was found that errors made by the keyboard and card-punch operators ranged from 0.004 to 0.2% per key stroke [3.2]. According to the estimates made in the early 1970s errors that occurred in manually kept radiotherapy records and manual execution of treatment planning might easily be as large as 1-2% per treatment field [3.23]. Therefore, in order to reduce errors made when executing procedures manually and to shorten the time of preparing the accelerator for operation, most manufacturers have developed microprocessor or minicomputer *systems of recording and verifying data or controlling the accelerator.* Among the purposes of such systems, one is to verify the actual data against treatment planning parameters, and then to record the implemented treatment procedure. These systems make it possible to record

(a)

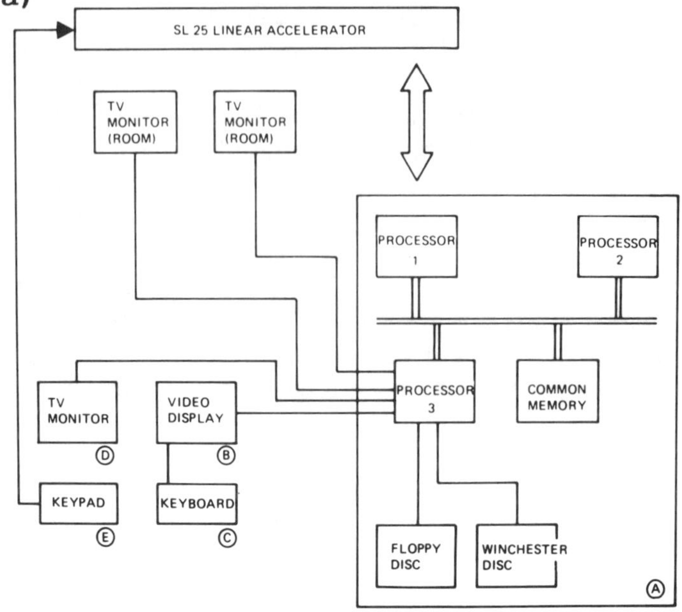

(b)

FIGURE 3.18 Microcomputer auxiliary service system for a radiation therapy accelerator (Philips SL 25): (a) — schematic, (b) — general view.

data from 100-300 patients treated within a certain period of time.

The main feature of the above systems is that they are only auxiliary with regard to the operator and never aspire to replace him entirely. The final decision as to the patient set-up or the machine proper adjustment rests with the operator. This is because computer systems, although versatile in monitoring the accelerator parameters, are unable to provide equally accurate information on the patient positioning vis-à-vis the machine.

The demand for automated control systems has increased in the past few years with the development of *multi-field technology and multi-energy therapy*. All this has led to a considerable increase in the number of parameters used in a single treatment.

Figs. 3.18a and b represent a schematic and a view of a microprocessor control system Philips SL25. The system consists of a control cabinet operator console A which comprises processors, external memories, power supplies and a modem unit for remote monitoring of accelerator functions. The operator console proper includes a video display unit B, keyboard C, TV monitor D and a keypad E. The control cabinet also incorporates TV monitors D. The TV monitor and keypad are used to control the irradiation.

Three 16-bit INTEL processors are interconnected through a MULTIBUS system (Fig. 3.18a). Processor 1 has a 128 kB RAM memory and is intended for data communication with the accelerator. Processor 2 has a 256 kB RAM memory which collects data to control all the functions of the accelerator. Processor 3 has an identical 256 kB RAM memory to control the access to the hard disc and other peripherals. A common 512 kB memory is shared by Processors 2 and 3 when additional memory is required. A 35 MB hard disc is the main external memory in which all programs, patient and accelerator data are stored. An 1 MB floppy disc is used for storing operational and test programs during the installation of the accelerator.

Each patient's data are entered from the keyboard. They include patient identification and non-mechanical parameters such as type of radiation, energy and monitor units. Mechanical parameters can be entered manually or transferred from the accelerator after the patient has been correctly positioned. The treatment prescription is thus stored on a hard disc for future

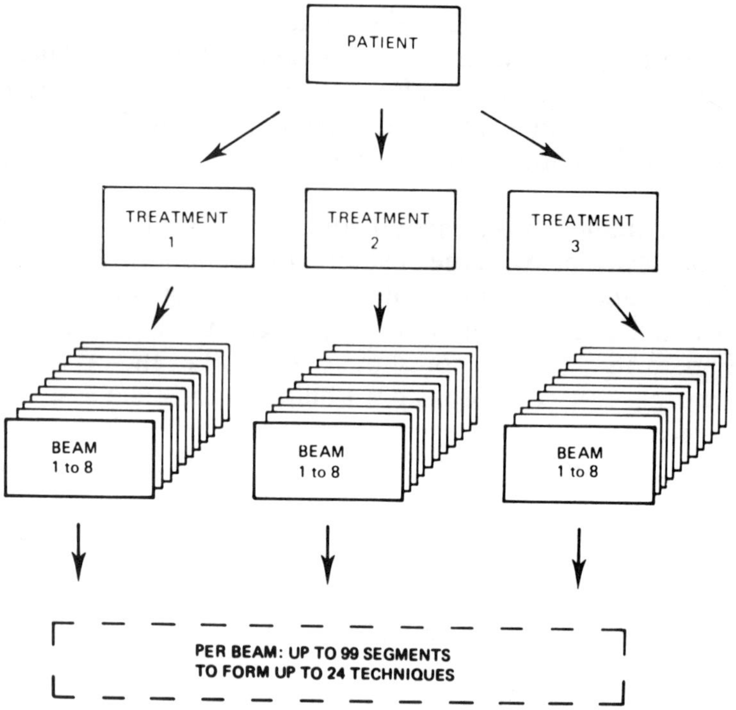

FIGURE 3.19 Structure of treatment prescription.

treatment sessions, or can be erased if it is not to be used in the future. Up to 250 patients can be recorded and assessed.

A treatment prescription can be formed of up to three beam (irradiation) configurations (Fig. 3.19). Each beam configuration may be used to irradiate a certain portion of the tumor and can consist of up to eight stationary beams. Several beam segments can also be combined in a single beam. These segments include, for example, various energies (cf. Fig. 3.20), wedge filter positions, etc. A certain arrangement of segments defines a *treatment technique*.

When a patient comes for the next treatment session, the operator calls up his prescription from the hard disc by typing in

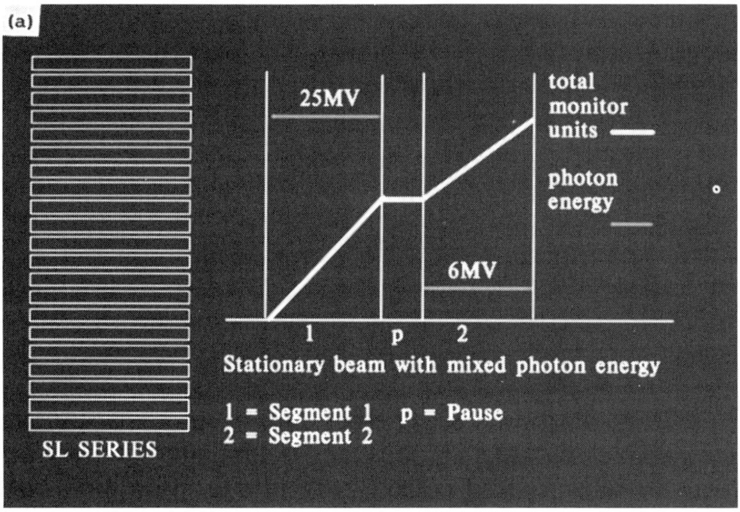

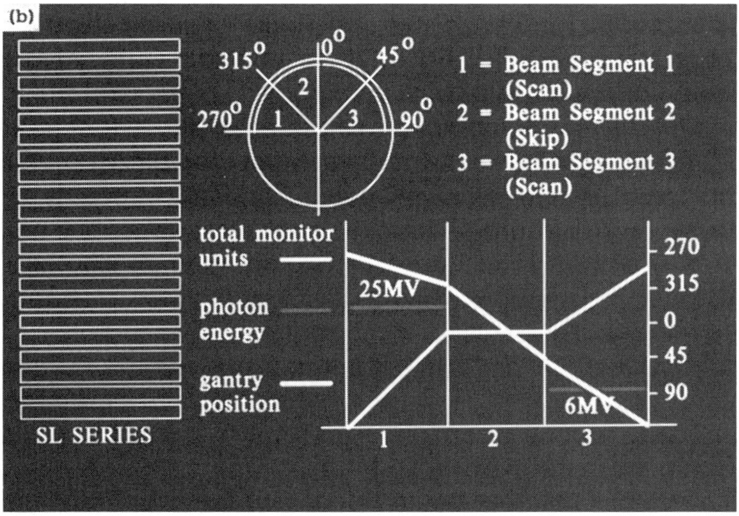

FIGURE 3.20 X-ray multi-energy therapy: (a) — time cycle in stationary treatment, (b) — arc treatment (scanning + skip).

the patient's identification symbol. After selection of the treatment to be given, the prescription data for the first beam are displayed on the operator's monitor and, simultaneously, on two TV monitors in the treatment room. The accelerator nonmechanical parameters are automatically set according to the prescribed values. The next step is to set mechanical parameters, such as the gantry angle, diaphragm rotation angle, field size and table parameters in accordance with the prescription within the predefined tolerances. For all mechanical movements eight different groups of tolerances can be defined. The code which determines the treatment prescription defines, at the same time, a combination of all tolerances for each of the selected beams.

The disc also contains all the special events, such as updated treatment prescriptions, activation of the override function, irradiation interruption either manually or automatically, and premature treatment termination.

The basic system, shown in Fig. 3.18a can be extended by adding special functions which make it possible, for example, to implement *multisegmental rotational therapy* (Fig. 3.20) and other nontypical procedures. In this are included procedures for dual exposure check radiographs. In standard procedures, a single exposure is envisaged in irradiations using a specific beam configuration.

3.1.2.5 Multi-energy Accelerators

General requirements. A characteristic feature of the present-day radiation therapy is that ever more precise treatment fields are generated with varied energies and types of radiation employed. According to the recently published data for a large city area (Philadelphia, USA) on caseload and beam preferences, 71% of patients were treated with low energy X-rays and 23% with high energy X-rays. Electron therapy was applied in a total of 18 % of patients, of which 12% were treated with *mixed therapy (electrons and X-rays)* and the remaining 6% with electrons only [3.21]. It would be most convenient for both the therapist and the patient if one could have a *one-session radiation therapy* by means of one accelerator. That is why, accelerator designers have long been trying to build a machine which would operate over a wide energy range and have a capability of fast switch over from X-ray to electron modality or vice versa [1.17, 3.1, 3.20 and 3.21].

The accelerator capability for *multienergy X-ray radiotherapy* is shown in Fig. 3.20a. In time period 1, the target volume is irradiated with a 25 MV stationary beam. The subsequent pause P is necessary for the accelerator switch over, after which the target volume is again irradiated with a 6 MV beam. In a case like this, when one has a wide energy range, it is possible to obtain a treatment field that corresponds to any energy between 6 and 25 MV (cf. Fig. 1.10).

Fig. 3.20b presents an accelerator operational cycle of a technique referred to as a *dual segment Skip-and-Scan technique*. In time period 1, a 25 MV beam irradiates the angle segment from 270° to 315°. In time period 2, the beam skips angle segment from 315° to 45°, and the next irradiation is started with a 6 MV beam, through the angle from 45° to 90°. The whole cycle is controlled automatically by a system shown in Fig. 3.18.

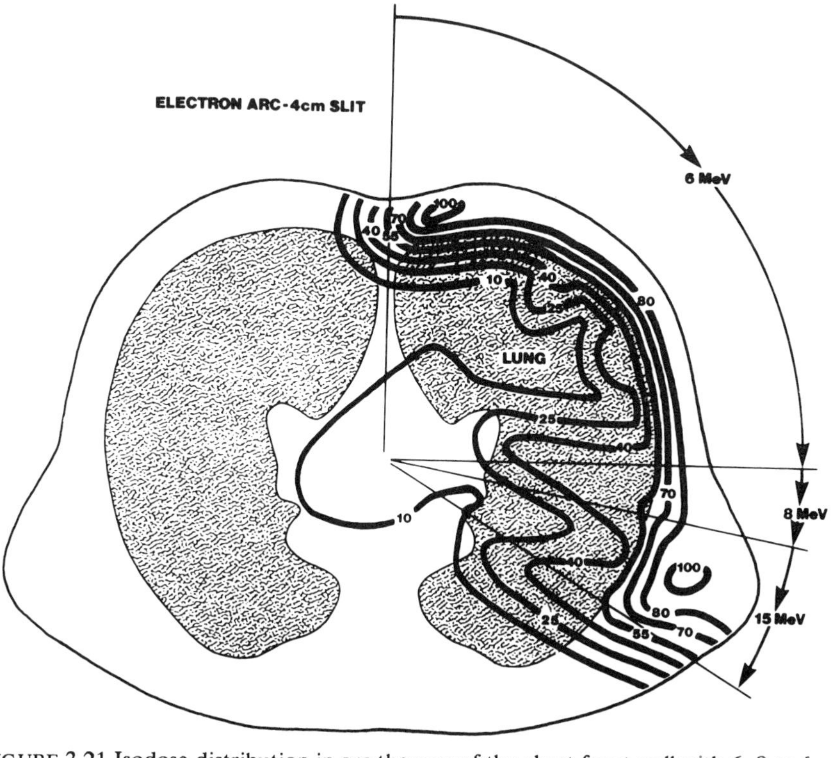

FIGURE 3.21 Isodose distribution in arc therapy of the chest front wall with 6, 8 and 15 MeV electrons (after F. Voogt [3.24]).

Fig. 3.21 is an example of a *multi-energy electron therapy*. Patient's chest wall is irradiated with the use of an arc technique. The initial body segment is irradiated with a 6 MeV beam, whereas all the subsequent segments with increasing wall thickness are irradiated successively with 8 MeV and 15 MeV beams. This means that sequential beams with three energies must be produced along a single arc.

Method of energy control. Whereas energy switch is a relatively easy thing to accomplish in electron modality due to the very small currents involved, the *X-ray modality, with beam currents higher by two orders of magnitude than those in electron modality,* involves considerable difficulty. The large energy switch in the X-ray mode load usually results in beam energy degradation, i.e. deterioration of the beam monoenergetic character. A detailed discussion of these problems is given in [1.17].

The conventional method of changing the accelerator output energy consists in varying the rf power supplied to the accelerating structure. In addition to varying the power, the output energy may also be changed by varying the beam load current in the accelerating structure. This causes the working point on the curve representing the variation of electron beam energy with beam current (Fig. 3.8) to shift in accordance with Eq. (3.2) (Sec. 3.1.2.3). The advantage this method has is that it is simple and reliable, due to which it has found wide application in accelerators produced by many manufacturers, e.g. the Varian Clinac-18 and Clinac-20, Siemens Mevatron 74 and Mevatron 77, AECL Therac 20 (CGR Saturn), and Philips SL75/14 and SL75/20. The method, unfortunately, also has a disadvantage: very rapid degradation of the beam energy spectrum leads to beam power reduction. In electron therapy, application of energy defining slits in the beam transport system may be helpful in this case. However, in X-ray therapy the use of this technique is much more problematic [3.1].

The method of varying the rf power in the accelerating structure in order to change the accelerator output energy has been employed with some success in Philips SL75/20 machine which generated two beams, 8 and 16 MV, that is with limited energy separation. In a Philips accelerator (SL25) X-ray beam energies attained were as high as 6 MV and 25 MV together with electron energies between 4 and 22 MeV. Both accelerators employed traveling wave accelerating structures.

In Varian Clinac 1800 machine an *energy switch* has been introduced. Its schematic is shown in Fig. 3.22. The accelerating structure, 1.4 m long, is divided in two sections: 0.4 and 1.0 m long. The shorter section bunches electrons and the longer one is an accelerating section. Both sections are coupled together by the energy switch whose construction is discussed in more detail in [3.24]. The control is performed by a side cavity with adjustable coupling, located between the rf coupler and the accelerating structure. The accelerator is intended for generating two X-ray beams with energies of 6 and 18 MV and five electron beams with energies of up to 20 MeV. The buncher section, which is always supplied with 0.4 MW regardless the X-ray output energy, operates under optimum conditions which ensure that the accelerating gradient is 10 MVm^{-1}. In the buncher, electrons are bunched and accelerated to 4 MeV. If a 18 MV beam is desired, the accelerating field in the second section is induced by rf 20 MW power, which results in the accelerating gradient of 14 MVm^{-1}. On the other hand, when a 6 MV X-ray beam is generated, the above gradient is reduced to 2 MVm^{-1},

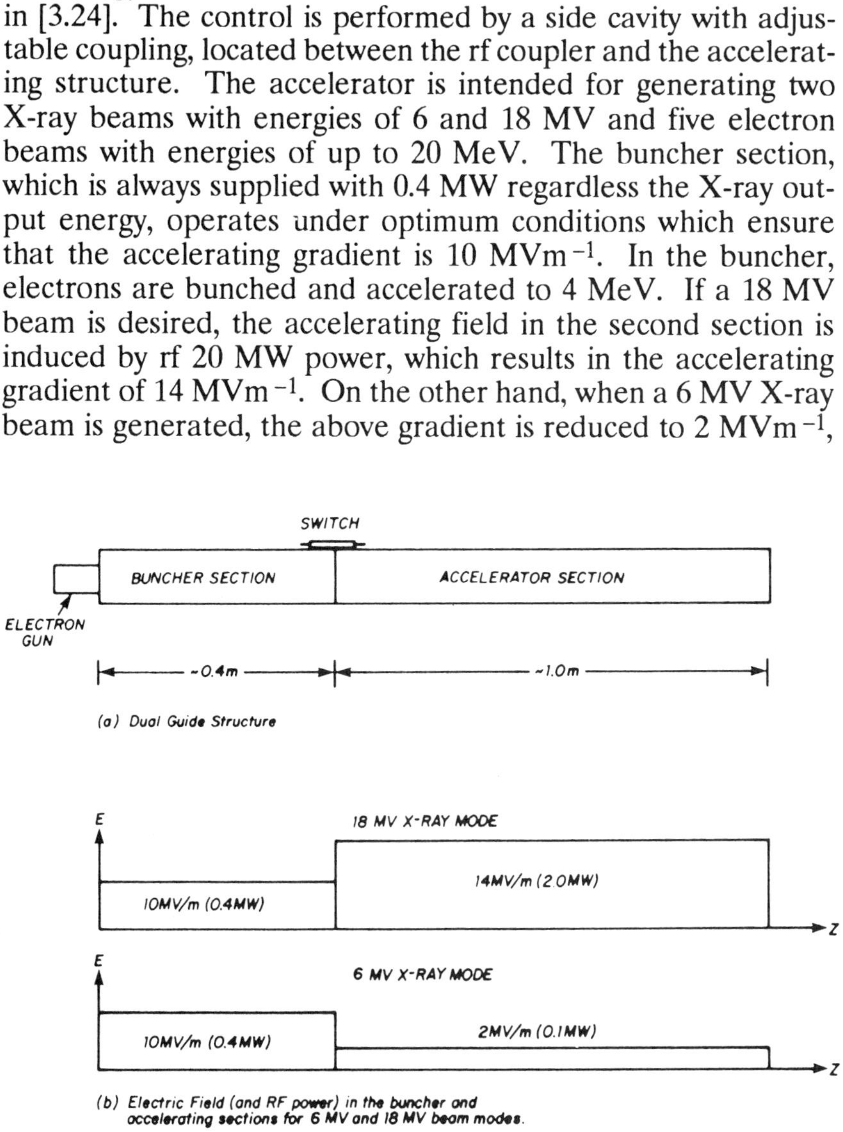

FIGURE 3.22 Varian Clinac 1800 accelerator with 6 MV and 18 MV energy switch.

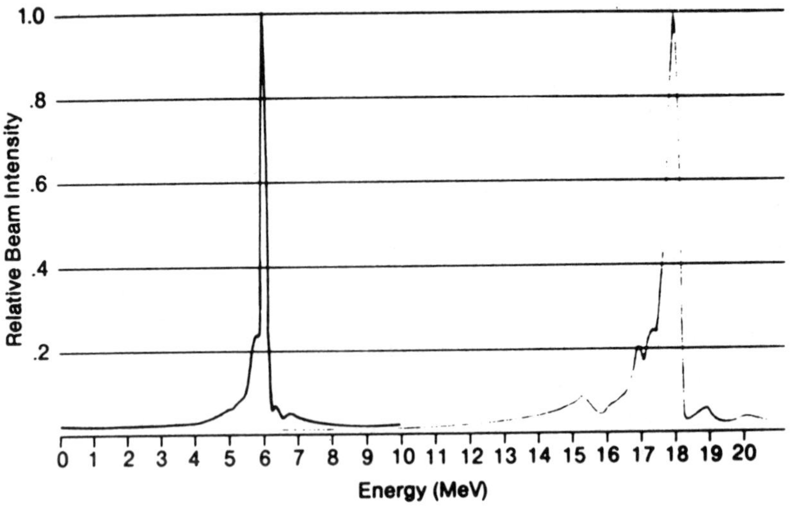

FIGURE 3.23 6 MeV and 18 MeV energy spectrum (Varian Clinac 1800) obtained with a magnetic spectrometer with 1% energy slit width (after E.Tanabe and R.W. Hamm [1.17]).

which corresponds to the power of 0.1 MW. Both beams are highly monoenergetic (3% FWHM), as measured with a magnetic spectrometer equipped with 1% energy slits (Fig. 3.23).

The Varian Clinac 2500 uses a 1.9 m structure which features X-ray beam separation in the range between 6 and 24 MV.

The Siemens Mevatron MD and Mevatron KD radiation therapy accelerators use a different design to control the X-ray beam energy, which consists of detuning the rf signal. The accelerating structure operates off-resonance, and as a result, output energies change. The Mevatron MD produces 6 MV and 10 MV X-ray beams, whereas the Mevatron KD machine, besides a range of electron beam energies, may generate five different dual energy photon combinations with a maximum separation of 6 MV and 20 MV.

Table 3.5 lists parameters of some multi-energy accelerators. In addition to linacs, it also includes microtrons (Sec. 3.3).

In view of the rapid development of multienergy X-ray therapy, as well as that of using the dual modality, both X-rays and electrons, multi-energy accelerators have become a viable alternative for oncological hospitals, as compared with the older

TABLE 3.5 Fundamental Parameters of Radiation Therapy Multi-Energy Accelerators.

Manufacturer and accelerator type	X-ray energies [MV]		Electron energies [MeV] and their number (in brackets)
	low	high	
Varian Clinac 1800	6(8)	10 or 15 or 18	6 to 20 (four)
	8	15 or 18	
Varian Clinac 2500	6	24	6 to 22 (six)
Siemens Mevatron	6	10 or 15	5 to 14 (six)
Siemens KD S	6	10 or 15 or 18	6 to 18 (six)
Siemens KD	6	10 or 15 or 18	6 to 21 (six)
Philips SL 15	6	10	4 to 15 (six)
Philips SL 18	6	15	4 to 18 (seven)
Philips SL 20	6	18	4 to 20 (eight)
Philips SL 25[1)]	6	25	4 to 22 (nine)
BBC Dynaray CH 20[1)]	6	16	5 to 20 (seven)
Mitsubishi ML-15M	6	10	6 to 15 (four)
Mitsubishi ML-20M	6	10	6 to 20 (five)
Jefremov LUER-20M	6	18	5 to 20 (six)
Microtrons			
Scanditronix MM 14	6	14	6 to 14 (four)
Scanditronix MM 22	6	up to 22[2)]	3 to 22 (nine)
Scanditronix MM 50	5	up to 50[3)]	2,5 to 50 (twenty)

1) Traveling-wave accelerating structure
2) Three X-ray energies between 6 and 22 MV
3) In basic version seven X-ray energies between 5 and 50 MV and three additional energies optionally.

single X-ray energy only machines. The cost of a multi-energy machine is \$ 500.000 higher. On assuming a 10 year lifetime the differential cost per year is \$ 50.000. Installation and operational costs for a new unit are comparable. However, the dual-energy accelerator has higher maintenance costs. Again, if this differential cost is \$ 20.000/year, the total cost increase for a dual-energy accelerator as compared with single X-ray energy unit would be \$ 70.000 per year. In the U.S. health care system, it corresponds to the revenue obtained from only 22 patients a year. The cost-effectiveness of the installation of a multi-energy accelerator which results in a 20% increase in the patient throughput due to increased treatment capabilities, cannot be questioned. It is true that multi-energy units have been shown to have slightly higher treatment time lost (2 to 4%) as compared with single X-ray energy machines (1 to 2%), but it is a small price to pay for most hospitals for greater treatment capabilities [3.1].

3.1.2.6 Multi-pass Systems

In an effort to design an accelerator which would be compact and short, attempts have been made to develop *multi-pass systems* in which one accelerating structure could be used twice or three times to accelerate the beam. Fig. 3.24 shows a standing wave *double pass accelerator system* to be used by AECL Therac 25 [F10] a few years ago. Electrons are produced by an electron emitter off the structure axis and are injected into the structure electrostatically. The electrons are initially accelerated to 13 MeV in the first pass, i.e. from right to left (Fig. 3.24). The standing wave structure is 1.2 m long and is powered by a 2.6 MW magnetron operating in the S-band. Next, the beam is extracted by a system of three focusing quadrupoles and is directed to a magnet which plays the role of a magnetic mirror. It bends the beam through 180° and reflects it back into the waveguide (from left to right), where it gains an additional energy of 12 MeV. In this way, the beam output energy at the treatment head may be as high as 25 MeV.

By adjusting the beam phase during the second pass, the output energy may be varied between 5 and 25 MeV. Therac 25 accelerator can also produce a single 25 MV X-ray beam. The electron beam is deflected in the treatment head by a scanning

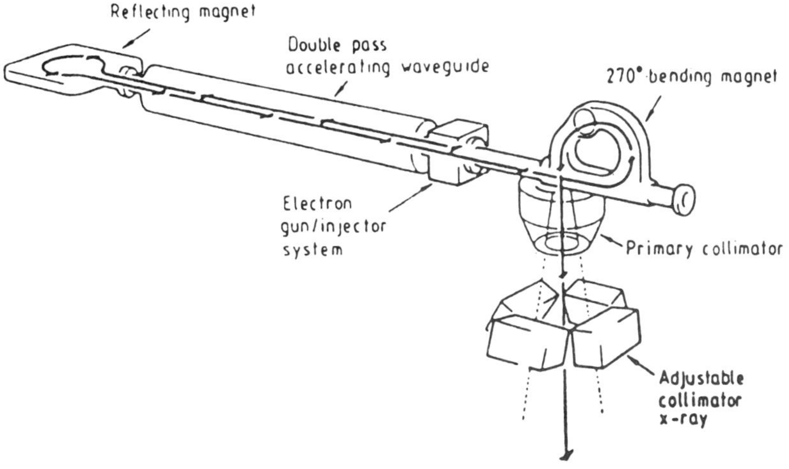

FIGURE 3.24 AECL Therac 25 recirculating (double-pass) linear accelerator [3.1]

system described in Sec. 3.1.5. The construction of the above accelerator is described in more detail in [3.25, 3.26]. The energy spread of the 25 MeV spectrum is 1% (FWHM).

3.1.3 The Beam Transport and Bending Systems

Beam focusing and stabilization. The magnetic system of beam transport incorporates: (1) electromagnet coils and control units which are mounted on the accelerating structure (cf. Fig. 3.14), (2) focusing quadrupoles (cf. Fig. 2.24) outside the structure, and (3) beam bending magnets. The electron beam which travels from the gun (source) through the structure to the X-ray conversion target or scattering foil is directed, focused and transported by the above systems.

The need for high degree of beam focusing results from the demand that the beam trajectory should be only slightly tilted, i.e. the effect on the penumbra should be small. It follows from Fig. 3.25 that, if the radius of the beam of accelerated electrons striking the X-ray target is r (usually r $\approx$ 1 mm), an X-ray beam emitted gives rise to penumbra of the width a in the plane of rotation. Since this beam is usually collimated by a diaphragm consisting of movable jaws, X-rays tangential to the vertical

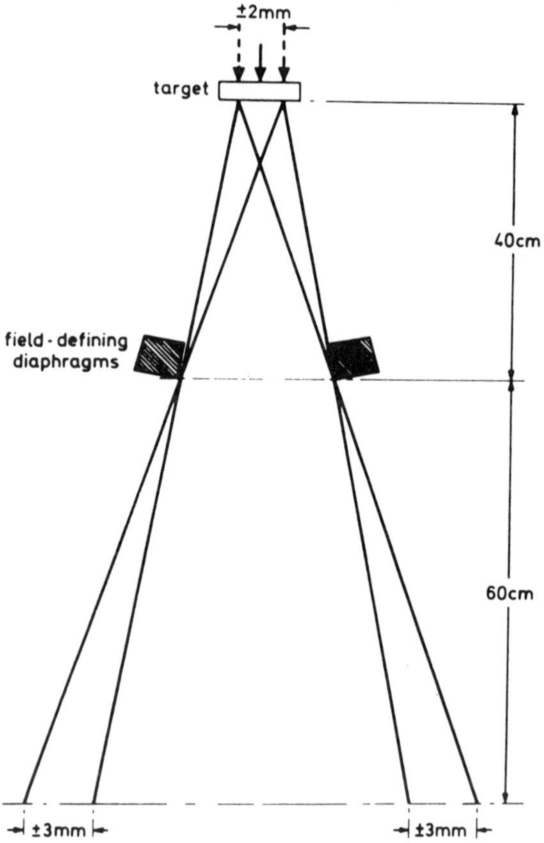

FIGURE 3.25 Geometrical and physical penumbras in an X-ray beam.

edges of the collimator partially penetrate the jaws and give rise to the penumbra whose total width is usually about 6 mm. The better the focusing the smaller the effect of the penumbra.

Since accelerators producing X-ray beams are, as a rule, equipped with flattening filters of conical shape (Fig. 3.38) what is of special importance is that the beam position and the beam angle with the X-ray target be stabilized.

This situation is depicted in Fig. 3.26 (cf. Fig. 3.4). If the electron beam axis coincides with the flattening filter axis (Fig. 3.26a), the X-ray beam, having passed the filter, has, as a result, a symmetrical flat field characteristics. The electron beam may, however, show instability of two kinds. It may depart from the central trajectory through the angle θ, called *angular divergence,* and it may also undergo the lateral positional departure Δx, called *displacement.* In both cases (Fig. 3.26b and c) the X-ray field becomes strongly asymmetrical for higher values of θ and Δx. At present, all radiation therapy accelerators for routine applications use the isocenter system (Fig. 3.5). For low energies of 4 to 6 MV, the treatment head can have the *in-line configuration* with a short standing wave structure (Fig. 3.5d). In this design of the head, the bending magnet is not necessary and, what is more, even focusing coils can be dispensed with. In the accelerators which employ the above design, the target is also

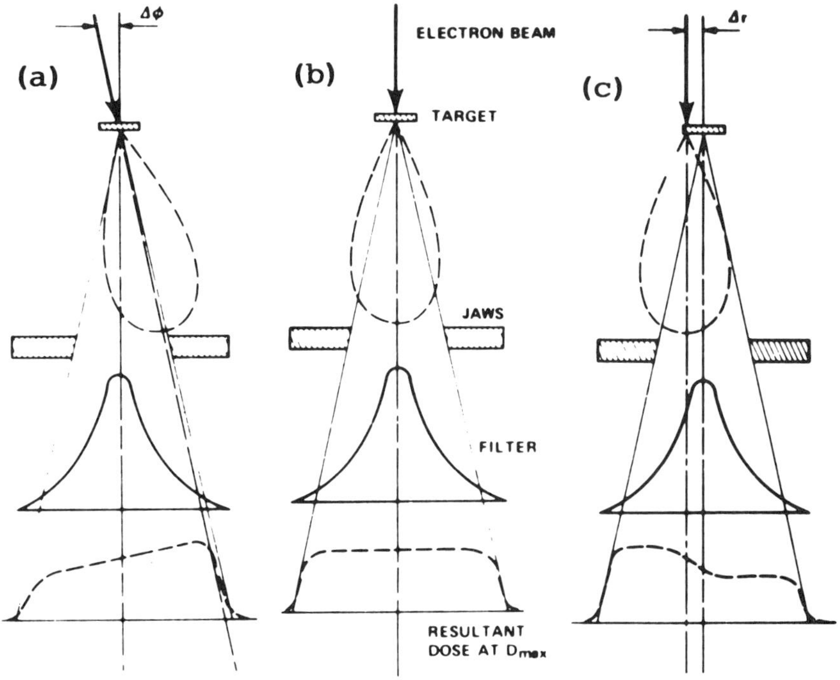

FIGURE 3.26 Effect of instability of position of an electron beam incident on X-ray target on the D_{max} distribution: (a) — correct position, (b) — beam angular divergence from the axis initially orthogonal to target, (c) —radial displacement x (after C.T. Karzmark [3.27]).

bombarded with low energy electrons. As a result, in order to obtain equivalent X-ray penetrability, higher peak energies must be used than those in accelerators with bending magnets.

Instabilities may be caused either by changes in beam energy due to rf power or frequency variations or by various other factors, such as mechanical strain resulting from gantry rotation. Influence of temperature and the earth's magnetic field or other external fields cannot be neglected.

The main reason why 90°or 270° bending magnets have been employed, as already mentioned in Sec. 3.1.2.1, is the attempt to accommodate long accelerating structures in the in-line configuration, or in other words, developing a rotational unit for higher energies and low-placed isocenter (Fig. 3.5b and c).

Bending magnets. Terminology. The problems concerning the trajectories of particles moving in a magnetic field have been dealt with in Secs. 2.4.1 and 2.4.2. The efficiency of bending the trajectory is determined by a parameter called *magnetic rigidity* (Eq. 2.19) For example, a 1.5 T magnetic dipole (Fig. 2.24) will bend a 25 MeV electron beam with a radius of 5.66 cm (Fig. 2.25). A 5 MeV electron beam can be bent with the same radius by a magnetic field of only 0.32 T.

The phenomenon of energy dispersion of the beam has also been discussed in Sec. 2.4.2. In radiation therapy accelerators, the dispersion of the beam at the accelerating structure output is often as high as 10%.

Typically, a *bending magnet* (Fig. 2.25) has two functions to perform: *it bends the beam and separates particles with various momenta* (with various energies). Both these functions are interconnected. When discussing bending magnets, there is a tendency to emphasize one or the other of these functions depending on the actual need. In the past, linear accelerators used to be equipped with only one dipole. At present, units are used which incorporate two or three magnets (Fig. 3.29) that better separate the above two functions. The electron beam which enters the beam transport system has a given energy spectrum. Besides, individual particles may exhibit some geometrical divergence and lateral displacement. In general, the task of the beam transport systems is to deliver the beam in the form of parallel and coaxial rays in a well-defined energy range. What is meant by "delivery" is that the electron beam should be delivered onto the X-ray target or the scattering foil.

In the field of bending magnets employed in radiation therapy accelerators, definitions and concepts are used which are analogous to those used in the field of optics. The term *achromatic system* is taken to describe a magnetic system which defines a particle beam, originally parallel, characterized by energy dispersion. The *nonachromatic magnet* plays a role of a prism that disperses light in basic colours. In *zero dispersion systems,* sometimes called *singly achromatic,* the particles travel in divergent trajectories after reaching the focal point. In achromatic systems, sometimes called *doubly achromatic,* the particles travel as a parallel beam after being focused. The term *double focusing* is taken to mean focusing both the transverse and radial components at the same point along the central axis. The image of a doubly focused beam is called *stigmatic*. The term *energy focus* is used for the focus of momenta associated with their previous dispersion, whereas the term *spatial focus* is used for the transverse and radial focusing of particles, associated with particles which, after focusing, are divergent or displaced with respect to the central ray (central axis). The terms *angular departure* and *lateral positional displacement* Δx have already been defined (Figs. 3.4 and 3.26).

90° bending magnets. Bending magnets employed in medical accelerators are shown in Fig. 3.27. Fig. 3.27a depicts a simple 90° beam bending magnet analogous to one shown in Fig. 2.25. It can also play a role of a magnetic analyzer since it separates the beam with respect to the energies of individual particles. That is why it may be used as an energy analyzer when a slit emitting particles in a selected energy interval is used (Fig. 3.27a). If individual input rays are shifted in parallel to each other, the beam at the magnet output is characterized by a considerable angular divergence, whereas a divergent input beam will still be divergent leaving the magnet. The magnet of this kind is referred to as an *nonachromatic magnet.* In order to obtain achromatism, the system must employ two or three partial magnets. Although 90° bending magnets featured all the advantages of simplicity of construction and small dimensions, which made it possible to locate the isocenter not so high above the floor, those magnets do not find application in present day accelerators because of the disadvantages mentioned above.

270° bending magnets. A simple example of a magnet of this kind is shown in Fig. 3.27b. It is single and achromatic, being characterized by zero dispersion and a uniform magnetic field. An input beam of an energy spread of up to ±10% is focused at one point on the X-ray target. Particles with higher energies have larger trajectories in the magnet (dotted line in Fig. 3.27),

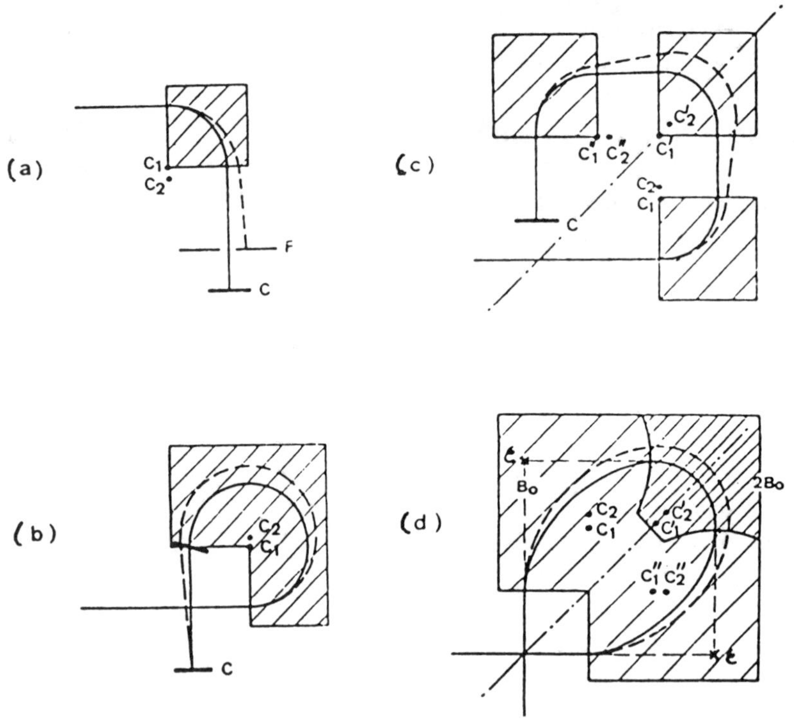

FIGURE 3.27 Beam bending magnetic systems: (a) — 90° magnet, (b) — triplefocal 270° magnet, (c) — multisectional 270° magnet, (d) — Pretzel-type 270° magnet (achromatic magnetic mirror)

whereas particles with lower energies move in orbits whose radii are smaller than the radius which corresponds to medium energy particles. This magnet also focuses through the same points all parallel entry paths over a range of axial displacements in two dimensions. Hence the title a *triple focusing system*. The beam, after traversing the focal point, still remains divergent. Similarly, changes in medium energy bring about a change in the angle

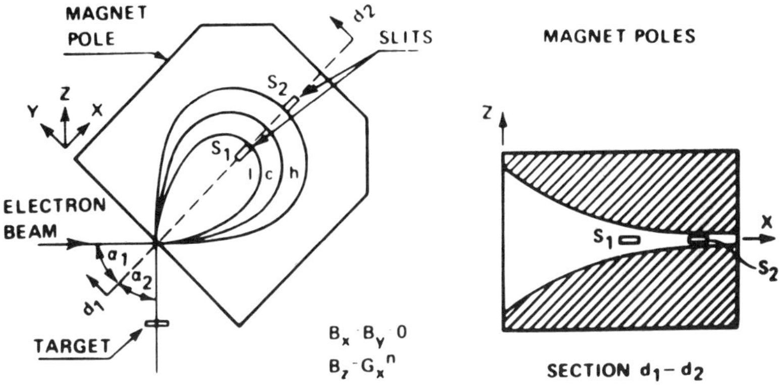

FIGURE 3.28 Pretzel-type magnetic system 270° doubly achromatic.

through which the beam strikes the X-ray target. Thus, the X-ray field becomes asymmetric.

Good achromatic properties can be achieved in a 270° magnet which consists of three magnetic systems separated by gaps with no magnetic field present (Fig. 3.27c). The disadvantage of this design is its complex magnetic system.

The Varian Clinac 18 accelerator is equipped with a three-sectional 270° magnet. The energy defining slit is located in the central section whose edges can give rise to bremsstrahlung since they constitute a target-like body for electrons which do not penetrate it. The bremsstrahlung radiation, however, owing to this location of the central slit, does not reach the patient. From the measurements with the magnetic analyzer it was found that the 6 to 18 MeV electron beams in Clinac 18 accelerators are characterized by the dispersion ranging from ±0.4 MeV to ±0.7 MeV (FWHM) [3.23]. In accelerators of the type described above the energy slit is adjusted so that it can let electrons pass with energy variations ranging ±3%. As already mentioned in Sec. 3.1.2.4, accelerators are equipped with devices for energy stabilization coupled to dosimetry systems (Fig. 3.14). Devices of this kind are intended for rf power control and, hence energy stabilization. In the Clinac 18 machine [3.23], energy spread values in the open type control mode were found to be ±1% over one treatment session and ±5% over one day, whereas those in the closed loop control mode were ±0.1% and ±1%, respectively.

Fig. 3.27d represents a system which can be implemented with only one magnet whose construction is shown in Fig. 3.28. It was developed in 1963 [3.28] and was called a *pretzel or achromatic magnetic mirror* since the particles which pass through it appear at the output as if reflected from the y-z plane. Such a system has been employed in the BBC Dynaray accelerators. The magnetic field increases along the x-axis. Individual orbits which correspond to electrons with different energies, are characterized by a reproducible shape so that all particles coalesce at the output regardless their momenta. Therefore, there is no energy dispersion and the bending is achromatic. Energy slits S_1 and S_2 which define the output energy range are located in the symmetry plane. Varian Clinac 2500 accelerators also incorporate magnets of this type.

Slalom-type systems. It has already been mentioned that the height of the magnetic bending system is a critical element in the attempt to achieve the lowest possible location of the isocenter. The vertical distance between the X-ray target and the main orbit of the bent electron beam can be taken for comparison. Fig. 3.29a is a schematic of a three-sectional 270° achromatic bending magnet in which the above distance is about 9.0 cm [3.21].

Since the actual beam consists of parallel rays, drift spaces or low magnetic field regions are required between individual sections to achieve clear cross-overs in the electro-optical system. Similarly, a small vertical distance can be obtained in the beam transport system which incorporates three in-line magnetic dipoles, as shown in Fig. 3.29b. The system of this kind, employed, e.g. in Philips accelerators [3.29] is often referred to as a *slalom-type bending magnet*. The first 45° bending magnet (Fig. 3.29b) is an energy analyzer which produces a separated beam at the output. Before the beam enters the next 45° bending magnet it passes through an energy slit S which selects medium-energy particles. Beam focusing in two orthogonal planes is started in the second 45° magnet. Next, the beam enters the third magnet which bends it vertically. The beam spot size at the X-ray target is less than 2x2 mm^2, whereas the input beam width and height are 5 mm [3.21].

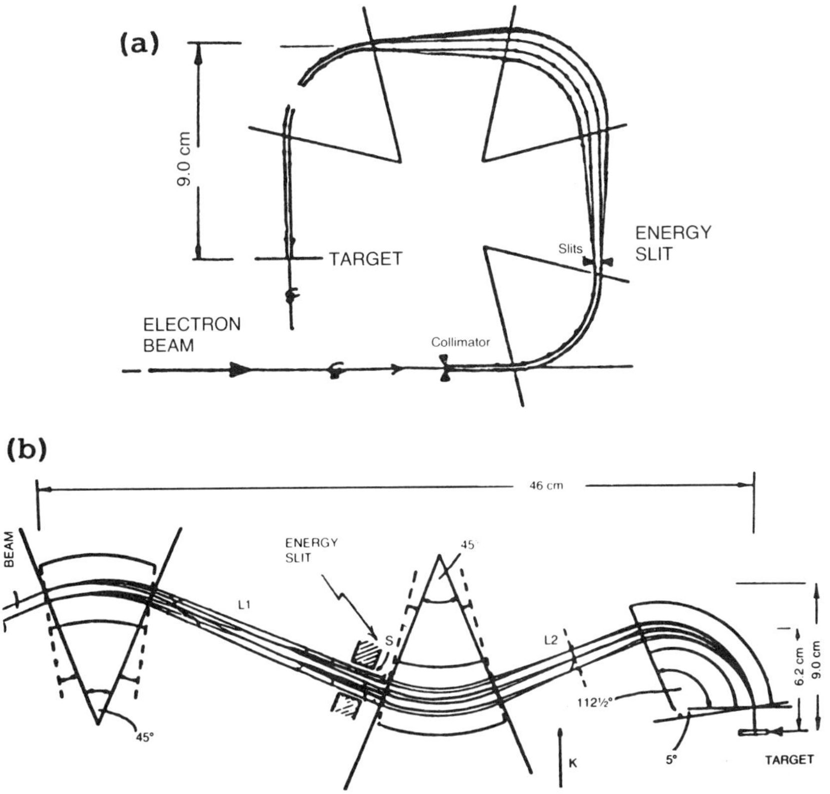

FIGURE 3.29 Three-sectional 270° magnetic systems: (a) — circular magnet, (b) — linear (slalom type) 45°/-45°/112° magnets (solid lines are trajectories of particles with ±6% energy dispersion (after C. Nunan [3.21] and J.I.M. Botman et al. [3.29]).

Both the three-sectional systems are equivalent, with respect to their vertical dimension. However, the system in Fig. 3.29a must be mounted inside the treatment head, whereas the online type shown in Fig. 3.29b, can be located along the accelerating structure. Because there is more free space in the treatment head, the jaws can thus be made of lead instead of tungsten as in the previous design [3.21].

The in-line system depicted in Fig. 3.29b is useful in bending electrons of energies between 2 MeV and 30 MeV. The input beam can have energy dispersion of 10% and the initial angular divergence of ±1 mrad [3.29].

The output energy dispersion in the well executed 270° bending systems is about ±3%, whereas that in the 90° systems may be as high as ±10% [3.30]. The energy slits in the 270° systems are located inside, those in the 90° systems are usually located at the magnet output, i.e. near the X-ray target.

3.1.4 Treatment Heads

3.1.4.1 Basic Systems

The design of a radiation treatment head varies considerably with the type of the beam produced. Low energy, i.e. 4 to 6 MeV, accelerators are intended exclusively for X-ray therapy; the beam of accelerated electrons is, in this case, *converted to bremsstrahlung radiation* and is not used directly in electron therapy. Varian Clinac 6/100 (Fig. 3.30) provides an example of a treatment head in accelerators of the above type. It also exemplifies an isocentric in-line treatment head in which the accelerating structure is incorporated in the head, which eliminates the need for beam bending magnets.

In this design, electrons are extracted from the structure through a vacuum-tight window and then are incident on the X-ray generating target. Next, the photon beam passes through a primary collimator with a conical aperture and strikes a flattening filter. The focal spot, where X-rays are produced, is only 2 mm in diameter and, therefore, the penumbra is small and the beam profile sharp (Fig. 3.65). The task of the accelerator is, however, to produce a flat 32x32 cm^2 irradiation field and the task of the flattening filter is to ensure that the nonuniformity for large size fields is less than ±3% (Sec. 3.1.6).

Below the flattening filter there is a system of ionization chambers for stabilizing beam parameters and measuring the dose and dose rate. The beam impinges on a *collimator which has two pairs of movable jaws* that are motorized and can define the dimensions of the treatment field. The treatment head shown in Fig. 3.30 is also equipped with an *optical light system* which projects the preset field onto the isocentric plane. It also includes an *optical distance indicator* that permits determination of the target-skin distance.

The accelerating structure proper (Fig. 3.30) and the space around the tungsten X-ray target are shielded to ensure protection against radiation. The treatment head rotates through a 360° with the axis of rotation (isocenter) located 123 cm from the floor. The distance between the target and the isocenter is 80 cm.

In accelerators which produce alternately X-rays and electron beams, the design of the treatment head must be slightly more complex. Fig. 3.31 shows, by way of example, the treatment head of the Philips SL25 accelerator. It is a multi-energy machine which generates two X-ray beams with energies of 6 MV and 25 MV (Table 3.5), as well as electron beams with nine energies ranging from 4 to 22 MeV. In the photon modality, the head operates in a system shown in Fig. 3.31a and is equipped with two rotating primary collimators: a high-energy collimator 2 and a low-energy collimator 6. The function of filter 3, on which the beam is initially incident, is to "harden" the radiation, i.e. to

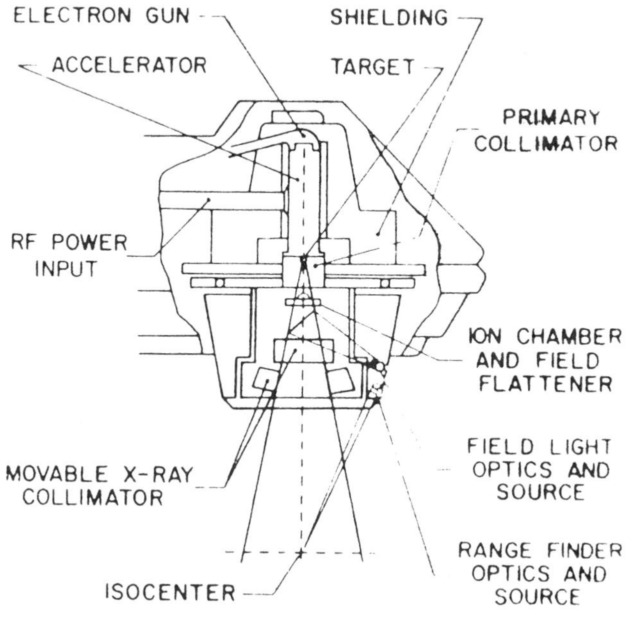

FIGURE 3.30 Schematic of a treatment head of the Varian Clinac 6/100 accelerator for X-ray therapy.

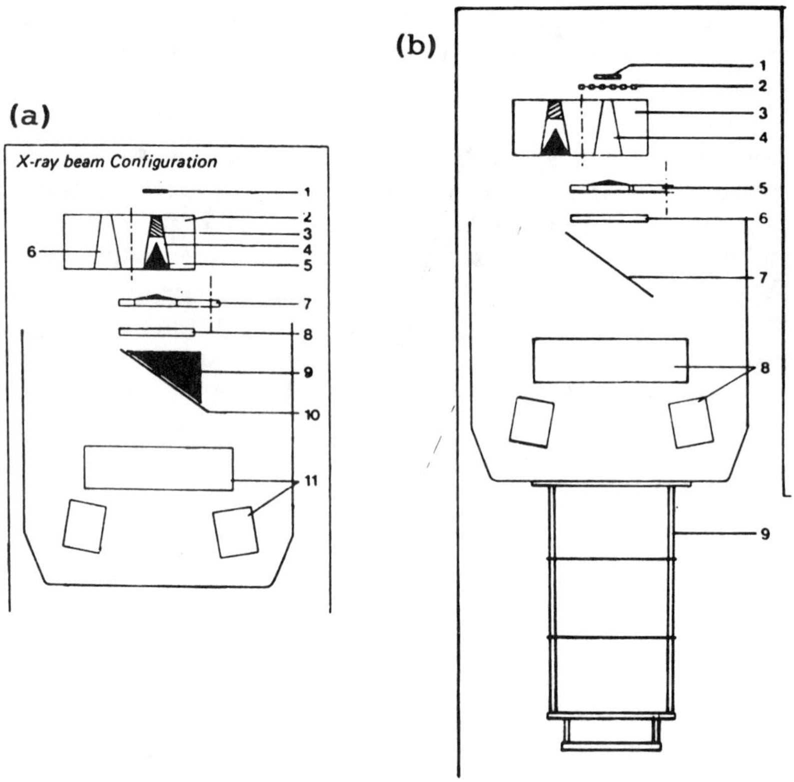

FIGURE 3.31 Schematic of a treatment head of the multi-energy Philips SL25 accelerator: (a) — X-ray treatment modality, (b) — electron treatment modality (description in the text).

suppress its low-energy component. Then the beam strikes the low atomic number flattening filter 5, through the high-energy filter proper. If, however, a low-energy X-ray beam is generated by the accelerator, instead of passing through the rotating primary collimator 2, the beam is incident on collimator 6 and on the low energy X-ray flattening filter 7.

A double thin walled ionization chamber 8 coupled to the dosimetry system is located below the collimators and filters. A motorized wedge filter 9 to shape the field profile (Fig. 3.40) is mounted under the chamber. Again, a Mylar mirror 10 incorporated in the optical field projection unit is located below filter 9. Finally, a movable collimation diaphragm 11, the collimator jaws, is mounted in the lower part of the treatment head.

The operation of the head in the electron treatment modality is shown in Fig. 3.31b. The beam passes through a vacuum-tight window 1 and strikes a high Z *primary scattering foil* which gives a broad beam profile (Fig. 3.37). Next, it is transmitted through a set of collimators 3 located on a rotating carrousel and the low-energy X-ray mode collimator 4 (cf. collimator 4 in Fig. 3.31a). Then it is incident on the second flattening foil 5 made of low Z material. Both these foils make up a scattering system which, in fact, is a filter which flatens the beam profile (Sec. 3.1.4.3). As in the X-ray modality, the electron beam passes through ionization chamber 6, Mylar mirror 7 and collimation X-ray jaws 8. Finally, it strikes a collimating aperture or cone 9 at the treatment head output. This aperture defines the treatment field size (Fig. 3.51).

3.1.4.2 Conversion of Electrons to X-rays

Conversion mechanism and its spectrum. If a charged particle is being decelerated in an external field, i.e. in the field of the nucleus, or only changes its direction, it emits electromagnetic radiation called *bremsstrahlung radiation,* which literally means "braking" or deceleration radiation. In practice, this radiation is often called X-ray radiation or simply X-radiation. Bremsstrahlung radiation, abbreviated to BR, is of particular importance in the case of electrons. The process in which the energy of electrons is transformed into the energy of X-ray photons, is called *electron-to-bremsstrahlung conversion.*

As an example of BR, Fig. 3.32 gives a conversion spectrum of 10 MeV electrons: A is a calculated curve and B is the actual spectrum obtained from the conversion at a 3 mm thick tungsten target (Linatron accelerator). Unlike the characteristic spectrum, this curve is continuous. Although the most probable energy of the primary electrons is 10 MeV, after conversion, the number of photons whose energies are close to 10 MeV is relatively small. Another feature of BR spectrum is a relatively large contribution of energies much lower than those of the primary particles. For example, a ^{60}Co source typically used for medical and industrial irradiations has two gamma energies of 1.17 and 1.33 MeV (mean energy used in calculations is

1.25 MeV). In order to obtain energetically equivalent BR, the target must be bombarded with 3 MeV electrons.

Targets used in radiation therapy accelerators to produce bremsstrahlung radiation are usually those of the *transmission type* (Figs. 3.30 and 3.31). The electron beam is incident on target front side surface and X-rays are emitted from the opposite side parallel to the primary beam direction. In this design, the depth dose distributions are strongly dependent on the target thickness. Fig. 3.33 shows depth dose distributions (ion chamber current vs. depth) for 0.52 to 2.00 cm thick lead targets (Fig. 3.33a) and 3.8 to 8.0 cm thick aluminum target (Fig. 3.33b). The measurements were made in a water phantom. For thin targets, the curves have peaks, especially for Al. These peaks are due to the fact that some electrons are converted to

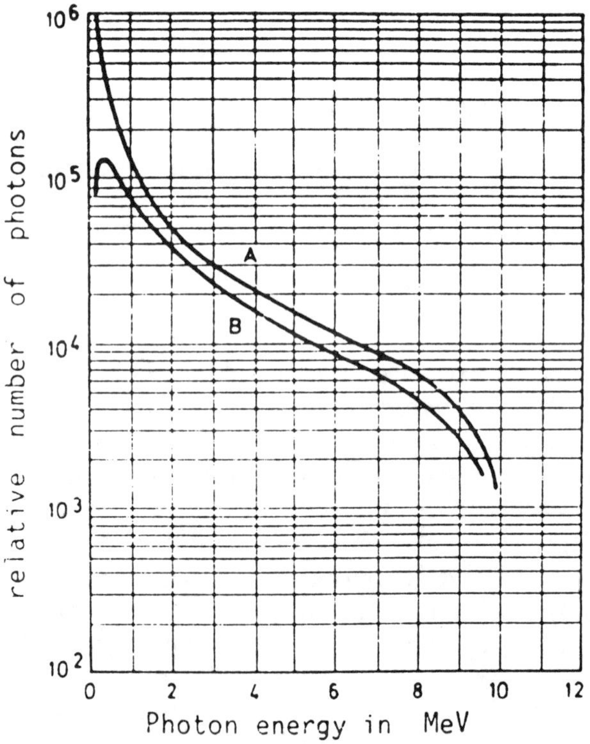

FIGURE 3.32 Bremsstrahlung spectrum from conversion of 10 MeV electrons: (a) calculated curve, (b) spectrum obtained from 3 mm thick tungsten target [3.31]

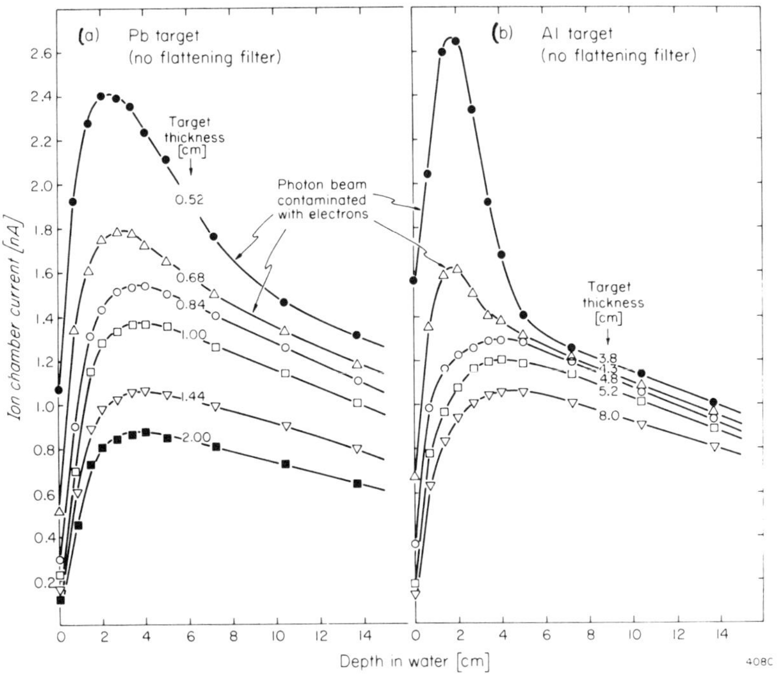

FIGURE 3.33 Ion chamber current vs depth dose for 25 MeV electrons (circular field of 10 cm diameter): (a) — lead target, (b) -- aluminium target (after E.B. Podgorsak et al. [3.32])

X-rays and some are transmitted through the target and thus increase the dose in the phantom. These latter electrons cause medically unwanted *contamination* of the X-ray beam. The greater the target thickness the less sharp is the depth dose maximum due to the smaller contribution of the transmitted electrons. When the target thickness of about 1 cm of lead and about 5 cm of aluminium is reached, this contribution is equal to zero and the initial peaks are completely removed.

In order to remove the unwanted electrons which contaminate an X-ray beam, either filters made of thin foils or magnetic methods are used. The thinner the target, the harder the X-rays emitted, i.e. the beam becomes more penetrating.

X-ray angular distribution. The bremsstrahlung radiation induced by high-energy electrons is emitted within a relatively narrow forward cone around the directions of primary electrons. The cone width becomes narrower with increasing energy. Fig. 3.34 shows the distribution of bremsstrahlung radiation for a thick target bombarded by 8 to 30 MeV primary electrons. Since the purpose of the treatment is to produce large fields, 30x30 cm^2 or even 40x40 cm^2, with nonuniformity less than ±3% (Fig. 3.65), the above figure can justify the use of flattening filters. These filters lead to lower angular nonuniformity at the cost of lower total intensity.

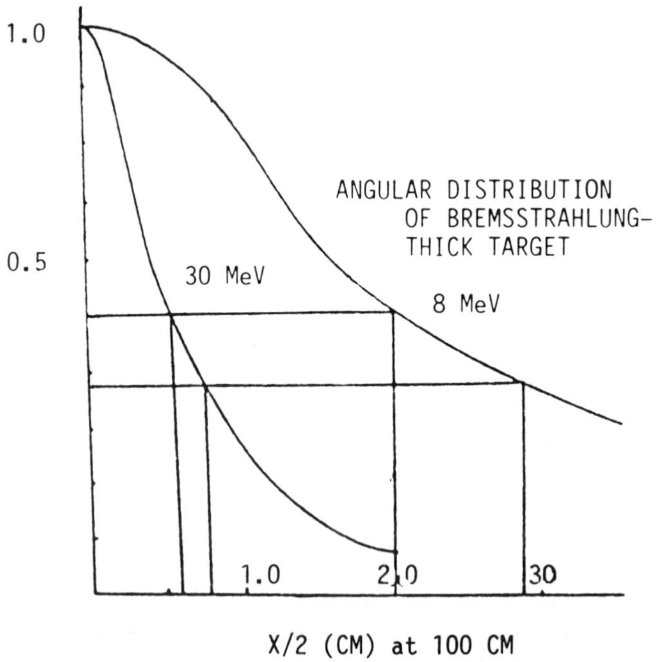

FIGURE 3.34 Angular distribution of bremsstrahlung for 8 and 30 MeV electrons incident on a thick target.

Conversion efficiency. The intensity of the bremsstrahlung initiated by high-energy electrons is dependent on the electron beam intensity and energy as well as on the *conversion efficiency.* This latter quantity is given by the ratio of the total power of the X-ray beam to the total power of the impinging electron beam.

The flux of electron-produced photons in the energy range between 3 and 5 MeV is given by

$$\Phi_{rad} = 5.77 \cdot 10^{-4} Z E_k^2 \tag{3.6}$$

where Z is the mean atomic number of the target material; E_k is the kinetic energy of electrons (MeV); Φ_{rad} is the unit energy flux of BR (MeV per electron). Let us assume that the electron beam intensity is 0.15 μA, which corresponds to the flow of $9.36 \cdot 10^{11}$ e s^{-1}. For electron energy E_k = 3.85 MeV and an Al target we get Φ_{rad} = 0.111 MeV per electron, i.e. the total X-ray energy flux for I = 0.15 μA will be $\Phi = 1.04 \cdot 10^{17}$ eV s^{-1}, or 1.68 J s^{-1}. For the above parameters of the electron beam the energy flux of the primary electrons is $\Phi_{el} = 3.61 \cdot 10^{18}$ eV s^{-1}, or 0.58 J s^{-1}. Thus only 2.97% of the electron energy becomes transformed to BR energy.

As seen from Eq. (3.6), the conversion efficiency depends on the atomic number Z of the target material. Fig. 3.35 shows that the photon conversion efficiency for an rf electron linear ac-

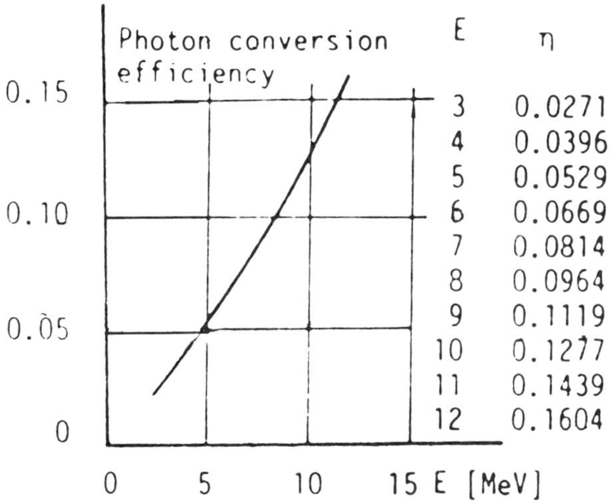

FIGURE 3.35 Photon conversion efficiency vs. incident electron energy (Cassitron accelerator).

celerator of the Cassitron type for 3 MeV electrons is 2.7%, while the efficiencies for energies of 6 and 12 MeV are 6.7% and 16.0%, respectively.

This is the reason why the electron beam intensity in bremsstrahlung production must be higher than that in the electron therapy mode by two orders of magnitude. In both cases, of course, the electron beam intensity gives rise to an identical treatment dose. Conversion efficiency is of fundamental importance in nontherapeutic applications of X-rays, namely in medical sterilization (Sec. 8.2).

Target design. Initially, targets were made from heavy metals, such as tungsten, gold, platinum and copper which is a good heat conductor. Early investigation in which comparison was made between the quality of radiation produced by a 16 MeV betatron and that by a 25 MeV linear accelerator showed that the percent depth dose distribution was almost identical for both machines, but, when this was examined in more detail, it was found that one of the reasons for the above identity was that the betatron employed a thin tungsten target, whereas the linac used a thick copper target. The contribution of low-energy X-ray photons was greater in the copper target. For this reason, in present-day designs, targets are used for which the spectrum generated corresponds to that obtained for thin targets. The term *thin target* is taken to mean a target which is a few mm thick, whereas a *thick target* is that several cm thick.

To obtain thin target characteristic, *layer-type targets* are currently employed. For example, a target may consist of a 2 mm thick gold layer, 6 mm thick copper layer, 2.5 mm water layer (cooling) and 0.5 mm thick stainless steel layer. When a target of this kind is bombarded with a 21 MeV electron beam, most of the electrons of energies less than 9 MeV are stopped in the copper layer whose conversion efficiency is much lower than that made of gold. That is why the X-ray spectrum is more similar to that of a thin layer than to that of a thick layer.

To produce detailed data on photon beam characteristics calculated spectra can be obtained by modelling the radiation machine head in a Monte Carlo simulation of the radiation transport from the target to a surface in treatment distance [3.137, 3.138 and 3.145]. The knowledge of the output photon energy spectra is essential for investigations of treatment head design and for certain dose calculation algorithms [3.146].

3.1.4.3 Flattening Filters

Electron beam filters (scattering foils). Like an X-ray beam (Fig. 3.34) an electron beam is also spatially nonuniform. Fig. 3.36 shows the 5, 10 and 20 MeV beam intensity measured as density variations on a radiographic film placed just below the primary collimator output. It can easily be seen that the beam becomes focused along the main axis as its energy increases. Therefore, in order to obtain a large uniform treatment field, it is essential that the spatial distribution of electrons be first made uniform or flattened. This is done in practice by means of *scattering foils.* For this purpose both *single foils systems,* called *primary foils* (Fig. 3.37a) and *double foil system* incorporating a primary foil and secondary foil (Fig. 3.37b) are employed. A *secondary foil* is placed under the primary foil at a distance between a few and some dozen cm. In the head shown in Fig. 3.31 this distance is 10 cm.

A primary foil is the scattering foil proper. It is made of a high Z material. Its thickness is limited by the maximum permissible electron energy losses and the permissible intensity of the bremsstrahlung radiation produced and is usually chosen so that

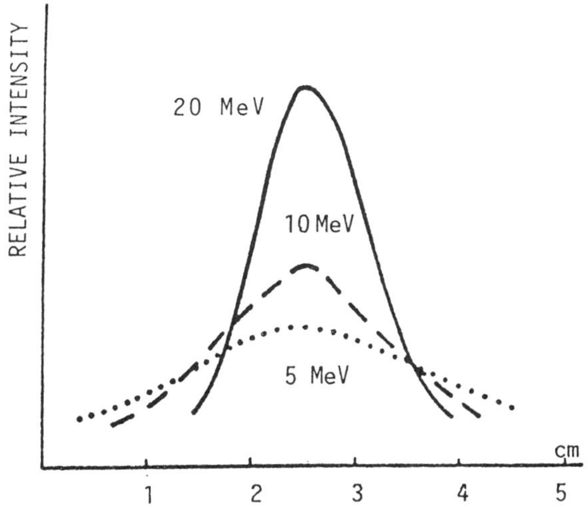

FIGURE 3.36 Spatial distribution of 5, 10 and 20 MeV electrons measured below the primary collimator.

the field intensity fall-off near its edges does not exceed 5% relative to the intensity on the axis. Single foil systems can create satisfactory *field uniformity (flatness)* if the field dimension (diameter) is not greater than 10 cm and the electron energies are less than or equal to about 10 MeV. A single foil is usually not sufficient to fully encompass the energy range. For this reason, a system consisting of several foils of various thickness, placed on a rotating feeder, is used.

In systems with a double foil, the secondary foil is usually made of a low Z material. This helps to minimize the bremsstrahlung radiation, electrons in the foil being absorbed rather than scattered. The foil thickness is profiled, decreasing (from the beam axis) towards the edges (cf. Fig. 3.31b), which requires accurate centering of the foil relative to the beam.

The primary distribution of the beam intensity (Fig. 3.36) is much broader for lower energy electrons than that for high-energy electrons. In addition, low-energy electrons are absorbed in the secondary foil much strongly than high-energy electrons are. Thus the primary foil thickness required for high-energy electrons is thicker. For a lower energy electron beam the primary foil is thinner, and as a result, the contamination of the treatment field with bremsstrahlung is smaller.

A double-foil flattening field for electrons is usually equipped with several primary and secondary foils. The selection of foils with appropriate thicknesses and profiles is done automatically. When a certain electron energy is chosen, rotating feeders with foils on them are set automatically in order to optimize the scattering system.

The advantage of the dual foil technique is that the treatment fields obtained are highly uniform (flat), especially for beams with energies higher than 15 MeV, and for large fields of 15 cm in diameter and larger. The disadvantages, however, are: (1) beam energy degradation, and (2) bremsstrahlung radiation, especially at energies higher than 20 MeV. Therefore, in generating treatment fields with high-energy electrons scattering foils are now being replaced by scanning techniques (Fig. 3.41).

X-ray filters. The choice of the shape and material of a *flattening filter for an X-ray treatment field* is determined by the requirements of (1) high uniformity of the field within the treatment volume, (2) increase in the penetrability (hardness) of the radiation as compared with the radiation emitted from the target, and

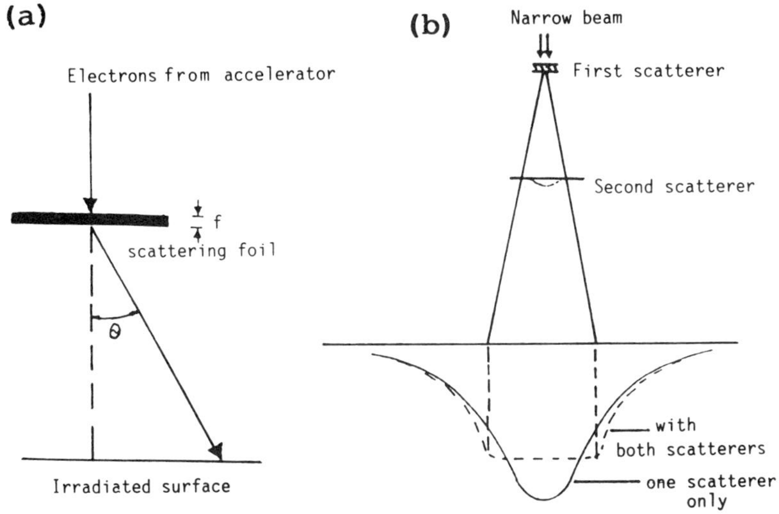

FIGURE 3.37 Scattering foil systems: (a) — single foil, (b) — double foil and comparative curves.

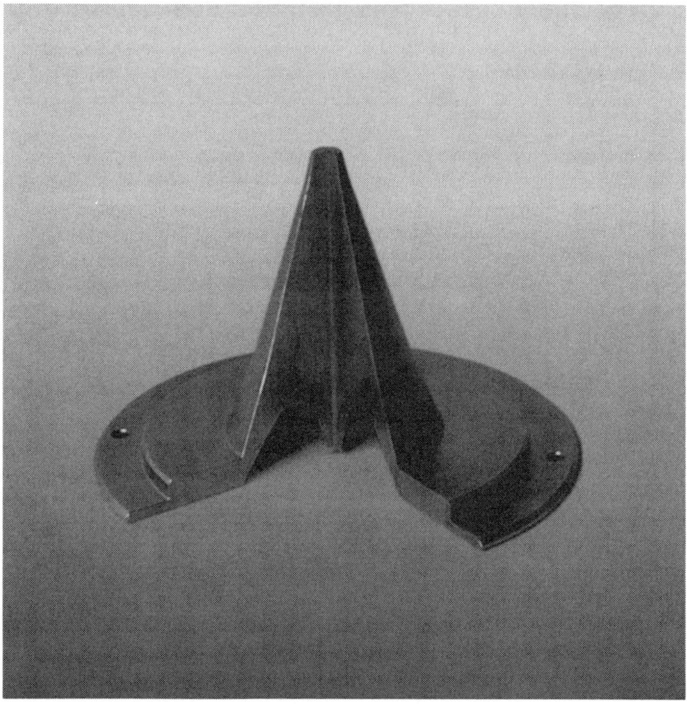

FIGURE 3.38 View of a flattening filter for X-ray field (Siemens).

(3) minimization of the possible generation of neutrons. Filters made of low Z materials are suitable for reducing low-energy X-rays; they ensure that the radiation used in therapy has a larger "hard" component. The low Z materials also feature low neutron emission. However, filters made of aluminium would have to be too large (in the case of high energy X-rays) to accommodate in the treatment head.

As a compromise, a double layer filter is usually employed, which consists of an aluminium and lead core. For example, if a 39 mm flattening filter is used for a 21 MV X-ray beam, photons traveling close to the beam axis are incident on a 2 mm thick Al layer and 37 mm thick layer of lead. The amount of lead layers decreases linearly, whereas that of the aluminium increases as the distance from the axis increases; the opposite ratio of the thickness is obtained at the edges of the maximum field size of 37x40 cm^2 [3.34].

Instead of layer-type filters, *solid filters* employed are made from homogenous materials, such as steel, which is a compromise. Fig. 3.38 shows a half section of a filter of this kind (mfd. by Siemens). It is obvious that a high degree of precision

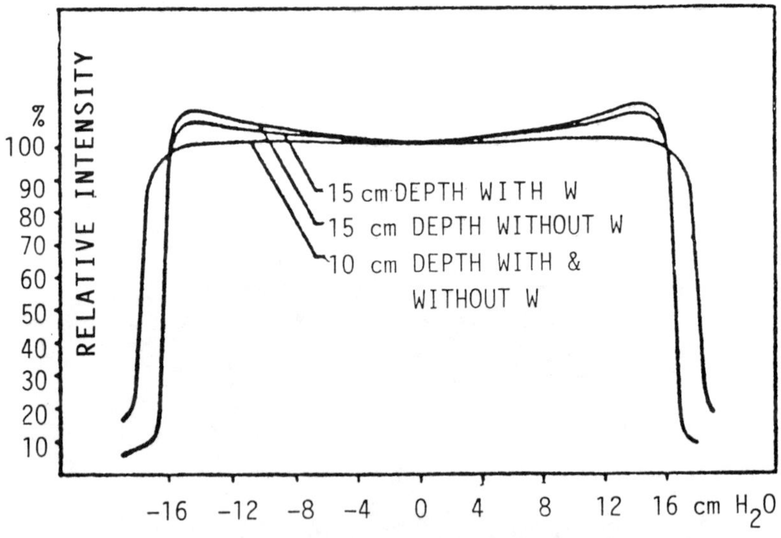

FIGURE 3.39 4 MV X-ray field profiles in a water phantom at 1, 5 and 10 cm depth, with and without an additional tungsten filter; field size 35x35 cm^2 (after L. Taumann [3.39]).

is required to align the filter's axis with the beam's axis (cf. Fig. 3.4). Homogeneous filters increase the "hardness" of central X-rays more than that of photons at a greater distance from the axis. As a result, the nonuniform energy distribution may be acceptable for high-energy photons, however, it gives rise to serious difficulties for low-energy photons.

A 4 MV central X-ray beam gives a maximum dose in a phantom at the depth of about 10 mm. As the energy of photons decreases, the maximum shifts closer to the surface layers. In this case, the skin dose increases, which may cause severe skin reaction (erythema) that can limit the total dose delivered to the treatment volume. This harmful effect can be alleviated by using an additional tungsten foil filter. Depth dose measured at 10 cm and 15 cm depth with and without the tungsten filter are shown in Fig. 3.39. The effect of the tungsten filter is more pronounced for a dose which corresponds to a smaller depth.

X-ray flattening filters in their central part absorb 50 to 90% of the beam intensity. That is why, in some designs, they are replaced with a beam scanning system (Fig. 3.41).

Wedge filters. Filters of this kind are placed in the main beam in the way shown in Fig. 3.31a (element 9). They are, in fact, absorption wedges which modify the beam profile (Fig. 3.40). Fig. 3.40a shows how the wedge filter angle is defined. The depth-dose curve is determined for the radiation field of dimension F and standard measurement depth. The wedge filter angle is formed between the straight line from measurement points for dimensions $\pm F/4$ and the line parallel to the entrance plane.

Wedges are supplied by the manufacturer as a system consisting of four wedges: 15°, 30°, 45° and 60°, included with the accelerator system. Maximum size of the modified field is usually 20x20 cm^2 (Siemens). The wedges can be introduced into the beam either manually when setting the accelerator for treatment or by means of a motor drive which can also operate during irradiation.

A design using a single motor driven 60° filter is also used (Philips, Fig. 3.31a). The effective angle of the filter action is dependent on the ratio of the treatment time with the wedge to that without the filter. When implementing the treatment plan, the computer system which controls the accelerator automat-

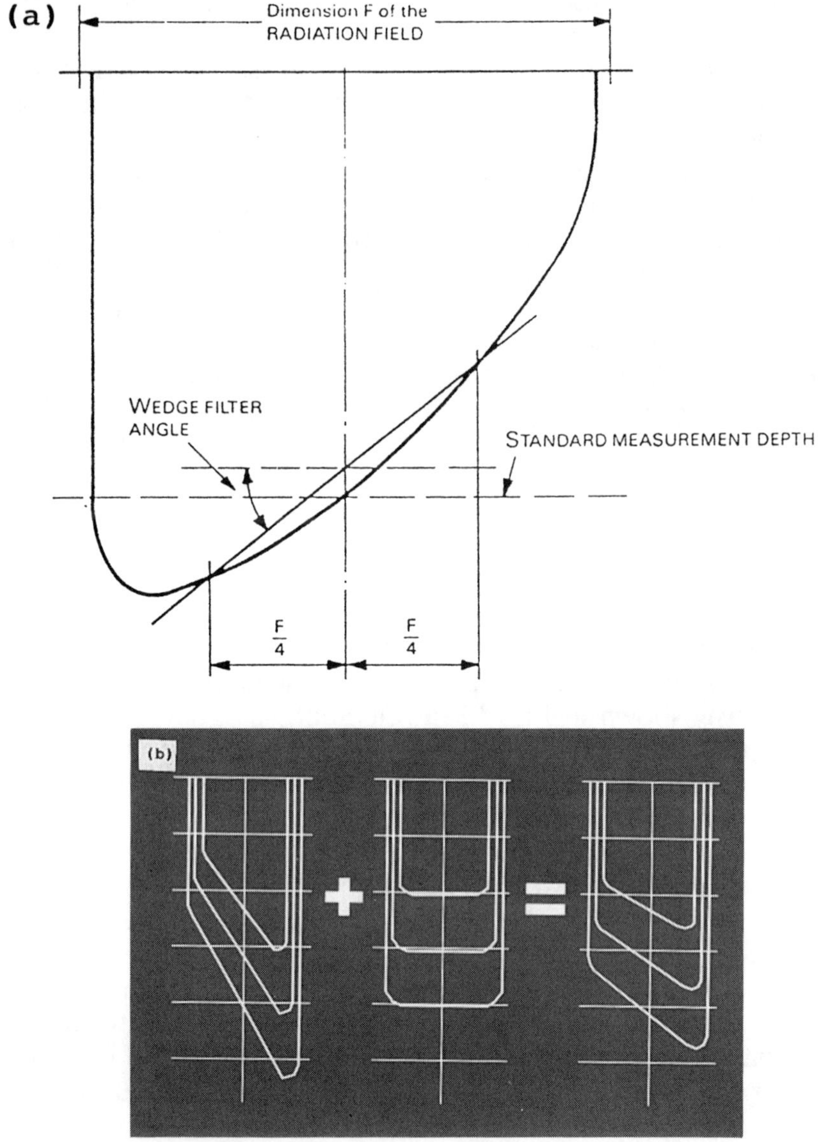

FIGURE 3.40 Wedge filter: (a) — definition of wedge filter angle (after 976 IEC Standard), (b) — modification of an X-ray field.

ically calculates the duration of the stage without the filter as well as the treatment time with the 60° wedge filter. When the first stage (without the filter) is completed the motor drive automatically introduces the filter during the second treatment stage. This is an essential solution in designing a fully automated irradiation process.

The scanning method. In the designs described heretofore, the beam of accelerated electrons would leave the accelerating structure in the form of a *narrow (pencil) beam,* and would subsequently be applied to the conversion target or onto a scattering foil without changing its direction and fixed position. However, in some radiotherapy accelerators, especially those which are intended for operation in high energy ranges, the stationary beam system is replaced with a *beam scanning method.* The scanning technique is usually implemented with two deflecting magnets (Fig. 3.41) that cause the beam to deflect in two orthogonal planes. In the system shown in Fig. 3.41, one of the magnets is placed in front of the bending magnet and the other behind it.

In the Therac 25 accelerator, which produces a 25 MV X-ray beam and electron beams of energies ranging from 5 to 25 MeV, the scanning magnets are computer-controlled. One of two rasters can be selected, whose patterns are shown in Fig. 3.42. Typically, the pulse repetition rate is 120 pulses per second for

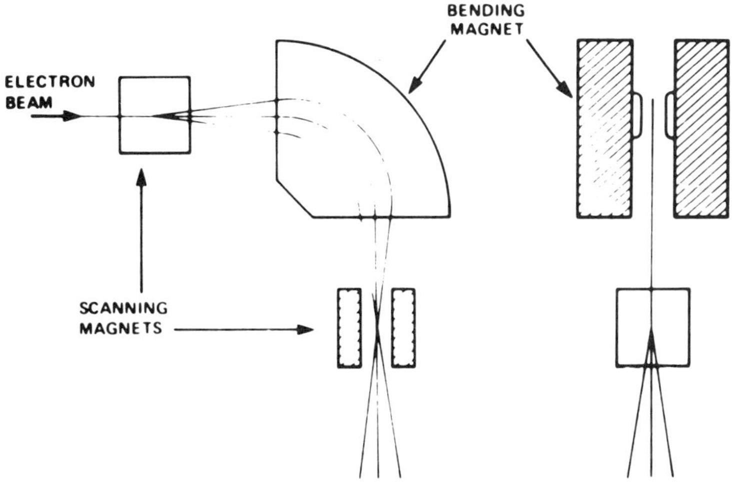

FIGURE 3.41 A schematic of a high-energy electron beam scanning system.

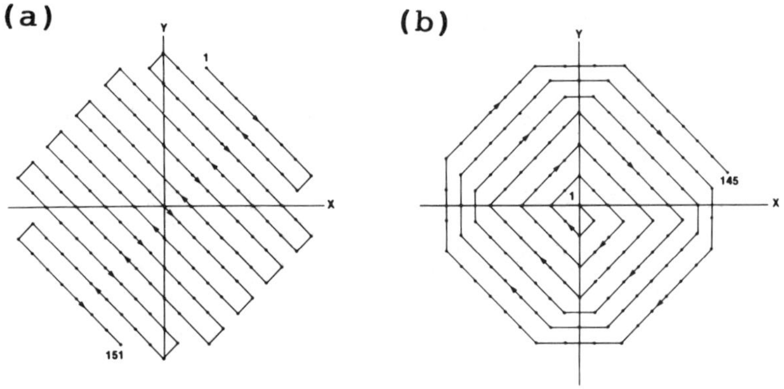

FIGURE 3.42 Scanning patterns of a beam in an AECL Therac 25 radiotherapy accelerator: (a) — raster pattern, (b) — spiral pattern (after P.O'Brien et al.[3.35]).

the dose rate of 250 cGY min^{-1} and the single pulse duration is 3 ms. The example shown in Fig. 3.42 represents the repetition rate of 150 Hz, the total single field scanning time thus being about 1 s.

After each 3 ms pulse the beam is deflected to another position represented by single dots in Fig. 3.42. The raster shown in Fig. 3.42a is similar to a TV raster and is usually used for high-energy beams, whereas the spiral pattern (Fig. 3.42b) is preferred for lower energies. The number of grid points in the net and the raster spacing in both directions is selectable and can be adjusted for each electron energies in order to optimize the flatness and the symmetry of the beam. Such symmetry is preserved if the dose per pulse is constant and the beam-on time corresponds to an integral number of raster scans.

In electron therapy, the scanning technique makes it possible to obtain a low level of beam contamination with bremsstrahlung due to the absence of scattering foils. The beam field may be programmed so that it matches specific requirements. If, however, the scanned beam is converted to X-rays, initial field flatness is achieved so that even a relatively thin flattening filter is quite sufficient.

The drawback that the scanning technique has is that it is rather more complex than the stationary method and involves several dosimetry problems (Sec. 4.1.6.3). This is specially true for phantom measurements.

It was previously feared that the scanning technique may be somewhat dangerous for the patient who could be exposed to a narrow (pencil) beam when the scanning system fails. These fears are not warranted because first, the accelerator is always equipped with a device that switches the beam off in case of any malfunctions in the scanning system, and, second, the beam shape is actually quite large. According to measurements reported in [3.35], when a 0.05 mm nickel foil window was used, the half-width (FWHM) of the 6 MeV electron beam with the deflection system off is 26 cm; it is still 6.8 cm when the energy is increased to 25 MeV.

3.1.4.4 Collimators and Applicators

X-ray collimation diaphragms. In a radiotherapy X-ray accelerator, a *collimator diaphragm* is included in a typical collimation system. As already indicated in Fig. 3.31a, the diaphragm consists of two pairs (upper and lower) of *movable collimation jaws.*

Fig. 3.43 shows the main operation modes of a collimation diaphragm. In the *symmetric mode,* both jaws of one pair are shifted symmetrically relative to the beam axis, and, as a result, the field is also always symmetrical to this axis (left-hand side of Fig. 3.43), whereas in the *asymmetrical mode,* the jaws of the pair (the upper pair in the right-hand side of Fig. 3.43) may move independently, i.e. each jaw can be shifted over the full range from the left to the right edge of the field. In the most recent designs, all four jaws of both collimation pairs can be shifted independently, and, therefore, a rectangular collimated treatment field can be obtained in a part of the full field.

The fact that the beam can be collimated in an asymmetrical (independent) mode greatly enlarges the flexibility of radiation therapy, particularly that for carcinomas of the breast (supraclavicular and tangential fields) and for Hodgkin's disease. Fig. 3.44 provides examples of such fields: A and B are tangential fields, and C is an anterior field.

Collimation jaws are made from materials that strongly absorb X-rays, that is mainly from lead or tungsten. They may be set manually prior to the treatment session, or controlled automatically. The latter mode of control becomes specially important in dynamic therapy. The term *dynamic therapy* is used to

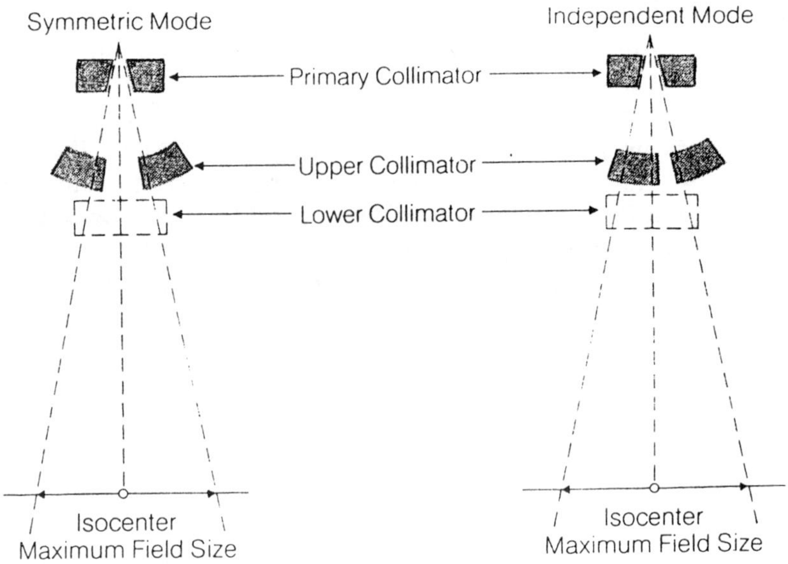

FIGURE 3.43 Symmetrical and independent movement of collimation jaws of an X-ray beam collimating diaphragm.

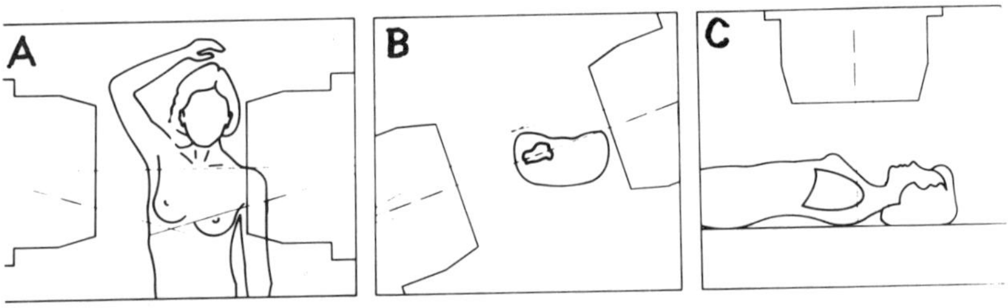

FIGURE 3.44 Examples of isocentric treatments: A — breast, chest wall and peripheral lymphatics, A and B — tangential fields, C — anterior field.

cover the method of treatment in which, during the beam-on cycle, one (or more) of the treatment parameters can be controlled such as, for example, the energy (Figs. 3.20a and 3.21), the effective value of the wedge filter angle (Fig. 3.40) or the geometric field size. Previously, the rotational or arc therapy was implemented only by means of open fields and of those employing wedge filters. Now, however, thanks to the new method of dynamic therapy, it is possible to combine shaded fields that are essential in the irradiation of irregularly shaped tumors with arc therapy. By using computer control, independent jaws can be shifted so that variable asymmetric fields can be obtained as a function of the gantry rotation. An example of an automatically controlled movement of the jaws in an asymmetrical (independent) mode is presented in Fig. 3.45 which shows a dose distribution obtained for a combined treatment of the nasal-pharyngeal cavity. A 20 MeV electron field in the frontal direction was used through a mask that shielded the eye region. Then two opposite 8 MV X-ray fields were made perpendicular to the electron field. During irradiation, one of the jaws was kept at a constant distance of 4 cm from the X-ray beam axis, whereas the other jaw moved in a preset (program-

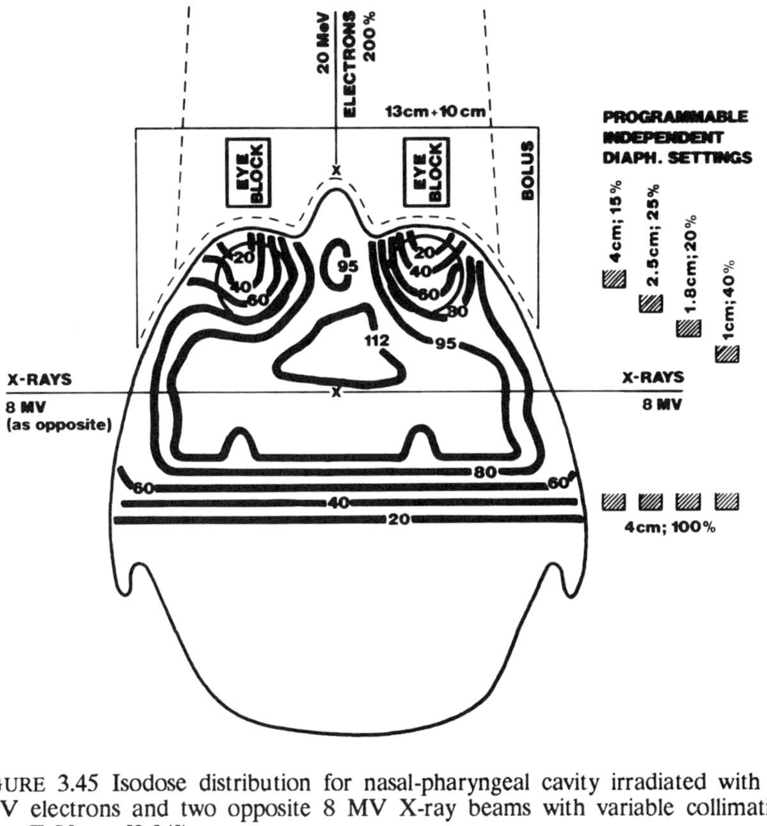

FIGURE 3.45 Isodose distribution for nasal-pharyngeal cavity irradiated with 20 MeV electrons and two opposite 8 MV X-ray beams with variable collimation (after F. Voogt [3.24]).

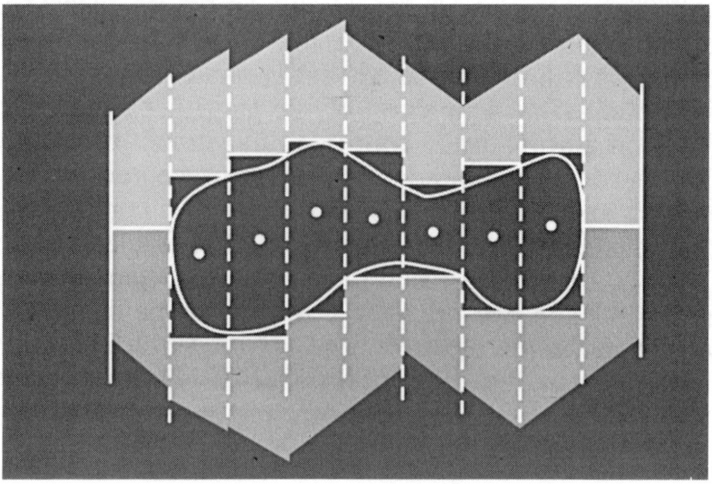

FIGURE 3.46 The principle of conformation therapy

mable) fashion: during the first 40% of the full cycle it was subsequently shifted at the distance of 1.0 cm and then at 1.8 cm (20%), 2.5 cm (25%) and 4 cm (15%), respectively.

Independent four-jaw collimation jaws may play an important role in *conformation therapy* in the future. In radiotherapy of malignant tumors it is important to give a homogeneous dose to the tumor, while avoiding as much injury as possible to the surrounding healthy tissue. In ordinary rotational therapy or the multifield stationary irradiation the high dose region will usually include a considerable amount of healthy tissue, because of the irregular shape of the target region. If the beam is just wide enough at all positions of rotation to cover only the desired region, a minimal amount of irradiation will be delivered to the surrounding healthy tissue. The concept of changing the field size during treatment in rotational therapy, so-called *conformation radiotherapy*, was introduced by S. Takahashi (1960). In order to plan and perform this sophisticated radiotherapy technique, the integrated systemization of radiotherapy from the CT, simulator and MR to the computer-controlled linear accelerator with conformational device is very important.

The principle of conformation therapy, which is in fact a *3D therapy*, is that the treatment field is divided into individual sections (Fig. 3.46), and then each section or, strictly speaking, its center has a separate treatment plan. During treatment, the y – jaws are shifted in such a way that the irradiation covers only a given section (Fig. 3.47), whereas the x – jaws are adjusted so that the beam is matched to the lateral dimension of the section.

Conformation therapy is, as yet, the domain of clinical research centers [3.132 and 1.133] since it requires sophisticated computer systems to provide treatment on a routine basis. This is probably less true of the dynamic therapy, mostly because the manufacturers are not too willing to produce such complex systems. Recently, however, some manufacturers have expressed greater interest to become involved in the research carried out by research centers [3.36]. This could lead to a more widespred use of dynamic therapy, and at a later date, conformation therapy, which has many advantages in quality assurance (Sec. 3.1.6.6) as it can create a homogeneous high-dose region in treatment fields with avoidance of injury to the surrounding tissues. Sophisticated radiotherapy is easily planned and performed. The introduction of conformation therapy with the use of X-CT and computers aims at decreasing the unwanted injury surrounding target areas and at improving the local control rate

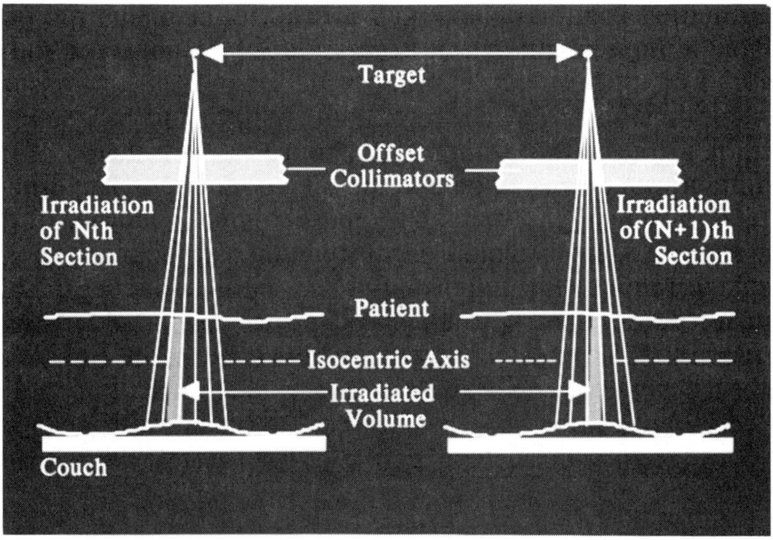

FIGURE 3.47 Sequential irradiation in conformation therapy with independent movements of collimation jaws (after F. Voogt [3.24]).

compared with the conventional conformation technique. Clinical evaluation of computer-assisted conformation therapy should be done in many hospitals and its clinical efficacy must be clarified [3.133].

Shadow filters and individual compensators, integrated systems. The previously described collimation diaphragms make it only possible to shape rectangular fields. To overcome this drawback individual apertures or filters, sometimes referred to as *shadow filters or shadow blocks*, are used. Some manufacturers supply shadow tray assemblies as standard equipment (Fig. 3.48). Shadow blocks made on order for each patient are placed between two transparent support plates made from organic glass or acrylate. The material from which these filters are made is usually a low melting alloy, Lipowitz metal, composed of 50% bismuth, 27% lead, 13% cadmium and 10% tin. This metal marketed under a variety of trade names, including Cerrobend and Ostolloy, has already gained widespread use for field shaping in photon and electron therapy and is replacing lead. The metal may be remelted after use (melting point 96°C) and molded into new field defining configurations.

The demand for individual beam modifiers for the present-day X-ray therapy is very high. According to [3.38], in a radiation therapy center where 30 to 40 patients are treated daily with X-rays (on average 2.25 static fields are used per one patient, which gives 65 to 90 fields per day), 90% of all fields employ individual patient beam modifiers. Since the weight of each block may be up to 15-20 kg (a block is meant to be a single collimating element or a shadow filter) it was essential that an automatic transport system should be designed, which would place the blocks into a block holder in the treatment head output [3.38]. In view of the large number of individual beam modifiers their identification has become an important problem. The holder or tray assembly shown in Fig. 3.48 is provided with a mechanical code in the form of seven holes in the support plate. In this way, up to 110 individual beam modifiers can be identified and then, with corresponding identification tags (numbers), stored in the computer that controls the accelerator operation.

Shadow blocks are intended for shielding organs and body critical regions against excessive irradiation. In addition, *compensating filters (compensators)* are also used, which are in the

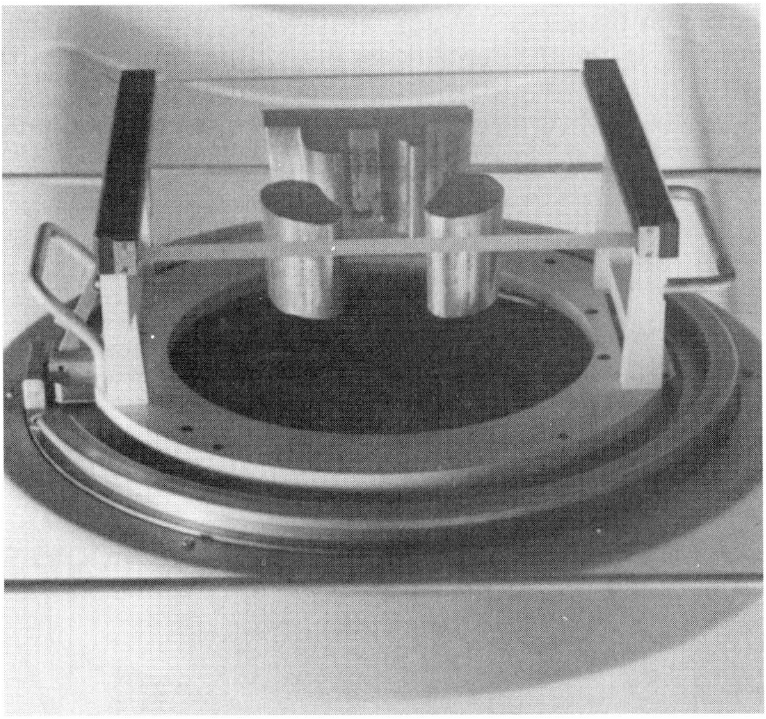

FIGURE 3.48 Individual shadow blocks in X-ray therapy (Philips Shadow Tray Assembly)

form of masks or forms made from metals or plastics. They serve to compensate for the patient inhomogeneities in the target field. They are particularly essential in the treatment of relatively large fields since the basic equipment, such as collimation diaphragms and wedge filters, are not sufficient to account for both the radiobiological and anatomical features within a single field.

The low energy electron beams from many medical accelerators exhibit a pronounced build-up of dose below the patient surface. At 6 MeV, for example, the surface dose can be as much as 30% lower than the maximum dose. Such beams are of limited clinical usefulness, since in practice it is usually necessary to treat a superficial disease uniformly from the surface down to some pre-selected depth. When using low-energy electrons for irradiation of inhomogeneous tissues, account must be taken of the considerable difference in the stopping of electrons in tissues of different density, e.g. soft tissues, lungs, bones etc. If not, the depth doses may differ considerably from those calculated. Fig. 3.49 demonstrates the differences in the stopping of 8 MeV electrons in a phantom simulating a chest wall of 2.5 cm thickness and lung tissue with 1 cm *bolus* for

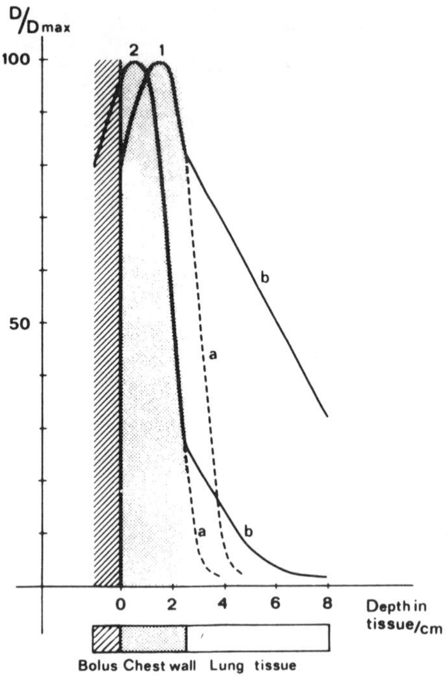

FIGURE 3.49 Central depth dose from 8 MeV electrons: in a 2.5 cm thick chest wall and lung tissue with (curve 2) and without (curve 1) 1 cm bolus compensation for build-up effect and thickness of the chest wall.

compensation of build-up and thickness of the chest wall (curve 2a + b) and without bolus (curve 1a + b). Stopping of the electrons in lung tissue is very slight. Consequently, their range is increased, which may result in intensive irradiation of large lung volumes causing pneumonitis and fibrosis together with insufficient dosage of the superficial tissues and the chest wall.

This can be avoided by using electrons of a lower-energy or by adjusting the thickness of the chest wall with an external absorber (bolus) of suitable thickness for the electron energy used.

Patients with head and neck lesions have been typically treated with low energy megavoltage beams, such 4 MV or even 10 MV photon fields. To improve the dose distributions $\approx$1.2 cm thick polystyrene slabs are then placed at 5-15 cm from the body surface [3.110]. Such compensators are also called *beam spoilers.*

An important innovation which has facilitated the use of individual modifiers is a system called *field-integrated dose modification* [3.39]. Field shaping blocks (shadow blocks) and other dose modifiers are combined into a single unit. The basis for the construction of such a system is the knowledge of the patient's anatomy and of the area of the main treatment field together with partial treatment fields which should receive a different dose relative to the fragments of the open part of the field. The radiotherapist determines doses for all areas of the field with regard to a reference plane, and then defines the primary modifier transmission values for the different partial regions. Other basic parameters are the radii of individual treatment field sectors and the depths relative to the reference plane as well as beam parameters. By taking into consideration the effect of other factors, calculations are then made [3.39] and the results are subsequently transferred to a *computer-controlled cutting machine* which produces the moulds and block forms to make separate modifiers. The modified system of modifiers is carefully checked with X-rays, and other tests are made before it is accepted for use.

In *dynamic therapy*, a computer-controlled accelerator can produce various treatment fields with the corresponding weighing coefficients both for single fields and various beam angles. In order to do this, the table, gantry and the collimation diaphragm must be controlled with precision, and the system of determining doses must be highly accurate. Dynamic therapy is still in its early stage of development, far from being applied routinely. Apart from the unconventional method of treatment planning, which is being now under intensive development, one of the issues that is now of greatest importance for dynamic therapy is the design and implementation of a system of security which would eliminate unpredicted radiation accidents during the treatment session. This calls for the construction of complex computer methods with highly developed verification systems.

Multi-leaf collimators. The purpose of *multileaf collimators* is to replace one pair of collimation jaws with a system of multiple leaves, as depicted in Fig. 3.50a. Each leaf can be separately shifted by means of a separate small servo motor over full range from the beam axis to its edges. Since about 95% of all treatment fields are not longer than 30 cm, when the raster is used and the leaf width is 1 cm, 30 pairs of leaves driven by 60 individual motors would have to be used.

In early designs, several to several tens of 2-3 cm wide leaves were employed, which made it possible to independently collimate fields with maximum size of 20x20 cm^2 [3.30]. At present, designs are known whose spatial resolution at the isocenter is 1 cm, and thus fields up to 30x25 cm^2 can be collimated [3.43].

Multi-leaf collimators are now used in X-ray [3.41], electron [3.42] and neutron [3.43] therapies. In generating stationary fields, collimators of this kind speed up the treatment session since the information on patient's individual fields can be stored in the controlling computer's memory. In arc therapy, sequential field shaping is also facilitated. This means that the implementation of the dynamic or conformation therapies is made easier. What this means in practice is that the highly complex task of treatment planning in these conditions consists of (1) determining the 3D tumor area using computed tomography and (2) planning radiotherapy to optimize the dose distribution in this area, and finally — at the last stage — (3) combining control of the accelerator movement with that of the radiotherapy table and the multicollimator.

Fig. 3.50b is a schematic of the program which adjusts the multileaf collimator during electron arc therapy [3.42]. The collimator is made up of 18 independently controlled 4 cm wide brass vanes which all together form 9 pairs. The dynamic range of motion of the vanes, symmetrically mounted on each side, provides a variable aperture width projected to isocenter of 2 cm minimum to 8 cm maximum. The length of each vane projected to isocenter is 4 cm yielding a maximum field length of 36 cm at isocenter. Individual vane velocity is 1 cm s^{-1}, which ensures full range movement from minimum to maximum aperture in less than 2 s. Since vane position adjustment between succesive arc increments is usually less than 1 cm, the transition

(a)

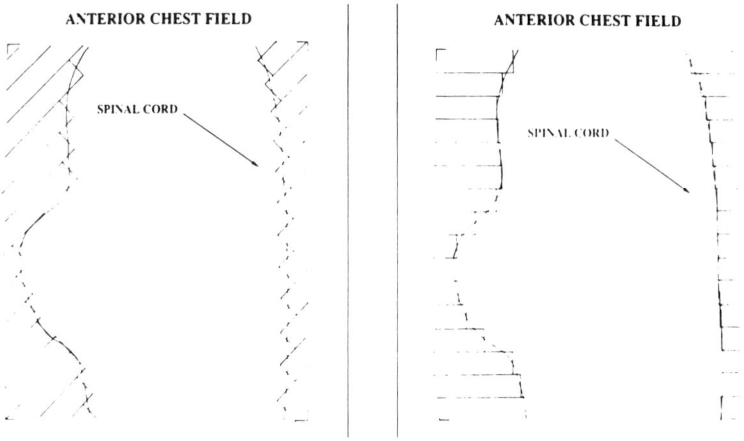

(b)

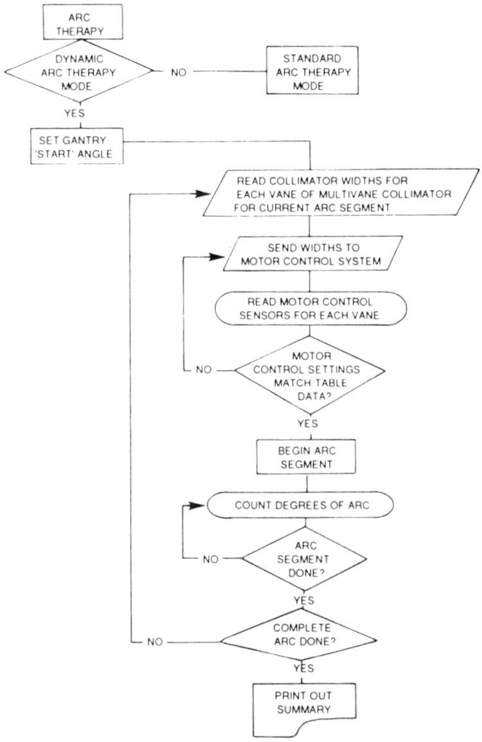

FIGURE 3.50 The multi-leaf collimator: (a) — principle of operation, (b) — flow chart showing a control program for a multi-leaf used in electron arc therapy (after D.D.Leavitt [3.42]).

time between vane position is less than 1 s, which corresponds to gantry rotation of less than 1°. In regions where the patient contour is changing rapidly, individual vane adjustments can be made for angular increments as small as 5°.

Electron applicators. Devices, referred to as *applicators,* are designed for shaping electron beam profiles. They may be made in the form of openwork aluminium strips and then are called *aperture applicators.*

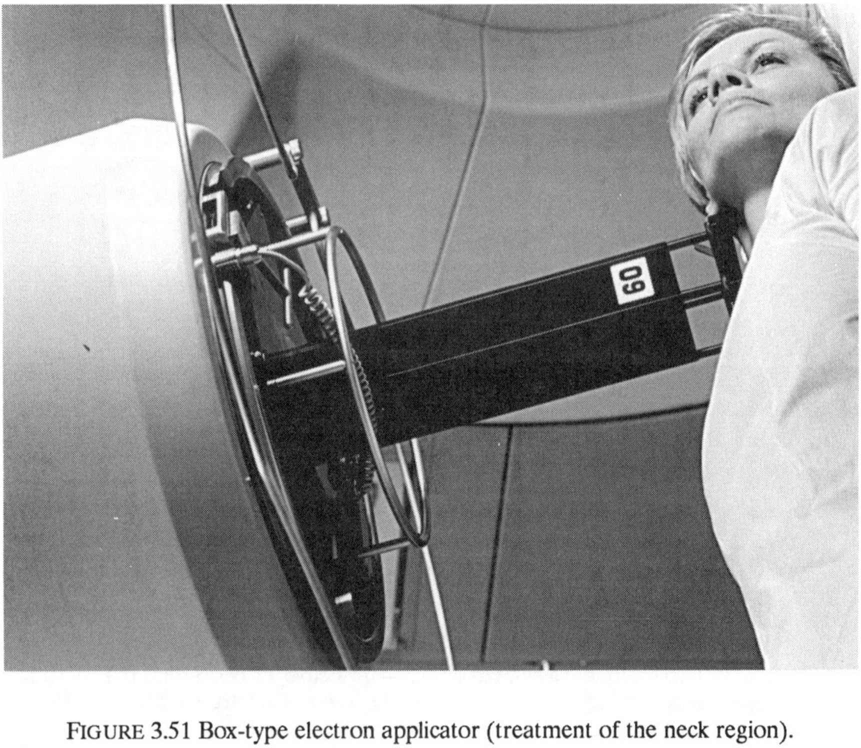

FIGURE 3.51 Box-type electron applicator (treatment of the neck region).

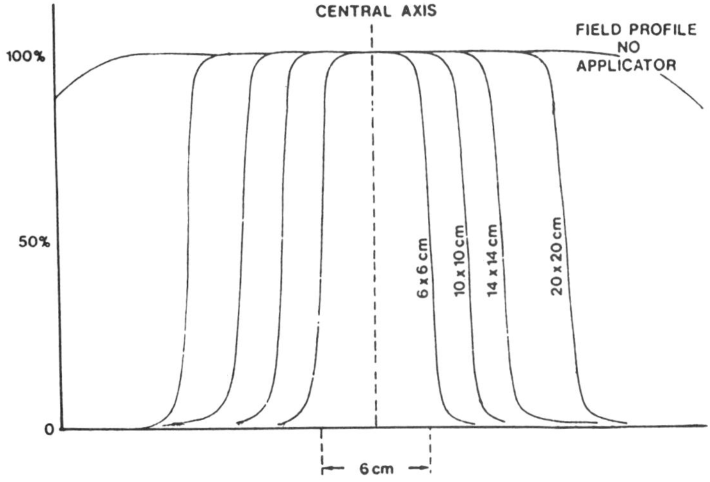

FIGURE 3.52. 14 MeV electron beam profiles obtained with 6x6 cm to 20x20 cm applicators (Philips SL).

Fig. 3.51 shows a frequently used *box-type applicators* which are employed for collimation of beams whose cross-section is square, rectangular or circular. The basic set includes five applicators (inset) with apertures of 6x6 cm, 10x10 cm, 15x15 cm, 20x20 cm and 25x25 cm. Spatial slits are made in the lower part of the applicators, which make it possible to inspect the irradiated field visually. Collimation inserts for treating irregularly shaped fields can be introduced into frames located below the slits. Each applicator is supplied with a set of different inserts.

In some designs, the applicator bottom frame is mounted on a spring support equipped with a safety switch. When the frame is struck the accelerator is immediately switched off. Applicators like those shown in Fig. 3.51 are typically placed at a distance of 5 cm between the applicator and the patient.

Applicators of the aperture or box type are not suitable for the arc or rotation therapies since they may pose a *real danger to the patient*. In rotation therapy the applicator is most frequently removed and the insert proper is mounted at the treatment head output.

The effect of the applicator on the beam profile is shown in Fig. 3.52. Here, a 14MeV beam without an applicator and that with rectangular 6x6 cm to 20x20 cm applicators are depicted. The main applicator advantages are a sharply defined beam profile and the lowest possible level of beam contamination with bremsstrahlung produced also in the applicator walls.

Some manufacturers supply electron applicators with an automatically controlled aperture which permits variable collimation of the beam. Applicators of this kind are equipped with movable walls, the positions of which can be adjusted with electric motors. In this way, a single applicator can produce any square or rectangular field from 7x7 cm^2 to 20x20 cm^2 (Philips) or 6x6 cm^2 to 20x20 cm^2 (Siemens).

3.1.5 Radiotherapy Patient Support Tables and Auxilliary Equipment

Radiotherapy patient support tables. Patient support tables can be divided into *ram-type construction tables* (Fig. 3.6) and those equipped with a *pedestal or a rectangular jack* (Fig. 3.53). The tables of the former type are simple in construction and permit the patient to be placed at a short distance above the floor. Thus large treatment fields can be obtained (Fig. 3.6) and the table can be removed from under the accelerator to allow the patient to be irradiated directly on his hospital bed. Tables of this kind, however, are seldom being used now.

A typical present-day radiotherapy patient support table is equipped with a support system as in Fig. 3.53. This construction permits angular rotation through $\pm 20°$ about the vertical axis through the isocenter. Three linear movements are possible: in the vertical direction over the range from 53 to 170 cm between the floor and the table top, and in the longitudinal direction over the range from +43 to −117 cm. The center of the central beam, that is the center of the accelerator output, is taken to be a zero point. The third movement is in the lateral direction and the table top can be moved over the range of ± 25 cm.

All the table movements are driven with electric motors with maximum speeds ranging from 3 cm s^{-1} (lateral) to 6 cm s^{-1} (longitudinal). The isocentric rotation has a maximum speed of up to 300° min^{-1}. The table can be set manually with a cranking

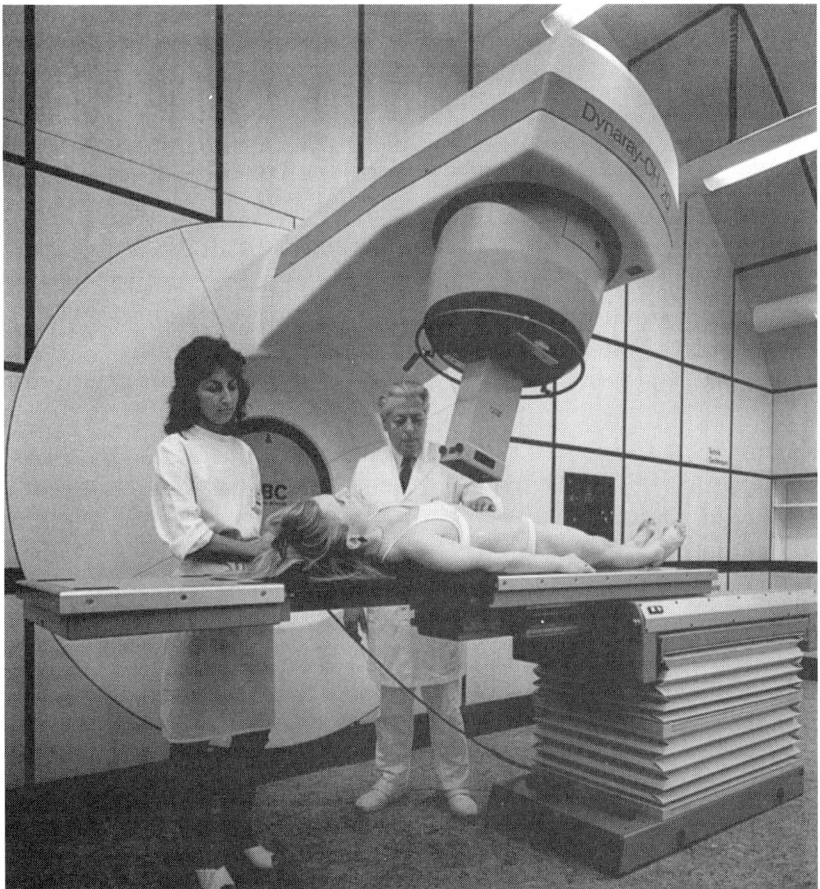

FIGURE 3.53 View of a radiotherapy table for routine treatments (BBC Dynaray Type 120P)

handle. It can also be moved automatically and manually controlled from a control console. In the automatic mode, the table movements are monitored with sensors which provide information on the exact position of the table for verification and therapy planning systems. This is particularly suitable for dynamic therapy.

In other designs, table top positioning can be carried out by means of four parallel built-in controllers, two at each side of the table. Two additional infrared controllers permit manual remote control, which is provided in order to stop the table movement if the operator becomes indisposed.

As for the table top, it is usually made of different segments. At one end it is provided with a headrest with a variable angle of inclination of up to 20°. In the middle part (directly under the accelerator head in Fig. 3.53) exchangeable frames can be used. A hard plastic table top finish can be replaced with a carbon fibre panel which is highly transparent as it absorbs less than 1% of photons. This panel is employed when irradiating from the side or obliquely from below. A *tennis racket type frame* can also be used when large (50x50 cm^2) fields are irradiated. A frame with a mylar foil is sometimes used instead of the tennis racket type frame. When carbon fibre or tennis racket frames are employed, they can have a film cassette placed directly under them.

The edge of the table top is shaped so that various accessories such as lamps, a microphone or leg supports for gynecological irradiations can be mounted on the table.

In order to achieve the required degree of accuracy during set up and beam-on stages the tables are usually equipped with an electric system to provide continuous speed control, motor braking and electromagnetic table support disengagement. The headrests can be mounted at both ends of the table top. Some tables are provided with an additional rotation through ±180° about another vertical axis through the table support center.

Like the ram-type tables (Fig. 3.6), *pedestal-type tables* can be moved beyond the beam range and thus the patient can be irradiated on his hospital bed. The pedestal-type table are usually heavy, their weight ranging between 400 and 1500 kg. For installation, they require a load-bearing structure built in the room floor. The depth of the pit necessary to accomodate this structure is usually between 20 and 30 cm.

Apart from typical tables for routine therapy (Figs. 3.6 and 3.55) surgical tables are used for the purposes of intraoperative therapy (Fig. 3.54).

Tables of this kind feature a much greater range of patient positions which can be adapted to a given treatment (Sec. 3.4.1).

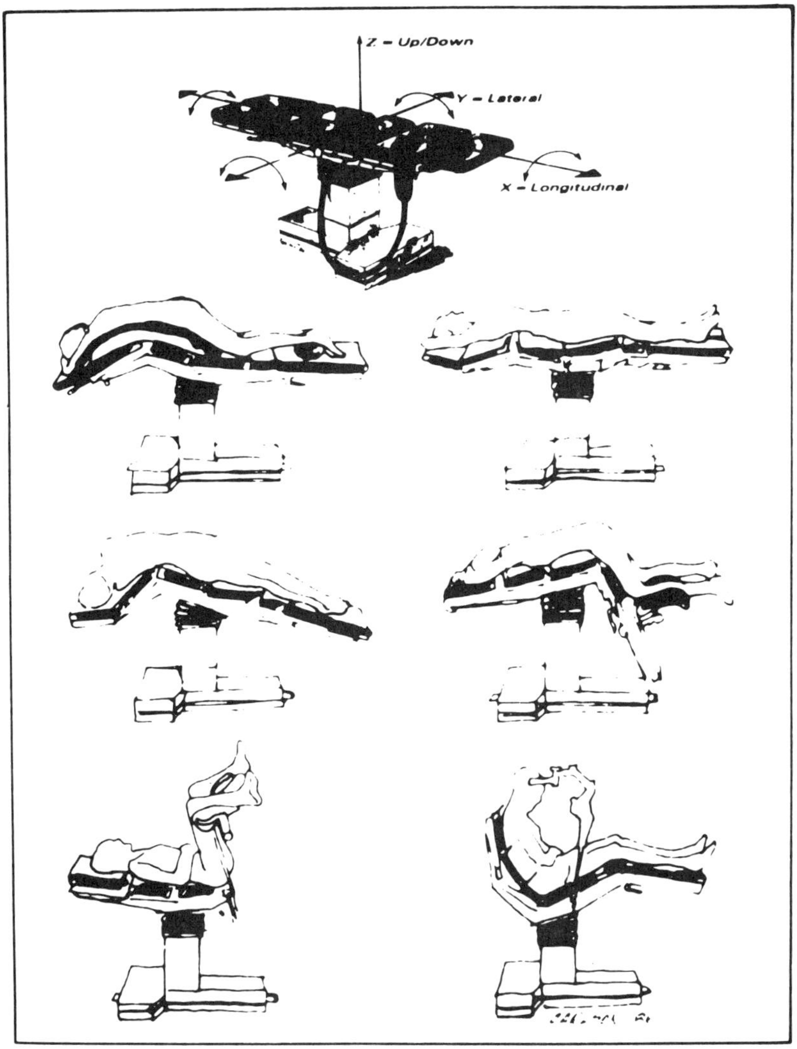

FIGURE 3.54 A surgical table for intraoperational irradiations (Siemens Mevatron ME)

Patient positioning systems. Currently, optical systems using lasers are employed for patient positioning, that is for setting-up his position on the radiotherapy table. Fig. 3.55 presents a typical *three-laser system* in which individual 1 mW beams are generated by a 632.8 nm wavelength laser. The first beam located in front of the treatment head defines the longitudinal plane, and two side beams define the patient's lateral plane. The line width at up to 5 m is less than 2.5 mm. The position of the longitudinal line, which "cuts" the patient's body in two symmetrical halves, may be adjusted within ±100 mm at 5 m. For the lateral line, the adjustment range is ±150 mm.

In addition to three-beam systems, *four-beam systems* are also in use: the fourth beam is generated by a source mounted in the ceiling close to the treatment head. Like in the system shown in Fig. 3.55, the longitudinal plane is defined by a single beam, whereas the lateral plane is determined by three crossing beams which project a light spot 4 mm in diameter with the adjustment range of ±76.2 mm (Gammex [F13]).

Systems of this kind are also equipped with a device which detects the patient's movements. For this purpose, reflecting markers are positioned on the patient's body. They reflect the laser beam light to detectors which are placed on the accelerator head. If the patient moves the reflected light disappears and the operator is signalled or the radiation is automatically switched off if the patient's movements exceed permissible limits.

Immobilization devices. An adequate immobilization method is an important guarantee in radiation therapy of high quality. Apart from precision in dose calculation and field determination, high demands are made on reproducible radiation sessions. An adequate immobilization and fixation of the patient by means of an individual mould or cost, results in an exact starting position for each radiation session. One of these moulding materials is Orfit®. This raw material (High Molecular Weight Thermoplastic Polyester) has been tested on toxicity by the FDA (Food and Drug Administration) in the United States. In industrial applications it has been found to be non toxic and there is no evidence of harmful side effects. Orfit becomes moudable at 55°C. At this temperature it becomes transparent, highly elastic, easy to mould and slighty adhesive, which makes it easy to adhere to itself and to other materials. In order to save the skin-sparing effect as much as possible, a thickness of 2 mm with maximum perforation is used (Catharina Hospital, Eindhoven). Orfit can be remoulded thanks to its "elastic mememory". This means that the material regains its original shape after reheating in water or with a heatgun and can be remoulded at any time.

Optical projection of the field. An optical system is usually built in the treatment head, which permits treatment field projection on the patient's skin surface (Figs. 3.13 and 3.30). The concurrency obtained for both fields, that is the optical and radiological fields, is usually better than ±2 mm. Depending on the design, the field dimensions are given in cm and projected as in Fig. 3.56.

Additionally, an optical distance indicator is also built in the treatment head, which determines the position of the isocenter and the target-skin distance. The accuracy of this measurement is usually better than ±2 mm over the distance to the target of 80 to 125 cm (Siemens).

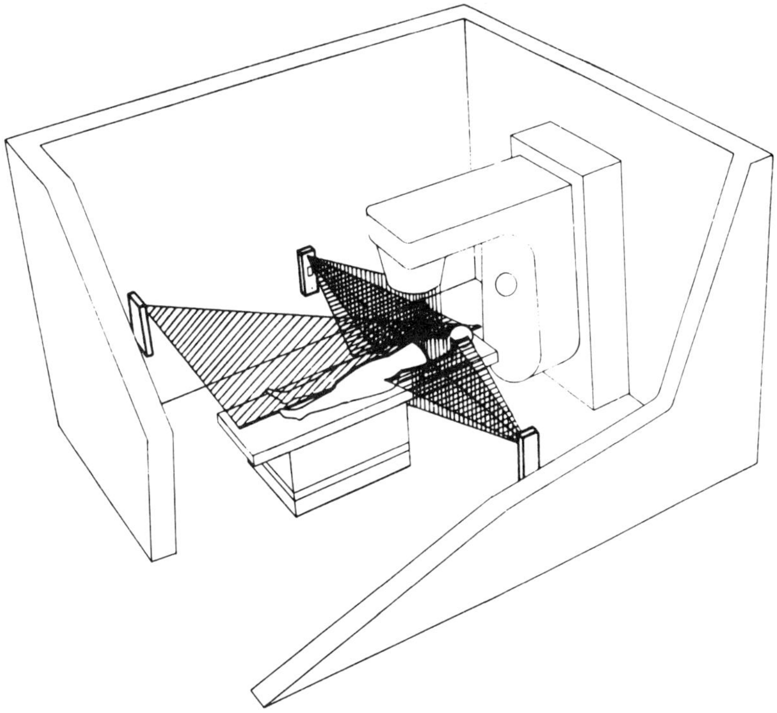

FIGURE 3.55 A three-laser system for patient positioning (Nuson A/S [F12]).

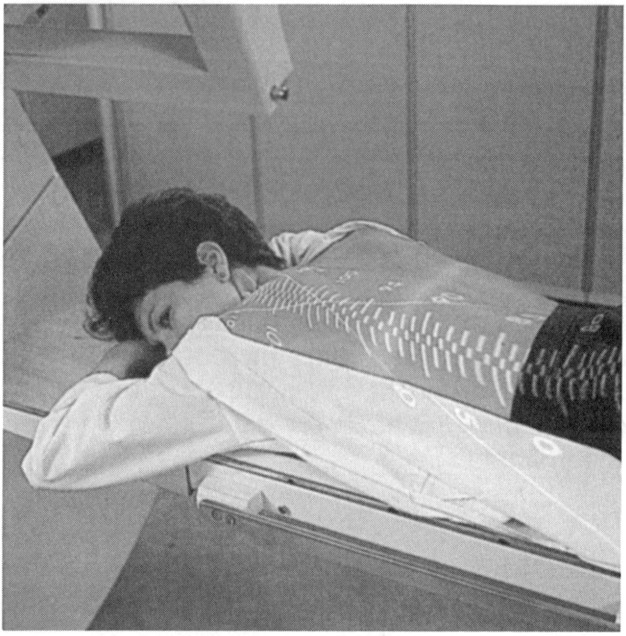

FIGURE 3.56 Optical projection of the beam field on the patient's skin.

Radiographic systems. These are used for verification of the treatment field relative to the target position prior to or during irradiation. Conventional verification of beam size, shape and position is based on portal films exposed weekly during the treatment, with visual comparison against a simulator X-ray films.

For this purpose *radiographic films (portal films)* with metal screens are placed in typical cassettes. The radiographic technique of beam field projection and that of the patient's position by means of portal films is cumbersome mainly because of the need of photochemical processing, switching off the beam to have the cassette removed as well as of the frequently poor contrast and high costs. For these reasons, work has been done to improve the portal imaging by using digital processing of radiographs

[3.44] and to replace the radiographic method with on-line methods using other types of detectors.

Over the last decade, megavoltage imaging systems have evolved from laboratory prototypes to clinically practical units. Over this period, real-time megavoltage imaging has been demonstrated with a variety of detector techniques including: (1) mechanically-scanned, linear arrays of solid state diodes or scintillation crystals, (2) rotating and/or translating liquid ionization chambers, (3) electronically-scanned liquid ionization chambers, and (4) camera-fluoroscopic systems which collect and focus the light mirror-lens of fiberoptic systems. The development of several of these systems has progressed to the

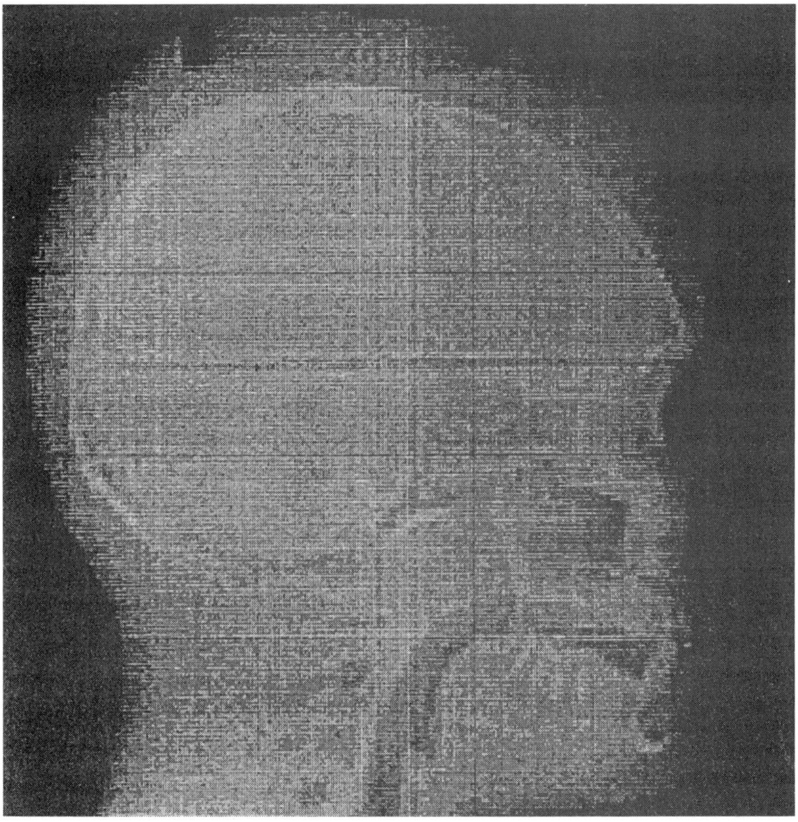

FIGURE 3.57 Image of a Rando/Alderson phantom head obtained with a multi-wire ionization chamber 128x128.

point where first generation imagers have now been introduced to clinical service [3.139 through 3.143, 3.149]. Finally, the advent of thin-panel solid state technologies offers the promise of an extremely powerful second-generation imaging system. With 10^5 to 10^6 imaging pixels consisting of pairs of hydrogenated amorphous-silicon photodiodes and transistors, 30 by 30 cm^2 imaging panels can now be created with prospects for 60x60 cm^2 panels later this decade. Such devices offer the prospect of compact, very highly radiation damage resistant systems capable of being read out in real-time at high frame rates [3.143].

Among designs of this kind is an innovative device developed by Asea Brown Boveri in which a film is replaced with a multi-wire ionization chamber filled with a special liquid carbohydrate. Inside the chamber, cover plates are lined with an array of 256 parallel electrode strips. The upper electrode array is arranged at 90° to the lower array. When a polarizing high voltage is applied to one strip of the lower electrode array, the resulting ionization signals from all the strips in the upper array are collected individually. As a result, the whole chamber is scanned to form an image of over 65,000 pixels, corresponding to the area of 32.5x32.5 cm^2.

After computer processing, the signals from the multi-wire chamber are transferred to a video monitor. Fig. 3.57 shows an image of a phantom head obtained with the above system, called Dynaray ID, using only half the resolution, a 128x128 array. When improved, this technique can be used in the future for verifying dynamic therapy.

X-ray diagnostic systems. As already mentioned, the quality of the radiographs which localize the beam and the patient's positions obtained directly with the use of a high-energy accelerator beam is often unsatisfactory even when some auxilliary devices, such as lead screens, are employed. If critical organs must be shielded (e.g. the spinal cord) in the treatment field by means of shadow blocks, the verification of the position of these blocks with portal films may be difficult.

To overcome the problem of improving the quality of radiotherapy portals two solutions are usually adopted. Some manufacturers provide their high energy 18 MV linear accelerators with a low-energy (6 MV) unflattened beam to be used only to obtain portal films whose quality is better than that taken with 18 MV X-rays. Still better results can be achieved by

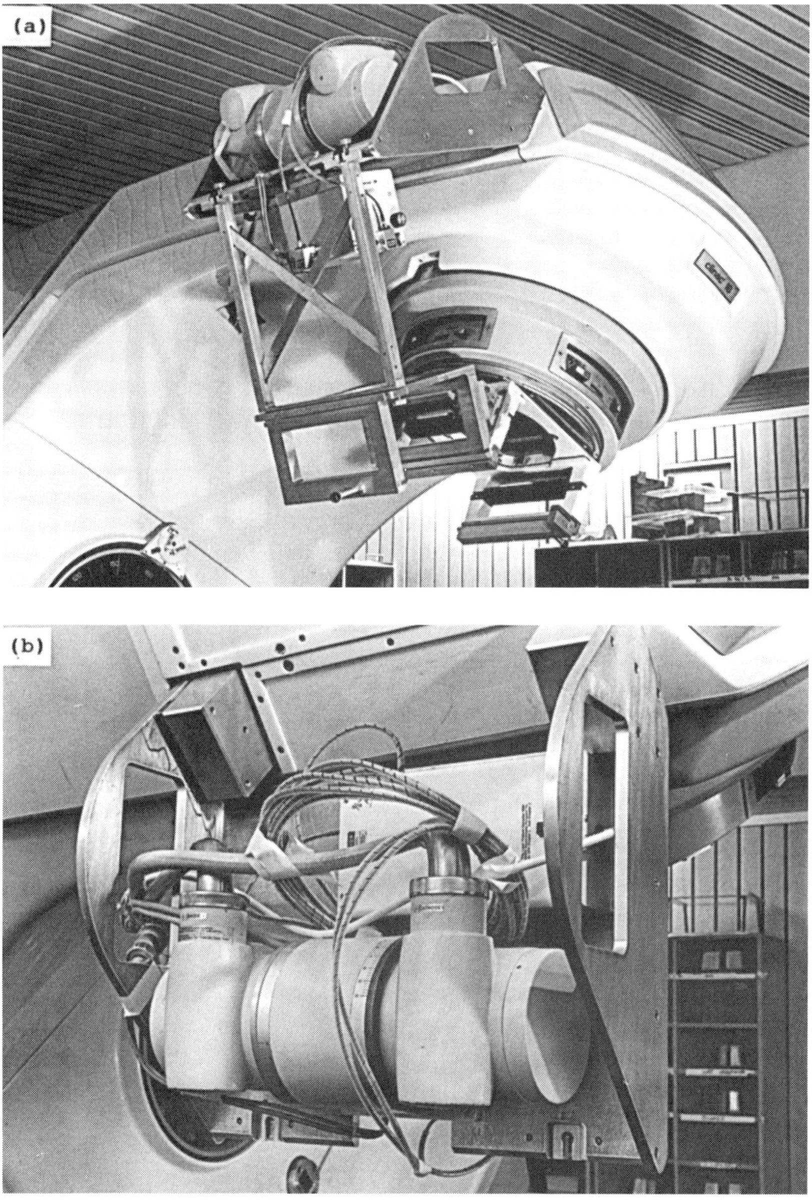

FIGURE 3.58 X-ray assembly mounted on a 10 MV accelerator head: (a) movable unit which can be mounted on the accelerator head, (b) X-ray tube in the unit (after P.J. Biggs et al [3.46]).

mounting an additional diagnostic X-ray unit (Fig. 3.58) directly on the side or in the head of the treatment unit. In this way, the patient's position and the localization of the shielding blocks can be verified. The coincidence between the mechanical, 10 MV beam and diagnostic isocenters is about 1 mm.

The design using an additional X-ray unit for localization purposes, shown in Fig. 3.58, is also an example of how the users themselves can improve the commercially available accelerators.

3.1.6 Fundamental Accelerator Quality Parameters

3.1.6.1 Methods of Presenting Beam Characteristics

A treatment beam, as a spatial structure, is characterized by defining its *main axis, or central axis* and two longitudinal and orthogonal planes (Fig. 3.59).

The parameters that are used to characterize the beam are: (1) the *percent depth dose curves along the central axis*, and (2)

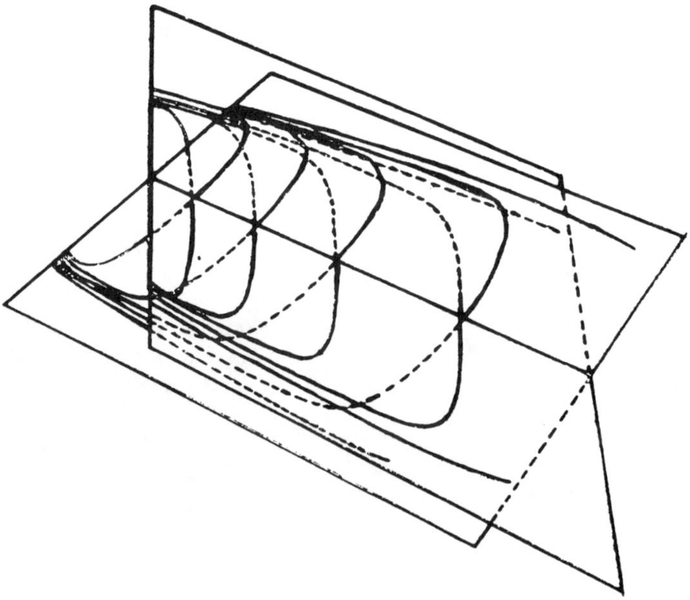

FIGURE 3.59 The central axis and two main (longitudinal) planes of the accelerator beam.

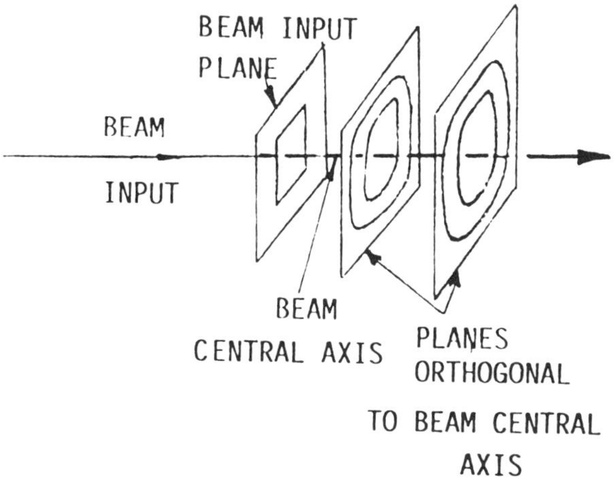

FIGURE 3.60 Planes orthogonal to the beam central axis.

the *dose distributions for a single treatment depth.* These parameters are usually determined for water as a standard scattering medium. The dose distribution is represented in the form of isodose curves which join points with identical percent depth dose values relative to the maximum dose on the beam central axis. Practically, planes in which isodoses are drawn are orthogonal to the field and either pass through or are parallel to the beam central axis (Fig. 3.59). Examples of dose distributions have already been shown in Figs. 1.7 and 1.8.

In addition to the longitudinal distributions, the beam may also be characterized by *transversal distributions in orthogonal planes* (Fig. 3.60). Examples of isodose distributions for orthogonal planes have been shown in Figs. 1.11, 1.14, 3.21 and 3.45. Typically, isodose distributions feature curves which correspond to the maximum values (100%) down to the minimum value which is usually that of 10%, graded in 10% or 20% increments.

The dose in the irradiated medium is a sum of doses from the primary and scattered radiation. The percent depth dose depends, therefore, on the parameters which affect both these components and which are (1) radiation energy, (2) depth of the medium, (3) type of radiation, (d) source-skin distance (SSD), (4) size and shape of the field, and (5) chemical composition of the scattering medium. That is why both the percent dose along the beam central axis and the isodose distribution constitute the main parameters that characterize a given type of a radiotherapy accelerator.

3.1.6.2 Beam Energy Distribution

The electron internal beam, or, in other words, the beam close to the accelerator output window but behind the bending magnets and slits for energy determination has a certain energy distribution which can be characterized by the following parameters (Fig. 3.61a): *maximum energy* E_{max}, *most probable energy* E_p, *mean energy* E_m and *energy spread* Γ in terms of the Full Width Half Maximum (FWHM). When the beam is passed through the output window and through other materials on its way to the phantom surface, its energy diminishes and the energy spread increases, which is shown in Fig. 3.61b. Curve 1 represents energy distribution close to the output window where the mean energy is approximately equal to the most probable energy. Curve 2 represents energy distribution of an electron beam which has penetrated a scattering foil intended for ensuring uniform spatial distribution (Sec. 3.1.4.3).

The scattering system is, in this case, optimized. If this is not the case, curve 3 is obtained. It features a much greater energy spread and the mean energy E_m is considerably lower than the most probable energy E_p because there are far more electrons that have undergone multiple scattering or those which have contributed to beam contamination.

The increase in energy spread shown in Fig. 3.61b (curve 3) results in a lower dose gradient (Sec. 3.1.6.3) and consequently impairs the beam quality. Therefore, a tendency in designing present-day radiotherapy accelerators is to minimize the energy spread Γ, or to achieve a *beam as monoenergetic as possible.*

The beam energy measuring device provided with the accelerator usually measures the *electron internal beam energy.* For this reason, this device, together with the energy selectors on the control console, should be calibrated in terms of energy which does not depend on the kind of materials in the beam

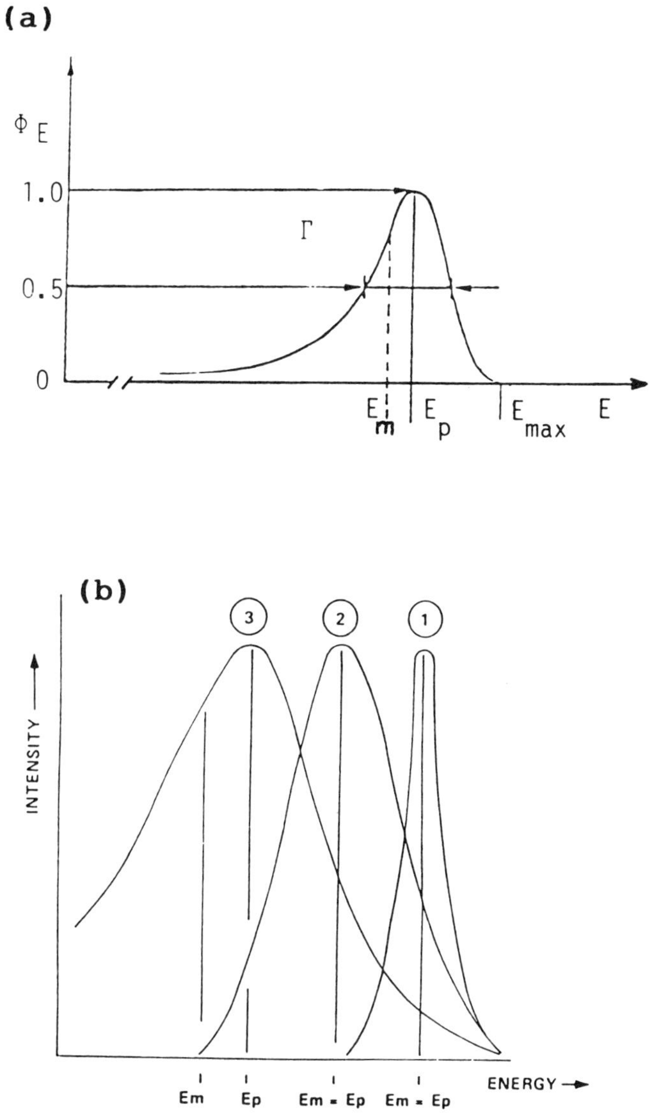

FIGURE 3.61 Beam energy distributions: (a) — typical, (b) — energy spread.

path. The proper energy used in this case is the most probable energy E_p.

3.1.6.3 Depth Dose Distribution

A typical *percent depth dose distribution* is shown in Fig. 3.62. It can be described with the following parameters: (1) level of *maximum absorbed dose* D_{max}, *surface dose* (defined at 0.05 cm depth) D_s, and dose D_x corresponding to the X-ray background. In addition, the *half value depth* R_{50} is also used, defined as the depth of the 50% dose point and is related to the beam mean energy E_m (Fig. 3.61a). *The therapeutic range* R_{85} is defined as the depth of the distal 85% dose point. The third distance parameter is *practical* or *projected range* R_p, defined as the depth of the point of intersection of the X-ray background and the extrapolated range; it is related to the beam most probable energy E_p (Fig. 3.61a). The practical range is the range concept in most frequent use for correlation with the electron energy.

Apart from the above doses and ranges, the depth dose distribution can also be described by means of gradient G defined as

$$G = \frac{R_p}{D_{max} - D_z} \frac{dD}{dz_{max}} \tag{3.7}$$

The above gradient determines the slope of the depth dose distribution. The greater it is the closer to each other become ranges R_{85}, R_{50} and R_p. From a known depth absorbed dose distribution, the dose gradient can be obtained simply by extrapolating the steepest section of the dose distribution, which is used to obtain the practical range, until it intersects the photon background D_x and the level of maximum absorbed dose D_{max}. The dose gradient is now obtained by dividing the practical range by the distance between these two intersections. This measure of the steepness of the absorbed dose distribution is a dimensionless number which varies slowly with eletron energy since the increase in practical range with energy almost balances the decrease in absolute slope of the absorbed distribution.

The *therapeutic range* R_{85} is that depth interval of the absorbed dose distribution which ideally should coincide with the target volume [3.112]. For a given electron energy, the width of this interval depends on the desired uniformity of absorbed dose distribution in the target volume. It is well known that differences as small as 5%-10% in the absorbed dose in some cases can cause a considerable change in the difference between the fraction of patients cured and that subsequently suffering complications, a fact which normally makes it desirable to strive for a very small dose variations inside the target volume.

The uniformity of the absorbed dose distribution perpendicular to the beam can sometimes be improved by slightly increasing the field size. Similarly, the distribution of absorbed dose parallel to the beam can often be improved by increasing the electron energy. However, both these methods result in a larger irradiated volume, a result which is generally not acceptable. If the same variation in absorbed dose is accepted, perpendicular and parallel to the beam, the variation of absorbed dose in the target volume should be less than 10%, since most accelerators are specified with a homogeneity of ±5%. Such a small variation with depth is generally unrealistic since a compromise has to be found between good uniformity and a small irradiated volume. The therapeutic range has therefore been defined as the depth interval within which the absorbed dose exceeds 85% of its maximum value. For most high-energy electron beams, the therapeutic range extends from the surface to the 85% depth dose, since the surface dose is often higher than 85% of the maximum dose [3.112].

The *surface dose* D_S (Fig. 3.62) in a volume irradiated by a therapeutic electron beam is generally of great interest since it is of importance for the degree of skin sparing obtained. The depth of the radiation-sensitive layers below the epidermis is generally on the order of 0.5 mm, a depth which is also accessible for accurate absorbed dose measurements with many detectors (Sec. 4.1.2). The *surface or skin dose*, D_S, is thus defined here as the ratio of absorbed dose at a depth of 0.5 mm to the maximum absorbed dose. D_S is often about 90% or less because of the buildup of the fluence of primary and secondary electrons and in most therapeutic beams also because of the contamination by oblique electrons and photons.

The *photon background*, D_X, is defined as an extrapolation of the tail of the absorbed dose distribution (Fig. 3.62) back to the practical range. The photons are either present as a contamination of the incident electron beam or generated in the irradiated medium itself.

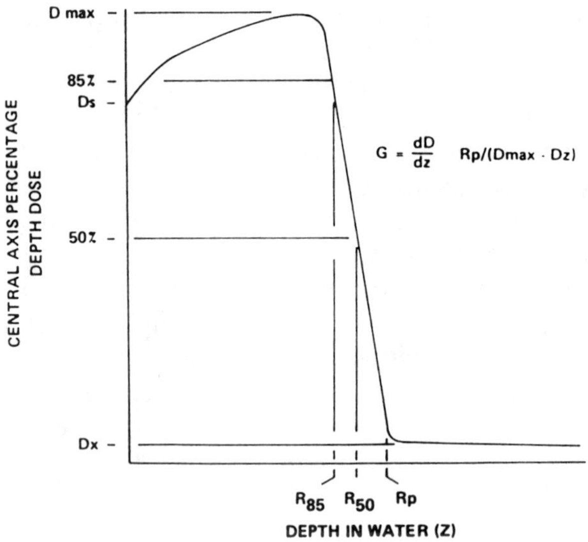

FIGURE 3.62 Depth dose distribution for an electron beam.

Actual depth dose distributions as a function of energy have already been shown for electrons in Fig. 1.12, and for X-rays in Figs. 1.3 and 1.10.

3.1.6.4 Treatment Fields

Field flatness. The *flattened area* of square X-ray fields is defined by straight lines joining points on the major axes and diagonals of square fields as shown in Fig. 3.63a. After 976 IEC Standard (Table 3.3) maximum field flatness is defined as the ratio of the maximum absorbed dose (averaged over not more than 1 cm^2 anywhere in the radiation field) to the minimum absorbed dose (averaged over not more than 1 cm^2 anywhere in the flattened area). The maximum deviation of the absorbed dose from the reference value in the reference plane perpendicular to the axis of the beam in the standard measurement depth of 10 cm — within the flattened are — is described in Fig. 3.63a and Table 3.6. The guideline values of the 976 IEC Standard for the maximum deviation are for radiation fields less or greather than 30x30 cm^2 6% or 10% respectively.

TABLE 3.6 Definition of the Flattened Area of a Radiation Field in the Standard Measurement Depth and Guideline Values of the 976 IEC Standard (Table 3.3) for Acceptable Ratios D_{max}/D_{min} of the Profile of the Relative Absorbed Dose (Fig. 3.63b).

Field dimension F of square radiation field	Dimensions defining the flattened area d_m	d_d	Guideline for D_{max} / D_{min}
5 cm - 10 cm	1 cm	2 cm	1.06
>10 cm - 30 cm	0.1 F	0.2 F	1.06
>30 cm - maximum	3 cm	6 cm	1.10

The definition of the *nominal radiation field* is based on the distance of the points of 50% of absorbed dose at the radiation beam axis along each of the major axes of the field. The test should be performed in a water phantom (Sec. 4.1.3) by measuring profiles of relative absorbed doses along the major axes and the diagonals of the radiation field as seen in Fig. 3.63b. Shown in the figure are only halves of the profile typical for a small radiation field (left) and for a large radiation field F>30 cm (right). All profiles lie within the permitted limits (Table 3.6). The maximum deviation of absorbed dose at points symmetrically displaced from the radiation beam axis and within the flattened area at the standard measurement depth of 10 cm for photons (cf Fig. 4.6) is a specific performance characteristic (3%). A similar requirement holds for the dependence of absorbed doses at each point within the flattened area on the angular position of the gantry and the beam limiting system, for nominal energies less, equal or greater than 30 MeV: 3% or 4%, respectively.

The explanatory diagram for flattness of the electron fields is shown in Fig. 3.64. In this case, the flattness is described as the ratio of the highest absorbed dose (averaged over 1 cm^2) anywhere in the radiation field at the standard measurement depth with respect to the absorbed dose on the radiation beam axis at the depth of dose maximum. For the purpose of this requirement the geometrical field defined by an applicator is projected into the phantom parallel to the radiation beam axis (Fig. 4.6). The phantom measuring position can be seen also in

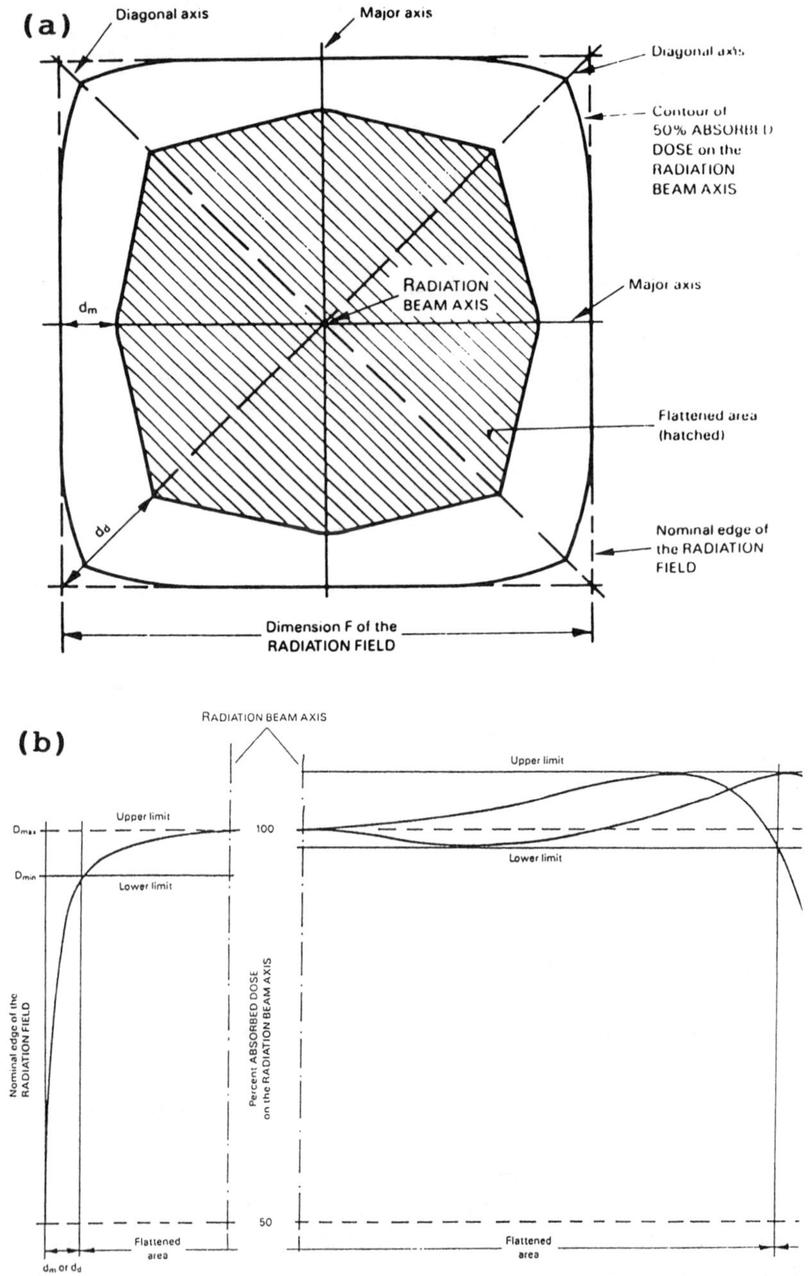

FIGURE 3.63 Radiation field parameters: (a) — flattened area (shown hatched) within the radiation field, (b) — examples of profiles of absorbed dose along the major axes or diagonal axes (after 976 IEC Standard, Table 3.3).

Fig. 4.6. Standard measurement depth S is defined as S = P:2, where P is penetrative quality. After IEC 976 Standard for each set of test conditions the following parameters shall be given:

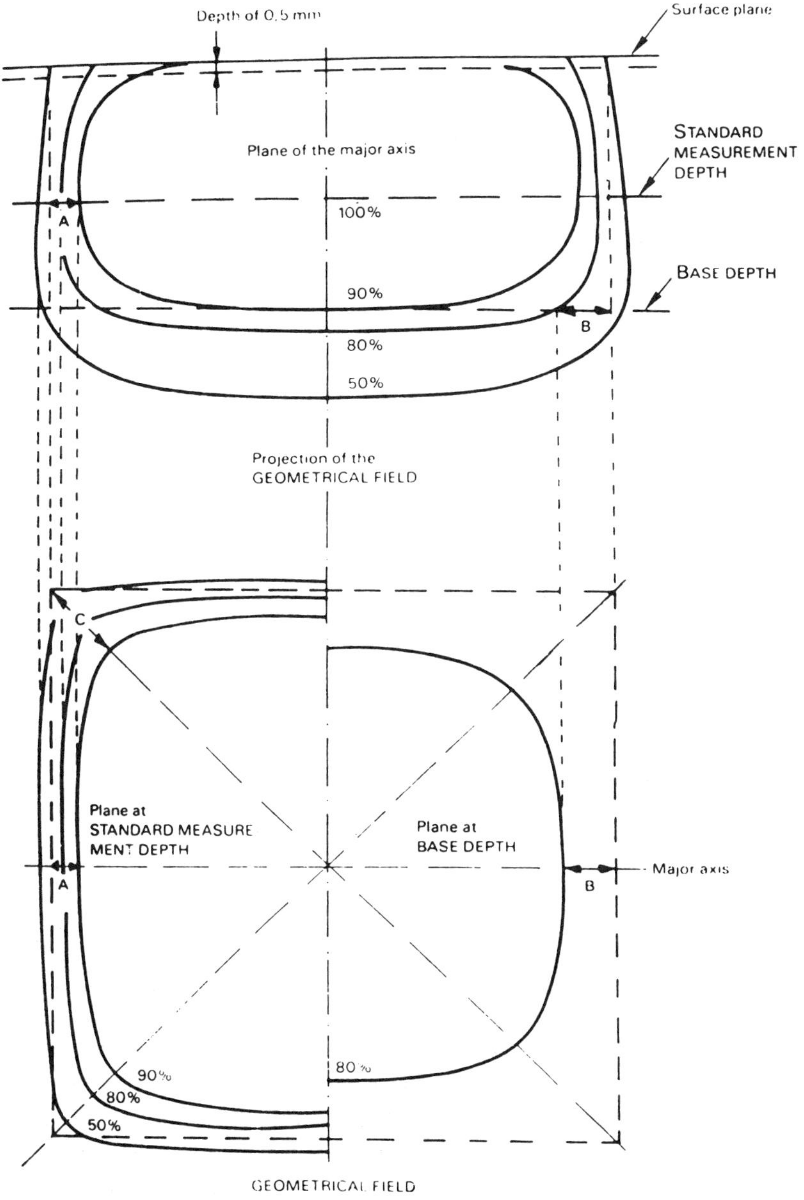

FIGURE 3.64 Explanatory diagram for flatness of the electron field (after 976 IEC Standard).

(1) the maximum distance A (Fig. 3.64) between the 90% isodose contour and the edge of the projection of the geometrical field on both major axes at the standard measurement depth,

(2) the maximum distance B between the 80% isodose contour and the edge of the projection of the geometrical field on both major axes at the base depth (90% isodose depth, Fig. 3.64),

(3) the maximum distance C between the 90% isodose contour and the corner of the projection of the geometrical field on the bisectors of the corners at the standard measurement depth; C is defined along the diagonal axes, for non-square radiation fields not along the bisector.

For an electron beam the area of flatness is thus limited by the point at which the relative dose is reduced to 95%. This criterion is a compromise between the highest possible degree of flatness and the suitable field size.

A useful measure of the field flatness can also be provided by the coefficient of flatness adopted in the recommendations of the Scandinavian Protocol (NACP). It is defined as the ratio of the area restricted by the 90% isodose contour to the field whose area is determined by the 50% isodose curve. According to these recommendations, the coefficient of flatness should be not less than 0.8 for fields 10x10 cm or larger. The higher the above coefficient, the better is the coincidence of the fields which correspond to the 90% and 50% isodose contours.

Border field. The quality of the treatment field is also determined by the width of the border field. Due to lateral scattering of electrons in tissues, the distance between the 90% and 10% isodose curves increases with depth, the rate of the increase also depending on the beam's initial energy. The criterion for the field flatness already implies a requirement for the part of the field border which is inside the geometric field. The remaining part between the geometric field boundary and the 10% isodose contour is strongly dependent on the shielding properties of the applicator and the mask (frame). In order to find a proper measure of the *"quality of collimation"*, the border field for the highest energy beams should be considered at a depth near the surface, where significant divergencies in isodose distributions are not yet present. The distance between the geometric field boundary and the 10% point on the beam profile should not exceed 5 mm depth for a 20 MeV electron beam.

Stray radiation regions. The conventional presentation of treatment fields by means of isodose patterns does not include any information about the region where the dose is less than 10% of the maximum value. However, these regions are important from the point of view of the radiation protection of organs outside the useful field. The scope of the protection depends on the shielding properties of the applicator and mask (frame). Among the most radiosensitive organs are the eye lens. In the routinely used treatment plan with, for example, 60 Gy (6000 rad) applied over 4 weeks, the threshold value which causes visible changes in the eye is typically 5% of 60 Gy or 3Gy. It is, therefore, essential that the beam profile at 0.5 cm depth be observed and the distance be determined between the

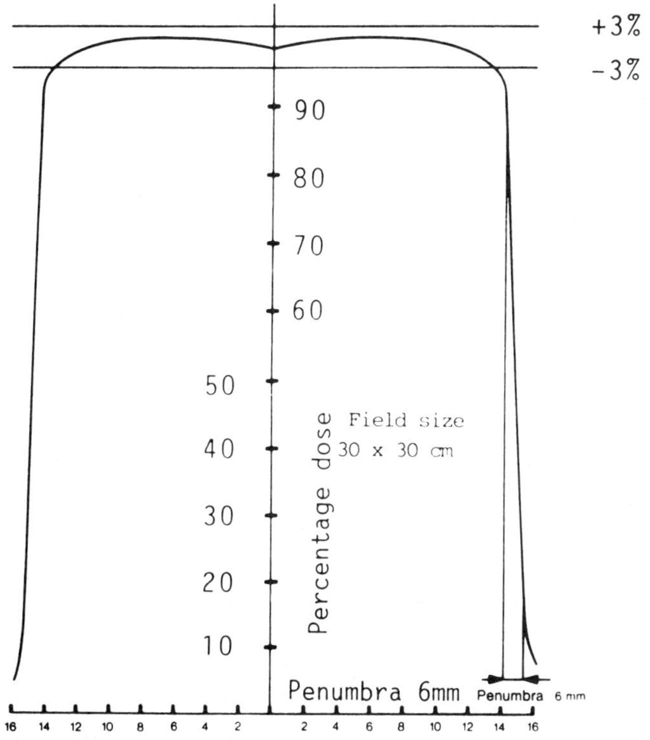

FIGURE 3.65. 8 MV X-ray beam profile in a water phantom at 5 cm depth, 30x30cm field.

geometric field boundary and the point beyond which the dose value is 5%. This distance is taken as a measure of the stray radiation distribution. If one wants, however, to have an additional margin of safety the determination of the 2% isodose contour is also desirable since it defines the outer boundary of the safety region with stray radiation.

Effect of penumbra. The slope of the edge of the beam profile is largely dependent on penumbras, whose formation was discussed in Sec. 3.1.3 (Fig. 3.25). The method of determining the effect of penumbra on the beam characteristics (isodose pattern) is shown in Fig. 3.65. The *penumbra* is defined as a distance between the points corresponding to the 80% and 20% isodose patterns at 10 cm depth. Typical penumbra values depend on the type of the accelerator and usually range between 6 and 9 mm. These values, for a given machine, increase with the increase in field size.

Following 976 IEC Standard the width of the penumbra shall be given for radiation fields of 5 cm x 5 cm, 10 cm x 10 cm and for the maximum radiation field for both X-rays and electron radiation.

Beam contamination. The radiotherapy beam should be *pure*, i.e. uncontaminated with radiation of another kind. This means that an X-ray beam should not be contaminated with electrons, and an electron beam should be free from X-rays. Both types of beams, in the case of higher energies in particular, can be contaminated with the highly unwanted neutron radiation.

X-ray beam contamination with electrons. One feature of X-ray treatment fields is that the depth of the maximum depth dose, D_{max}, becomes smaller with increasing field size. An example of this effect for a 25 MV X-ray beam and the field size change from 4x4 cm^2 to 28x28 cm^2 is shown in Fig. 3.66. By changing the field size, for the same value of beam energy, the effect of build-up is markedly diminished. The initial uncertainty of whether this effect is due to the beam contamination with soft X-rays or electrons has been dispelled by experiment [3.48 through 3.50]. By using an additional magnet to free the beam from electrons and employing filtering foils, it has been shown unequivocally that the contamination effect is caused by electrons [3.30 and 3.48]. The degree of contamination increases markedly with field size and X-ray energy. A detailed discussion on the production of secondary electrons by an X-ray

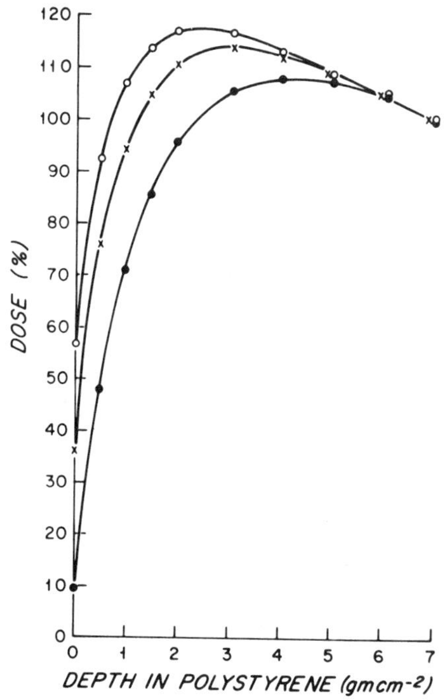

FIGURE 3.66 Depth dose curves for 25 MV X-ray beam (Clinac 35) in a polystyrene phanton: o – 38x28 cm^2 field, x – 12x12 cm^2 field, and • – 4x4cm^2 field (after J.P. Biggs and C.C. Ling [3.48]).

beam in materials with a various Z's and in air is given in [3.49 and 3.50].

The unwanted effect of secondary electrons can be easily reduced, if not eliminated, with the use of high Z foil filters. So far, this technique has not found application in commercially produced accelerators.

Bremsstrahlung photon contamination. Bremsstrahlung is present in electron beams because it is emitted by the various components of the treatment head, such as the accelerating structure output window, scattering foils, collimator edges, ionization chambers and applicators.

Photon contamination leads to the undesirable irradiation of tissues outside the treatment field (D_x dose in Fig. 3.62). Typi-

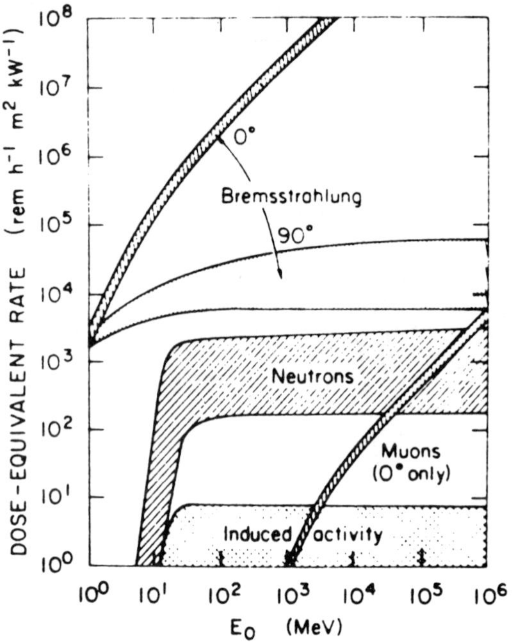

FIGURE 3.67 Dose-equivalent rates for various radiations produced by a 1kW electron beam, as a function of electron energy (after W.P. Swanson [3.56]).

cally, the contamination contribution to the total electron beam intensity is 1.0-3.0% for the energy range of up to about 15 MeV. Since the bremsstrahlung intensity is a function of energy (Sec. 3.1.4.2), this contribution may be even as high as 5% for energies higher than 15 MeV. Bremsstrahlung contamination dose can be defined from the residual tail on the depth dose distribution beyond the electron range (Fig. 3.62).

Neutron contamination. In electron radiotherapy accelerators neutrons are produced in the (X, n) and (e^-, n) reactions, which have energy thresholds higher than 5 MeV. The neutron production yield is strongly dependent on energy up to about 25 MeV. With higher particle energy, the emission rate increase is lower. This is depicted in Fig. 3.67 which shows dose equivalent rates for various types of radiation produced by a 1 kW high-energy electron beam. Although, typically, the *neutron fluence* in a radiotherapy accelerator is some 1000 times lower than the electron flux, neutron contamination is of some importance because the neutron RBE is 10 to 15 higher than that for electrons [3.51 through 3.62]. The main sources of neutrons are components of the treatment head such as the target, collimators, and the flattening filter. Fig. 3.68 presents the results of measurements and calculations of neutron fluence per absorbed treatment dose as a function of electron energy. Points are representative of several accelerators. The upper solid curve corresponds to the case in which all neutron-producing components of the treatment head are made of tungsten. The lower dotted curve corresponds to the case in which these components are made of copper. Giving neutron fluence in terms of the unit absorbed dose considerably helps in making the subsequent shielding calculations essential in radiation protection (Sec. 4.3.2) since neutrons are present not only in the patient's body but also in the treatment rooms.

In electron therapy, the electron beam intensity is usually less than 1% of that used when producing X-rays, thus electron beam contamination with neutrons is generally small, its effect being negligible. On the other hand, in X-ray therapy and for energies higher than 10 MeV (Figs. 3.67 and 3.68), the neutron dose contribution in the total therapeutic dose must be carefully analyzed and taken into consideration. Using Monte Carlo simulations with a typical treatment head geometry [3.59], it has been shown that a copper target, a steel flattening filter and a tungsten collimator give a total neutron yield of 1.6×10^{10} neutrons per cGy of peak photon dose. The neutron yield is approximately isotropical and is not significantly attenuated by the treatment head shielding. Most of the neutrons (82%) are produced in the collimator material and the remainder are produced mainly by the target (12.5%) and the flattening filter (5.7%). The neutrons that are produced are fast, their spectrum approximately that of the fission neutrons. In the above case, the effective energy of the neutrons in the treatment room is about 0.4 MeV. Using the fluence-to-dose-equivalent conversion, recommended by the NCRP Report No. 51, i.e. 7.2×10^{9} cm^{-2} Sv^{-1}, the calculated dose-equivalent rate at 1 m from the target is 2×10^{-3} Sv neutrons per peak photon Gy or 0.2% Sv Gy^{-1} [3.59].

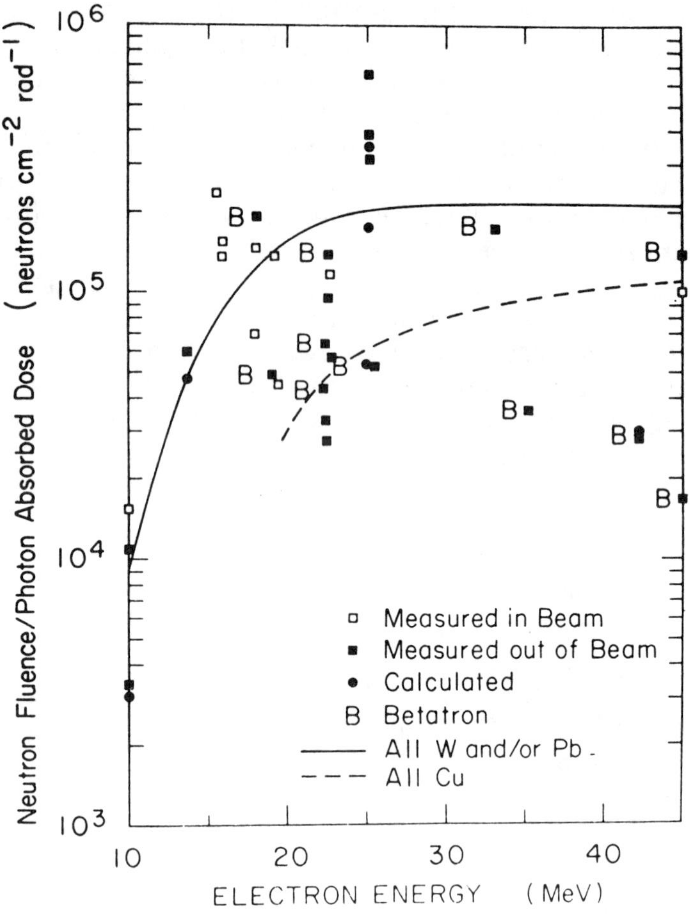

FIGURE 3.68 Neutron fluence per 1 cGy of treatment dose as a function of electron energy (after W.P. Swanson [3.57]).

The conversion from neutron fluence to dose equivalent is highly dependent upon a knowledge of neutron spectrum, and in studies of neutron measurements around medical linear accelerators, values of 2.5 to 20x10^9 cm^{-2} Sv^{-1} were used.

It follows from [3.54] that, for a typical total treatment dose of 6000 rad and beam energy of 20 MV, the neutron dose in the treatment field will be 120 rem. This is true for a case in which the neutron contamination contribution in the primary beam intensity (SSD = 100 cm) is 0.2%. The dose absorbed by the patient within the field will be only 12.0 rad, but the absorbed dose equivalent will be 120 rem since the RBE for the neutrons with the above energies is 10. The patient irradiation outside the treatment field for the neutron contamination of only 0.02% will give an absorbed dose of 1.2 rad corresponding to the equivalent dose of 12 rem [3.54].

Calculation of neutron shielding requires knowledge of the neutron emission from the treatment unit. According to the measurements on a large number of accelerators from various manufacturers, the neutron emission for 10 MeV electrons are of the order of 10^8 n rad^{-1} of photon dose at isocenter [3.62]. For the upper energy of 25 MeV used in present-day linear accelerators, this value becomes as high as 10^{10} to 10^{11} n rad^{-1}.

The *thermal neutron fluence* from a fast neutron source such as a therapy accelerator, placed in a cavity with thick concrete walls, may be expressed as

$$\Phi = \frac{k\,Q}{S} \tag{3.8}$$

where Φ is thermal neutron fluence rate (n cm^{-2} s^{-1}); Q is the fast neutron emission rate (n s^{-1}); S is the inside surface area of the cavity (cm^{-2}), and k is a constant equal to 1.26 ± 0.10. The above relationship was shown to be true both for thermal neutrons and for scattered fast neutrons produced in typical radiotherapy rooms [3.62]. Eq. (3.8) may thus be used for calculating shielding parameters for such rooms. In accordance with Eq. (3.8) if the emission of fast neutrons as primary particles is expressed in relation to the radiation treatment unit (Q in n rad^{-1}), the fluence, Φ_n is being expressed in n cm^{-2} rad^{-1}.

The values of the permissible neutron contamination levels recommended by the IEC regulations are given in Sec. 3.1.6.5, whereas the problem of shielding radiotherapy rooms is discussed in Sec. 4.3.2.

3.1.6.5 Radiation Protection

Over thirty years ago in the early applications of radiotherapy accelerators, it was established, probably by analogy to recommendations for diagnostic X-ray units, that the *level of stray radiation* emitted by the accelerator should not exceed 0.1% of the dose at isocenter. In addition to all components of the scattered radiation, this level also included the dose from neutron radiation appropriately normalized for its RBE. The level so established turned out to be quite problematic, especially when it was found that the contribution from the scattered radiation in the patient's body was much greater than the stray radiation from the accelerator. The chronological development of requirements and regulations including the American, Canadian and IEC regulations for the radiation safety of radiotherapy accelerators for energies of 1 to 50 MeV is presented in [3.63].

In the late 1970s, twenty national standardization committees, including the United States, adopted the IEC 601-2-1 document

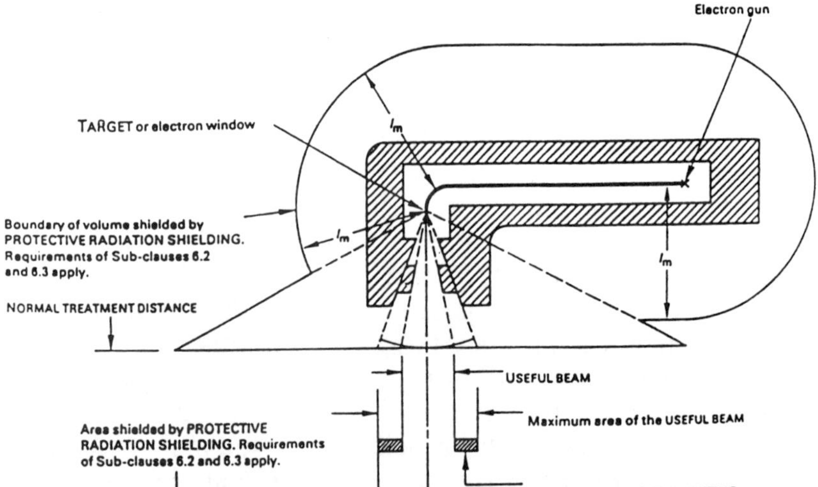

FIGURE 3.69 General requirements for radiation safety in radiotherapy accelerators according to IEC recommendation (Document 601-2-1, Table 3.3)

(Table 3.3) on radiation safety of medical accelerators in the energy range of 1 to 50 MeV.

These regulations provide some limits for the protection of the patient against stray radiation in the useful beam and against radiation outside the useful beam (Fig. 3.69).

Protection of the patient against stray radiation in the useful beam. For stray X-radiation during *electron irradiation* the absorbed dose on the axis of the beam at a depth 100 mm farther than the practical electron beam range (ICRU Report 21) shall not exceed the limits shown in Fig. 3.70a. The measurement shall be made in a phantom (Fig. 4.6), the incident surface of which shall be at the normal treatment distance (NTD) as defined by the applicators and normal to the axis of the beam (Fig. 3.69). The incident surface of the phantom shall have dimensions at least 5 cm larger than those of the radiation field. The depth of the phantom shall be at least 5 cm greater than the depth of the measurement.

At the largest field size available, the absorbed dose at the surface on the axis of the beam during *X-irradiation* shall not exceed the limits shown in Fig. 3.70b. The measurement shall be made using a phantom (Fig. 4.6), the incident surface of which shall be at the normal treatment distance (NTD) from the source X-radiation and normal to the axis of the beam, and using an instrument which allows extrapolation to the surface absorbed dose. The incident surface of the phantom shall have dimensions at least 5 cm larger than those of the radiation field. The depth of the phantom shall be at least 5 cm greater than the depth of measurement. All beam modifying devices which are removable without use of tools, except field-flattening filters, shall be removed from the useful beam.

Protection of the patient against radiation outside the useful beam. Each item is specified separately: (1) leakage radiation through beam limiting devices, (2) leakage radiation (excluding neutron radiation) outside the maximum useful beam and (3) neutron leakage radiation outside the maximum useful beam.

(1) Adjustable or interchangeable beam limiting devices shall be provided. For X-radiation, they shall attenuate the radiation so that the absorbed dose in the area shielded by the beam limiting devices at the normal treatment distance does not exceed 2% of the absorbed dose on the axis of the beam measured at an equal distance from the radiation source. Measurements

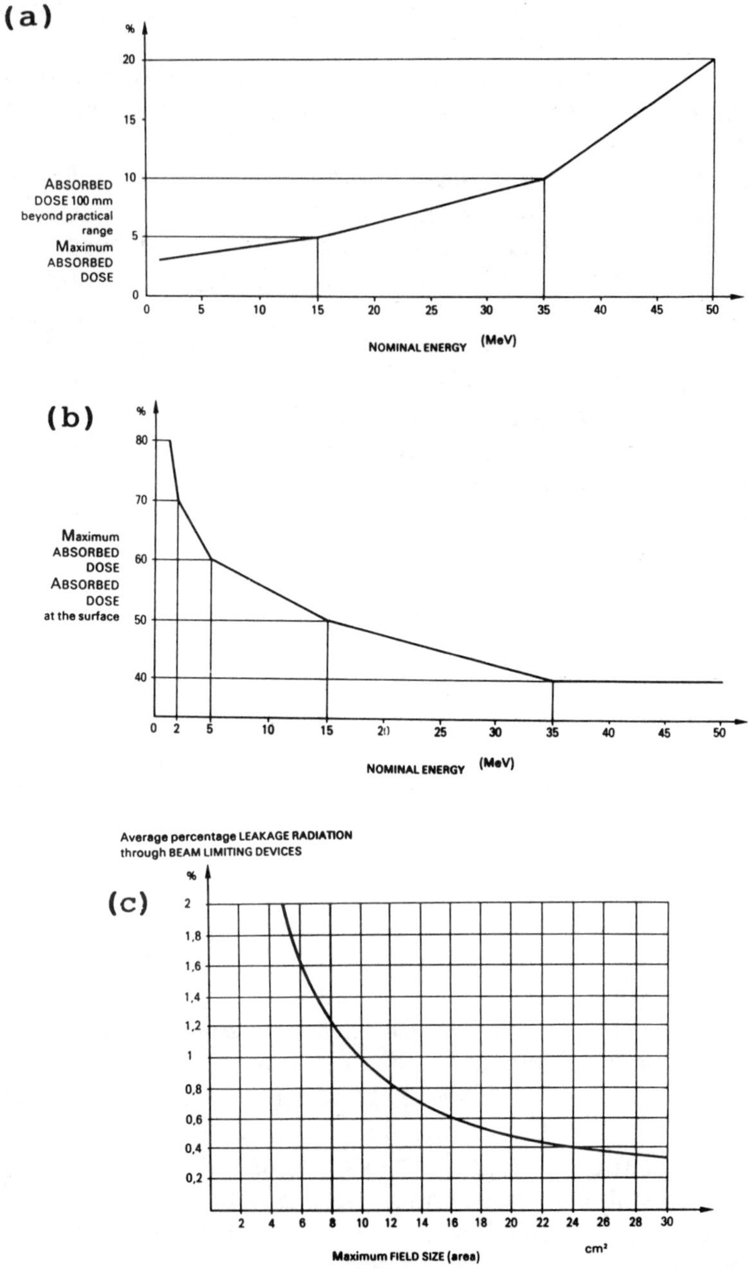

FIGURE 3.70 IEC limits: (a) — of stray X-radiation during electron irradiation, (b) — of absorbed dose at the surface during X-irradiation, (c) — of leakage radiation through beam limiting devices (large fields) (after IEC Publication 601-2-1)

of leakage radiation through the beam limiting devices shall be made with any residual aperture closed by at least two tenth value layers of a suitable absorbing material.

For electron irradiation, adjustable or interchangeable beam limiting devices shall be also provided. They shall attenuate the electron radiation, including X-radiation generated by electrons incident on the beam limiting device and other parts of the radiation head, so that the following limits apply. The absorbed dose in a plane perpendicular to the axis of the beam at the normal treatment distance shall not exceed:

(a) an average of 2% of the absorbed dose on the axis of the beam at NTD. This limit shall apply in the area between a line 4 cm outside the 50% isodose contour and the periphery of the maximum available geometrical field size;

(b) a maximum of 10% of the absorbed dose on the axis of the beam at the NTD. This limit shall apply in the area between a line 2 cm outside the 50% isodose contour and the periphery.

For the equipment in which the maximum geometrical field size exceeds 500cm^2 at the NTD (large square fields), the product of the average absorbed dose due to leakage radiation through the beam limiting devices and the area shielded by the beam limiting devices shall not exceed one-tenth of the product of the maximum absorbed dose on the axis of the beam and the area of the useful beam for a field size of 10 cm x 10 cm. A graph of average percentage leakage radiation versus maximum field size is shown in Fig. 3.70c.

These recommendations concern leakage radiation through collimator jaws (cf. Fig. 3.25). This problem is of particularly great importance when small fields are irradiated with accelerators with large maximum field size. For example, if the leakage through the collimation jaws is 5%, then for irradiating a 5 cm rectangular field in an accelerator with a maximum field of 35 cm, the treatment dose received by the patient would amount only to some 30% of the total dose due to the leakage through the jaws. The total dose is meant here as a product of the dose and the irradiated volume. That is why the above mentioned IEC recommendations provide that the dose due to the leakage through the jaws should not exceed 2% of the maximum therapeutic dose.

(2) The equipment shall be provided with a protective radiation shielding which will attenuate the radiation so that the fol-

lowing conditions are satisfied. In a plane circular surface of radius 2 cm centered on and perpendicular to the axis of the beam at the TNT (Fig. 3.69) and outside the area of the maximum useful beam, the absorbed dose, due to leakage radiation (excluding neutron radiation) shall not exceed a maximum of 0.2% and an average of 0.1% of the absorbed dose measured at the point of intersection of the axis of the beam and the plane surface. The values refer to measurements which have already been averaged over 100 cm^2. Except in the area defined above, the absorbed dose due to leakage radiation (excluding neutron radiation at 1 m from the path of the electrons between the electron gun and the target or electron window (Fig. 3.69) shall not exceed 0.5% of the maximum absorbed dose on the axis of the beam at the NTD.

(3) The levels for *doses from neutrons* (Sec. 3.1.6.4) have been established separately. The absorbed dose due to neutrons outside the maximum useful beam measured over the plane defined in (2) (see also Fig. 3.69) shall not exceed a maximum of 0.05% and an average of 0.02% of the absorbed dose in the useful beam measured at the point of intersection of the axis of the beam and the plane surface. These values refer to measurements which have already been averaged over 200 cm^2. Except in the area defined above, the absorbed dose due to neutrons measured at 1 m from the path of the electrons between the electron gun and the target or electron window shall not exceed 0.05% of the maximum absorbed dose on the axis of the beam at the NTD.

Radiation safety for persons other than patients. After IEC Standard 602-2-1 it shall be possible to lock the accelerator into the *stand-by state* (for definitions see Table 3.4). When the interlock is removed the accelerator is in the *preparatory state*. The preparatory state shall be indicated at the treatment control panel. A special circuit shall be provided to allow external interlocks to be connected, for example, for room doors. There shall be an indication at the treatment control panel that this special circuit is open or closed. This special circuit together with the interlocks shall be combined in a *system of interlocks*. Irradiation shall not be possible unless the system of interlocks is satisfied and all selection procedures have been completed. When these conditions are satisfied, the equipment is in the *ready state*.

The ready state shall be indicated at the treatment control panel and it shall be possible to transmit this indication to other locations. An audible indication of the ready state shall be given at the accelerator in the treatment room. Access to electrical circuits shall be provided so that it is possible to interlock the equipment from other locations against progress to the ready state, for the protection of persons in the controlled areas (Table 3.4).

In the stand-by and preparatory states the absorbed dose rates, which may arise from, for example, radioactivity originally present or induced in the equipment, or from field emission, shall not exceed during a period greater than 10 s immediately following termination of irradiation a value of:

0.2 mGy per hour at any readily accessible place 5 cm from the surface of the equipment;

0.02 mGy per hour at 1 m from the surface.

Measurements shall be averaged over an area up to but not exceeding $10cm^2$ at a distance of 5 cm from the surface or 100 cm^2 at a distance of 1 m from the surface.

The American recommendations differ from those of the IEC by being more stringent about the stray radiation both in the patient's plane and outside it, whereas Canadian recommendations are more stringent about the permissible level of the leakage radiation through the collimation jaws [3.63].

3.1.6.6 Quality Assurance

Introduction. Periodic quality control, usually referred to as *quality assurance* (QA) is an important element of maintenance of a radiotherapy accelerator and is essential in ensuring high standard of operation in accordance with the requirements presented in Sec. 3.1.1.

It is clear that the entire safety and quality assurance (QA) for radiotherapy includes all links of a long chain connecting [3.113]:

— the tumor diagnostics of a patient (cancer comprises a group of 50 to 150 diseases, the number depending on the degree of diagnostic refinement),

— the decision for radiotherapy (alone or combined with other treatment methods),

— the localization of the target volume and the relevant organs at risk,

— the choice of an adequate radiotherapy equipment and treatment method,

— radiation safety aspects (for the patient and employees) of the equipment,

— functional performance of radiotherapy equipment including test procedures, clinical dosimetry and treatment planning,

— simulation of the treatment,

— optimization of the medical prescription (target absorbed dose, spatial absorbed dose distribution, fractionation scheme etc.), influenced by individual radiobiological parameters of the patient,

— the patient adjustment and realization of the treatment,

— the verification procedures for treatment planning and performing,

— the patients examination accompanying the treatment and correlated possible changes of the treatment parameters (i.e. shrinking field technique) and

— the consequent and accurate follow-up analysis.

Obviously, a comprehensive QA program in radiation therapy has also both clinical and physical components but, because this book deals with radiotherapy accelerators, only the physical and technical aspects will be addressed here, and a small but important part of the above chain will be mentioned.

Quality assurance in radiation oncology may be defined as those procedures that ensure a consistent and safe fulfilment of the dose prescription to the target volume, with minimal dose to normal tissues (Sec. 1.1.2, Fig. 1.6) and minimal exposure to personnel [3.114]. The most significant factors affecting the ability to deliver the correct tumor dose (Sec. 3.1.1.1) include: exact dose calibration, accurately determined depth dose, off-axis dose characteristics, wedge and block factors, etc., that will be obtained during commissioning of a radiotherapy accelerator. In addition, certain other data will be obtained during an *acceptance test* for such machines, including measurements of the mechanical, radiation-beam and radiation protection specifications. A good QA program will, therefore, monitor each of these parameters in order to ensure that the machine is performing within specifications and that the dose is being accurately delivered. The protocol for these QA tests should recommend the equipment to be used, the frequency of measurement, the techniques to be followed, and suggested performance criteria. Obviously, the equipment and techniques should be as simple as possible in order to reduce the time effort and cost of such test. The frequency is chosen depending upon the likelihood of a change and the effect upon treatment if a change should unexpectedly occur. Finally, action and notification levels must be established [3.114].

Quality assurance in radiation oncology has lately been defined also in terms of the 6 P's (Rush-Presbyterian — St. Luke's Medical Center, Chicago [3.147]). The first P is the Prescription for the patient, as prescribed by the radiation oncologist, and checked by the clinical physicist. The prescription contains total dose, daily fraction, number of fraction per week, total number of days, and NSD or TDF. A uniform nomenclature should be used to indicate where the dose is to be administered. The second P symbolizes the Physical Dosimetry of the radiation beam (Sec. 4.1). This is performed by the physicist. It is essential that accuracy and consistency be maintained in the hospital treatment beam with regard to output, beam energy, beam flatness, optical distance indicator, and applicability of the inverse square law when using extended treatment distance. A Lucite phantom (Sec. 4.1.3) can be used to measure doses at 5 cm and 12 cm deep along the beam axis and to measure the energy. The phantom can also be used to measure the output of electrons using the 1.5 cm and 3.0 cm depth. Output check intervals are weekly, biweekly or sometimes daily. The beam flatness of the different beam is checked using film dosimetry. The obtained results [3.147] indicate that the electron output fluctuates more than the photon output and sometimes more by 5% of the nominal value. The third P, Planning of treatment for the patient (Sec. 4.2), is again accomplished by team effort only. The end results of this planning lead to calculation of the treatment time or monitor units to deliver the daily dose for the treatment. The chart of every patient receiving radiotherapy is double-checked always within a couple of days of initiation of treatment. Checking accuracy of all physical parameters, verifying all computations, and rectifying any existing errors is the clinical physics responsibility. In St. Luke's Medical Center experience 5% of the patient charts checked over the last two decades comprising over 20,000 patients contained errors of greater than 5% in calculated daily dose, treatment time, or monitor units.

The fourth P is the Precision in dose delivery. The study of monitoring the patient set-up parameters using a computer shows that errors do occur and of the order of 3%. The fifth P is the Pattern of Treatment outcome. The sixth P is the Plan for the future.

International QA Standards. A good QA program for radiotherapy must be flexible, that is, it must account for various local requirements. Differences in national traditions for implementation of regulations and standards, in organization of medical services, in financial resources and in the level of qualifications and science in diverse countries cause divergent interests. For this reason, despite international and national standardization institutions (Sec. 3.1.1.2), particularly those responsible for dosimetry protocols, as yet, there no internationally recognized QA protocol for radiotherapy, and the existing national recommendations vary from country to country (Table 3.10). However, numerous studies have been conducted for some time now on the international standardization of the above QA protocol. Table 3.7 lists the IEC nonmandatory guidelines on geometrical and dosimetrical tolerances for application of absorbed doses to patients by photon therapy with electron accelerators [3.113].

For some of the functional performance characteristics, e.g. isocentre sphere, through which the radiation beam axis passes for all gantry angles, the achievable accuracy of the test method is on the same order as the tolerance value to be measured. In the IEC 976 performance standard it was decided not to combine measuring uncertainty and tolerance range in one number in the hope of smaller uncertainties of measuring procedures in the future.

The root mean square of all geometrical partial tolerances add up to about ±4 mm (Table 3.7) and of all dosimetrical partial tolerances to about ±4% which is approximately comparable with dates which are in discussion between therapists as desirable values for medical reasons [3.113] (cf Sec. 3.1.1.1, 4.1 and 4.2). It should be reminded that the tolerance values of Table 3.7 are only those correlated to deviations of the functional performance characteristics and they do not include additional deviations caused by the other links of the therapy chain.

TABLE 3.7 Survey on Geometrical and Dosimetrical Tolerances for Application of Absorbed Doses to Patients by Photon Therapy with Electron Accelerators According to the (non-mandatory) Tolerance Values for Functional Performance Characteristics of the Guidelines, IEC Report 977 to the IEC Standard 976 (Table 3.3), (after J. Rassow [3.113]).

A. Affecting precision of isocentric positioning of a selected point in the target volume

Functional performance characteristics	Geometrical tolerance along field length X and field width Y	Geometrical tolerance along depth z
° Light field indication of the centre and the edges of the radiation field	±2mm	–
° Indication of the radiation beam axis	±2mm	–
° Radiation beam axis relative to the isocentre	±2mm	±2mm
° Indication of the position of the isocentre	±2mm	±2mm
° Indication of the distance from the isocentre along the radiation beam axis	–	±2mm
Root mean square	±4mm	±3.5mm

B. Affecting precision of planned absorbed dose delivered to a selected point in the target volume

Functional performance characteristics	Dosimetrical tolerance*	
Dose monitoring system		
° Reproducibility	±0.5%	
° Proportionality	±2.0%	
° Dependence on the angular position of the gantry and the beam limiting system	±1.5%	
° Stability throughout the day	±1.0%	
° Stability (from day to day) throughout the week	±1.0%	
Root mean square	±2.9%	±2.9%
° Stability of absorbed dose due to the stability of penetrative quality of ±1% (e.g. 6 MV X-rays at the standard measurement depth of 10 cm)		±0.4%
° Stability of field flatness with the angular position of the gantry and the beam limiting system		±1.5%
Root mean square		±3.3%

* The flatness tolerance of 6% (for radiation fields larger than 30cm x 30cm :10%) within the flattened area is to be added, if not included in the dose distribution within the treatment planning procedure. This is to be done in the form of a symmetrical tolerance value ±3% resp. ±5%.

For safety assurance, IEC standards are prepared for radiation, electrical and mechanical safety (Table 3.3 and Sec. 3.1.6.5), which are in updated form now combined in the newest draft-standard [3.115]. This standard contains definition of safety characteristisc, requirements with quantitative tolerance limits, test procedures and test conditions and a discrimination between the necessity of a type test and a site test, depending on the possibility of a change of this characteristics between individual machines of the same type. *Type test* is a test by the manufacturer of one or more devices or equipment made to a certain design to show that it complies with certain criteria. *Site test* is a test of each individual device or equipment after installation to establish compliance with certain criteria [3.115].

The complexity of technical constructions makes it impossible to find in all cases test procedures general and concise enough to be applicable to all types of equipment. Some fault conditions are disastrous, e.g. absorbed dose rate excess, and can only be simulated for testing the related safety device. Therefore, test grades are defined as follows [3.113]:

— Inspection of the analysis of equipment designs as related to the specified radiation safety provisions.

— Visual inspection or functional test or measurement of the equipment. The test procedures are specified in the standard and based on operating and fault condition states achievable without interference with the circuitry or construction of the equipment.

— Functional test or measurement of the equipment. The test procedures are specified by the manufacturer in accordance with the principle stated in the standard and may require interference with the circuitry or construction of the equipment [3.113].

Caused by a disastrous accident the new version of the IEC safety standard [3.115] contains an additional new requirement that under any fault conditions of absorbed dose rate excess the excess absorbed dose at the normal treatment distance shall be less than 4 Gy. That means for equipment capable of both X-radiation and electron radiation that the shutting-off of the radiation beam will need to be completed within a time of the order of one millisecond resp. before the next pulse of radiation.

Safety and QA for radiotherapeutical equipment is a very important part of optimizing the radiotherapeutical results without accidents, but one should not neglect that many other links in a long chain mentioned above have also crucial influence on the final results (cf Sec. 4.1 and 4.2), and physical and technical aspects should not be over-estimated only because they are comparatively easily measurable and/or because they are a comfortable excuse for idleness in other fields [3.113].

Individual QA Protocols. The following two US programs adapted to local conditions provide a good example of specific QA protocols. They have been based on the long experience and legally effective guidelines [3.15, 3.116], the first one having been developed and implemented at the Mallincrodt Institute of Radiology, Washington University School of Medicine [3.64 and 3.65]. All the control procedures were divided in four different groups depending on their frequency:

(1) Daily checks, performed each morning by the therapy technician who operates the machine. These checks include : (a) machine operating parameters, (b) beam output stability, (c) beam symmetry by means of a meter built in the machine, (d) laser isocenter indicator (visual check), (e) optical distance indicator (visual check), and (f) audio-visual patient-operator communication. Depending on the accelerator type and on the heating up period, etc., the above procedures require 10 to 30 min [3.64].

(2) Weekly checks performed during lunch hour or in the evening by a dosimetry technician or a physicist. These checks include: (a) congruence of the radiation field with the optical indicator field by examining a radiograph, (b) electron beam output stability, (c) electron beam bending magnet parameters settings.

(3) Monthly checks performed by a dosimetry technician or physicist on specific Friday afternoons reserved for a QA program. These checks include: (a) visual inspection of the mechanical parts of the accelerator including the blocking tray assembly and treatment aids, (b) test of safety devices, (c) optical range finder, (d) mechanical and digital indicators, (e) congruence of the radiation field with the optical indicator field by densitometry, (f) isocenter, (g) beam symmetry, (h) dosimetry system stability and its rad/MU scale, (i) beam energy (percent dose).

(4) Annual calibration performed on a weekend by a qualified medical physicist. These tests include: (a) alignment of

the therapy beam, localizing optical system and collimator, (b) dose monitor linearity, (c) dose monitor end of scale effect, (d) beam uniformity as a function of gantry angle, (e) dosimetry system calibration (rad/MU), (f) field size dependence, (g) dependence of dose calibration on accelerator orientation, (h)

TABLE 3.8 Daily, Weekly, Monthly and Yearly Quality Assurance Checks (after P.R. Almond and B.E. Bjarngard [3.114]).

Daily

Gauges

Interlocks

Lamp tests

Sidelights

Optical Distance Indicator (ODI)

Output Constancy check for fixed beam and arc where applicable

Weekly

Flatness/Symmetry with Film

Light-Radiation Field Congruence

Cross-Hair Alignment

Monthly

Emergency Off Switches and Interlocks

Gantry and Collimator Angle Indicators

Field Size Indicators

Energy/Depth-Dose

Mechanical Distance Indicator

Annually

Calibration at the gantry angle

% Depth-dose for several field sizes

Dose-rate for one field size

Output factors

Check inverse square law

Wedge factors

Tray factors

Monitor linearity and end effect

Flatness/symmetry in water for several field sizes and with film

Output check and Flatness/Symmetry Check at 0°, 90°, 180° and 270°

Machine Alignment

Focal spot position

Jaw symmetry

Definition of isocenter by intersection of axes of collimator, gantry and couch

Stability of gantry arm and bearing under rotation

Couch rotation and table top sag

depth dose factors in relation to tissue-phantom ratios, (i) off-axis field parameters, (j) wedge filters transmission, and (k) blocking filters attenuation.

The results of the above checks are recorded in a special notebook which subsequently serves as a valuable resource to estimate the operation of each device separately.

A QA protocol, generally regarded as one of the best US programs, has been presented by J.L. Horton in his textbook [3.117]. Table 3.8 lists the daily, weekly, monthly and yearly QA checks recommended by Horton, ACPM [3.116] and AAPM [3.15] [see 3.114]. It should be pointed out that some flexibility should be applied to this table.

For example, Horton recommends that most of the tests listed in Table 3.8 as annual checks, be done quarterly, i.e. every 3 months. A policy for the precise frequency at which each test is done should be developed at each institution. This table should be considered, therefore, only as a recomendation of a reasonable QA program.

Table 3.9 adopted from the ACMP [3.116] summarizes these tests. This table not only includes the parameter under investigation but also the main reason for making the test and the suggested performance criteria. If the measurements fall outside the criteria, the parameter should be adjusted so as to come into compliance If the parameter is significantly outside the criteria and/or is impossible to adjust to the required value, it will be necessary to call the authorized manufacturers' service to make the required adjustments, unless in-house expertise is available.

This in-house expertise will generally consist of the medical physicist and/or hospital engineer. With a good preventive maintenance program in which the parameters are measured and adjusted on a regular basis as outlined here, linear accelerator can be kept running in good operating condition. These tests and adjustments are generally simple to learn and easy to implement. When such a program is neglected and problems are not fixed as they arise, the machines can deteriorate, and authorized manufacturers' service will be required, which is often difficult to obtain.

A thorough discussion of the problems of the quality management model for radiation oncology physics has recently been presented in [3.134].

TABLE 3.9 Quality Assurance Tests (after P.R. Almond and B.E. Bjarngard [3.114])

Parameter	Reasons for making the observation	Suggested performance criteria
External Beam Machines		
1. Dosimetry		
a) Central axis dose calibrations (A)	Calibration of the radiation beam using the local standard beam	The stated accuracy is ±3.4% for beams other than Co-60, ±2.5% for Co-60 (6)
2. Photon Only Electrical Machines		
a) Constancy checks for megavoltage machine dose per monitor unit along the central axis (D)	Checks for the possibility of change in calibration due to some change in the electronics of the unit	±3% of calibrated dose (1)
b) Constancy checks for conventional machine dose per unit time (W)	Checks for the possibility of change in calibration due to some change in the electronics of the unit.	±3% of calibrated values (1)
c) Depth dose (M)	Checks for changes in the beam characteristics.	±2% of calibrated values (1)
d) Beam uniformity (W)	Checks for changes in the beam characteristics.	±3% of calibrated values (1)
3. Electron Beam Equipment		
a) Dose calibration (A)	Calibration of the electron beam using the local standard system.	±4% (1)
b) Beam uniformity (W)	Check for changes in the beam.	±5% (1)
c) Depth dose (M)	Check for changes in the beam.	±3 mm at 80% (1)
d) Dose per monitor unit constancy check (D)	Check for changes in the linac or beam.	±3% (1)
e) Dosimetry reproducibility and linearity (A)	Check for changes.	Document
4. Geometry		
a) Field positioning aids		
1) Light beam and radiation field (W)	Check for coincidence.	Within 3 mm (1)
2) Mechanical distance indicator, lasers, SSD lights (M)	Change in alignment aid affects dose to patient.	±2 mm variation from correct distance (1)

3) Scale readouts (M)	Check for any change in readout	±1 unit (1)
b) Machine Alignment		
1) Focal spot position (A)	Check for any change.	Document
2) Jaw symmetry (A)	Check for any change.	Document
3) Definition of isocenter by interaction of axes of collimator, gantry and couch (A)	Check for any change	±2 mm (1)
4) Stability of gantry arm and bearing under rotation (A)	Check for any change.	±2 mm (1)
5) Couch rotation and table top sag (A)	Check for any changes.	Document
5. Treatment Accesories (A)		
a) Wedges and compensators	Check basic parameters.	Document
b) Field shaping blocks (A)	Check basic parameters.	Document
6. Emergency Off System (M)	Check basic parameters.	Document

(1) AAPM Report No. 13 [3.15], (6) SCRAD Report,
(A) Annually (M) Monthly (W) Weekly (D) Daily

Differences of QA Protocols. Table 3.10 shows some recommendations included in various protocols [3.69] of control frequencies for various parameters of accelerator dosimetry systems. Each QA program depends on the use of a given set of recommendations. For example, if the Varian Clinac 1800 accelerator were to be checked according to the IEC recommendations, 3100 individual checks would have to be made annually, whereas using the German norm DIN 6847/5 the number of such checks would only be 1900 per year [3.70].

In this situation, the QA programs of a given accelerator model will vary from installation to installation. In developing programs, the users employ recommendations from one of the Dosimetry Protocols [3.71 through 3.73]. Table 3.11 lists the requirements for radiation therapy QA checks developed in Finland on the basis of the Nordic Association of Clinical Physics (NACP) Protocol which does not specify any individual control regulations, rather it provides some principles and recommendations. The program and QA criteria included in

TABLE 3.10 Frequency of QA Checks as They Affect Dose Monitoring Systems (after T.J. Godden [3.69])

Test	ISPM	AAPM	NACP	ICRU	IEC
Output calibration					
X-ray	D	D	W		W
electron	D	W	W	D-W	
Output stability					
during day			3M		M
moving beam therapy			3M		M
Reproducibility (X + e)		M	M	3M	6M
Proportionality (X + e)		M	M	3M	6M
Output variation with					
gantry position	M		3M		6M
gantry rotation (arc)	M	12M			6M
Electron applicator factors	3M			6M	

(D = daily, W = weekly, M = monthly, 3M = three monthly, etc.)

Table 3.11 are now being implemented by all nine radiotherapy centers in Finland [3.74].

Figs. 3.71a and b show the results of measurements of dosimetry stability of two multi-energy X-ray accelerators with energies of 6 and 24 MV and 10 and 18 MV respectively. Both accelerators were calibrated in agreement with the recommendations included in the AAPM Protocol (Report No. 13, 1984 [3.15]), according to which the permissible dosimetry calibration error may be 2%, including the calibration error of the dosimetric chamber and the instability of the accelerator proper. It follows from the measurements in Fig. 3.71 that the calibration error is mainly dependent on the stability of accelerator operation. The mean calibration values may be slightly different from the expected values for which the machine has been calibrated. The systematic difference of this kind may occur because there exists a calibration drift in the accelerator. The changes shown in Figs 3.71a and b, however, lie within the range set by a QA criterium.

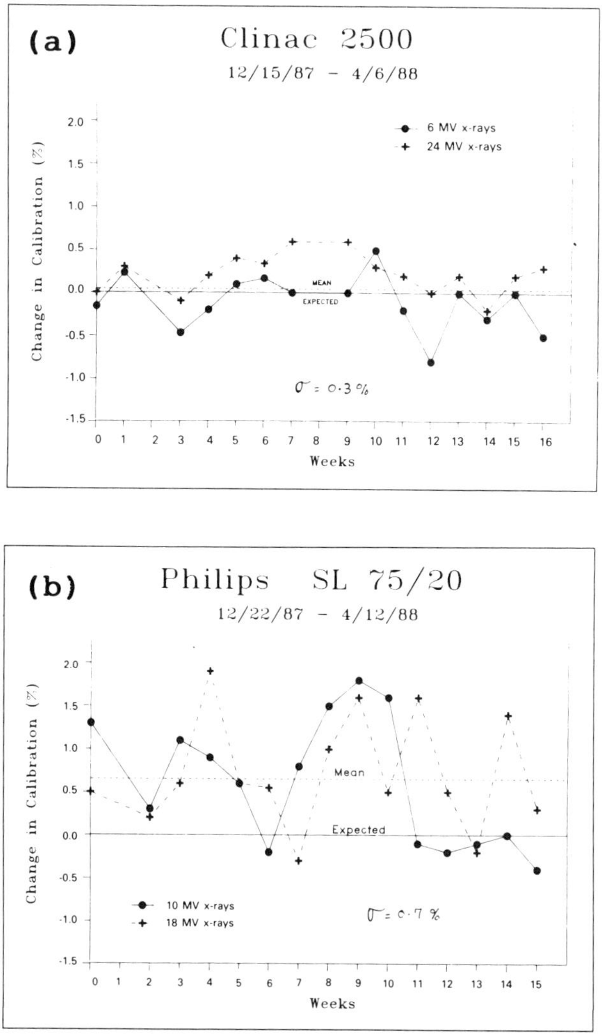

FIGURE 3.71 Weekly change in dosimetry calibration for multi-energy accelerator: (a) — Varian Clinac 2500, (b) — Philips SL 75/20 (after F.M.Khan [3.73]).

TABLE 3.11 NACP Recommendations for QA Checks Used in Finland (after P.Ruotsalainen [3.74]).

Check	Frequency	Accepted Range (±)
Relative dose	1 day	3%
Agreement of light field & display	1 day	3 mm
Agreement of optical &mech. SSD	1 day	2 mm
Agreement of light & rad. field	1 week	2 mm
Flatness & symmetry of rad. field	1 week	3%
Constancy of energy	1 week	X 1% e 4%
Agreement of optical & mech. axis	1 month	2 mm
Scales and displays	1 month	
Absolute dose	6 months	2%
Depth dose curves, profiles	6 months	
Penumbras	6 months	< 8 mm
Factors of dose monitor (stability, linearity, dependence on gantry angle)	1 year	
Dose dependence on field size, wedges	1 year	1%
Stability of rad & mech. isocenter	1 year	2 mm
Operation of locks, limits, lights & emergency system	1 day - 1 year	

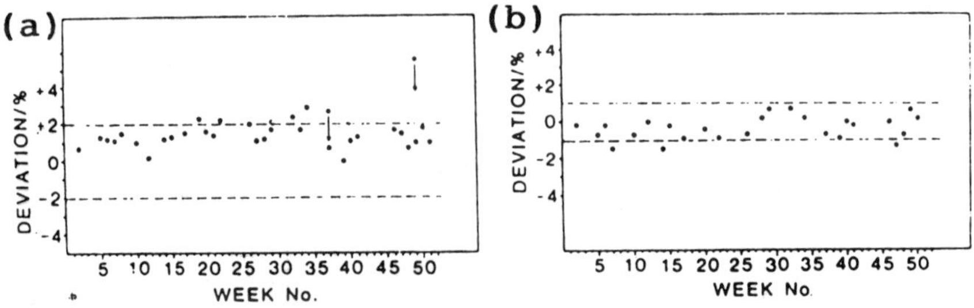

FIGURE 3.72 Calibration stability for: (a) — dose monitor, (b) — beam energy (Varian Clinac 4/100) (after O. Mattson [3.71]).

Fig. 3.72 represents the results of stability measurements of a dose monitor (Fig. 3.72a) and energy stability (Fig. 3.72b) carried out in accordance with the NACP Protocol for the Varian

Clinac 4/100 accelerator. According to NACP, the monitor calibration stability should be better than 2%. The arrows in Fig. 3.72a indicate that recalibration with a potentiometer was necessary in weeks 37 and 49. The deviation in week 49 was caused by a failure in the power supply. The results of the energy stability indicate that, except for some values, the results of checkup measurements lie within the normal range of stability of ±1%, which conforms to the NACP Protocol.

A separate problem is that of commissioning measurements carried out when the accelerator is set in operation after installation. These measurements usually cover a 6 month period. Their description is given in [3.66]. Detailed QA programs are also used when new types of accelerators are put on the market. New machines are carefully examined at one of the well-known medical centers for promotion purposes [3.67 and 3.68].

3.1.7 Survey of Commercial Radiotherapy Accelerators

Fundamental parameters of some of the radiotherapy rf linear accelerators available on the world market in 1992 are presented in Table 3.12. The first group of machines comprises accelerators with energies of 4 to 6 MeV. They are generally intended solely for X-ray therapy, and do not incorporate electron beam extraction; the only exception is the Siemens M 6740 machine, in which, in addition to a 6 MV photon beam, 5 to 12 MeV electron beams can also be employed.

The basic group of accelerators is made up of multi-energy machines, that is those which make it possible to use photons of at least two energies and electrons with 4 to 7 different energies during a single treatment (Sec. 3.1.2.5 and Table 3.5).

The dose rate is usually controlled over a wide range and in accelerators with lower energies, e.g. 4 to 6 MeV, its maximum value for an X-ray beam ranges from 200 to 300 cGy min^{-1}. In higher energy accelerators, the upper limit of the X-ray dose rate is as high as 400 or even 500 cGy min^{-1}. On the other hand, for an accelerator intended only for producing electron beams, such as Siemens ME, the dose rate upper limit is 900 cGy min^{-1}.

The maximum size of the photon fields, 40x40 cm^2, is standardized in all types of accelerators. The field size of a treatment field is determined with pre-collimators and collimating diaphragms (cf. Sec. 3.1.4.4). As for electron fields, their maximum dimensions are usually slightly smaller and do not exceed 20x20 cm^2 or 25x25 cm^2. The dimensions and shapes of individual fields are determined by the parameters of applicators used (Sec.3.1.4.4).

Typically, the rotation angle of the treatment head is 370°, that is irradiation can be carried out with a certain margin or overlap. Among the accelerators listed in Table 3.12 only the Siemens ME machine features a limited rotational movement about the vertical axis of up to ±115°. It is specially intended for intraoperative electron therapy. Maximum rotation values for all types are 1.0 rpm (rev min^{-1}) and minimum values are 0.1 or 0.2 rpm. In some designs, the speed is controlled continuously up from the value close to zero.

At present, maximum energies in any of the high-energy accelerators do not exceed 25 MeV, although in the 1970s machines were manufactured in which the maximum energy was as high as 50 MeV.

Thanks to the high dose rate of both electron and X-ray beams, the linear accelerator throughput can be 50-70 patients daily or even more. The throughput is, of course, not only dependent on the type of radiotherapy procedures, but also on the kind of tumors irradiated.

3.2 Betatrons

The principle of betatron operation has been discussed in Sec. 2.5 (Figs. 2.30 and 2.31). At the time when betatrons were most popular, that is in the early 1970s, there were about 200 accelerators of this type the world over, mostly in the West-European countries and in Scandinavia. Due to the reasons given in Secs. 1. 1 and 2.5, betatrons ceased to be produced in the second half of the 1970s. However, they were very durable and some of them are still in operation.

Over the course of betatron development ever increasing energies were being achieved. At the beginning, the betatron upper energy was that of 25 MeV (Allis-Chalmers). It was, however, unsuitable for deep therapy except for the irradiation of tumors in the region of the neck. Therefore, in the early 1960s, production was started on betatrons with a maximum energy of 35 MeV (Asklepitron 35 MeV, Brown Boveri). In 1965, the betatron maximum energy was already 42 MeV

TABLE 3.12 Basic Parameters of Some Radiotherapy rf Linear Accelerators

	TYPE	RADIATION		ENERGY, MeV		DOSE RATE X-RAY $cGy\ min^{-1}$	X-RAY FIELD SIZE, cm	HEAD ROTATION ANGLE	HEAD ROTATION SPEED, rpm	SAD cm
		X	e⁻	X	e⁻					
Varian	Clinac 4	•	–	4	–	250	0x0 to 40x40	360°	0.1 to 1.0	100
Varian	Clinac 6	•	–	6	–	70 to 200	0x0 to 40x40	360°	0.1 to 1.0	100
BBC Brown Boveri	Dynaray D 4/6	•	–	4 to 6			max 40x40			
Siemens	Typ M6300	•	–	4	–	250	max 40x40	370°	0.2 to 1.0	100
Siemens	Typ M6700	•	–	6		300	max 40x40	370°	0.2 to 1.0	100
Siemens	Typ M6740	•	•	6	5 to 12	250	max 40x40	370°	0.2 to 1.0	100
Varian	Clinac 1800	•	•	6+10[1)]	6 to 20	400				
Varian	Clinac 2500	•	•	6+24	6 to 22	100 to 500	0x0 to 40x40	365°	0.2 to 1.0	100
Siemens	ME[2)]	–	•	–	6 to 18	300 to 900[(e−)]	3)	±115°	0.2 to 1.0	100
Siemens	MD 63-6700	•	–	4+ 6	–	250[4)]	0x0 to 40x40	370°	0.2 to 1.0	100
Siemens	KD S 67-7860	•	•	6+18	6 to 18	300[4)]	0x0 to 40x40	370°	0.2 to 1.0	100
Siemens	KD 67-8067	•	•	6+23	6 to 21	300[4)]	0x0 to 40x40	370°	0.2 to 1.0	100
Philips	SL 18	•	•	6+15	4 to 18	25 to 400	max 40x40	370°	0 to 1.0	100
Philips	SL 25	•	•	6+25	4 to 22	25 to 300	max 40x40	370°	0 to 1.0	100
Mitsubishi	ML - 4M	•	–	4	–	50 to 150[5)]	max 30x30	390°	0.5	80
Mitsubishi	ML - 20M	•	•	6+10	6 to 20	200 to 1000[5)]	max 35x35	390°	0.2 to 0.8	100
Jefremov	LUER-20M	•	•	6+18	5 to 20	60 to 300[6)]	2x2 to 40x40	360°	0.1 to 1.0	100
Jefremov	LUER-40M	•	•	8+18	5 to 40	150 to 300	2x2 to 40x40	360°	0.1 to 1.0	100

1) 15 or 18 MV; 2) dedicated for intraoperative irradiations (Sec. 3.4); 3) depending on the electron applicator size; 4) for electron beam the dose rate is 300 $cGy\ min^{-1}$ (optionally to 900 $cGy\ min^{-1}$); 5) R min^{-1}; 6) for electron beam 60 to 500 $cGy\ min^{-1}$

TABLE 3.13 Principal Parameters of Some Radiotherapeutic Betatrons

Manufacturer, country	Type	Emergent radiation		Energy range, MeV	Dose rate, Gy/min		Irradiation field, cm	Angle of rotation
		e^-	X-rays		e^-	X-rays		
Allis-Chalmers, U.S.A.	25 RTM	•	•	10-25	6(1m, 25 MeV)	1.4 (1m, 25 MeV)		
Brown Boveri (BBC), Switzerland	Asklepitron 35	•	•	0-35	10	0.75 (1m)		±105°
Brown Boveri (BBC), Switzerland	Asklepitron 45	•	•	3-45	5 (1.1 m)	0.80 (1.1 m)	from 3x3 to 20x20	±120°
Techsnabexport, U.S.S.R.	B5M-35	•	•	7-25	5	0.50 (1 m)	from 20x20 (1 m)	±90° to 18x18
Zakladna rozvoje uranoveho prumyslu, Czechoslovakia	Chiratron	•	•	4-19	1.50 (65 cm)	0.25 (65 cm)		±120°
Siemens, Federal German Republic	18 MeV	•	•	4-18	10 (50 cm)	2 (50 cm)		
Siemens, Federal German Republic	42 MeV	•	•	5-42	15	3	from 3x3 to 20x20	±100°

(Siemens), and later 45 MeV betatrons were manufactured (Brown Boveri, Asklepitron 45). Table 3.13 lists the basic parameters of some of these machines.

All the betatrons in Table 3.13 were intended for both X-ray and electron beam production. The radiation dose rate was usually 25 to 140 R min^{-1}, that is it was much lower than that currently achieved in linear accelerators (Tab. 3.12).

In betatrons, the dose rate was strongly dependent on the dimensions of the treatment field, decreasing rapidly with decreasing field size. The weight of the betatron electromagnets increased approximately in proportion to E_{max}^3. For example, Siemens 18 MeV machine weighed 400 kg, the Siemens 42 MeV betatron weighed 6000 kg and the treatment head of the BBC Asklepitron 45 weighed 10t ! The heavy weight together with large dimensions made it difficult to design a movable treatment head essential for rotational (arc) therapy. Betatrons, therefore, were not able to use the full treatment head rotation about the axis of the patient, rather they employed a pendulum-like movement in the full maximum arc of ±100° (Tab. 3.13).

On the other hand, the reason why betatrons became quickly popular was the fact that they considerably surpassed units used in conventional X-ray therapy. For example, the 5-year survival rate in patients who had their urinary bladder tumor irradiated with 200 kV by X-ray therapy was 14%, whereas in X-ray therapy using a 35 MV betatron it was as high as 55%. Betatrons have played an important role in the development of radiation therapy, at present, however, they are only important from the historical point of view. The main reasons why they have been outdistanced by linear accelerators were their low dose rates, large dimensions and weight, and, hence difficult maneuverability.

3.3 Microtrons

The principle of operation of microtrons has been discussed in Sec. 2.8 (Fig. 2.43a and b). In addition to linacs, they constitute another type of radiotherapy accelerators produced commercially. Their prevalence is, however, smaller: they constitute about 1% of all the linear accelerators installed at present.

Microtrons used in radiation therapy are being produced by the Swedish company Scanditronix [F14]. Having developed an earlier version of the MM 10 type (10 MV photons and 6 to 10 MeV electrons), the current production is being focused on multi-energy accelerators (Tab. 3.5). The M22 type machine is a circular microtron (Fig. 3.73), equipped with a 2.2 m diameter magnet. The accelerating structure (resonator) is supplied with a 2MW magnetron at 3 GHz. The number of accelerating orbits may be as high as 42. In each orbit, the electrons gain an additional 0.535 MeV. They can be extracted from successive orbits 10 through 42; nine beams can be used, their energies ranging from 5 to 22 MeV. The beam extracted from the microtron is passed through a transport system to a rotating gantry with a treatment head comprised in one unit. In this way, one microtron can supply two gantries (Fig. 3.74). Provision is also made for the third beam which can be used for experimental purposes [3.75].

The MM 22 microtron has two therapeutic photon beams whose energies are within the range of 5.4 to 21 MV [3.76]. In addition, ten electron beam energies from approximately 3 to 22 MeV are also provided. The tenth energy is obtained by the energy degradation of the beam extracted from orbit 10 in the graphite-gold scattering foil. The maximum photon field is $37x40\ cm^2$, and the electron field is $20x25\ cm^2$ [100-cm source-

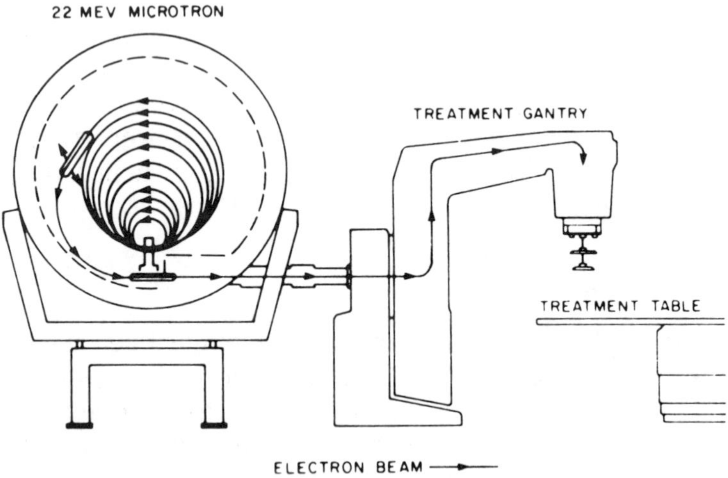

FIGURE 3.73 Schematic diagram of the MM 22 radiotherapy microtron (mfd. Scanditronix)

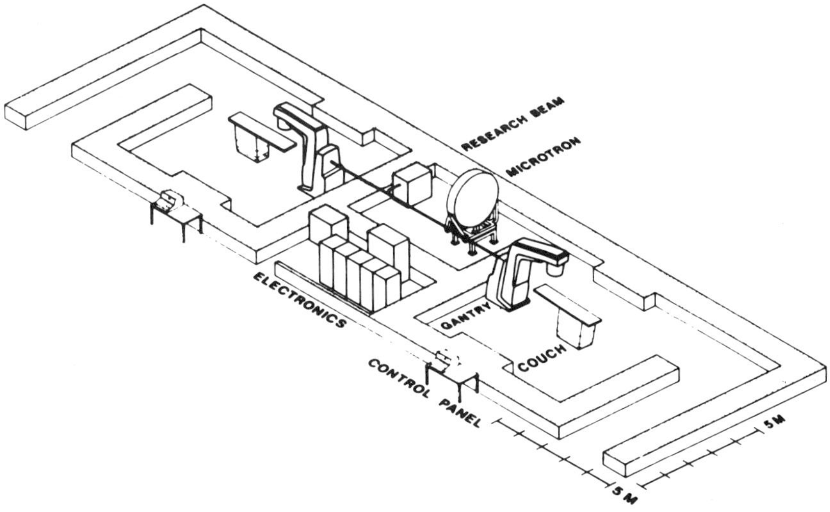

FIGURE 3.74 Radiation therapy unit equipped with a microtron supplying two rotational gantries (after A. Brahme and D. Reinstad [3.71]).

axis distance (SAD)]. The results of studies on the quality of X-ray beams are presented in [3.76] and those on the quality of electron beam are discussed in [3.77].

The MM 50 microtron utilizes a *recirculating accelerating structure* (Fig. 2.43b). Electrons are injected into a linear accelerator which accelerates them to 5 MeV. Two bending magnets, which change the beam direction by 180°, are mounted on both sides of the accelerating structure. Thus the beam can be passed through the same structure several times. At the output of the MM 50 microtron electrons are obtained whose energies range from 2.5 to 50 MeV (20 values) and the electron beam dose rate is regulated over the range of 0.5 to 5 Gy min^{-1}. The machine can also produce photon beams within the range of 5 to 50 MV (7 energy values) with the beam dose rate in the range equal to that of electrons. In order to obtain a flat field, a beam scanning system (Sec. 3.1.4.3) is employed. Thick flattening filters where the beam quality is degraded and which makes the adjustment of beam flatness very critical are thus avoided. The helium filled treatment head makes it possible to use very pure and unscattered electron beams. In the MM 22 microtron, scattering foils and flattening filters are used for the same purpose.

The treatment head in the MM 50 microtron is equipped with a multi-leaf collimator, placed in helium atmosphere, which consists of 64 independent elements. This collimator can therefore be used for both electrons and photons giving sharp beam collimation, except possibly for low energy electron where scattering in the air between the treatment head and patient surface may be too large. This computer-controlled system permits convenient shaping of irregular fields.

At present, the MM 50 microtron is a unique radiotherapy accelerator which makes it possible to produce beams with energies higher than 25 MeV. The reasons presented in Sec. 3.1.6.5, among others, explain why the value of 25 MeV is currently regarded as the upper limit of the routine radiotherapeutic range both for electron and X-ray beams (cf. Tab. 3.12).

A distinct advantage of microtrons is that their beams are monoenergetic. The beam energy spread (FWHM) is not greater than 0.1 MeV [3.77]. This has a good effect on dosimetric patterns of both electrons and photons [3.76 and 3.77]. Since the accelerating structure is located outside the treatment head and the gantry (Fig. 3.73), the head can be of a compact type. However, the area taken by the treatment room must be larger (Fig. 3.74).

3.4 Dedicated Systems

3.4.1 Accelerators for Intraoperative Radiation Therapy

The technique of *intraoperative radiotherapy* (IORT) makes it possible to directly treat with radiation tumors surgically exposed or regional lymph nodes. The radiation treatment is carried out after the tissues above the neoplasm have been removed and when the lesion itself is unresectable or only partially resectable. The general rationale for using IORT is to improve the therapeutic ratio of local control versus complications. This objective can be achieved by: (1) decreasing the volume of the irradiation boost field with precise visual localization of deep-seated tumors during operation, (2) excluding from the radiation field all or some of the dose limiting sensitive structures by operative mobilization, shielding, or the use of appropriate electron beam energies, and (3) therefore, increasing the effective dose to the tumor volume.

This technique was initiated over 25 years ago [3.78] and since then it has generated an ever increasing interest [3.79 through 3.83]. To irradiate an open surgical field, 6 to 20 MeV electron beams are usually employed at present since their sharply defined range, compared with bremsstrahlung radiation, ensures that the underlying tissues are protected. Physics aspects

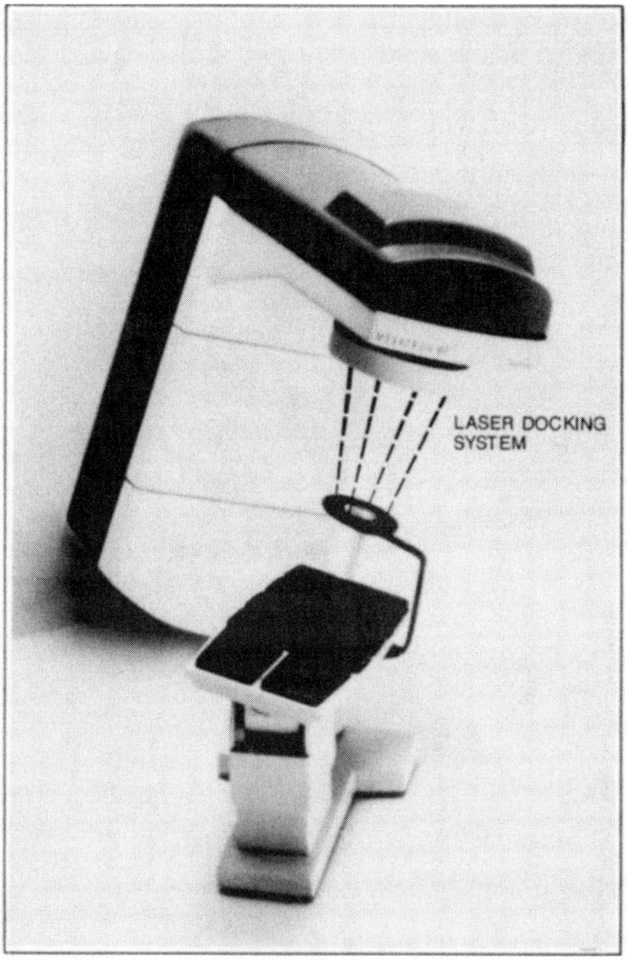

FIGURE 3.75 View of a Siemens ME electron accelerator for intraoperative radiotherapy.

of intraoperative radiotherapy are presented in [3.78, 3.84 and 3.85]. Typically, the doses used are about 1500 cGy [3.78 and 3.86].

Compared with conventional radiotherapy treatment, intraoperative procedures involve additional problems of sterility, radiation protection and patient positioning. In some medical centers, anesthesized patients are transferred from an operating room to the radiotherapy department and then transferred back to the surgical suite for closure of the body cavity (National Cancer Institute, Mayo Clinic and Massachusetts General Hospital). Another approach is to convert the radiotherapy room into an operating room (Howard University, Harvard) [3.80]. An idea has been also put forward of having the operating room adjacent to the therapy room and providing a lead door between them, which would enable the patient to be quickly moved without having to transport him over large distances. It should be noted that the patient is anesthesized and both the surgical and anesthesia teams have to leave the room during

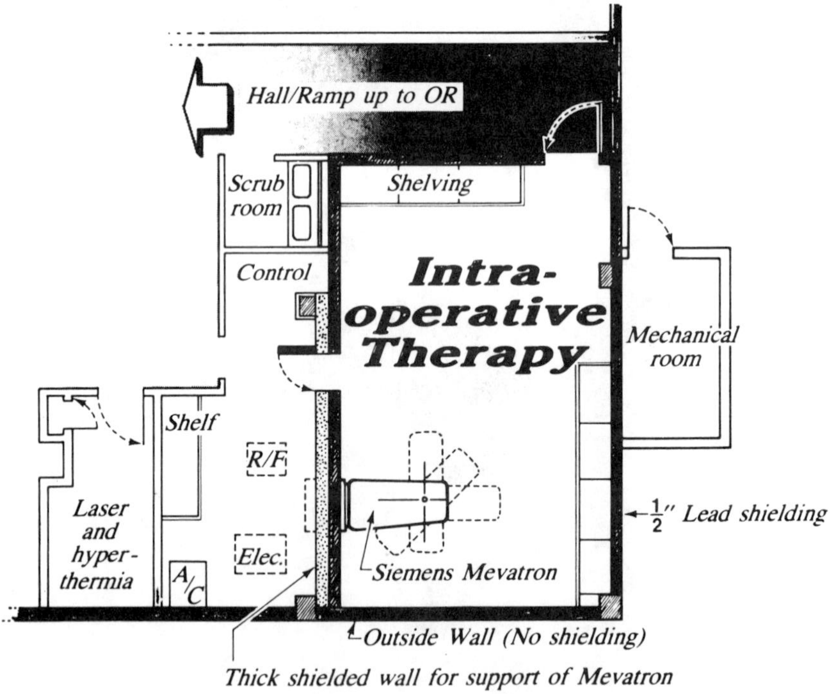

FIGURE 3.76 Placement of the Siemens ME accelerator for intraoperative radiation treatment in an operating room (University of Texas M.D. Anderson Hospital).

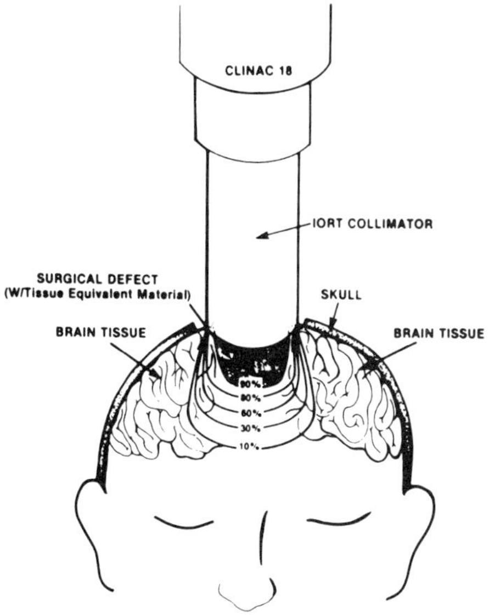

FIGURE 3.77 View of an intraoperative radiation therapy system for brain tumors used at Howard University Hospital (Washington, USA) (after A.L.Goldson and M.Y.Halyard-Richardson [3.86]).

treatment. Therefore, additional monitoring of the patient's condition (arterial blood pressure, respiration, etc.) by an anesthesiologist on a multichannel oscilloscope is required.

In view of the ever increasing interest in intraoperative radiotherapy, an accelerator has been recently developed which is intended for such purposes. The Siemens ME machine, shown in Fig. 3.75, is intended for direct installation in the operating room (Fig. 3.76). This design serves its purpose best in a dedicated facility because it eliminates the patient's transportation and is cost-effective (about $ 1.2 million).

The Siemens ME accelerator delivers five electron energies within the range of 6 to 18 MeV with a maximum field diameter of 15 cm. The dose rates selected can be 300, 600 or 900 cGy min^{-1}. Applicators and other field shaping elements can be autoclaved.

The accelerator operates in conjunction with a surgical table already shown in Fig. 3.54. A detailed description of the system is given in [3.87]. The facility situated at the M.D. Anderson Hospital (University of Texas) is discussed in [3.88].

Intraoperative radiation treatment is often used in patients with primary malignant tumors of the brain [3.86]. Fig. 3.77 shows a system used for such treatment at Howard University Hospital (HUH) in the USA. The Varian Clinac 18 accelerator is equipped with a collimator which permits direct irradiation of the brain tissue through a hole in the skull. It is seen from Fig. 3.77 that the collimator is in contact with the tissue. Part of the surgical defect is filled with saline-saturated gauze to act as a tissue equivalent material both to maintain dose homogeneity and to prevent the irradiation to deeper normal brain tissue. Typically, the electron energies used in cases of this kind are 9 to 18 MeV and the single intraoperative dose fractions are 1500 cGy at the dose rate of 500 cGy min^{-1} [3.86].

3.4.2 Total Body Irradiation Systems

Total body irradiation (TBI) as well as *half body irradiation* (HBI) or *total nodal irradiation* (TNI) has been used for over fifty years since it was initiated by Chaoul and Lange in 1923 [3.89]. The physics fundamentals of TBI are presented in [3.90], whereas dosimetric problems are discussed in [3.91 and 3.118]. Typical examples of TBI include irradiation of patients with acute lymphoblastic leukemia prior to bone marrow transplantation. Using the Seattle method, a single fraction dose of 10 Gy is used; the irradiation lasts for about 2.5 h (dose rate about 7 cGy min^{-1}) [3.92]. In preparation for bone transplants a single fraction radiation dose is 500 cGy and a dose rate is 50 cGy min^{-1}; multiple fractions amount to 1400 cGy [3.90].

The above examples require that the patient be irradiated with treatment fields as large as 50x200 cm^2 and relatively low dose rates. These requirements cannot usually be met by most of the existing radiotherapy accelerators installed in typical radiation therapy establishments. Even if the patient positioning is such that he is placed on a special table close to the floor (Fig. 3.6), fields larger than 70x70 cm cannot be achieved. For the fields required in TBI therapy the distance SAD should be from 2 to 4m. For such large values of SAD, the conventional radiotherapy accelerator should have its treatment head mounted horizontally, the dimensions of the treatment room permitting. Fig. 3.78 shows example of this design employed at the Zentrum Radiologie in Freiburg (FRG) [3.92]. Here, the patient is placed on the diagonal of the room with a 4.5 m side and the SAD is 3 m. The patient is positioned on his back or in a chair.

Among the serious difficulties involved in the TBI therapy is that of achieving dose distribution uniformity. The main reason why nonuniformities as great as $D_{max}:D_{min} = 1.5$ are encountered is that the body has an irregular shape and the tissues, e.g. the lungs, are inhomogeneous. The nonuniformities can be compensated for by using special 3D masks (compensators), designed and shaped from styrofoam with an automated cutter [3.93 and 3.119]. In order to protect critical regions of the body, e.g. the lungs, lead screens are used [3.92].

Total body irradiations are usually carried out by means of X-ray accelerators with energies within the range of 4 to 15 MV [3.94 through 3.96, 3.119].

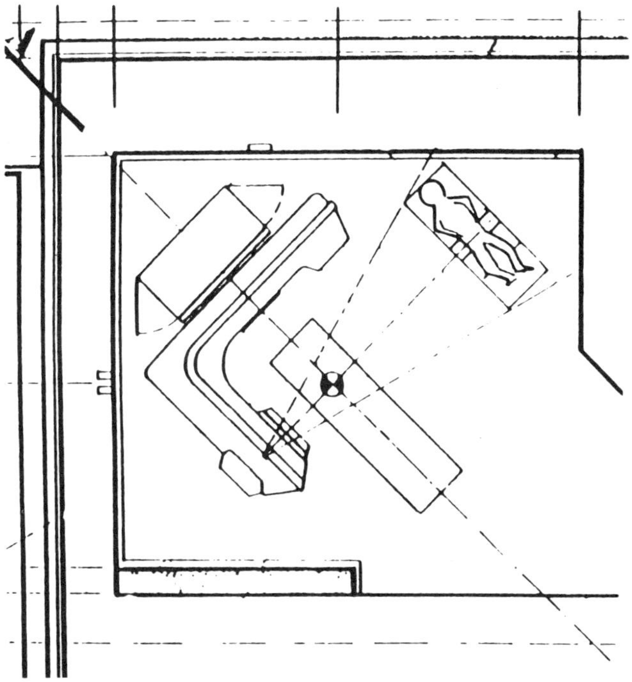

FIGURE 3.78 A system for total body irradiation using Varian Clinac 4, SSD=300cm, installed at the Zentrum Radiologie (Freiburg, FRG) (after K.Kuphal et al. [3.92]).

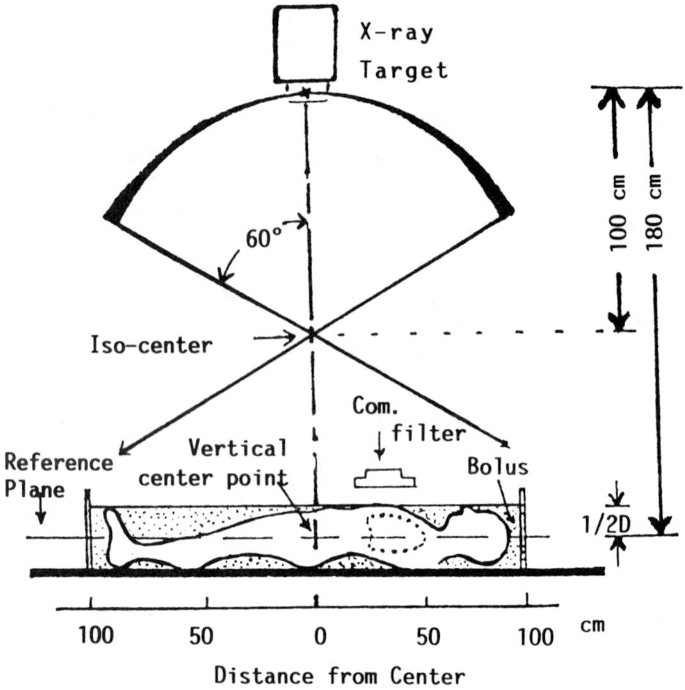

FIGURE 3.79 A system for total body irradiation using a pendulum beam, installed at the Tokyo Women's Medical College (Japan) (after H.Hirabayashi et al. [3.95]).

When large SSD distances of the order of several meters cannot be obtained, the design shown in Fig. 3.79 may be useful. It was developed at Tokyo Women's Medical College [3.95] and consists in placing the patient in a comfortable prone or supine position. The treatment head rotates through a pendulum angle of 60° relative to the vertical axis. Thus large field size of more than 60x200 cm^2 can be obtained and the beam moves over the entire length of the patient from head to foot and back several times during one treatment. The dose uniformity over the long axis of 200 cm is within 3%. It is also possible to use a mask to compensate for the variable body thickness and special filters which protect some selected regions like the lungs, for example, to keep the dose below 800 cGy [3.95]. Optimization of a compensator of this kind has been discussed in [3.120].

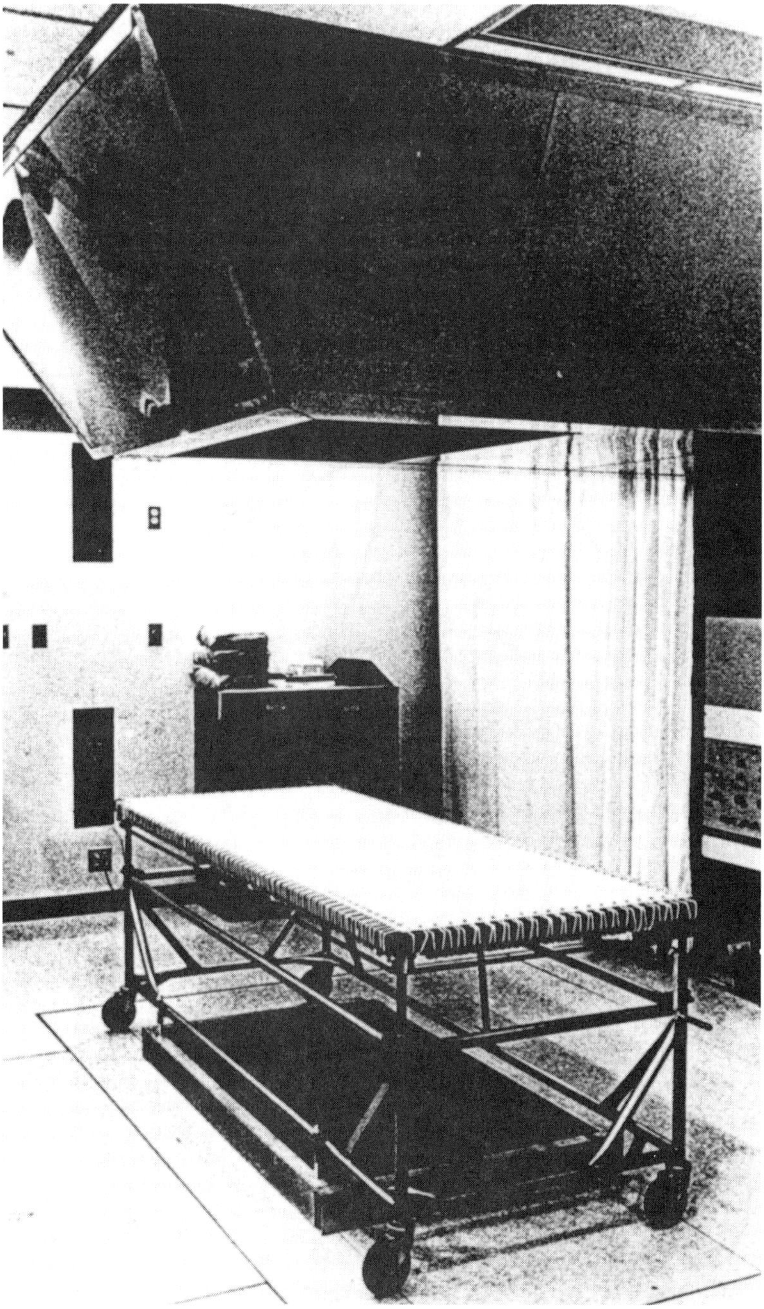

FIGURE 3.80 View of a total body irradiation facility at Harvard Medical School (USA) (after W.R.Lutz et al. [3.96]).

Recently, a dedicated facility has been developed at Harvard Medical School (USA), intended exclusively for total body irradiations of large-field treatments [3.96]. The view of the facility is shown in Fig. 3.80. Two old 4 MV linear accelerators have been utilized for this purpose. One of them was mounted on the floor, and the other was suspended from the ceiling in such a way that the maximum distance between the treatment heads was 410 cm. Both beams are, therefore, vertical. The patient is placed on a carriage with stretchers and the treatment field is 80x220 cm^2. The flattening filters ensure that the field uniformity is within $\pm 3\%$ over the area of 70x200 cm^2 (in air). The dose rate in the middle plane, at the distance of 205 cm from each head, may be controlled over the range between 0.05 and 0.8 Gy min^{-1}; each source being responsible for 50% of the preset value. The positions of both treatment heads in the vertical plane may be changed to ensure higher dose rates. Each head can also be moved over the range of 120 to 205 cm relative to the middle plane, that is the patient's plane. At the minimum distance of 120 cm each head produces beam dose rates of up to 1.1 Gy min^{-1} [3.96].

3.4.3 Systems for Radiosurgery

Radiosurgery is a focal brain irradiation technique which delivers a single high dose of the order of several thousand cGy to local intracranial structures or lesions with diameters of a few mm to a few cm [3.97 and 3.98]. The doses are usually delivered in a single fraction to a very small volume of 0.5 to 8 cm^3 [3.99]. Prior to irradiation, the region is carefully localized with techniques generally used in brain surgery (stereoencephalotomy) such as computed tomography (CT) or angiography. The technique of radiosurgery is used not only for treatment of small tumors, such as pituitary adenomas, but also in certain non-neoplastic conditions, such as inoperable arterio-venous malformations (AVM). The main feature of these techniques is the requirement of exceptionally high spatial accuracy of 1 mm and radiation volume dose accuracy of $\pm 2\%$. A rapid treatment field fall-off outside the target volume is also required.

Radiosurgery was introduced in 1951 by Leksell [3.97] who initially employed 200 to 300 kV_p X-ray beams, and later used cobalt sources. In the most recent design of this kind, a radiosurgery unit (Gamma knife) has 201 ^{60}Co sources, each of 1040 GBq distributed over a segment of hemisphere, the resulting beam being focused at a sharply defined focal point [3.100 and 3.101]. The disadvantage radioisotope dedicated units have is that they are relatively expensive (typical installation costs for building and equipment of about $ 5 million [3.97]) so their availability is low (Stockholm, Buenos Aires and Sheffield).

Radiosurgical treatment can also be implemented with proton or helium ion beams accelerated in accelerators such as synchrocyclotrons or isochronous cyclotrons (cf. Sec. 5.1.3.4). Radiosurgery with heavy charged particle beams was started in the late 1950s in Uppsala by Larsson et al. [3.121], in Berkeley by Lawrence et al. [3.122] and in Boston by Kjellberg et al. [3.123, 3.124] with proton beams from cyclotrons, and recently progressed to helium ion beams obtained from a 230 MeV synchrocyclotrons [3.125 and 3.126]. The technique is based on advantageous Bragg peak dose distributions from energetic heavy charged particles (Sec. 5.1), which, with multiple field irradiations (up to 12), give a sharp dose fall-off outside the target volume. However, its main disadvantage is the dependence on high-energy cyclotrons or synchrocyclotrons as a source of radiation, making it very expensive and therefore not generally available.

Both techniques, the one using focused cobalt beams and the other using heavy charged particles, are thus theoretically quite well suited for radiosurgery and have proven their effectiveness on several thousand patients by now [3.127]. However, they are not widely used because they are very expensive. The capital and operating costs for this type of equipment can hardly be justified by a single institution, which, even with a large catchment area, can at best treat only several dozen patients per year. For this reason, work has been going on for some time on the possibility of using popular and low-cost, medium energy range 4 to 10 MV, electron radiotherapy machines in radiosurgery [3.97 through 3.99, 3.102 through 3.105].

Fig. 3.81 shows Varian 6/100 accelerator adapted to radiosurgical treatment [3.99]. It produces a 6 MV X-ray beam collimated with a supplemental *external collimator.* Internal collimating diaphragms, mounted in the accelerator head (Sec. 3.1.4.4), are not suitable for fields which are small since a diaphragm placed at 70 cm from the isocenter (cf. Fig. 3.25)

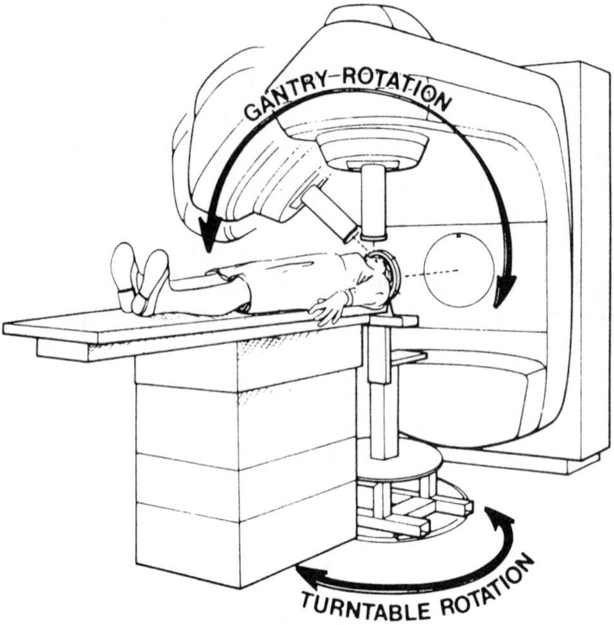

FIGURE 3.81 A system for radiosurgery using Varian 6/100 accelerator and a supplemental collimator, Harvard Medical School (after W.Lutz et al. [3.99]).

produces relatively large penumbras. In addition, using typical collimators it is difficult to accurately fix a small circular field.

The supplemental collimator, seen in Fig. 3.81, may be equipped with eight exchangeable collimation Cerrobend® inserts. The distal end of the collimator is 23 cm from the isocenter and the diameter of the circular collimator opening may vary from 12.5 to 30.0 mm as projected to isocenter.

The patient is positioned on a typical treatment table but his head is in the *Brown-Roberts-Wells (BRW) stereotactic system* for localization of the irradiated volume. The BRW frame is also used in computed tomography to develop a treatment plan. This target volume can then be irradiated from any direction with spatial accuracy within 2.4 mm. Better accuracy can be obtained with the angiography localization. For collimators with medium opening diameters the dose fall-off at 1.0 cm outside the irradiated field is lower than 20% of the target dose.

Radiosurgical techniques are characterized by the particular approach they take toward fulfilling the goal of delivering the prescribed dose to the target with a sharp dose fall-off outside the target volume (Fig. 3.84). Theoretically, the optimal dose distribution is attained if the beams are aimed toward the target center through a large number of surface points distributed uniformly over the upper hemisphere of the patient's skull. This distribution of beam entry points is called 2π geometry (Fig. 3.82a), because the beam-entry points are distributed essentially over a half sphere. In 2π geometry all beams except those in the transverse plane enter the upper hemisphere and exit the lower hemisphere. The beams in the transverse plane enter and exit the same plane, resulting in a large number of parallel-opposed beams in that plane. For parallel-opposed beams, the dose to a given point outside the target represents a superposition of contributions from the beam approaching the target (entrance beam) and the beam that already passed through the target (exit beam). Therefore, parallel-opposed beams should be avoided in radiosurgery because they degrade the dose fall-off outside the target in the plane containing the beams. In the proposed 2π geometry radiosurgical technique, beams in transverse plane are avoided by limiting the beam entry points to a spherical segment of 160°x160° in the upper hemisphere, as shown in Fig. 3.82a.

The isodose distributions resulting from a 160°x160° technique are optimized in terms of sharpness of the dose fall-off outside the target volume, and their shape is essentially isotropic, that is, the isodose surfaces are spherical providing identical dose fall-offs in any radial direction outside the target volume. The 160°x160° technique represents an ideal situation, which can be only approximated in practice because of the constraints imposed on the pattern of beam-entry points by the equipment design. The patterns of beam-entry points in current clinically used photon beam radiosurgical techniques are shown in Fig. 3.82b-h [3.150].

Two models of the gamma unit are available: in one the sources are distributed evenly over 160°x60° sector of the source core; in the other, the sources are distributed on a ring containing the transverse plane and extending 36° into the upper hemisphere. The points of beam entry on the patient's head for the gamma unit models are shown in Figs 3.82b and c.

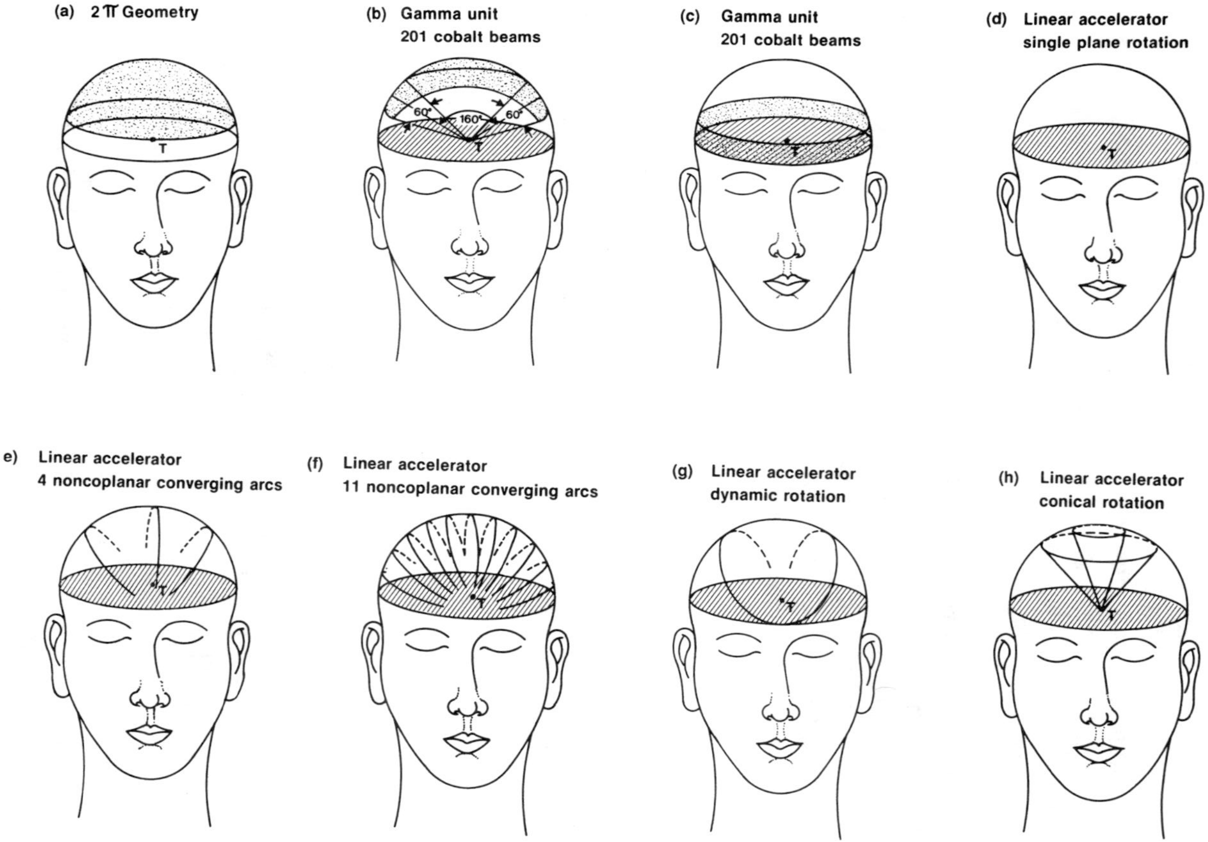

FIGURE 3.82 Beam-entry patterns on a patient's skull for various radiosurgical techniques (after E.B. Podgorsak [3.150]).

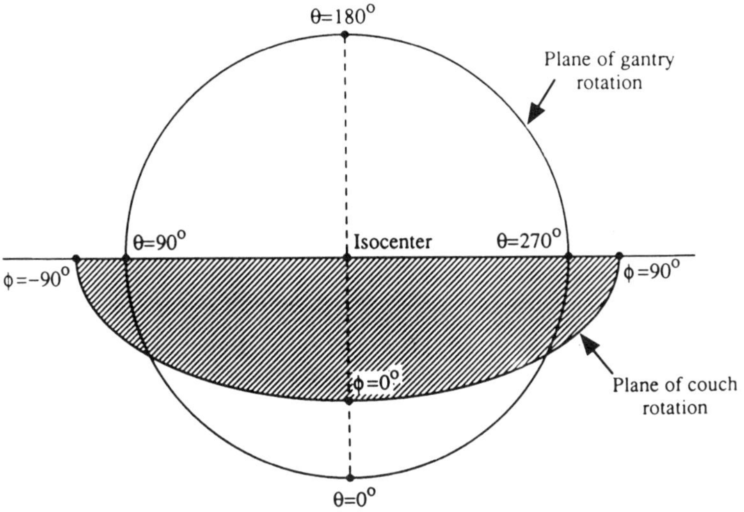

FIGURE 3.83 Planes of motion of the gantry and treatment couch or chair in linear accelerator-based radiosurgery (gantry rotation angle θ and couch rotation angle Φ) (after E.B. Podgorsak [3.150]).

Four methods can be used to perform radiosurgery by means of a linear radiotherapy accelerator: (1) *single plane rotation* [3.98], (2) *multiple noncoplanar converging arcs* [3.99, 3.103], (3) *dynamic rotation* [3.97, 3.127 through 3.130] and (4) *conical rotation* [3.151]. Each technique is characterized by a particular set of individual rotational motions of the linac-gantry or patient from given start to stop angles. The convention used for gantry angle θ and couch angle Φ is shown in Fig. 3.83.

The first technique, similar to standard radiation treatment, is to employ rotational (arc) irradiation in a single plane (Fig. 3.82d). The patient is placed supine on a stationary couch ($\Phi = 0°$), and the gantry rotates from $\theta = 0°$ to $\theta = 360°$ during treatment. This method is rather simple and provides a guarantee that the dose falls off rapidly in the direction orthogonal to the plane of rotation. In the latter plane, however, the dose fall-off outside the treatment field is slower due to the superposition (overlapping) of the infinite number of parallel and opposite beams. Since the intracranial structures are often situated in the vicinity of highly radiosensitive regions which have to be protected, the method of rotational irradiation in a single plane does not give satisfactory results in the majority of radiosurgical procedures.

Better results are obtained with the use of *multiple convergent noncoplanar arcs* (University of Miami) [3.106 through 3.108]. The principle of this method is shown in Fig. 3.82e and f. The treatment area is localized stereotactically and is placed in the isocenter. It is then irradiated using several arc movements (4 to 11) over segments shorter than 180° in order to avoid situations in which the treatment field would be irradiated with parallel and opposite beams. This method can be implemented in a system like one shown in Fig. 3.81 by rotating the table between successive arc irradiations.

The dose fall-off outside the target volume obtained with the method of multiple convergent noncoplanar arcs is comparable with similar parameters achieved with the previously mentioned radioisotope system and with heavy ion beams.

Further development of radiosurgery methods with linear accelerators has been provided by the method referred to as the *dynamic rotational technique* (McGill University, Montreal, Canada). An isocentric linac is used, with the patient in the supine position on the treatment couch [3.150]. The stereotactic frame is attached directly to the couch, which is immobilized by a simple pressure plate. The main feature of dynamic rotation is the continual simultaneous rotation of the gantry and couch during treatment. The gantry rotates 300° from $\theta = 30°$ to $\theta = 330°$ and the couch rotates 150° from $\Phi = 75°$ to $\Phi = -75°$. Thus, each degree of couch rotation corresponds to 2° of gantry rotation. The beam entry trace on the patient's skull (Fig. 3.82g) has a peculiar pattern (baseball seam) but always lies in the upper hemisphere, causing all beam exit points to lie in the lower hemisphere. Thus, although all beams intersect in the target volume and the gantry travels almost a full circle (300°), there never is a parallel-opposed beam situation to degrade the steepness of the dose fall-off outside the target.

In a *conical rotation technique* the patient rotates on a special treatment chair from $\Phi = 0°$ to $\Phi = 360°$, while the gantry is stationary at a given angle θ between 90° and 180°. The stereotactic frame is attached to a pedestal, which in turns is mounted to the base plate of the patient support assembly. Up to three gantry position (θ = 100°, 120° and 145°) are used for a typical treatment, resulting in coaxial circles for beam-entry traces in the upper hemisphere and conical irradiation patterns, as illustrated for two gantry positions in Fig. 3.82h [3.150].

In Fig. 3.84 are compared the calculated and measured dose profiles in several directions through the target volume in a humanoid head phantom for single-plane rotation [curves (a) and (b)] and the dynamic curves radiosurgical procedures [(c) and (d)]. Again the field diameter is 1 cm and beam energy 10 MV. Calculated profiles are shown as solid and dashed curves, measured profiles as data points [3.131].

Single dose delivery of 5000 cGy takes about 15 minutes for patient setup on the machine and about 30 minutes for the target dose delivery, typically in five full dynamic rotations from

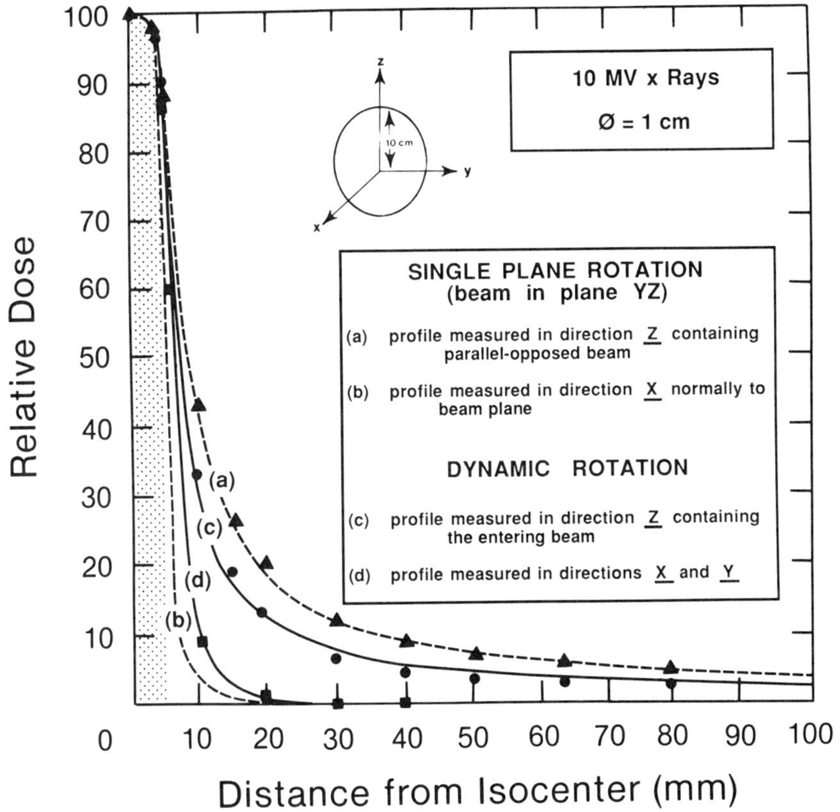

FIGURE 3.84 Dose profiles in several directions through the center of a humanoid head phantom for single plane rotation [(a) and (b)] and for dynamic rotation [(c) and (d)], (after B. Pike et al. [3.131]).

the respective couch and gantry start-to-stop angles. The staff required for a single high-dose irradiation for AVM includes a neurosurgeon, a radiation oncologist, a physicist, an engineer, and a radiotherapy technologist. The fractionated dose deliveries take about 10 to 15 minutes for patient setup on the machine and 5 minutes for the target dose delivery in one single dynamic rotation. The fractionated treatments are usually given by two physicists and a technologist, and occasionally involve a radiation oncologist.

Table 3.14 summarizes other parameters of dynamic stereotactic radiosurgery [3.128]. Dynamic rotation is a practical and inexpensive alternative to radiosurgery with the gamma unit or with proton beams as it gives almost the same precision in dose delivery and yields similar dose fall-offs outside the tar-

TABLE 3.14 Dynamic Stereotactic Radiosurgery (after E.B. Podgorsak [3.128]).

isocentric linear accelerator	—	photons (4-20 MV)
radiation field	—	circular (diameter 0.5 - 3 cm)
moving field therapy	—	continuous and simultaneous gantry and couch rotation; gantry from 30° to 330° and couch from 75° to -75°
stereotactic frame	—	for target localization with DSA, CT and MRI, treatment set up, and patient immobilization
3-D treatment planning system	—	for calculation of 3-D isodose distributions directly superimposed onto CT, MRI, or DSA images
lesions treated	—	AVM, acoustic neuroma, cavernous angioma, low grade glioma, cranyopharin gioma
dose prescribed	—	4000-5000 cGy in one irradiation, 100% at isocenter 90% at edge of target
dose to critical organs	—	target — 100% 5000 cGy thyroid — 0.2% 10 cGy breast — 0.06% 3 cGy gonads — 0.02% 1 cGy

TABLE 3.15 Technical Aspects of Linac-Based Radiosurgical Techniques Implemented Clinically in the 1980s (after E.B.Podgorsak [3.150])

Technique	Location	Patient Position	Radiation Field	LINAC Energy (MV)	Gantry Rotation	Couch of Chair Rotation	Beam-Entrance Pattern
Single-plane rotation	University of Miami, Miami	Supine	Circular	10	Full rotation 0° – 360°	Stationary $\Phi = 0$	Fig. 3.82d
Noncoplanar converging	Hospital Espanol, Buenos Aires	Sitting	Circular	10	Arcs 130° – 250°	Stationary, several positions	Similar to Fig. 3.82f
Noncoplanar converging	City Hospital, Vincenza	Supine	Square	4	5 to 10 arcs, 50° – 170° 190° – 310°	Stationary, 5 to 10 positions between 90° and -90°	Similar to Fig. 3.82f
Noncoplanar converging	DKFZ, Heidelberg	Supine	Circular	15	11 arcs 20° – 160° 200° – 340°	Stationary, $\Phi = 90°$, 75°,....15°, $\Phi = -15°$, -30°,....-75°	Fig. 3.82f
Noncoplanar converging	Joint Center, Boston	Supine	Circular	6	4 arcs, 50° – 310° 40° – 140° 220° – 320°	Stationary, $\Phi = 0°$ $\Phi = 90°, 45°$, $\Phi = -45°$	Fig. 3.82e
Dynamic rotation	McGill University, Montreal	Supine	Circular	10	1 arc, 30° – 330°	Rotation, $\Phi = 75°$ to $- 75°$	Fig. 3.82g
Conical rotation	Emory Clinic, Atlanta	Sitting	Circular	6	Stationary, 100°, 120° 145°	Full rotation, $\Phi = 0° – 360°$	Fig. 3.82h

get volume. However, when considering a linac for use in radiosurgery, one should keep in mind that requirements for accurate radiosurgery are far more stringent than those applied to standard radiotherapy, since the target volumes in radiosurgery are much smaller and the dose is delivered in a single irradiation [3.128].

The dosimetric characteristics of minibeams used for radiosurgery are very different from those of the radiation beams used in classical radiotherapy [3.135]. Stereotaxic radiosurgery with a linac provides an extreme challenge in the calculation of 3-D dose distributions as normally performed on a digital computer, i.e. the calculation of isodose surfaces by interpolation from doses calculated on a Cartesian grid [3.136].

At present, more and more centers became involved in linac-based radiosurgery. By 1990 some 40 centers in North America alone were offering this treatment modality or seriously considering doing so [3.152]. It is likely that this number will double by the end of 1991 [3.150]. The technical features of linac-based radiosurgical techniques implemented clinically in the 1980s, are summarized in Table 3.15.

CHAPTER 4

Accelerator Laboratories for Routine Therapy

4.1 Dosimetry of Accelerator Therapy Beams

4.1.1 Basic Dosimetry Problems

The basic sequence of *dosimetry procedures* implemented during accelerator beam therapy is depicted in Fig. 4.1. The first stage is *absolute calibration of a dosimeter*, which most often is an ionization chamber or, less frequently, a semiconductor, thermoluminescent detector, or dosimetric film (Sec. 4.1.2). The second stage consists in *determining the absorbed dose at a well-determined reference point* in a homogeneous standard medium, such as water or, sometimes, a plastic phantom (Sec. 4.1.3). In the third stage, the *radiotherapy field distribution* is determined, that is the absorbed dose distribution is measured at an arbitrary point in the phantom.

Before stage four is started, data on patient topography must be taken into account. These are obtained from such diagnostic techniques as radiography, X-ray computer tomography, MRI tomography, etc. The main point in stage four is to determine the *absorbed dose distribution in a particular model of a patient.* All the procedures implemented in this stage, including the biological effectiveness of the type of the radiotherapy beam used, are generally called *radiotherapy planning* (Sec. 4.2).

The fifth stage in the sequence of dosimetry procedures is to perform *patient irradiation* proper, that is to deliver a dose to the target (tumor) volume, monitored or measured by an internal accelerator dosimetry system (Sec. 3.1.2.4).

Direct *patient dose monitoring* is of particular importance, since it makes it possible to detect systematic errors made in

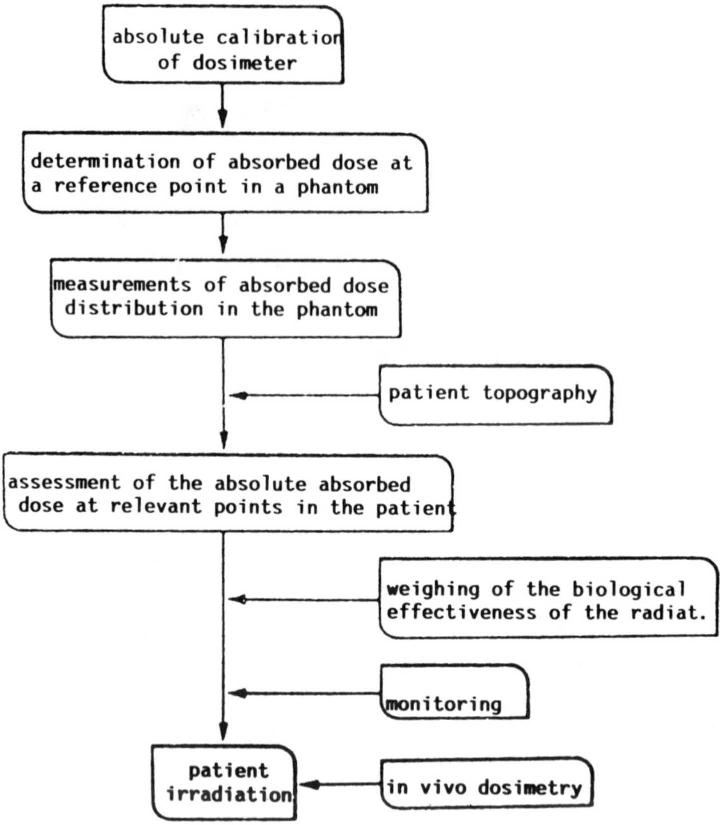

FIGURE 4.1 Sequence of dosimetry procedures in radiotherapy (according to B.J.Mijnheer [4.1])

radiotherapy planning and in some irradiation techniques used. In addition, random errors in dose calculation, patient positioning and accelerator operation can also be exposed. What is important is to monitor the dose level inside critical organs or close to them. However, direct measurement of the dose in the tumor volume, except for some special cases, is not feasible. Instead, in practice, surface doses on the patient skin are determined with ionization chambers, semiconductor or thermoluminescent detectors.

According to the recommendations included in some dosimetry protocols (NACP 1980, [3.18]), the dosimetry of patient doses should be carried out at two *safety levels*. At the first level, systematic errors in radiation therapy planning and irradiation procedures should be detected. This includes measurements for each radiation field during first irradiation sessions as well as those made when the patient anatomy is changed.

The second level incorporates measurements carried out during the whole series of irradiations, with the aim of detecting any errors in setting up irradiation parameters and in accelerator output values.

Sec. 3.1.1 has already provided a list of basic clinical requirements to keep the accuracy of the patient dose within ±5% or even ±3.5% [3.6]. The main trend in the development of the present-day radiation therapy dosimetry is to make a detailed analysis of errors at all stages of measurements (see Table 3.2) and, consequently, to minimize them. For this purpose, clinical dosimetry procedures are being standardized with the aid of dosimetry protocols and codes of practice described earlier in Secs. 3.1.1 [3.3 through 3.18 and 4.1].

A detailed discussion of the fundamentals of clinical dosimetry is beyond the scope of this book. The reader interested in basic problems of clinical dosimetry is referred to the textbooks [4.2 and 4.107 through 4.114]. The dosimetry of radiotherapy electron beams is described in [1.30], whereas the main clinical accelerator dosimetry issues are presented in an interesting overview [4.6]. In what follows, we shall deal only with the fundamental measuring equipment and some methods employed in the dosimetry of accelerator beams used in routine therapy. The main emphasis will be placed on the features, such as higher beam energies and the pulse-like nature of radiation resulting from the beam time structure (Sec. 2.3.4), which set them apart from the parameters of conventional dosimetry. In addition, specific measuring requirements will also be presented, resulting from the present-day radiation therapy planning methods and the tendency to have all the dosimetric procedures fully automated and integrated with planning procedures (Sec. 4.2).

In view of the well-known dependence of field parameters on the field size, accelerator dosimetry techniques have witnessed a division into *small field dosimetry* [4.7 and 4.8] and *large field dosimetry*. The term *small field* has not been universally defined; it is, however, usually taken to describe square fields whose side is between 5 and 8 cm. It is clear that small field dosimetry must also account for measurements of the smallest possible fields with diameters or sides sometimes as small as 1 cm.

4.1.2 Dosimetric Detectors

Calorimetric dosimeters. Among various dosimetric techniques, only the *calorimetric method* provides an absolute measurement of the dose. All other methods are secondary and are referred to a reference value, most often an ionization measurement in air.

A *calorimeter* consists of: (1) a medium thermally insulated from the environment, which absorbs radiation energy, and (2) a thermometric system that determines the temperature rise of the absorber. Recommendations on the choice of a proper absorbent, which is usually graphite or water, are given in [4.2]. Apart from *graphite or water calorimeters,* those of mixed design, e.g. polystyrene-water systems, are also used [4.3]. Because of the difficulty in use, calorimeters are not employed in routine dose measurements. At present, they are being used exclusively for standardization and comparative determinations [4.5 and 4.6].

Liquid dosimeters. Liquid dosimeters are water solutions of compounds in which chemical transformations are linearly related to the radiation energy absorbed. The basic dosimeter of this type is the *Fricke dosimeter* which contains air saturated water solution of 0.4 mol/l sulphuric acid, its density corresponding to that of the human tissue, and 0.001 mol/l $Fe(NH_4)(SO_4)$ and 0.001 mol/l NaCl. Ionizing energy radiation absorbed in the solution oxidizes the two-valence iron to three-valence iron. The Fe^{3+} ions resulting from this reaction cause considerable increase in the spectral absorption of the solution, especially visible in the ultraviolet region of about 305 nm. By measuring spectrophotometrically absorption changes for the above wavelength, a molar extension coefficient is determined, which is then used to calculate the absorbed dose. The accuracy of the measurement may be as high as $\pm 1\%$. Dosimeters of this type are often employed in national dosimetry systems as calibration dosimeters. In accelerator dosimetry they are only used for comparative measurements [4.6] in which high accuracy over a wide range of dose rates is of great importance (cf. Fig. 4.2).

It should be noted that in the present-day multi-energy radiotherapy accelerator (Sec. 3.1.2.5) doses can be produced of up to 200 Gy in about 0.5 s [4.6] and consequently, measurements in the range of high dose rates require detailed knowledge of the detector pulse characteristics and of the radiation system used (Sec. 4.1.5).

Ionization chambers. These are detectors which are at present most often employed in routine accelerator dosimetry because of the wide range of the dose and dose rate measurements that they can provide (Fig. 4.2), as well as of the good energy characteristics they exhibit.

Depth dose measurements usually require *thimble-type chambers* with active volumes ranging from 0.1 to 1.0 cm^3. Their walls

TABLE 4.1 PTW-Freiburg [F17] Ionization Chambers for Dosimentor System (simultaneous measurements of exposure and exposure rate of high-energy photon and electron beam)

Chamber	Type	Energy Range[1]	DI4 Display[3] Resolution Gy	Range min. Gy	Range max. Gy
0.1 cm^3	M23323[2]	140 kV ... 40 MV	0.1 m	10 m	999.999
0.125 cm^3	M233642	140 kV ... 40 MV	0.1 m	10 m	999.999
0.3 cm^3 rigid	M23332	140 kV ... 40 MV	0.1 m	10 m	99.9999
0.3 cm^3 flex.	M233641	140 kV ... 40 MV	0.1 m	10 m	99.9999
1 cm^3	M23331	70 kV ... 40 MV	0.1 m	10 m	99.9999
30 cm^3 cyl.	M23361	70 kV ... 40 MV	0.1 μ	10 μ	999.999 m
30 cm^3 flat	M233612	50 kV ...150 kV	0.1 μ	10 μ	999.999 m
1000 cm^3	M2337	60 kV ... 40 MV	0.1 μ	10 μ	99.9999 m
electron	M23343	2 MeV ... 40 MeV	0.1 m	10 m	99.9999

1) For ^{137}Cs and ^{60}Co radiation "free in air" with build-up cap.
2) "Sealed" chamber, definition: change of air measuring volume cannot be detected for 8 hours in the range from +10°C to +38°C.
3) Dose integrator DI4 (type 77511) with 6 digit display, 1 digit corresponds to 10^{-14}C).

are most often made from graphite or plastic (nylon, A-150 plastic and organic glass) or less often from aluminum. A detailed review of chambers of this type is given in [4.9]; parameters of a set of chambers used together with the Dosimentor dosimetry system (manufactured by PTW-Freiburg [F17]) are listed in Table 4.1.

Each dosimetry protocol usually provides a detailed list of requirements for chambers used in absorbed dose measurements. *Cylinder-type chambers* are usually recommended for photon beam measurements,but the AAPM protocol [3.14] also accepts *parallel plate chambers.* Below are presented some requirements for chambers used in photon beam determination [4.115].

The ICRU 14 recommendations [4.11] state that the internal diameter of the chamber should not exceed 1 cm and the chamber walls should be made of low Z material and be thin. The NACP 1980 protocol [3.18] states that the active volume should range from 0.1 to 1.0 cm^3, whereas the diameter of the air cavity should be between 4 and 6 mm, its length being not greater than

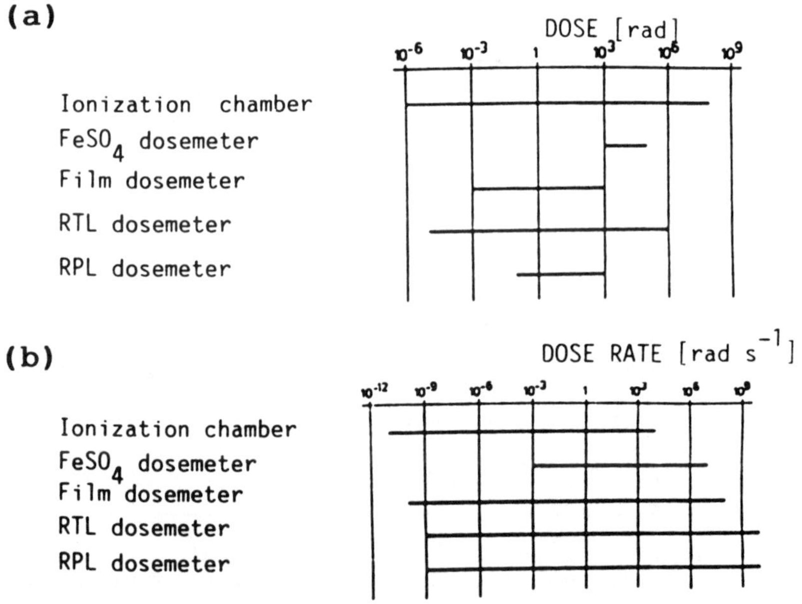

FIGURE 4.2 Measuring ranges of some dosimeters: (a) — doses, (b) — dose rates.

25 mm. The chamber walls should be made from graphite, tissue, water, or air-equivalent material. The wall material should conform to that of the build-up cap and the central electrode. The total thickness of the chamber and the build-up should be 0.45 ± 0.05 g cm^{-2}. The AAPM 1983 protocol [3.14], on the other hand, states that the internal diameter of the chamber should also not exceed 1 cm, and the chamber walls should be made from a material equivalent to the measuring phantom. Finally, the latest dosimetric protocol issued by the IAEA in 1988 [4.12] states that the chambers should be of cylindrical type and their walls made of low Z material conforming to that of the central electrode. The chamber wall thickness should not be greater than 0.1 g cm^{-2}, the recommended value being about 0.05 g cm^{-2}. The cavity and the electrode diameters should not exceed 7 and 2 mm respectively, whereas the cavity length should not be greater than 25 mm. The total thickness of the build-up cap and the chamber wall should range from 0.4 to 0.6 g cm^{-2}, the recommended value being about 0.45 g cm^{-2}.

As for ionization chambers used in determining electron beam parameters, the ICRU 21 [4.2] and AAPM [3.14] recommendations, apart from flat chambers, accept thimble-type chambers. The NACP protocol [3.18] recommends the use of flat chambers for $E_0 \leq 10$ MeV and thimble-type chambers for $E_0 \geq 10$ MeV. The IAEA recommendations [4.12] are similar, although both types of chambers are acceptable for the energy range of 5 MeV $\leq E_0 \geq$ 10 MeV. In most recent codes of practice the formalism to determine the absorbed dose in high-energy electron beams is based on an air kerma calibration of the ionization chamber and application of the Bragg-Gray equation [4.117 and 4.118].

Surface doses are measured by means *of flat chambers* [4.116] those using a *parallel plate system of electrodes,* equipped with metallized windows which also serve as an output electrode. The minimum depth of measurement is limited to the thickness of the window electrode which is typically of the order of 1 to 4 mg cm^{-2} [4.10]. For practical dosimetry it is often not possible to construct a sufficiently small detector with adequate sensibility. Therefore, if a cylindrical ion chamber is irradiated laterally in a phantom, the effective point of measurement has to be determined. This point is different from the geometrical center of the cavity [4.116].

Thermoluminescent detectors. TLD's constitute the second type of detectors employed in accelerator dosimetry both in phantom and in vivo measurements. Usually LiF is used as a detector. Its response is independent of absorbed dose rate up to 10^8 Gy s^{-1} [4.13]. According to [4.6], at about 350 Gy, saturation phenomena begin to limit response linearity. Several thermoluminescent systems have been developed for the purposes of clinical dosimetry. The TLD-20 system, manufactured by Therados company may serve as an example of such systems [F15]. The probe, in the form of long LiF rods, are enclosed in a thin, 1 mm teflon tube. The outer diameter of the probe is 1.5 mm, and its length — depending on the type — may be 4 cm, 8 cm or 15 cm. The measurements can be made over the range of 0.01 to 50 Gy (1 to 5000 rad). However, the linear response is limited to the interval between 0.01 and 5 Gy due to the nonlinear response of LiF.

Conventional methods used for the *spatial mapping of radiation field* using microdensitometric radiography and ionization chamber scanners (Sec. 4.1.4) utilize time-consuming mechanical scanning techniques to obtain two-dimensional mapping required. However, using a thermoluminescent phosphor, computerized digital video techniques offer a faster alternative for fields characterization.

An sensitive digital dosimetric system has been developed for the spatial mapping of small radiation fields using a CaF_2:Dy phosphor. A 1 mm thick, 25 mm diameter optically flat crystal is exposed to radiation and its subsequent thermoluminescence image is viewed by an *intensified charge injection device (ICID) camera* directly coupled by a tapered fibre-optic extended to the heated crystal. Video outputs are then digitized by a computerized real-time frame-grabbing system having image processing and analysis capability. Spatial resolution measurements of the system have indicated a resolution limit of 83 ± 4 μm. The system is capable of mapping small radiation fields and the edges of larger fields using absorbed doses as low as 0.2 mGy [4.119].

The introduction of a *phosphor plate reader* has presented an opportunity to develop a radiation therapy network system [4.120]. The plates can be used for portal imaging as well as simulation studies. Digital images (matrix size 512x512) from the plate reader can be displayed on video for review and verification. The images may be archived for record keeping. An entire network can be developed using relatively lower cost hardware devices of archive, workstations, and phosphor readers. The remaining challenges include interface of block cutting capabilities with the therapy workstation. Further integration with reports, patient charts and quality assurance procedures will increase the utility of digital radiation therapy management system (Sec. 4.2.4) [4.120].

Film detectors. Film is employed in dosimetric measurements and in profiling as well as in accurate positioning of radiotherapy fields. The present-day field analyzers (Sec. 4.1.4) are equipped with automated *densitometers* which permit fast readout of the film density after development.

The results of the analysis of the film method to determine relative doses for photon and electron radiation beams used in therapy are presented in [4.19]. For the Kodak XV-2 film placed in a polystyrene phantom (22 x 22 x 22 cm^3), the linear dependence between the film density and dose holds up to 25 cGy for 9 and 15 MV X-ray beams, and to 30 cGy for 4 MV photons and 6 to 20 MeV electrons. For X-ray irradiation (half-layers of 0.7 and 1.7 mm Cu) the linear range extends up to 10 cGy, whereas for Co-60 it is as high as 35 cGy. Fig. 4.3 shows the relationship between the film density and radiation energy for the above Kodak film at a constant dose. It may easily be seen that the dependence of the film sensitivity on energy is weak for 6 to 20 MeV electrons (Fig. 4.3b) and photons with energies from that of Co-60 to 15 MV. Compared with the sensitivity obtained for cobalt, the sensitivity for X-rays with the effective energy of 60 kV (half-layer of 0.5 mm Cu) increases by a factor of 8.

Semiconductor detectors. Semiconductors are used, for example, in the form of a *silicon diode.* Due to their high density, their sensitivity exceeds that of similar gaseous detectors by a factor of several tens of thousands, which implies that very small, quasi point-like detectors can be designed with the sensitive volume less than 1 mm^3.

Semiconductor detectors are specially well suited for very high dose rate measurements and, therefore, for accelerators equipped with scanning systems (Sec. 3.1.4.3). The dose from a single pulse may be more than 100 times higher compared with a scattered beam and may amount to several cGy. In a detector

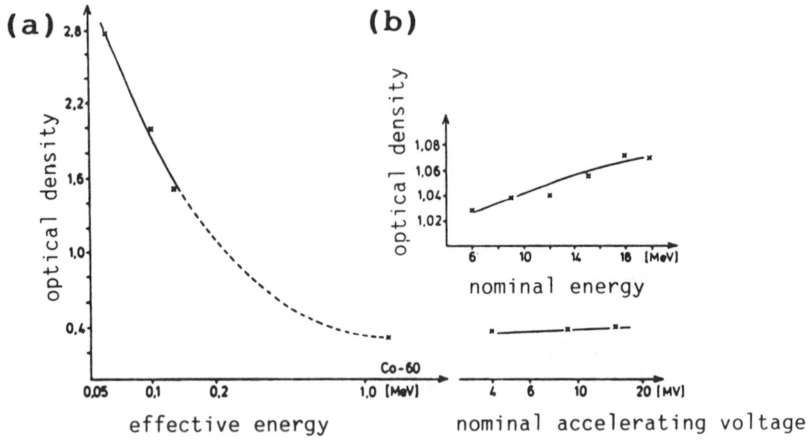

FIGURE 4.3 Dependence of Kodak XV-2 film density on energy at constant dose: (a) — photon, (b) — electrons (after J. Rostkowska et al. [4.19]).

which has a doping level of the order of 10^{15} per 1 cm^3, pulsed dose rate linearity is maintained up to about 30 cGy [4.14 and 4.15]. The disadvantage of semiconductor detectors is that their energy response is non-uniform, particularly in the 400 keV region.

The Therados company has developed a series of semiconductor detectors for use in field analyzers. With the sensitive volume of 0.2 mm^3, they allow measurement to be made with $\pm 1\%$ accuracy in 1 to 45 MV photon fields and 4 to 40 MeV electron fields. Detectors of this type can be used both for in vivo measurements and those inside phantoms. The thickness of the housing in which the detector is encapsulated is about 30 mg cm^{-2} and thus surface measurements can be carried out at the depths corresponding to this value.

Similar diodes are incorporated in the Victoreen Radocon II device employed in small field dosimetry [4.8]. The external diameter of the silicon diode is 2.9 mm and its length is 12.7 mm.

The applicability of semiconductor detector in the in vivo measurements has been thoroughly investigated [4.16 and 4.121], whereas their use in measurements inside inhomogeneous regions (inhomogeneities) is discussed in [4.17]. For electron beams, the presence of an inhomogeneity perturbs the dose distribution in the surrounding medium. These perturbations are due to different beam scattering and different electron stopping powers by media with different densities. Dosimetry of these perturbations is particularly important in computerized radiotherapy planning (Sec. 4.2).

Silicon diodes are well suited for in vivo photon beam dosimetry if a number of characteristics, such as energy and temperature dependence of the dosimeter response, are taken into account [4.121].

Other types of detectors. Intensive work is under way at present on the application of *quartz resonators as dosimeters* in clinical dosimetry [4.18]. By irradiating piezoelectric quartz SiO_2 used at present in millions of semiconductor devices one can obtain measurable changes in its resonance frequency over the dose range of 10^2 to 10^4 Gy. The dose is thus determined by measuring the frequency. However, no dosimeters which employ the above principle have been designed for practical use.

4.1.3 Dosimetry Phantoms

Dosimetry phantoms usually contain water, and less often plastic materials as an absorber.

Plastic phantoms. According to recommendations included in the NACP 1981 [3.18], AAPM 1983 [3.14] or ICRU 35, 1984 [3.12] materials, such as *polystyrene and polymethylmethacrylate (PMMA)* may be used in phantoms for measuring electron beams. Fig. 4.4 is a diagram of a *PMMA phantom* type 2967 (PTW-Freiburg [F17]) designed for the range from 70 kV to 50 MV photon radiation and 1 MeV to 50 MeV electron radiation. The phantom consists of PMMA plates with dimensions of 30 x 30 cm^2: 1 plate 1 mm thick, 2 plates 5 mm thick, and 29 plates 20 mm thick. Into some plates thimble chambers or a FeS dosimeter ampoule can be inserted without air gaps. The axis of the FeS ampoule is 7 mm or 13 mm under the plate surface depending on the direction. Into the twin plate (type 23343/12) the electron chamber type 23343 (Table 4.1) can be inserted. The reference point of the chamber is in this plate approx. 0.4 mm under the surface of the twin plate.

The PTW-Freiburg *solid phantom* type 2996 — according to ICRU report 23 — consists of a PMMA block with the dimensions of 20 x 20 x 12 cm^3 and an additional plate with the dimensions of 20 x 20 x 3 cm^3. In the block is a hole in which two

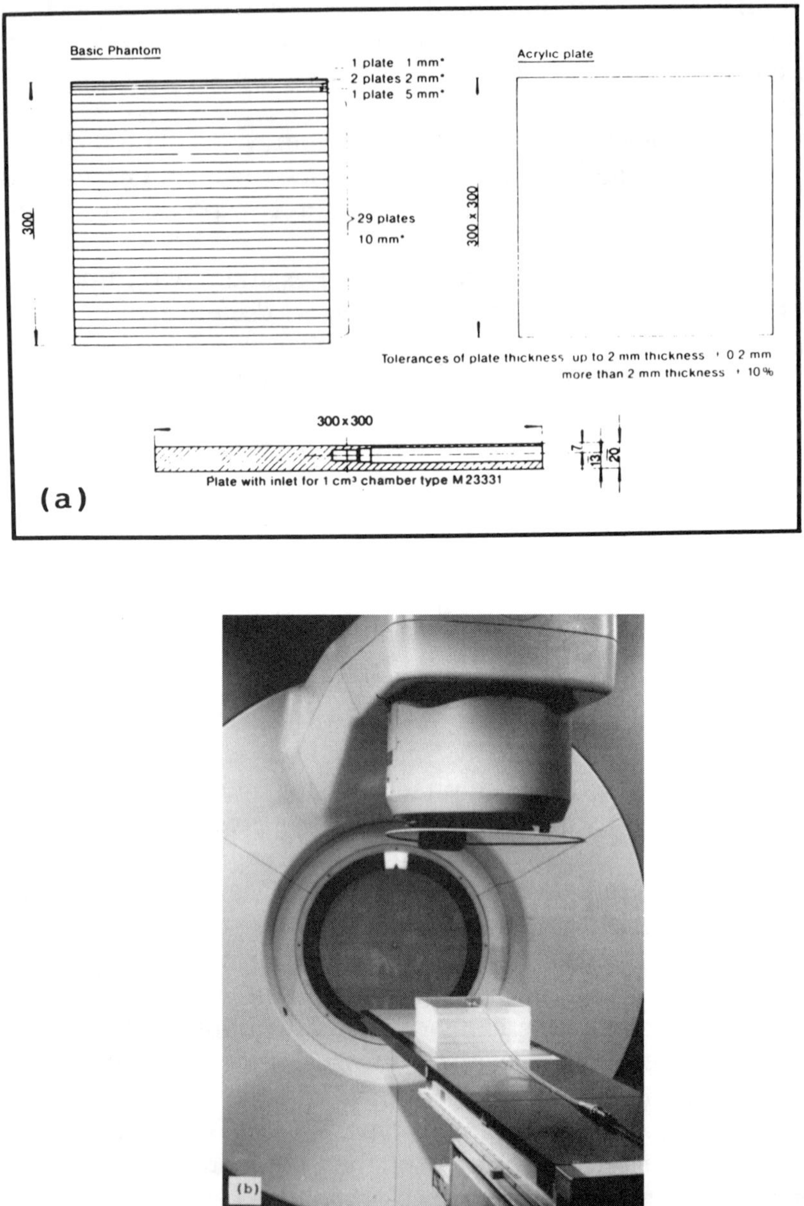

FIGURE 4.4 The PMMA phantom type 2967 (mf. by PTW-Freiburg): (a) schematic, (b) view of the phantom on the treatment table; tolerances of the plate thickness: up to 2 mm ± 0.2 mm, more than 2 mm thickness: ±10% weight of storage case with phantom 37 kg.

adaptors can be inserted without air gaps for different ionization chambers and FeS ampoules. This phantom can also be used for calibrations in the tip-to-tip method. By irradiating the phantom block from different directions and with the help of an additional block, measuring depths of 50 mm, 70 mm and 100 mm are possible. The weight of the phantom is 7 kg.

Besides polystyrene, which has a specific weight of $\rho = 1.037$ gcm^{-3}, or PMMA (plexiglass or metaplex), *acrylic resins* are also used. Radiation Measurements Incorporated (RMI) [F16] produces a water-equivalent absorption material called "Solid Water" and PTW-Freiburg [F17] — phantom material RW3 called "Goetingen Water". The RW3 is polystyrol with TiO_2 and has high water equivalence in the range from Cs-137 to 59 MV for photons and electrons. Solid Water can achieve calibrations within 1.0% of the true water dose. A detailed intercomparison of various materials in respect to water is given in [4.20].

RMI manufactures also tissue equivalent materials for bone, muscle, fat, breast, lung, and innerbone. These materials have absorption and scattering properties within 1.0% of living tissue and provide adequate simulations for electron and photon applications between 0.01 and 100 MeV [4.21].

The phantom shown in Fig. 4.4 has a block-type arrangement. In some designs, instead of blocks very thin plastic sheets are used, which make it possible to control the measured or scattering regions with greater precision.

A specific feature of measurements in plastic phantoms is that electric charge is stored in the phantom material. By interacting with the ionization chamber placed inside the phantom the presence of the charge can lead to serious errors in the readings [4.22 through 4.24]. As a result, the dosimeter readings may be increased by up to 10% [4.22]. This undesirable effect could be eliminated by the use of conducting materials since both polystyrene and methyl polymethacrylate belong to a group of good insulators.

Water phantoms. Water phantoms are usually cubicoid water containers with 10 to 20 mm thick transparent plastic walls. Their capacity ranges from 130 to 200 l depending on design. Their internal dimensions depend on the maximum field size which can be measured. In small and medium field dosimetry a certain *margin of safety*, usually 5 cm, must be maintained between the field and the container walls. The container is open at the top, which permits vertical measurements to be made in such a way that the beam passes through the water volume without entering the window. For horizontal measurements, the container is provided with horizontal or vertical slits equipped with 0.1 mm foil windows.

Fig. 4.5 a is a view of an automated PTW-Freiburg MP3 water phantom. It is made of 20 mm thick PMMA walls; its external dimensions are 694 mm (length), 596 mm (width) and 502.5 mm (height), the capacity being 200 l. The measuring probe can be moved automatically within the measuring region of 580 x 500 x 400 mm^3. To position the phantom for measurements, the container (Fig. 4.5a) is placed on a trolley equipped with an elevator which allows for vertical movement so that the water upper level, when the container is full, may be at a distance of 110 to 150 cm relative to the floor. An additional platform on three ball bearings mounted between the trolley and the container permits fine movement control relative to the trolley table top. In this way the container can be moved up and down, or horizontally, by ± 10 mm. Additionally, the container can also be shifted by $\pm 1°$ about its vertical axis.

During measurements the phantom is filled with water from an auxiliary cart equipped with a 220 l container and pump which provides 30 l min^{-1}, so that the time needed to have the container filled up to its maximum capacity is less than 7 min.

The probe can be moved in a 3D region by means of stepwise motors mounted outside the container. Its maximum regulated speed is 50 mm s^{-1}, whereas the accuracy of positioning in each direction is 0.1 mm. The motor steps can be varied between 0.1 and 1.0 mm.

At present, water phantoms usually operate together with large automated measuring systems, called *field analyzers* (Fig. 4.5b). They will be discussed in Sec. 4.1.4.

Examples of photon field profile measurements in water phantoms are presented in Figs. 3.39 and 3.65 and the depth response characteristics in a polystyrene phantom can be seen in Fig. 3.64.

Phantom position. Extensive measurements are made in the plane at a standard measurement depth within the phantom to simulate conditions at approximately average tumor depth. The distance from X-ray source to isocentre is defined as the *normal treatment distance* (NTD) for isocentric equipment (Fig. 4.6).

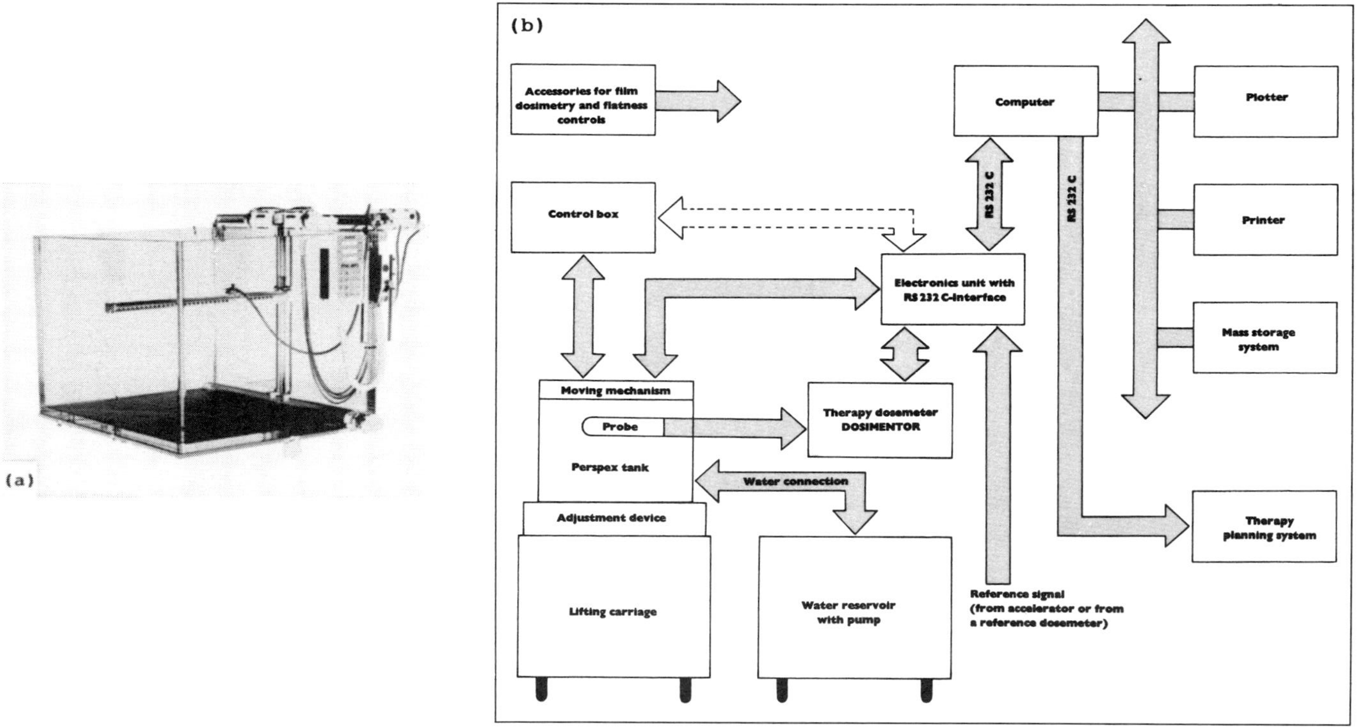

FIGURE 4.5 The automated MP3 Type PTW-Freiburg Radiotherapy Field Analyzer: (a) — view of water phantom, (b) — schematic diagram of the analyzer.

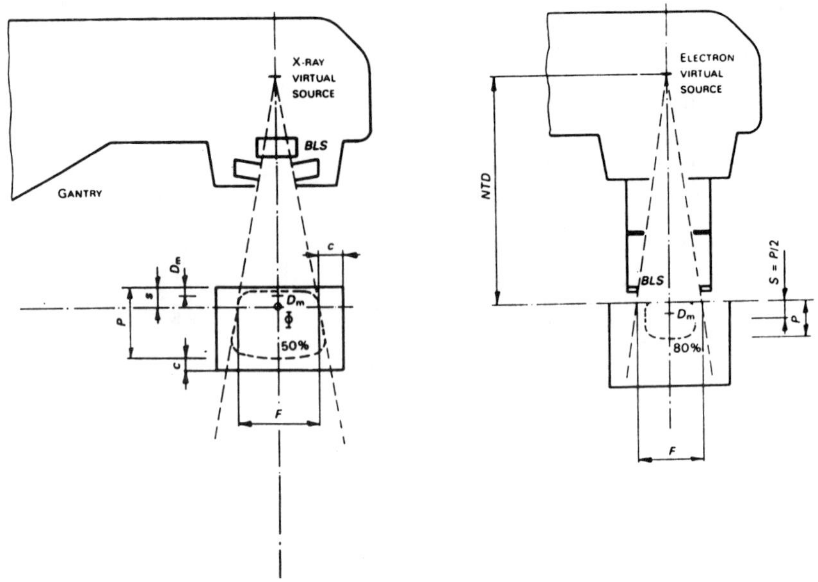

FIGURE 4.6 Phantom position (IEC Rep. 997, Table 3.3): Φ — X-ray isocentre, F — radiation field, P — penetrative quality, S — standard measurement depth, C — phantom clearance, D_m — depth dose maximum on the radiation beams axis, NTD — normal treatment distance, BLS — beam limiting system.

To simulate isocentric treatment techniques the phantom surface is positioned nearer to the radiation source from the isocentre by a distance equal to the standard measurement depth. This constitutes a more severe test condition than *SSD treatment technique* (patient surface at normal treatment distance) because the surface is closer to secondary radiation sources such as the beam limiting system, and because a greater solid angle from the radiation source is included by a given radiation field. For electron radiation, SSD treatment techniques are normally used and are simulated by positioning the phantom surface at the NTD specified by the manufacturer.

After IEC 977 recommendations for tests with very large radiation fields, a water phantom covering somewhat more than a quadrant of the radiation field may suffice in performing a test in each successive quadrant. However, when flatness and other parameters which depend on scattered radiation are tested, solid blocks of water equivalent material should be added to provide the appropriate contribution of scattered radiation. For dose measurements involving gantry rotation, a solid phantom or a water phantom attached to the radiation head may be used.

For X-ray radiation, the *standard measurement depth* S is specified as 10 cm, because this is approximately the average over all portals of the depth to the centre of the tumor for the majority of tumors in the body for the majority of patients. This constitutes a more severe requirement for flattened radiation fields than a lesser reference depth because of the reduction in beam penetration with displacement from the radiation beam axis. Tumors of the head and neck are typically at a lesser depth, averaged over all portals, than in the torso, but radiation fields are also typically smaller, for which the shoulders of dose profile tend to be lower than on the radiation beam axis. By specifying flatness at 10 cm depth, relatively flat isodose curves are obtained at typical depths of head and neck tumors.

For electrons the standard measurement depth is specified (IEC 977 Report) as half the penetrative quality P (Fig. 4.6, right) because this depth corresponds approximately to the depth of the centre of the tumor for the majority of the patients. This is in lieu of the discrete steps in reference depth which have been used traditionally for stepwise ranges of energy.

The depth dose distribution of the X-ray beams is dependent upon several factors other than the energy of the electron beam striking the radiation target of an accelerator. Similarly, the depth dose distribution of the electron beam is dependent upon several factors other than the energy of the electron beam striking the electron window (cf. Sec. 3.1.4). Therefore, instead of normal energy, the term *penetrative quality* is used as a more precise descriptor of the radiation beam produced by such accelerator. The penetrative quality P for electron radiation is specified at 80% absorbed dose because therapists typically specify the target volume to be within an isodose contour of at least 80% (Sec. 3.1.6).

4.1.4 Radiotherapy Field Analyzers

Manual measurement of 3D response characteristics is a cumbersome and time-consuming procedure. Therefore, more than twenty years ago design of special systems was initiated with the aim of developing automated measurements and processing of results obtained with the use of calibration phantoms presented in Sec. 4.1.3. These systems are now referred to as *radiotherapy field analyzers*. Their main purpose is to provide comprehensive information on field parameters such as depth dose distribution, beam profiles and isodose distributions for each axis and their combinations as well as diagonal scanning.

Typically, a sequence in the analyzer operation includes (1) positioning the detector at a given region of the phantom, (2) performing a dose measurement, and (3) normalizing the values obtained in order to eliminate the effect of beam intensity fluctuations on final results. This is usually done by employing a *standard detector* placed at a point inside the phantom whose parameters are in conformity with the recommendations of the dosimetry protocol used. In some designs, the accelerator internal dosimetric system can be utilized as a standard detector. In both cases the measurement is made in a differential mode.

The present-day radiotherapy field analyzers are complex measuring systems intended not only for automated measurements and processing of results, but also for feeding the results directly to the associated radiotherapy planning systems (Sec. 4.2). Among the analyzers available commercially and those often used in practice, mention should be made of such systems as the PTW-Freiburg Type MP3 [F17], Therados Type RFA-3-40 and Type RFA-3-50 (measuring fields of 40 x 40 x 40 cm^3 and 50 x 50 x 50 cm^3 respectively) [F15], Scanditronix RFA-300 (measuring field 49.5 x 49.5 x 49.5 cm^3) and AECL Medical Therascan (measuring field 40 x 40 x 40 cm^3) [F10].

We shall consider the operation of a radiotherapy field analyzer with the example of the PTW-Freiburg Type MP3 device, whose schematic diagram is shown in Fig. 4.5b. This analyzer operates together with the water phantom shown in Fig. 4.5a. The detector inside the phantom can be operated both automatically and manually. The detector readings are transferred by the Dosimentor therapeutic dosimeter to the control system equipped with a RS 232 C interface coupled to a computer. The control system is also supplied with readings from a standard detector or the accelerator internal dosimetric system. After processing, the results are fed by the computer to a printer and a graphic recorder and are stored in a large memory unit. They can also be directly conveyed to the therapy planning system which cooperates with the field analyzer.

The MP3 analyzer uses cylindrical chambers with sensitive volumes of 0.3 or 0.125 cm^3 as detectors. These chambers are made of PMMA doped with graphite, their walls being 0.75 mm thick or 90 mg cm^{-2}. Both chambers can be used for measurements with 0.1 to 45 MV photon beams or 5 to 45 MeV electron beams. Similar chambers manufactured by the PTW company are employed in the AECL Medical Therascan radiotherapy field analyzer.

Typically, present-day analyzers are also equipped with *optical densitometers.* When photon or electron field isodose distributions are determined with the aid of a film technique, the film placed between two plates can be, after development, analyzed with a scanning device incorporated in the phantom. For this purpose, for example, a glass plate with a film is placed in the empty water phantom and then, by using the mechanism which moves the detector in the x-y plane, the film density can be analyzed. In the MP3 analyzer, the control system (Fig. 4.5b) is coupled to an additional horizontal scanning system which permits the analysis of portal films with maximum dimensions of 350 x 350 mm. The same scanner can be positioned directly on the radiotherapy table and, equipped with an ionization chamber, can perform measurements of the field homogeneity in air, i.e. without the phantom.

The MP3 analyzer is provided with an additional device to check if the radiotherapy field is concurrent with the optical field projected by the optical indicator. For this purpose, the analyzer is fitted with a silicon photoelement probe with a sensitive area of 7 mm^2 and sensitivity of 50 nA lx^{-1}. This probe corresponds structurally to the ionization chamber and makes it possible to map the optical field in the recorder or the monitor screen and then compare the results with the shape of the radiotherapy field.

Once the analyzer is ready for operation, the time necessary for determining the response characteristics, e.g. a typical set of ten isodoses, is usually 10 to 20 min. That is why in the present-day radiotherapy practice, field analyzers are being increasingly employed more often in routine work, such as in accelerator quality assurance (QA) or in the examination of irregular field distributions with the use of individual filters as well as in many other applications.

4.1.5 Pulse and Energy Characteristics. Radiation Survey Meters.

The applicability of radiation survey meters in the detection and measurement of radiation in the vicinity of an accelerator must depend on the thorough knowledge of their pulse and energy characteristics. The survey meters are usually calibrated with Co-60 or Cs-137 radiations and the calibration radiation field is thus of a continuous type. On the other hand, when in use, the survey meter is often in an accelerator which produces a noncontinuous high-energy pulsed beam. The problems of the time structure of the beam generated by electron linear accelerators, which are by far most common in radiotherapeutic applications, have been discussed in Sec. 2.3.4. As a result of the pulsed nature of the beam instantaneous beam currents may exceed mean values by a factor of 100.

It is, therefore, clear that survey meters used for radiation protection during the work with medical accelerators should have their response independent of the beam time structure.

This is a problem which has been recently given more attention. For example, at the Stanford Accelerator Center (SLAC), a *pulsed X-ray facility* has been built which permits simulation of 360 ns to 5 μs electron accelerator pulses whose frequency can be varied from 60 to 360 Hz [4.26]. Fig. 4.7 shows the results of measurements made with this facility. The y-axis represents meter's readings, whereas the x-axis shows the true dose rate. Fig. 4.7a gives the response of the Radector II meter which is equipped with a pressurized ionization chamber filled with argon. The chamber was irradiated with 1 μs 100 Hz X-ray pulses. For a dose rate as low as 1 mGy h^{-1} the response is not fully linear, but the reason for that could be that the ionization chamber polarization voltage is too low [4.27]. Similar deviations from linearity can also be found in dosimeters equipped with scintillation detectors, in which the effect of photocathode overloading of the photomultiplier may be detrimental.

Fig. 4.7b shows the pulse response curves of the dose rate meter XETEX 303B intended for measurements in the vicinity of pulsed radiation sources. A calcium fluoride scintillator and a vacuum photodiode serve as the detector. The measurements (Fig. 4.7) involves 2 μs pulses with the repetition rate of 120 Hz. The survey meter's response is linear up to the dose rate of 40 mGy h^{-1}.

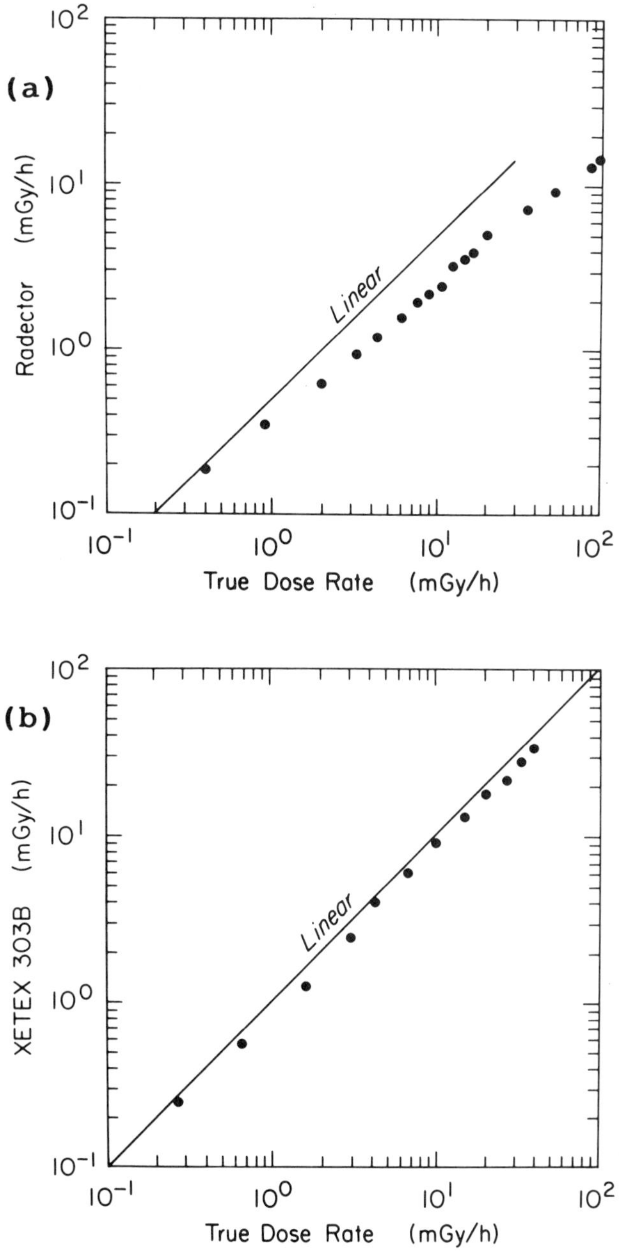

FIGURE 4.7 Pulse curves of radiation protection monitors: (a) — Radector-type device, (b) — XETEX 303B-type device (after R.C. McCall and N.E. Ipe [4.27]).

Equally important is the energy response in the high-energy range. The survey instruments calibrated with the use of Co-60 radiation sources are reliable up to the energy of about 1.25 MeV (mean value for Co-60), whereas for Cs-137 the calibration energy is 0.61 MeV. It has been found [4.28] that typical radiological survey instruments equipped with Geiger-Mueller counters and calibrated for Cs-137 are twice as sensitive for 6 to 7 MeV photons. The sensitivity of a meter with an ionization chamber was found to be only higher by about 20%, whereas that of instruments incorporating a NaI scintillator was only 30 to 40% relative to Cs-137 calibration energy. This is why measurements carried out with survey meters whose energy response is unknown may eventually lead to gross errors.

4.1.6 Some Fundamental Measurements

4.1.6.1 Determination of Energy and Quality of Accelerator Beams

Methods used in the determination of the energy and quality of accelerator beams can be classified in two groups: *laboratory techniques* employed mostly in research and *routine techniques* used in practical applications. The first group includes magnetic and activation methods for determining electron beam energies. The fundamentals of both these techniques are discussed in the classic paper [4.29]. A characteristic feature of the routine methods, on the other hand, is that the beam energy is determined indirectly, i.e. by introducing such parameters as the practical range, R_p (Sec. 3.1.6.3, Fig. 3.62) and the half-layer value R_{50}. The methods used in routine measurements usually result from individual dosimetry protocols. Typical measurements of this kind will shortly be discussed below.

Magnetic spectrometry. The magnets used in the beam energy analysis have already been discussed in Sec. 2.4, Fig. 2.25.

For $E \gg E_0$ (where $E_0 = m_e c^2$) in a uniform magnetic field

$$E = ec\,RB = 300\,RB \qquad [\mathrm{MeV\,T^{-1}\,m^{-1}}] \qquad (4.1)$$

where e is the electron charge, c is the speed of light, R is the radius of the beam path curvature and B is magnetic induction. The geometric resolving power of a magnetic spectrometer, which determines the resolution of two adjacent energies follows from the relationship

$$\Delta E/E = R / (S_1 + S_2) \qquad (4.2)$$

where S_1 and S_2 are the entrance and exit slit widths.

Magnetic methods are now seldom used for practical measurements of beam energies in clinical applications. In [4.30] a spectrometer is described for measuring beam energies above 35 MeV where the indirect method of energy determination with the R_{50} range is no longer valid.

Activation (threshold) methods. They involve the use of *threshold nuclear reactions* which feature a well-defined energy threshold within the range of measured energies. These techniques are suitable both for electron and bremsstrahlung beam energies in the corresponding reactions of (e^-, n) and (X, n) type. According to the ICRU 21 recommendations [4.2], several reactions with threshold values from 10.8 to 28.9 MeV can be used for calibration purposes. For example, in the reaction ^{63}Cu (X, n) ^{62}Cu (threshold value of 10.85 MeV) as a result of the activation of a Cu foil, the isotope of ^{62}Cu is produced. It emits 2.9 MeV positrons with 9.7 min half-life. By irradiating water two activation reactions can be induced: ^{16}O (X, n) ^{15}O and ^{16}O (X, 2n) ^{14}O with energy threshold of 15.67 and 28.91 MeV, respectively. In these case, 1.7 and 1.84 MeV positrons are emitted respectively and their half-lives are 124 and 70.5 s.

The disadvantage of the threshold techniques is that no reactions are available whose energy thresholds lie below 10 MeV and that the measurements themselves are cumbersome and time-consuming.

Routine methods of determining electron beam energies. In accordance with the considerations presented in Sec. 3.1.6 (Figs. 3.61 and 3.62), one of the most important parameters that describe an electron beam is the *mean electron energy,* E_0, on the phantom's surface. The value E_0 is determined from ionization measurements in the phantom along the beam main axis (Fig. 3.59). The ICRU 21 protocol [4.2] recommends that E_0 should be obtained from the relationship

$$\rho R_p = 0.521\,\overline{E}_0 - 0.376 \qquad (4.3)$$

where ρR_p is the practical range in g cm^{-2} in the material of ρ density. In (4.3) E_0 is given in MeV. Other protocols, e.g. IAEA [4.12], NACP [3.18] and AAPM [3.14 and 3.15] recommend that the mean electron energy be determined using the range R_{50}. Energy E_0 in MeV can then be obtained from the relationship

$$\overline{E}_0 = 2.33\, R_{50} \tag{4.4}$$

where R_{50} (cm) is the depth in the phantom at which the dose (NACP, IAEA) or ionization (AAPM) falls to 50% of the maximum ionization dose determined on the beam main axis. Relationship (4.4) is valid for water but it may also be used for phantoms which contain organic glass, polystyrene and A 150 material when respective correction coefficients have been introduced. The values of these coefficients given by the AAPM are not in agreement with those provided by the NACP due to the difference in the production and storage of the materials produced in the USA and Europe [4.31]. The IAEA gives a relationship between E_0 and R_{50} for ionization curves because by determining the value of R_{50} from the ionization curve, according to (4.4) as recommended by the AAPM Protocol, we are led to an error in energy determination since the coefficient 2.33 MeV cm^{-1} has been measured for the dose response [4.31]. The values of E_0 and R_p serve to obtain the mean electron energy for the depth Z in the phantom according to

$$\overline{E}_z = \overline{E}_0\,(1 - z/R_p) \tag{4.5}$$

This relationship is used for electrons with energies $\overline{E}_0 \leq 10$ MeV for the whole range of z, whereas for electrons with energies higher than 10 MeV it is only valid for small depth values. The IAEA protocol also provides the relationship $\overline{E}_z / \overline{E}_0$ as a function of z/R_p in the form of a table for energies of 5 MeV $\leq E_0 \leq$ 50 MeV.

Methods of determining the quality of an X-ray beam. In order to provide a characteristic of a bremsstrahlung beam, the ICRU, NACP and AAPM protocols use the term a *nominal accelerating voltage* which, in the manufacturer's list of accelerator parameters, corresponds to the maximum energy of emitted photons (ICRU 14 [4.11]), or is determined experimentally for each X-ray beam (NACP [3.18], AAPM [3.14]). The nominal accelerating voltage is determined from the parameter referred to as the *beam quality coefficient,* which is given in terms of ionization measured at 10 and 20 cm depth in the phantom:

— NACP: J10/J20 for SSD = 1m and 10x10 cm field;

— AAPM: J20/J10 for the source-detector distance of 1 m (or equal source-isocenter distance) and 10x10 cm field in the ionization chamber plane.

The AAPM recommendations accept the use of a phantom made from organic glass or polystyrene.

The IAEA protocol does not recommended the nominal accelerating voltage, rather it postulates the use of the radiation quality coefficient, called the *tissue-phantom ratio or TPR,* which is the ratio of doses at 20 and 10 cm depth in a water phantom. The measurements must be made using the geometry recommended by the AAPM protocol.

4.1.6.2 Measurements of Absorbed Doses

Recommendations on the measurement of absorbed doses included in dosimetry protocols are not identical; the differences in protocols have been thoroughly analyzed in [4.32]. The determination of the absorbed dose at the point of reference requires that the recommendations should specify some parameters such as the type of phantom, detector, the depth of the point of reference at which the standard measurements is made, the effective measuring point, perturbation coefficients [4.118], and, finally, a suitable selection of *stopping power coefficients* [4.122].

X-ray dose. As already mentioned in Sec. 4.1.3 X-ray dose measurement is made using a differential method relative to a well defined reference point in the phantom. The recommendations stipulated in protocols for ionization chambers have been analyzed in Sec. 4.1.2.

Most of the dosimetry protocols discussed here recommend that the absorbed dose should be measured in a water phantom, although the AAPM protocol also adopts the use of polystyrene and PMMA (organic glass) phantoms. According to the ICRU 14 recommendations [4.11], the reference depth should be 5 cm for beams within the energy range between the energy of Co-60 and the nominal energy of 10 MV, and 7 cm for those whose energies range between 11 and 25 MV. The NACP [3.18] protocol again recommends 5 cm depth for the nominal accelerating voltage of 1 to 10 MV and 10 cm depth for voltages between 10 and 50 MV. On the other hand, the AAPM recommendations [3.14] specify that the depth should depend on the diameter of the ionization chamber. For example, for chambers with an active volume of 0.6 cm^3 and diameter of 0.63 cm (Nuclear Enterprises) the depth should be 5 cm for the radiation energy between Co-60 up to 15 MV and 7 cm for the energy between 15 and 36 MV. In the IAEA [4.12] recommendations the following depths have been adopted: 5 cm for beams with energies ranging from that of Co-60 to $TPR^{20}_{10} = 0.70$; 17 cm for beams with $TPR^{20}_{10} > 0.70$, and 10 cm for beams with $TPR^{20}_{10} < 0.70$.

The measurements in a water phantom should be carried out in a sealing cap. According to the ICRU 14 recommendations [4.11], this should be a *build-up cap* made from a water-equivalent material because the ionization in the chamber is assumed to be caused by the electrons which are produced in the phantom and the cap. The error caused by lack of water equivalence depends on the material and thickness of the cap and that of the chamber walls. The cap should be made from 1 mm thick PMMA (NACP), or from less than 0.5 mm thick low Z material (AAPM), or from a PMMA tube with 0.5 mm wall thickness to match the ionization chamber.

Electron doses. As for detectors (Sec. 4.1.2), the ICRU and AAPM recommendations allow measurements with thimble-type chambers along with flat chambers. The NACP recommends the use of flat chambers for mean energies E_0 lower than 10 MeV and the use of thimble-type chambers for energies higher than 10 MeV. The IAEA recommendations are similar: both types of chambers are acceptable for the energy range between 5 and 10 MeV. Requirements for thimble-type chamber construction have already been presented in Sec. 4.1.2.

The depth of the reference point in the phantom depends on the energy of electrons on the phantom surface. And thus, according to the ICRU 21 recommendations [4.2], for the 10x10 cm^2 field and SSD = 100 cm, the depths should be as follows:

0.5 cm	for	$2 < E_0 \leq 5$ MeV
1 cm	for	$5 < E_0 \leq 10$ MeV
2 cm	for	$10 < E_0 \leq 20$ MeV
3 cm	for	$20 < E_0 \leq 50$ MeV.

The NACP protocol [3.18] recommends the following values:

R_{max}	for	$1 < E_{p,0} \leq 5$ MeV
R_{max}	(min 1 cm)	$5 < E_{p,0} \leq 10$ MeV
R_{max}	(min 2 cm)	$10 < E_{p,0} \leq 20$ MeV

for the field larger than 12x12 cm and for $E_{p,0} < 20$ MeV. R_{max} is the depth at which the dose is maximum, whereas $E_{p,0}$ (MeV) is the most probable energy at the phantom surface related to R_p (cm) by

$$E_{p,0} = C1 + C2\,R_p + C3\,R_p^2, \tag{4.6}$$

where C1 = 0.22 (MeV), C2 = 1.98 (MeV cm^{-1}) and C3 = 0.0025 (MeV cm^{-2}).

The AAPM recommendations [3.5] also use the term R_{max} for a wide parallel beam incident at a semi-infinite phantom. Practically, this is a 10x10 cm^2 field [4.31] unless it is shown experimentally that the field is too small and, when increased, the shape of the dose curve along the central beam changes. The central point of a thimble-type chamber at R_{max} should be such that the chamber wall facing the source embraces at least the 99% isodose. From the IAEA recommendations the following values are determined

R_{max}	for	$E_0 \leq 5$ MeV,
R_{max} or 1 cm	for	5 MeV $< E_0 \leq$ 10 MeV,
R_{max} or 2 cm	for	10 MeV $< E_0 \leq$ 20 MeV,

for the 10x10 cm field at a typical distance of SSD (e.g. 100 cm). R_{max} denotes the position of the effective measuring point rather than the center of the chamber as in the AAPM protocol.

Concerning the material of the phantom, the ICRU 21 protocol recommends water, whereas the NACP suggests water for energies E_0 higher than 10 MeV and plastic for lower energies. The AAPM protocol admits use of both water and plastic, while according to the IAEA recommendations plastic should be used for energies below 5 MeV and water for E_0 higher than 10 MeV; in the range between 5 and 10 MeV both water and plastic can be used.

A detailed discussion of the differences between various dosimetry protocols from the point of view of absorbed dose determination is given in [4.32]. Those problems have also been dealt with at the Third Conference of the European Society for Therapeutic Radiology and Oncology (ESTRO) [4.33]. In spite of the differences in the standard measuring devices recommended, or calibration methods and measuring procedures used and the different data concerning the fundamental physical parameters, the actual differences in the final results of measurements have turned out to be small. For example, for photon absorbed doses measured in water phantoms the differences usually fall within the range of ±1% [4.34 through 4.36].

4.1.6.3 Measurements in Scanning Systems

Some linear radiotherapy accelerators and microtrons are equipped with *beam scanning systems* which serve in place of flattening filters to ensure homogenous treatment fields (Secs. 3.1.4.3 and Figs 3.41 and 3.42). Undoubtedly, the advantage of the above design compared with scattering foil systems, is that the therapeutic range for the same nominal energies is extended, the photon contamination of the electron field is reduced, the surface dose is lower and the depth-dose curves, which are independent of field size for field sizes larger than 7x7 cm^2, are steeper. In addition, large homogenous fields of up to 30x30 cm^2 can be obtained and overdosage in the surface region near the field borders can be eliminated [4.37]. Unfortunately, these advantages are offset by some difficulties in dosimetry, which result from the very nature of the scanning system in which the electron beam is subjected to a simultaneous action of alternating magnetic fields in two mutually perpendicular directions (Fig. 3.41). As a result, the irradiated surface is scanned by an inhomogeneous electron beam whose diameter ranges from several to a dozen or so centimeters depending on the beam energy. At a given point, the dose is thus delivered to a well-defined region of the field (Fig. 3.42). Consequently, it looks as if a homogenous beam is simultaneously incident at the whole surface.

In order to ensure an accurate determination of a dose in a field with a scanning beam, sometimes dosimetric films are used. In addition, several electronic circuits have been developed to operate together with accelerators equipped with scanning systems [4.37 and 4.38]. By way of example, a system of this kind intended for operation with a CGR MeV Therac 20-Saturne accelerator will be presented below. A pulsed electron beam with a repetition frequency of 100 Hz is scanned with the frequencies of 0.615 and 4 Hz over a field area which is independent of the size of the therapy fields used. For the energies of 20 MeV and 6 MeV the field sizes are 60x60 cm^2 and 90x90 cm^2, respectively (SSD = 100 cm). Due to the beam divergence and the scattering in the exit window as well as in the monitoring ionization chambers and in the air, the half-value width of a single beam pulse in the isocenter plane changes from 8 cm for 20 MeV to 25 cm for 6 MeV electrons. During a scan half-cycle of 0.79 s, about 80 pulses are delivered to the scanned area. The statistical dose distribution in the area shows variations whose dispersion becomes averaged statistically with the number of scan half-cycles.

The second, even more important, contribution to statistical dose inhomogeneities results from the fact that the beam starts and stops at arbitrary points in the field. In electron therapy when the irradiation times are about one minute, statistical dose fluctuations of this kind can be neglected. However, dose measurements on phantoms require much longer times (Sec. 4.1.4). For the measuring time of 20 s, encompassing 12 full scanning cycles, the relative standard deviations are as high as 2%. They are reduced to about 1% only when 1.5 to 2 min times are used for the minimum of 60 full scan cycles. Besides, the measurements should be started at the same moment when the beam is switched on. It is clear that in this case, depth-dose distributions, dose profiles and isodose contours cannot be recorded in an automated water phantom. Dose distributions must be measured point-by-point in accordance with the beam direction (Fig. 3.42) by a single point measurement taking 20 to 120 s [4.37].

The dose delivered during one pulse to a small detector at the frequency of up to 6000 pulses per minute may be much higher than the calculated mean dose and that measured with a large-surface chamber.

The measuring arrangement used for monitoring a scanned beam is shown in Fig. 4.8 [4.37]. It uses a water phantom RFA3 (Therados) equipped with a silicon detector (active surface 1.5 mm dia. and 0.1 mm thick). The electron beam deflected in

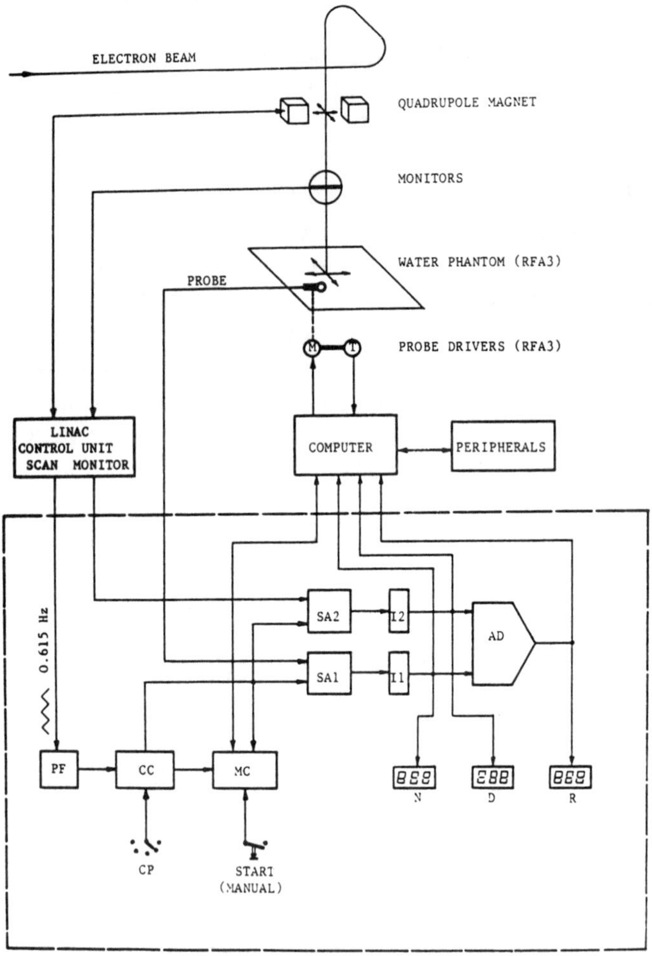

FIGURE 4.8 Schematic of a system for dosimetry measurements of radiotherapy fields with an accelerator equipped with scanned electron beams (after E.Ertan et al. [4.37]).

the x-direction (0.615 Hz) and y-direction (4 Hz) is delivered through an internal control unit to an auxiliary TS01 system intended for triggering and controlling the field analyzer. It is fed, after differentiation, to a pulse forming element PF which generates trigger pulses. The time between two successive triggering pulses is 1.58 s and it corresponds to the complete scan cycle. The number of such cycles, during which the system can be operative, may be programmed with a CP pre-selector determining the total measuring time by the number of pulses subsequently counted in the cycle counter CC. The control unit MC, coupled to the above units, activates the switched amplifiers SA1 and SA2. As a result, the currents from the point detector (probe) and accelerator dosimetric monitors are synchronously integrated during an exactly defined time interval. The outputs of both integrator I1 and I2 are coupled to an analog divider AD which ensures precise division of both charge values. The ratio of these charges, R, the readings N (detector probe charge) and the denominator D (dose monitor charge) are displayed on LED indicators.

Thanks to the fact that the measuring time is synchronously triggered with a lower scan frequency the minimum measuring times of single points can greatly be reduced.

Reproducibility for standard deviations not larger than 1% for a given measuring point may be achieved within 6.3 s.

In measurements carried out with scanning systems it is necessary to have the dosimetric measurement related to the full scan cycle or its multiple.

4.1.6.4 Some Monitoring Measurements

The rationale for and the method of employing monitoring measurements essential for good quality of performance of radiotherapy accelerators have already been presented in Sec. 3.1.6.6. In addition to measurements performed by means of devices incorporated in the accelerator or by those included in the standard equipment, several measuring instruments have been developed for comprehensive quality control. One such instrument is a monitor presented in [4.39]. Its detector consists of 49 ionization chambers with an air volume of 1 cm^3 each, arranged in a 7x7 matrix to cover a field size of about 30x30 cm^3. The detection unit measuring 34x50x12 cm and weighing about 9 kg is located in the treatment room. Each chamber incorporates a separate electrometer amplifier and the readings from the set of 49 amplifiers are processed by a computer situated outside the treatment room.

During monitoring measurements the detector assembly is irradiated with a dose of 300 cGy. The currents of all chambers are then integrated by individual amplifiers. The amplifiers' output voltages, proportional to the charge collected by individual chambers, are sampled and displayed by a computer on a video monitor. The final readings can be represented in the manner shown in Fig. 4.9. The assembly can display the readings from individual chambers as numbers for the whole 7x7 detector matrix (Fig. 4.9a) or as percentage deviations of in-

(a)

906	929	941	932	937	938	937
911	938	955	952	956	945	931
924	948	952	973	957	936	927
919	937	944	946	941	937	920
932	941	953	973	957	946	941
935	948	956	984	958	949	935
927	935	945	960	948	941	937

(b)

-5	-2	-1	-2	-1	-1	-1
-4	-1	0	0	1	-1	-2
-3	0	0	2	1	-2	-3
-3	-1	-1	946	-1	-1	-3
-2	-1	0	2	1	0	-1
-2	0	1	4	1	0	-2
-3	-2	-1	1	0	-1	-1

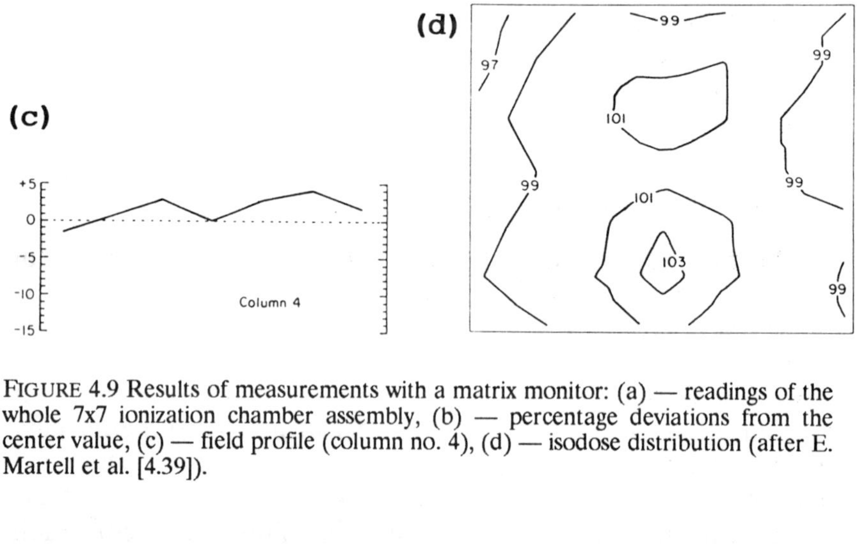

FIGURE 4.9 Results of measurements with a matrix monitor: (a) — readings of the whole 7x7 ionization chamber assembly, (b) — percentage deviations from the center value, (c) — field profile (column no. 4), (d) — isodose distribution (after E. Martell et al. [4.39]).

dividual measuring points relative to the center chamber values (Fig. 4.9b). A profile plot along any row or column of the matrix (Fig. 4.9c) can also be presented. In order to estimate the field flatness an *iso-intensity contour map* can be displayed as in Fig. 4.9d. The system can provide a reading of the water phantom calibration coefficient in terms of monitor units per 1 cGy.

The above systems ensure fast and comprehensive monitoring of the fundamental operational parameters of the accelerator.

4.2 Radiotherapy Planning and Management

4.2.1 Chronology of Development of Clinical Procedures

Fig. 4.10 presents a flow chart of typical clinical procedures employed in the diagnosis and radiotherapy of malignant tumors. Phase One consists of the diagnosis of tumors by means of such methods as X-rays, radioisotopes, computer tomography, ultrasound, etc. When all the necessary diagnostic examinations (together with supplementary tests) have been made and the malignancy has been classified, the decision on the appropriate form of treatment must be taken. After appropriate

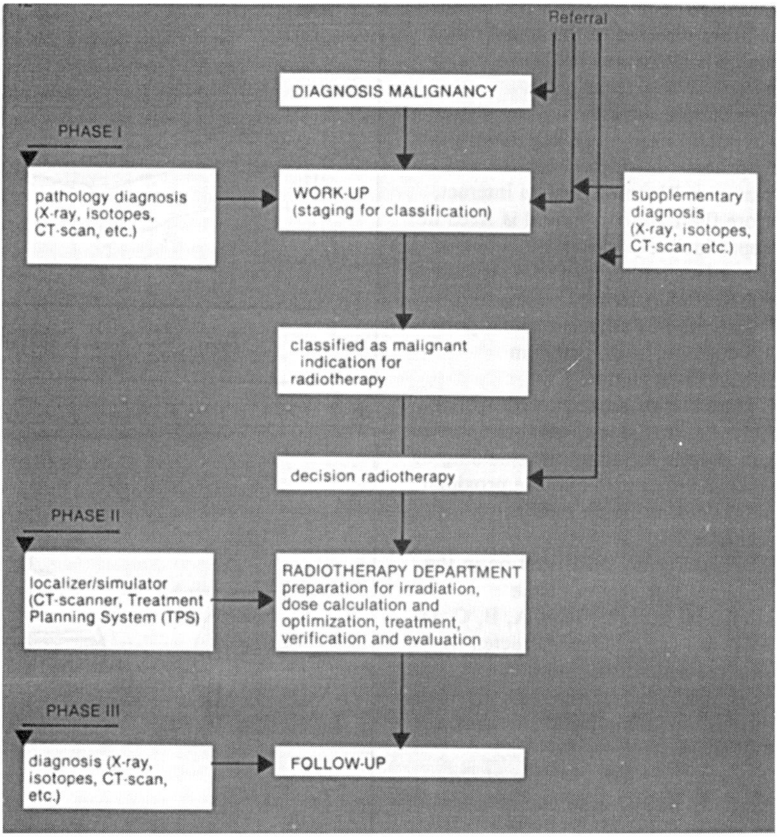

FIGURE 4.10 Clinical procedures employed in the diagnosis and therapy of malignancies.

radiotherapy has been decided upon, the patient is referred to a radiotherapy department and Phase Two begins. Before radiation therapy is started, the patient undergoes further diagnostic examination and the treatment planning is initiated. At this stage, the decision on the proper method of irradiation must be taken and the dose distribution must be calculated and optimized for the prospective radiotherapy fields. Next, the results of planning must be verified and the radiation treatment implemented. Phase Three (Fig. 4.10) is a follow-up which also includes X-ray, radioisotope, and CT-scan examinations to evaluate the treatment. In each phase it is, of course, mandatory to keep the patient's relevant records and data.

Until the mid-1970s, a large number of radiotherapy centers had been using manual methods of therapy planning and its optimization (Table 4.5). Since the late 1960s, the fast growth of

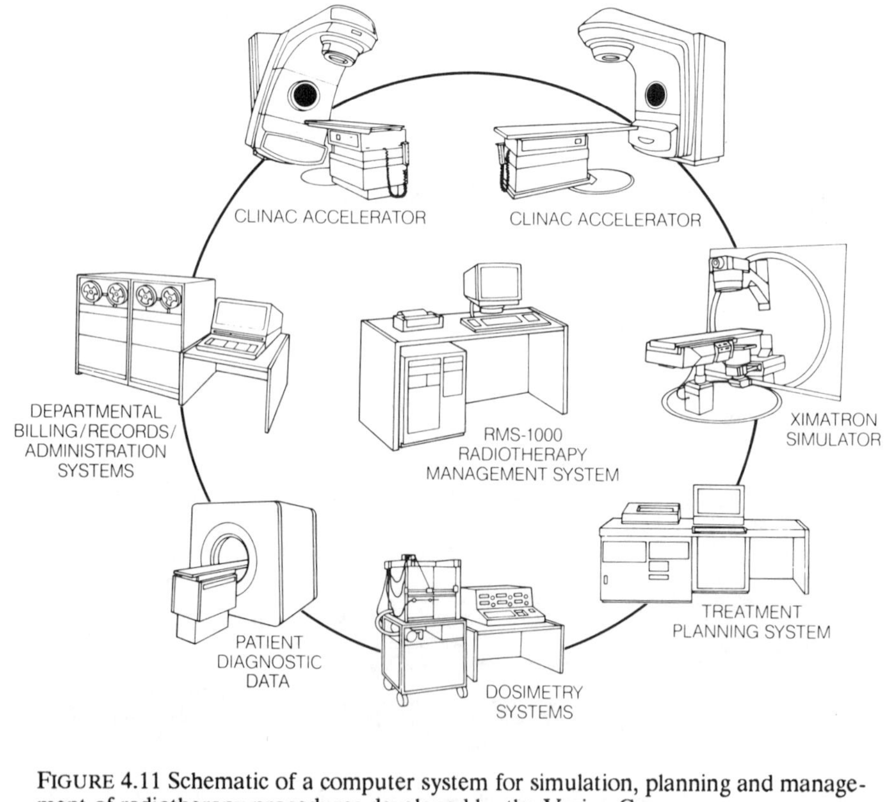

FIGURE 4.11 Schematic of a computer system for simulation, planning and management of radiotherapy procedures developed by the Varian Company.

computer techniques has brought about changes in the automation of several operations in therapeutic procedures, mainly in the process of the time-consuming therapy planning. In the 1970s, simulation techniques (Sec. 4.2.2) and computerized radiotherapy planning methods were introduced on a larger scale. Accelerator manufacturers developed dedicated operational systems which facilitated planning and implementation of the treatment process as well as the optimum control of individual stages of this process. Fig. 4.11 is a schematic representation of a modern system developed by the Varian Company, one of the top manufacturers of radiotherapy accelerators. It is a multi-accelerator system as some more sophisticated radiotherapy departments are now being equipped with two to four machines. Apart from the simulator, the system incorporates the therapy planning system proper which cooperates with a computerized dosimetry unit and is also fed with data from diagnostic instruments. The whole assembly is controlled by a RMS-1000 Management System (Fig. 4.11) coupled, together with other machines, to the hospital general computer system.

In the early 1980s the rapid development of computed tomography evoked positive response in the design of a new eneration of radiotherapy planning and management systems. New planning systems which integrated tomographic data were then born and new simulation techniques based on these data were developed. All this helped to increase the accuracy of planning, particularly that of electron therapy. Verification systems as well as patient localization and positioning techniques were also improved.

The 1980s also witnessed a development of a more comprehensive approach to all phases of the radiotherapy process. In addition to some 3D planning systems (Sec. 4.2.4) work was started on the design of large regional systems, such as, for example, the Scandinavian CART system (Sec. 4.2.4.7).

Since the early 1980s intensive work has been carried out — mostly in the USA — on a totally new generation of intelligent and knowledge based systems which would be able, in addition to computer aided decision making, to evaluate therapeutic decisions critically and draw the necessary conclusions. A concise description of this work is presented in Sec. 4.2.4.8.

4.2.2 Radiotherapy Simulators

The sequence of a present-day radiotherapy procedure implemented with an accelerator is as follows: simulation — planning — irradiation — checking the results — recording and filing. The instruments referred to as *simulators* are used for the first phase of the treatment procedure. Being a copy of the accelerator head and the treatment couch, they are intended for simulating the actual treatment. The view of a simulator manufactured by the Varian Company is shown in Fig. 4.12.

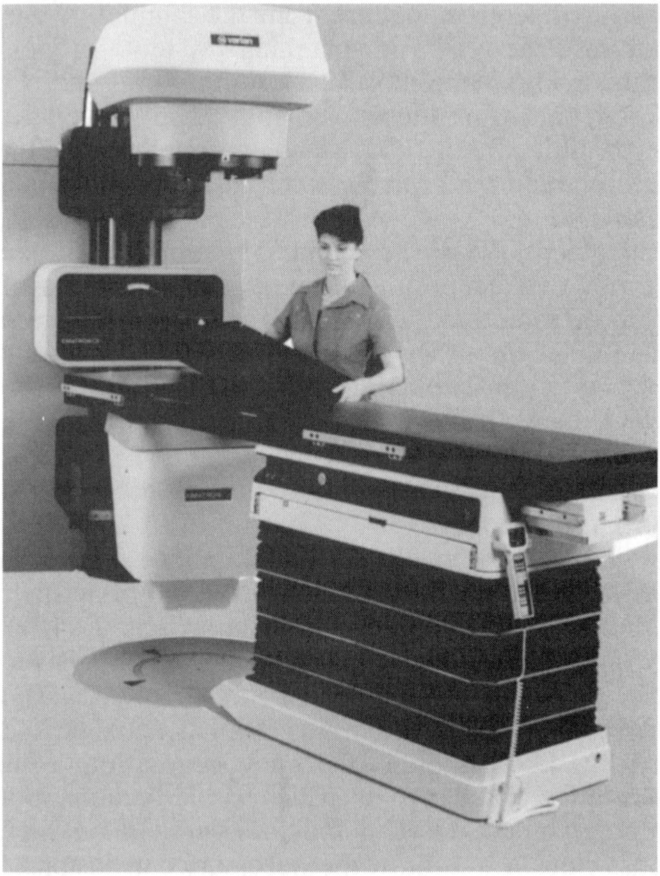

FIGURE 4.12 The view of a varian-Ximatron C-Series radiotherapy simulator (Focus-Axis-Distance, FAD 80 to 140 cm, 50x50 cm^2 max. field size).

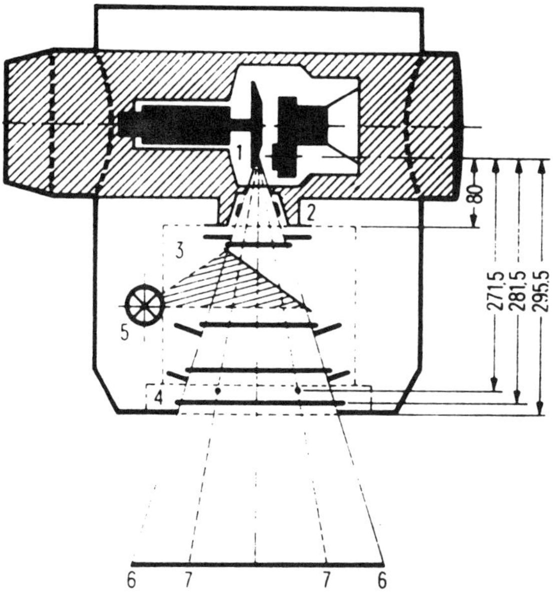

FIGURE 4.13 Construction of an X-ray head of a radiotherapy simulator (Siemens)

The rotating gantry of the same type as that of the accelerator is equipped with a head on which all the devices for beam modification, such as shadow filters and electron applicators (Sec. 3.1.4.4), can be mounted. The range of movement in the simulator head corresponds to that of the accelerator.

The useful beam is simulated with X-rays produced by a unit shown by way of example, in Fig. 4.13 (manufactured by Siemens). The X-ray beam generated by a tube with a rotating anode 1 is emitted through the oil cap 2, its maximum size being limited by a diaphragm 3 which ensures maximum field size 6 in the patient's plane through the rotating axis. Another measuring diaphragm consists of variable wires 4 which permit projection of the treatment field 7 in the same isocentric plane. The total field may also be projected optically thanks to a viewfinder 5.

A typical radiotherapy simulator is a *relatively complex kinematic system,* in which the number of virtual movements is as high as 18. The image of the simulated beam is usually viewed at an X-ray screen or in the *fluoroscopic mode* on a TV monitor. For image recording, radiographic films in conventional cassettes are also used. The accuracy of separate simulator movements is better than 1 mm.

The simulator should perform the following functions. It should: 1. represent the accelerator operational parameters such as the field size, source-isocenter distance, angular rotation about the main axis, etc., 2. represent the movements of the treatment couch, 3. represent the movements and scales of individual parts of the equipment used under real conditions, 4. represent optically the real dimensions of the radiotherapy field, 5. visualize the dimensions and direction of the treatment beam using a simulation X-ray beam, 6. localize coaxially the treatment area relative to the body contour.

Detailed requirements for the construction and operation of radiotherapy simulators are listed in [4.40 and 4.41]. The precision of simulation obtainable in various measuring conditions is analyzed in [4.42] and the integration of a simulator in a computerized radiotherapy network is discussed in [4.43].

Like accelerators (Sec. 3.1.6.6), simulators require a proper quality assurance program [4.44 and 4.45]. Quality control tools for such a program are presented in [4.46].

The main trend in the development of present-day simulators, apart from the improvement in computer control systems and in the X-ray image detector assembly, is the integration of the simulation functions with the functions of a CT scanner (Table 4.5) [4.47 and 4.48]. In the next few years, the simulators are expected to be equipped with a tomographic facility which will make it possible to obtain pictures of the quality comparable to that achieved with the second or even third generation CT scanners [4.48]. The patient will then be able to have his CT scan made in the position strictly corresponding to that during the treatment, which would greatly facilitate precise localization of the lesion irradiated. This technique would be of considerable importance in 3D treatment planning since the plan could be calculated with the patient in position on the couch and, if there are any corrections to be made as to the beam entry point, the new position could fed back to the simulator. The beam angle would then be automatically corrected and a new control radiograph taken would provide a true beam picture from the point of view of the source.

Digital simulators open up new possibilities of widespread use of multileaf collimators (Sec. 3.1.4.4 and Fig. 3.50). Individual collimator leaves could thus be digitally represented on the simulator screen. This allows the radiotherapist or operator to draw an outline around the tumor and have the leaves automatically form around it. This information could then be fed to the accelerator. This is, of course, only possible when the analogue technique of tumor localization, that of using X-ray films, has been replaced with digital presentation (cf. Sec. 3.1.5 and Fig. 3.57) of the port image. The two images would be automatically compared and the mechanical multileaf collimator would be set until its position is fully compatible with the position obtained from the simulator.

Table 4.2 shows the availability of simulation and treatment planning equipment in the US facilities [1.61]. The simulation process is a critical component assuring accuracy of treatment delivery, and every modern radiation oncology department should have a dedicated treatment simulator. Although significant progress in this area had been made in the U.S.A. between 1978-1986 (Table 4.2) nearly 20% of the facilities did not perform any simulation, and simulators, with both fluoro and radiograph capabilities, were available to less than 50% of the facilities in 1986. Improvement in this area is needed and 100% compliance should be the goal [1.61]. Treatment planning computers are also not available to more than 15% of the facilities in 1986, and a similar compliance should be the goal.

TABLE 4.2 Availability of Simulation and Treatment Planning Equipment in the US Facilities (after J.C.H. Chu et al. [1.61]).

	1978	1983	1986
No. Facilities	1093 (100%)	1083 (100%)	1150 (100%)
No. Facilities Performing Simulation	251 (23%)	749 (69%)	947 (82%)
Sim. with Fluoro	165 (15%)	343 (32%)	511 (44%)
No. Facilities with RTP computer		742 (68%)	953 (83%)

4.2.3 Radiobiological Fundamentals of Treatment Planning

A detailed discussion of the radiobiological fundamentals and planning methods used in radiotherapy is outside the scope of this book. The reader is referred to some source papers [4.49 through 4.56]. Here we shall only shortly mention some problems, mainly instrumental issues, with particular emphasis placed on future development in this field [4.56].

Fractionated therapy. The selection of an optimum schedule of *fractionation* still lies at the center of radiotherapists' interest. The criterion which is most commonly used for optimization is called the *Therapeutic Ratio,* the ratio of the dose that gives a 50% probability of local cure to the dose that gives a 50% probability of complications (Fig. 1.6). The following schedule of fractionation is accepted as a standard: *30 2 Gy fractions and 5 fractions weekly over 40 days.*

Since the dependence of the probability of cure and that of complications on the total dose is not unequivocal and is a function of fractionation, the 50/50 criterion is not unequivocal, either. The optimum *fractionation protocol* is one that gives the

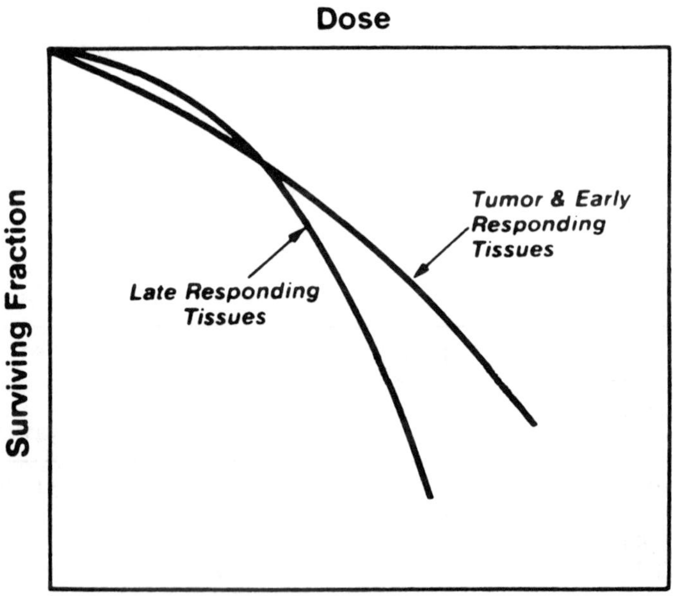

FIGURE 4.14 The dose response relationship for late-responding tissues in more curved than early-responding tissues.

TABLE 4.3 Fractionation Schemes (after C.G. Orton [1.80]).

Parameter	Type of fractionation				
	Conventional	Hyper-	Accelerated	Hypo-	Rapid
Gy/Fraction	1.7-2.5	0.8-1.3	1.6-2.2	>3	>2.5
Fractions/day	1	2-5	2-3	<1	1
Number of fractions	25-35	>50	25-35	<20	<25
Overall treatment time (weeks)	5-7	5-7	<4	5-7	<5
Total dose (Gy)	45-70	55-80	40-65	30-60	30-60

greatest probability of local cure, which is equivalent to obtaining a minimum number of surviving tumor cells [4.57 and 4.123].

Fractionation is based upon the 4 R's of radiotherapy: (1) repair, (2) reoxygenation, (3) repopulation and (4) redistribution [1.80]. *Repair* is represented by the shape of the shoulder of the cell survival curve at low doses. Since, in general, cell survival curves for the late reacting tissues of the body which are typically responsible for affecting tissue and organ tolerance, have a shallower shoulder than the curves for early reacting tissues and tumors (Fig. 4.14), low doses per fraction and low dose rates are of utmost importance when normal tissue damage has to be limited, whereas large doses per fraction might be more appropriate when rapid tumor response is vital. Various *fractionation schemes* are shown in Table 4.3.

Some creative combinations of these have also been tried with some success, such as *hyperfractionation* to a large field with concomitant accelerated fractionation to a small boost field. One of the most promising of recent modified fractionation studies is that of accelerated hyperfractionation in which 3 fractions per day of 1.4-1.5 Gy are given on 12 successive days for advanced head and neck and bronchial tumors. It is likely that this could become standard practice for the treatment of tumors with a short cell doubling time or when radiation-induced accelerated proliferation is suspected.

Reoxygenation of hypoxic regions of tumors is thought to be of major influence in keeping the doses needed to sterilize tumors below tolerance limits for many sites. Such reoxygenation takes time, so protraction of the course of radiotherapy is often necessary.

Repopulation of regions of tissues in which cell-damage has occurred is a two-headed sword. On the hand, in normal tissues it is desirable to allow repopulation to occur whereas, in tumors it is important to keep repopulation to a minimum. Hence, when normal tissue damage has to be avoided, prolongation of treatment schedules are often used when cells in these tissues are capable of repopulation. Conversely, for rapidly growing tumors overall treatment times are reduced.

Redistribution, which occurs when partial synchrony of cells into the resistant phases of the cell cycle is induced after each fraction of radiation, is another two-headed sword. If the next treatment fraction is delivered when these cells have cycled into the sensitive phase, then this is good if these are tumor cells but bad if they are of normal tissues. Not enough is known yet to utilize redistribution to advantage in radiotherapy. It is possible that, during this decade, certain hyperfractionation schemes may be devised specifically to take advantage of this phenomenon for some tumors in some sites [1.80].

Radiobiological models. One of the methods of determining optimum methods of fractionation is one that involves mathematical modelling of processing that take place in irradiated cell colonies. When ionization radiation interacts with matter some of its energy, defined by the absorbed dose, causes excitation and ionization of atoms. As a result, chemical and biological changes occur in living cells, which lead to transient or permanent transformations in the whole body. One of the models which describe the effect of radiation on living cells is the so-called *multiple-target model.* It represents a cell as a group of active centers to be hit by the radiation. For small doses only some targets become damaged. This leads to sublethal lesions in the cell which are then capable of regeneration with time. However, when a given threshold dose is exceeded lethal lesions occur, cells are unable to regenerate, and, as a result, they die. Apart from the absorbed dose, other factors, such as the type of the irradiated cells, the degree of their oxygenation, temperature, phase of the cell cycle and the type and spectrum of radiation, all this affect the type of injury.

For the multi-target model it has been assumed that the survival rate of a cell colony which is homogeneous from the point of view of its radiation sensitivity can well be described by

$$S/S_0 = 1 - \left(1 - e^{-d/D_0}\right)^n \tag{4.7}$$

where S_0 is the initial number of cells, S is the number of surviving cells, d is the radiation dose, D_0 and n are the parameters depending on the above factors and the degree of well-oxygenated cells.

The $\log(S/S_0)$ vs. radiation dose is referred to as a *survival curve.* Fig. 4.15 is an example of such a curve expressed by equation (4.7). It also provides explanation for the parameter

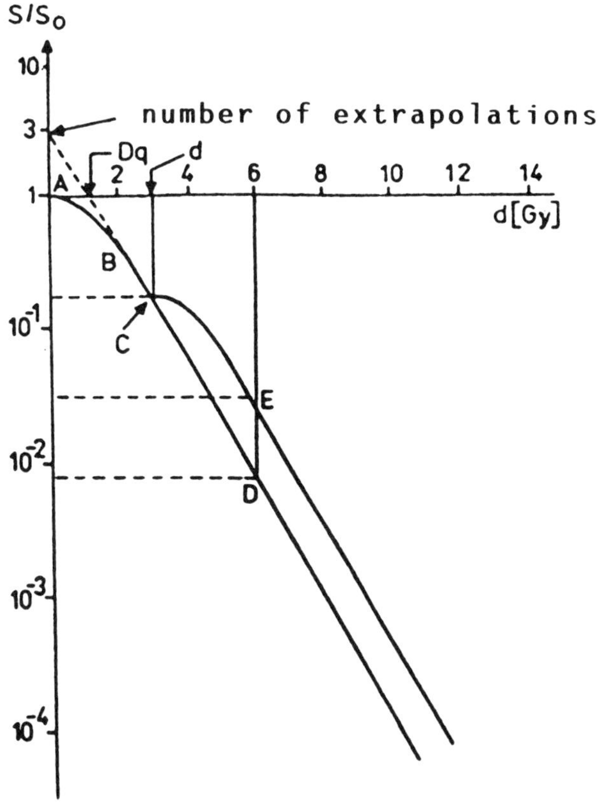

FIGURE 4.15 The cell survival curve: D_0 = 100 cGy, n = 3 (the number from extrapolations) and the quasi-lethal dose $D_q = D_0 \ln(n)$.

n, which is a number taken from extrapolations, and D_q, which is called the *quasi-lethal dose.* D_0 determines the slope of the rectilinear part of the survival curve. The section AB, or, in other words, the shoulder of the survival curve, represents the cumulation of sublethal injuries. The cells are able to repair the injuries with time. If cells are exposed to another irradiation right after the first one, this effect is represented by the section CD on the curve. However, if adequate time is left between irradiations for the cells to recuperate from sublethal injuries a new shoulder of the curve (DCE) is obtained.

Apart from fractionation, mathematical modelling of the processes occurring in irradiated cell colonies is of fundamental importance for the radiobiological problems of therapy planning. A detailed overview of the latest developments in this field is given in [4.58 and 4.124].

Methods of irradiation. Fundamental *techniques of irradiation* employed in radiation therapy are presented in Fig. 4.16. The *treatment volume,* which approximately corresponds to the *tumor volume,* is shown within the solid lines as small squares. According to the ICRU 29 recommendations [4.60], this volume is confined by a surface isodose curve which corresponds to the minimum dose to be absorbed by the tumor volume. In practice, the treatment volume is most frequently defined by the 80% isodose curve. The *real irradiated volume* is, however, larger (the dotted lines in Fig. 4.16). Following the above recommendations [4.60], the real irradiated volume is taken to mean a volume in which the absorbed dose has a significant value from the point of view of malignant cell radiation sensitivity, or strictly speaking, of the *cell tolerance.* This dose is usually expressed in percent (e.g. 50%) with respect to the absorbed dose in the treatment volume.

The methods presented in Fig. 4.16 have been classified in two groups: *conventional uniform beam irradiation* (Fig. 4.16a) and *unconventional non-uniform beam irradiation* (Fig. 4.16b) *techniques.* In general, the value of a given irradiation technique strongly depends on the extent at which the irradiated volume coincides with the treatment volume. The closer the shaded area approximates the area with a dashed contour the better the technique. From the analysis of the techniques presented in Fig. 4.16 it may be concluded that the methods using non uniform techniques provide a better match of these areas.

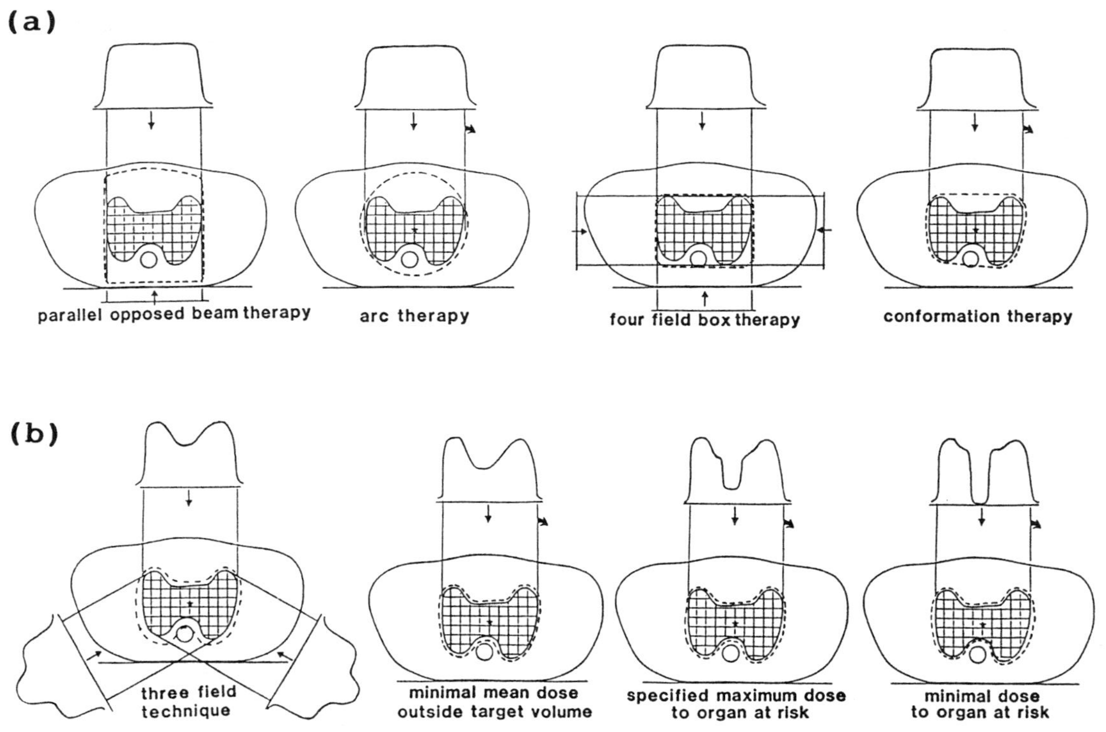

FIGURE 4.16 A list of the main radiotherapy techniques: (a) — conventional uniform beam techniques, (b) — nonconventional non-uniform beam techniques (after N. Ulso and A. Brahme [4.59]).

Conventional radiotherapy planning must give an answer to the question of the dose distribution in the patient when a given technique, that is, a given set of incident beams is employed. A precise answer to this seemingly simple question is a very complex problem.

Fundamental methods for photon beams. The *method of Cunningham* [4.64] provides the widest possible scope of operation and gives very good results over a full range of X-ray and gamma energies regardless the size and shape of the beam field or shields, wedge filters or other beam modifiers. In this method, the radiation in the scattering medium is mathematically divided into *primary radiation and scattered radiation.* The contributions from both kinds of radiation to the total dose are separately calculated with the use of T and S coefficients defined as

$$T(d, r_d) = \frac{D_F(d, r)}{D_A(d_A, r)}, \tag{4.8}$$

$$S(d,r_d) = T(d,r_d) - T(d,0), \tag{4.9}$$

$$T(d,0) = \lim_{r_d \to 0} T(d,r_d) \tag{4.10}$$

where $D_f(d,r)$ is a dose in the phantom at the depth d for a field with a radius r = SSD : (SSD + d); SSD is the source (conversion target) to phantom surface distance; $D_A(d_A,r)$ is the reference dose determined at the same point as $D_F(d,r)$.

If D_A is measured in air, coefficient T is referred to as the *tissue-air ratio* (TAR). For D_A measured in a phantom this coefficient is called a *tissue-phantom ratio* (TPR), but when d_A is the depth of the maximum dose the term *tissue-maximum ratio* (TMR) is used. For energies higher than 400 keV, it is most useful to employ TPR coefficients because of the difficulties involved in measuring the dose in air and because of the change in the maximum dose depth with field size. The scattering coefficients S are denoted, by analogy, as SAR *(scatter-air ratio),* SPR and SMR (Table 4.4).

TAR (d, 0) is extrapolated and it cannot be measured directly. For energies used in teletherapy, TAR is independent of the

TABLE 4.4. A List of Symbols and Parameters Used in Radiation Therapy Planning.

BSF	back-scatter factor
CF	correction factor
ETAR	equivalent tissue-air ratio
OF	output factor[1)]
PDD	percent depth dose
PSF	peak scatter factor
SAR	scatter-air ratio
SMR	scatter-maximum ratio
SPR	scatter-phantom ratio
SSD	source-to-skin (surface) distance
TAD	target-axis distance
TAR	tissue-air ratio
TMR	tissue-maximum ratio[2)]
TPR	tissue-phantom ratio
TSD	target-skin distance

1) defined as a ratio of the dose rate at d_{max} corresponding to maximum dose and the irradiated field to that for d_{max} and the reference field.
2) for depth d_A and the corresponding d_{max}.

source-to-skin distance. Therefore, in order to calculate the dose it is sufficient to know the depth dependence of TAR.

The primary and the scattered doses calculated at the depth d on the beam axis can be expressed by

$$D_P(d,r) = D_A(d_A,r) \cdot T(d,0) \qquad (4.11)$$

$$D_R(d,r) = D_A(d_A,r) \cdot \overline{S\,(d,r)} \qquad (4.12)$$

where S (d,r) is the resultant value of S determined with the *Clarkson method,* i.e. by summing up contributions from all sectors making up the radiation field. Thus

$$\overline{S(d,r)} = \frac{1}{2\pi} \sum_{i=1}^{\infty} \theta_i \; S\,(d,r_i) \qquad (4.13)$$

where N is the number of sectors, θ_i is the angle of the i-sector, r_i is the radius of the i-sector.

The *percent depth doses,* PDD, measured for the same energies but different therapy machines, exhibit small but significant differences, usually ±2%. These differences result from the differences in the spectrum of the radiation which reaches the phantom surface and from the different definitions of the radiation field size. The radiation spectrum is a complicated function of the material of the collimator and therapy unit head and of the geometry of the assembly.

From equation (4.8) the following expression for T can be obtained

$$T(d,r_d) = \frac{PDD(d,r)}{PDD(d_A,r)} \cdot \left(\frac{SSD + d}{SSD + d_A}\right)^2 T_A\,(d_A,r_A) \qquad (4.14)$$

where d_A is the reference depth, r_d, r r_A are field radii at the depths of d, 0, and d_A, respectively, for the same collimator aperture.For the depth d_A equal to that of the maximum dose, the coefficient $T_A\,(d_A,r_a)$ is referred to as a peak scatter factor (PSF) when TAR values are calculated, whereas for energies higher than 8 MV, $T_A\,(d_a,r_a)$ can be taken as equal 1.

In order to determine coefficients T and S the depth dose curve and coefficient T_A at d_A must be known. Moreover, the knowledge of values of the coefficients T and S over the full useful range of fields r=(0...60)cm and depths d=(0...30)cm can only be obtained if the values PDD at over 1000 points are known. This also includes the knowledge about the fields whose sizes exceed the maximum field which can be obtained for a given accelerator or any other radiotherapy machine. Therefore, to delimit the number of dosimetry measurements, PDD is expressed as a parametric function of two variables: field size and depth. For this purpose, mathematical models are used, e.g. the *Kornelsen-Young model* [4.61] which describes the PDD function outside the area of growth as a function of depth d and square side w

$$PDD(d,w) = 100\left[1 - \left(1 - e^{-\frac{d - d_m(w)}{Q(w)}}\right)^{M(w)}\right] \qquad (4.15)$$

where $d_m(w)$ is maximum dose depth as function of field size, $Q(w)$ is a parameter which describes the gradient of the exponential dose decrease with depth and $M(w)$ is a parameter which accounts for the influence of scattering effects on the deviations from a strictly exponential decrease of PDD. The maximum dose depth d_m can be measured directly, whereas parameters Q and M can be determined by matching the function PDD(d,w) with measuring points by the least squares method.

In the region of dose growth another model is used, that of Bomford [4.62], which utilizes two parameters instead of three employed in the Kornelsen-Young model. Thus we have

$$PDD(d,w) = PDD(0,w) + \left(100 - PDD(0,w)\right)\left(1 - e^{-\frac{5d}{d_m(w)}}\right) \quad (4.16)$$

PDD(0,w), as well as the dose itself in the growth region, may be measured relatively precisely by using, for example, an extrapolation chamber. It follows from equation (4.16) that already at the depth of 1/3 of d_m, PDD equals 80% and it only slightly depends on PDD(0,w). It is thus the maximum dose depth, d_m, which determines the value of PDD in the growth region and which can be measured by means of a standard ionization chamber. Computer programs for both models are given in [4.63].

In the further development of the TAR method, Sontag and Cunnigham introduced a *correction coefficient* CF. This method is referred to as the *equivalent tissue-air ratio method* (ETAR) [4.64]. Modification of the method was latter made by Iwasaki [4.64 and 4.66].

Now arises the question of determining the distribution of an optimum therapeutic dose which is taken to mean a dose characterized by the following features: 1. the dose gradient in the tumor area should be minimized, 2. the ratio of the absorbed dose in the tumor volume to the total dose absorbed by the patient should be maximized, 3. the total dose should be minimized, 4. the reference isodose corresponding to the dose necessary to destroy the tumor should have shape similar to that of the tumor, 5. the dose absorbed by the critical organs should be as low as possible, and 6. the volume of the maximum absorbed dose should be located precisely inside the tumor volume.

Fundamental methods for electron beams. Treatment planning in electron therapy is much more diversified and complicated than that in photon therapy [1.26]. For low-energy electrons and single straight-on fields the knowledge of the parameters on the central axis is sufficient for planning. Complications appear when the effects of the surface curvature, oblique beam incidence or beam rotation have to be taken into consideration. Therefore, for radiotherapy planning of deep-seated tumors with beam energies above 10 MeV, when tissue inhomogeneities, irregularly shaped fields and uneven surfaces are encountered it is necessary to use sophisticated computer-assisted calculation methods.

Fig. 4.17 shows a classification of models employed in radiotherapy planning with electron beams. Physical models are among the best approaches to this problem. They make use of diffusion (age)theory and of the multiple Coulomb scattering theory. A detailed overview of all models up to the late 1970s can be found in [4.67] and the work carried out in the 1980s is presented in [1.26].

In practical implementation of clinical calculations, the summation technique based on a *pencil-beam idea* is often used. In

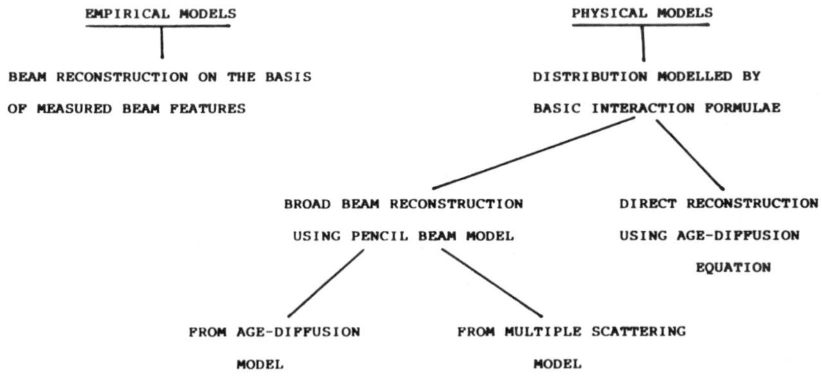

FIGURE 4.17 Methods used in electron therapy planning models (after S.C.Klevenhagen [1.26]).

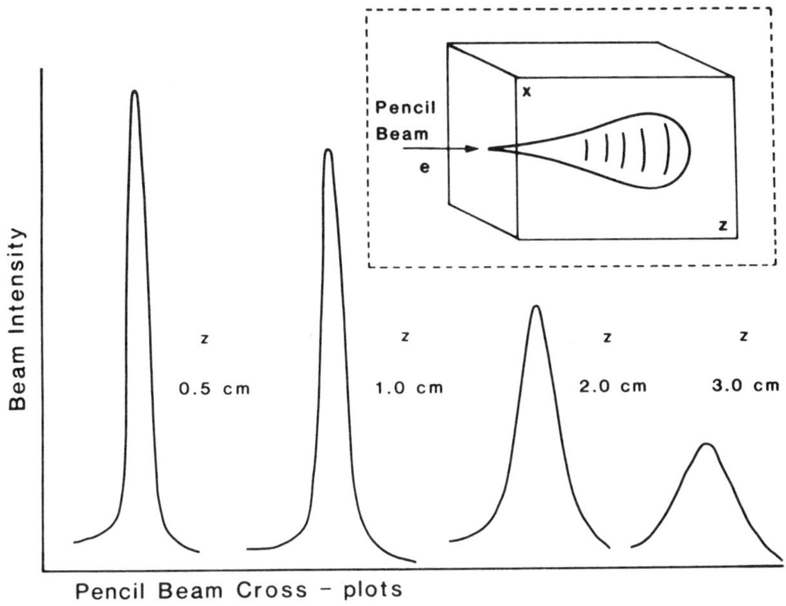

FIGURE 4.18 Experimental profiles for 10 MeV electron beams at 0.5, 1.0 2.0 and 3.0 cm (after S.C. Klevenhagen [1.26]).

this technique, the broad-beam is considered as a set of pencil-like beams contained in the electron field (Fig. 4.18). The energy distribution in the pencil beam is calculated from the multiple-scattering or diffusion(age) models. The effects of each pencil beam are then summed up within the entire field. This technique makes it possible to determine field distributions of arbitrary shape and diversified geometric conditions. It also allows for corrections of tissue inhomogeneities using data from the patient's CT scans. All these calculations can be carried out in 3D geometry.

The *pencil beam technique* has been developed specifically for radiotherapy accelerators. Fig. 4.18 shows profiles for a typical distribution in the central plane of a 2 mm beam of 10 MeV electrons in a water phantom. The curves obtained for various depths show that the resulting lateral distribution has the shape of a Gaussian probability function. The width of the curve is defined by the *beam lateral spread parameter*, σ, just as the

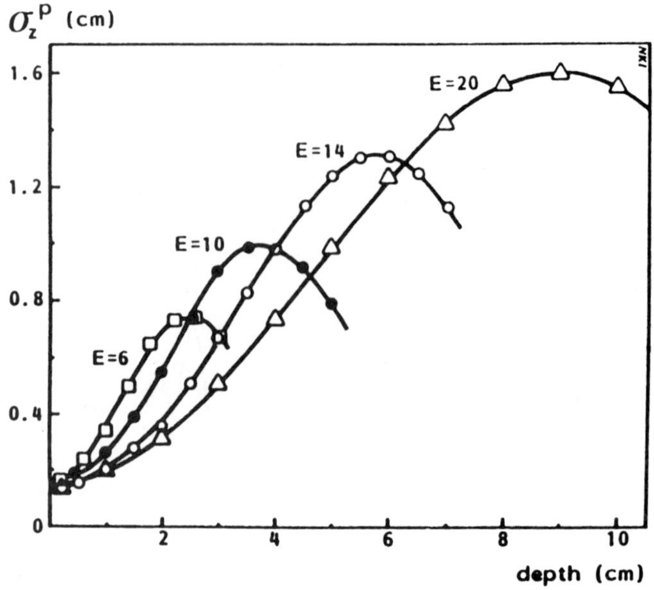

FIGURE 4.19 Electron beam spread parameter, σP(z), determined experimentally with applicator for a 10x10 cm field and for various energies and depths (after I.A.D. Bruinvis [4.70]); σ(z) values are derived values from off-axis to on-axis dose ratios at the edge of the field — these σ values are denoted by σP(z) (profile).

spread of experimental data is defined by the standard deviation. Therefore, if the distribution curve is narrow and has a sharp peak, σ is small, its value increasing with the increase in the distribution width. Fig. 4.19 shows that the lateral beam spread depends on depth: first it increases with increasing depth and then, after reaching a maximum, it decreases.

Thanks to the use of only a few parameters, the pencil beam technique is especially well suited for developing models of electron beams. For this reason, it is being now commonly employed in summation techniques for electron field planning.

By using the pencil beam technique several practical algorithms have been developed. In recent versions, they make it possible to account for the tissue inhomogeneities by making use of the CT-scan data on electron density determined for the patient body. A typical algorithm of this kind developed by Cygler et al. [4.69] has the following form

$$D(X,Y,Z = \quad (4.17)$$

$$= \left(\text{erf}\,\frac{WXZ/2 - X}{\sqrt{2}\,\sigma} + \text{erf}\,\frac{WXZ/2 + X}{\sqrt{2}\,\sigma}\right) \cdot \quad \text{(field's width)}$$

$$\cdot \left(\text{erf}\,\frac{WXZ/2 - Y}{\sqrt{2}\,\sigma} + \text{erf}\,\frac{WXZ/2 + Y}{\sqrt{2}\,\sigma}\right) \cdot \quad \text{(field's length)}$$

$$\cdot\, D_0(0,0,Z_{sk}) \left(\text{erf}\,\frac{WXZ/2}{\sqrt{2}\,\sigma}\right)^{-2} + \left(\frac{SSD + Z_{sk}}{SSD + Z}\right)^2 \quad \text{(depth dose C/A)}$$

This algorithm, based on the generalized theory of multiple scattering in Hogstrom notation, applies to regular fields WX x WY and to a given SSD. The dose $D_0(0,0,Z_{sk})$ refers to the central axis value; Z_{eff} is the effective depth, erf (error function) is the beam lateral spread function and W is pencil beam weight.

A detailed overview of the present-day models employed in electron field planning is presented, together with extensive bibliography, in [1.26].

Planning in 3D geometry. The methods of radiation therapy planning considered thus far are implemented with the aid of systems which will be discussed later in Section 4.2.4. They enable construction of comprehensive lateral dose distributions, making use of one degree of freedom of the accelerator, i.e. the head rotation angle about the isocenter. This approach restricts radiotherapy planning to *coplanar beams.* The present-day teletherapy machines, however, make it possible to use a larger number of degrees of movement. In addition, the rotation of the set of collimators and the therapy couch rotation can also be used. As a result, irradiation with *non-coplanar beams* can be implemented. *Methods of 3D planning* [4.71 through 4.77] aim at utilizing additional movements and developing optimized radiation therapy planning in the region of the chest, neck, head, esophagus and other anatomical areas. The recent generation of 3D planning systems will probably achieve the dose accuracy determination of 3%, which will constitute an important step forward compared with the existing 2D systems. Three-dimensional radiotherapy has been increasingly valued as a method which uses 3D planning techniques and other methods of irradiation control to ensure that all the spatial possibilities are exploited.

Thus 3D planning should enable precise dose determination in any plane. This is true both for doses in all standard lateral, longitudinal and circumferential planes, as well as in all oblique planes. In this way, the dose distribution along the tumor length or along the radiosensitive organ can be followed so that all the details of the resulting plan can be made more precise. The planning system which uses non-coplanar beams, apart from handling the necessary calculations, must provide the user with the conceptual information so that the sometimes unusual geometry of the plane involving the collimator and table movements can be represented in a verifiable, clear and feasible system.

Typically, in 3D planning, the approach adopted is that called *Beams Eye View* (BEV) schematically shown in Fig. 4.20. The patient's body sections are analyzed in such a way as if the observer's eye was placed at the source and the observer was looking along the beam axis [4.79]. Each projection corresponding to the 'view' for a given beam is shown on the screen of the monitor incorporated in the 3D system. The patient is then represented by a set of sections to match the beam's divergence and each section shows the target area and the surrounding normal tissues. The beam is represented by a rectangle which clearly visualizes the beam's dimensions and position, and thus enables estimation of the tumor area coverage by the beam.

By way of example, the method of coordinate transformation defining the accelerator movements, described in [4.76], will be

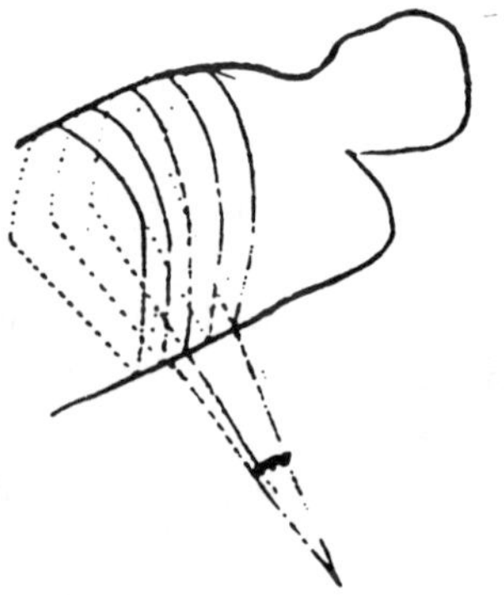

FIGURE 4.20 Beam's Eye View.

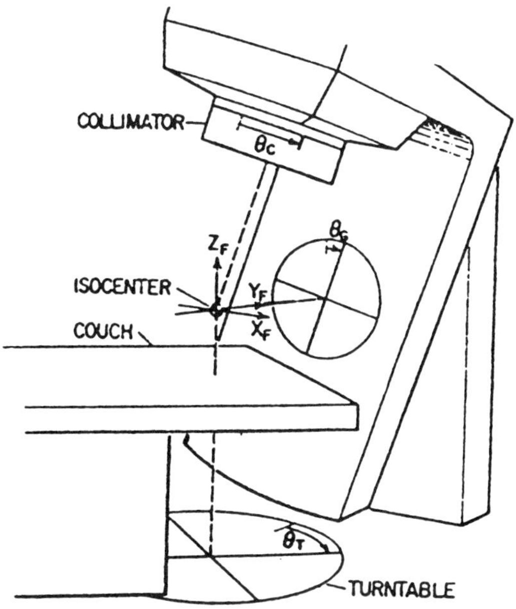

FIGURE 4.21 The accelerator kinematic system (after R. Siddon and B.E. Bjarngard [4.76]).

presented below. A radiotherapy accelerator (Fig. 4.21) may be described by three orthogonal coordinate systems: the *turntable T system* (Fig. 4.22a), the *head (gantry) G system* and the *collimator C system.* Each of the coordinate systems is obtainable from any of the others and from a fixed (F-system) by rotations.

The origin of the coordinate T-system (Fig. 4.22a), rigidly fixed to the treatment couch, is placed at isocenter. The turntable system is the most convenient form to have all the patient data specified in, for example, of multiple CT scans. Since individual points of dose calculation are usually related to the patient anatomy, the T-system is also convenient for presentation of such points.

The collimator C-system (Fig. 4.22b), fixed with the head (gantry) system shown in Fig. 4.23, also has its origin at isocenter. The C-system is most convenient for determining beam parameters such as the field size, primary profiles, penumbra, etc.

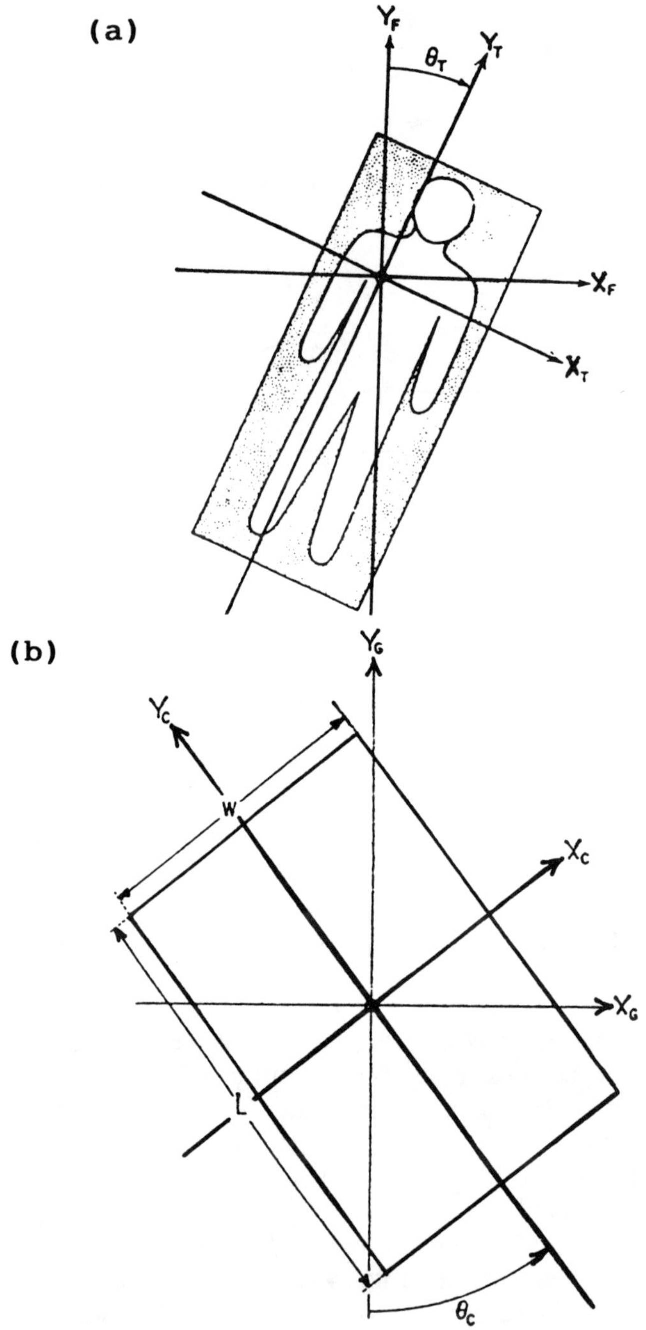

FIGURE 4.22 Schematic of: (a) — T-system, (b) — C-system (after R. Siddon and B.E. Bjarngard).

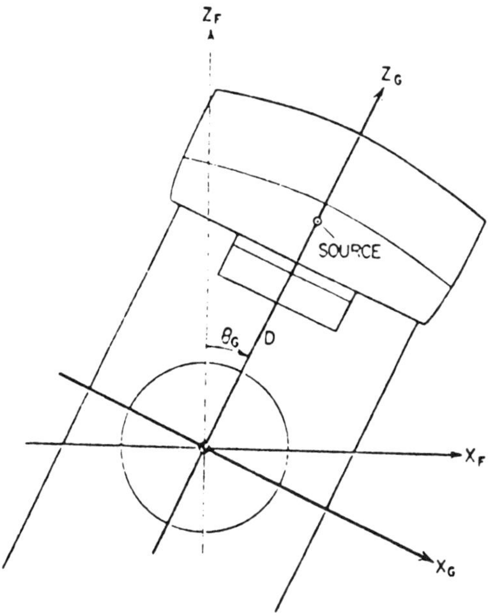

FIGURE 4.23 Coordinate systems of the collimator — head (gantry).

The transformation of data from the T-system to C-system and vice versa may be expressed as

$$[V]_C = R_Z(\theta_C)\, R_Y(\theta_G)\, R_Z(\theta_T)\, [V]_T \qquad (4.18a)$$

$$[V]_T = R_Z(-\theta_T)\, R_Y(-\theta_G)\, R_Z(-\theta_C)\, [V]_C \qquad (4.18b)$$

where $[V]_C$ and $[V]_T$ are the components of the vector V expressed in the C-system and T-system, respectively. The angles θ_C, θ_G and θ_T correspond to the collimator, gantry and turntable position angles, respectively. The rotation operators R_Z and R_Y may be expressed as 3x3 orthogonal matrices.

The following discussion will focus on the geometric problems encountered in 3D dose calculations [4.76]. The radiation dose at the point of calculation, located in space by the vector Q is expressed as primary and scatter components (cf Eqs. 4.8 through 4.10).

In the simplest case, the method of primary dose calculation will have to take into account three coefficients: 1. an attenuation coefficient which depends on the depth d of the calculation point, 2. an inverse square coefficient which depends on the distance r between the radiation source and the point of calculation, and 3. an off-axis coefficient which depends on the off-axis coordinates of the calculation point. The latter coefficient allows for such factors as beam profiles, wedge filters, etc. The calculation of the parameters d, r and the off-axis coordinates for a 3D system is a very tedious procedure but it can be simplified by introducing coordinate transformations as the example below shows.

The calculation point P within the patient's body contour is most frequently defined in the T-system of coordinates. In this system it has components $[P]_T$, whereas the radiation source, located by the vector S, is usually expressed in the C-system and has components $[S]_C$. In order to calculate the depth, d, it is necessary to specify the calculation point P, the source location S, and the patient location in the same system of coordinates. Since the calculation point P and the patient location have been initially expressed in the T-system, the source S (Eq. 4.17) has to be transformed from the C-system to the T-system:

$$[S]_C = R_Z(\theta_T)\, R_Y(\theta_G)\, R_Z(-\theta_C)\, [S]_C \qquad (4.19)$$

and then the depth calculation has to be carried out in the T-system. As the parameters P and S are expressed in the same system of coordinates, the distance r can easily be determined.

The *off-axis ratio* requires that the coordinates of the calculation point be determined with respect to the system covering the radiation field. Since this field is usually given in the C-system, the required coordinates are obtained by the transformation of the calculation point P (Eq. 4.17) from the T-system to the C-system:

$$[P]_C = R_Z(\theta_C)\, R_Y(\theta_G)\, R_Z(\theta_T)\, [P]_T \qquad (4.20)$$

The required value of the off-axis ratio is the expressed in the C-system.

The scatter dose may be given in terms of the weighted sum of the ratios of the scatter to primary components, the weighing factors being calculated with respect to local primary doses. The summation is carried out over the elementary fields contained within the field under consideration and located with the vectors Q. The summation over the field can most easily be done in the C-system of coordinates because this is the system which describes the field. In the C-system, the components of the vector Q are determined by $[Q]_C$. To calculate the local primary dose at Q the depth of the calculation point needs to be determined. For this purpose, the coordinates of the calculation point should be transformed (Eq. 4.18) from the C-system to the T-system, in which the patient is described:

$$[Q]_T = R_Z(-\theta_T)\, R_Y(-\theta_G)\, R_Z(-\theta_C)\, [Q]_C \qquad (4.21)$$

Next the depth calculations are made in the T-system. Another method for fast evaluation of patient set-up during radiotherapy is presented in [4.126].

The common feature of all the methods of 3D planning is the need to use mainframe computers.

Patient position control. The use of CT scanners has permitted precise localization of patient's internal structures. As a result, the present-day radiotherapy planning methods provide high accuracy in *tumor localization*. On the other hand, the new accelerators allow precise shaping of the therapy field. During the implementation of irradiation it is assumed that the patient is properly immobilized. In practice, this is not always so. Immobilization devices such as blocks, polypropylene casts or fiberglass thermo-mold casts are often inadequate and cannot eliminate the patient's movements on the treatment couch. The irradiation process is known to be often disturbed by the patient's uncontrollable movements, especially breathing. In the thorax area, breathing may lead to considerable changes in the patient's contour. For example, during moderate breathing the variations in the real SSD amount to ±1 cm. For deep breathing, they can be as large as ±2 cm. When irradiated with a 6 MV photon beam, a 2 cm change in the patient's thickness can cause change in the dose of 8 to 10%, whereas for 15 MV X-ray similar changes in doses range from 5 to 7% [4.80].

If radiotherapy planning is based on breathing averaged CT scans over the variations in the patient's outline, the resulting dose to the patient during treatment will be about the same as the calculated dose. In treatment planning CT "snapshots" are, unfortunately, also used, because CT scans are usually made in the inspirational phase of the breathing cycle. In such a case, the error resulting from the chest movements caused by breathing can lead to significant errors in dose delivered.

So far, no commercial systems are available which are equipped with a device to determine directly the patient's position. However, several attempts have been made to develop a system of this kind. An example of such a system is shown in Fig. 4.24. A line image is projected from a HeNe-laser on the patient's skin. At the same time, he is viewed by two video cameras at fixed angles relative to the light source. The cameras record the light reflected from the patient's contour surface. An analog-to-digital digitizer feeds this information to a microcomputer. The system measures 25 contour profiles per second. Changes in the contour can be determined on-line: if they ex-

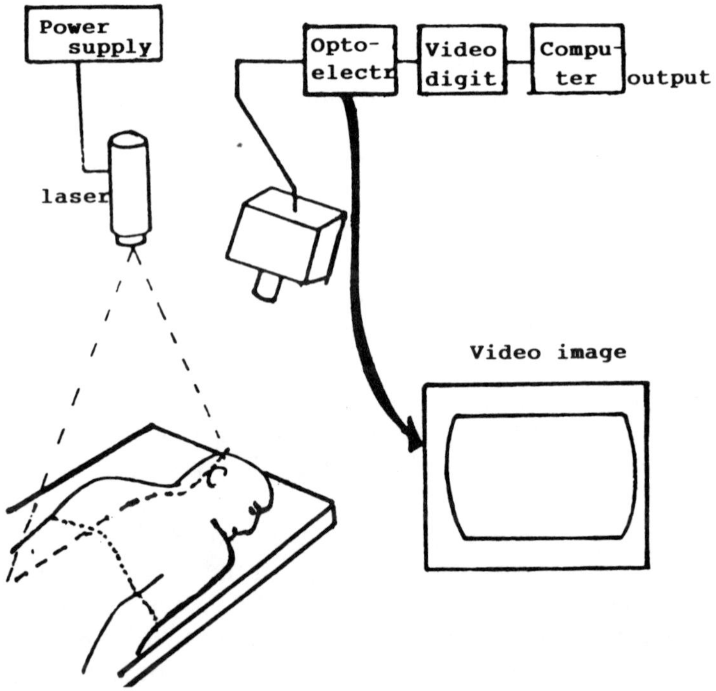

FIGURE 4.24 A system for monitoring patient's position (longitudinal profile) (after S. Peltoli [4.80])

ceed some predetermined limits the microcomputer turns the accelerator off and the irradiation is stopped.

The system shown in Fig. 4.25 can also be used to gate the accelerator so that the irradiation is performed in some selected phases of the breathing cycle. In this way, the technique of gated radiotherapy increases the accuracy of dose delivery. In addition, the patient who makes any rapid movements is safeguarded against overdosage [4.80].

4.2.4 Radiotherapy Planning and Management Systems

4.2.4.1 Individual Dose Calculation

Using the high-energy X-ray and electron beams at his disposal, it is possible today for the radiotherapist to heal a patient by destroying even deep tumors without irradiating the tissue above the tumor beyond their tolerance dose. In most cases the tumor is subjected to a 'crrossfire' of several fields of irradiation in order to reduce the dose imposed on the healthy tissue (Fig. 4.16). Thus it is clear that the skin dose of each individual field cannot be used as a reference, as was the case earlier with conventional X-ray equipment; differences in the anatomy of the patient would lead to the tumor being subjected to different doses. Consequently, the dose at the tumor itself or, even better, the dose distribution throughout the body, has to be determined. Such a distribution can, of course, be calculated manually, though this would take so long that it would not be possible to prepare individual irradiation plans for each patient. If radiotherapy is to be optimized on an individual basis, *computer-aided treatment planning* is therefore essential.

A fundamental problem which arises concerns the representation of the above-mentioned dose distribution throughout the body. Four dimensions are involved: the three physical dimensions plus the dose level. Only a *two-dimensional graphical representation* is possible on a sheet of paper. As a rule, the number of dimensions involved in radiotherapy planning is reduced by limiting the representation to a single plane. Although this plane is usually a cross-section through the patient for certain irradiation techniques it could just as easily be a longitudinal section or be freely oriented. The number of dimensions is reduced once more by the definition of the isodose. The simplest example of an isodose is given by the measurements carried out on an accelerator, using the water phantom. Such isodoses have to be corrected due to the uneven surface of the body and the deviations of the body-tissue densities from the unit density. All treatment planning systems therefore require two types of data: the *radiotherapy machine data* measured in the phantom and the *anatomical data* of the patient. Whereas the machine data have only to be entered once and possibly modified after major repairs, etc., the patient's data have to be entered individually. The input must therefore take place simply and quickly. As a rule, the patient's data are obtained from one or more sections through the patient. The sections themselves can be obtained by various means: for example, the outer body contour can be simply drawn employing the principle of the pantograph. Internal contours (organs and the tumor) then have to be drawn inside this outline by the doctor, to which end he can refer to radiographs taken from various directions with a scale, and make use of his own knowledge of anatomy. This method is, of course, prone to error, and large deviations from the actual position of the tumors and organs are possible. What is more, published mean values have to be used to determine the densities of the organs.

The advent of *computer tomography*, therefore, represented a significant advance: a computer tomograph (CT) shows a cross-section through the patient, and the position of the tumor and the organs, in particular the critical ones, can be determined with high precision. The densities can also be calculated to a high degree of accuracy from the CT numbers, representing the various electron densities.

4.2.4.2 Chronology and Trends in Development

The chronology of the development of treatment planning and execution methods are presented in Table 4.5 which summarizes the work at the Catharina Hospital (Eindhoven, The Netherlands) [4.125].

The aim of this work, in compliance with recent trends of integration of computer-aided functions and systems, was to develop a departmental network in order to fully exploit modern computer assisted information handling. A split-level radiotherapy approach was used, in which the complete radiotherapy process was divided into two categories: (1)

TABLE 4.5 Development of Processes of Treatment Preparation, Treatment Execution and Related Quality Assurance (Radiotherapy Department of Catharina Hospital) (after D. Baker et al. [4.125]).

BEFORE 1974	FROM 1974 UNTIL 1980
– Manual dose optimization by means of isodose charts for Cobalt-60 beams	Introduction of: – Treatment planning computer – Localizer/simulator (X-ray) – Linac (electron beams)
– No simulator – No fixation technique – No verification system	In-house development of: – Fixation techniques – A Body outline measuring device – A megavoltage radiographic system for portal verification
FROM 1980 UNTIL 1984	**AFTER 1984**
– Access to CT-scanners In-house development of: – Verification and recording system – Improved fixation techniques – A practical training programme for technicians	– Integration of CT-data in the dose planning process – Development of a new simulation process completely based on CT-data for high precision radiotherapy especially for electron beams – Ongoing improvement of fixation techniques and verification systems

routine procedures and equipment, and (2) high precision procedures and equipment.

The first category includes about 70% of the patients who are treated with conventional techniques (2D planning, conventional X-ray simulators) (see Table 4.6a). The second category includes about 30% of patients who require high precision procedures and equipment (3D planning, RT-adapted CT-scanners) (see Table 4.6b).

TABLE 4.6a Routine Procedure and Equipment (Catharina Hospital, Eindhoven, The Netherlands) (after M.A. Crommelin, D. Bakker and W. Dries [4.127]).

IMMOBILIZATION Accurate reproducible daily set-up	THERMOPLASTIC CAST 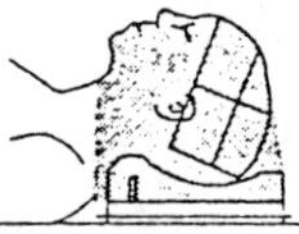
LOCALIZATION & SIMULATION Localization of target volume & simulation of proposed treatment portals	LOCALIZER / SIMULATOR 
DOSE PLANNING 2D - dose optimization for photon beams	2D - DOSE CALCULATION 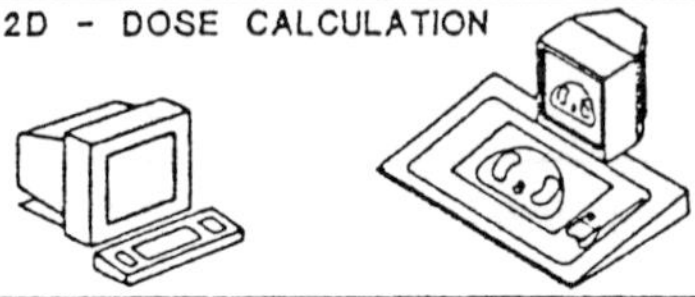
FRACTIONATION Efficiency - oriented treatment schedules	NORMAL TISSUE TOLERANCE $NSD = D \cdot N^{-0.24} \cdot T^{-0.11}$ $TDF = n \cdot d^{1.538} \cdot x^{-0.169} \cdot 10^{-3}$
TREATMENT - Photons : low / medium energy - Limited aids & appliances in static beams - Passive control : verification / recording - Assisted set-up	LINAC 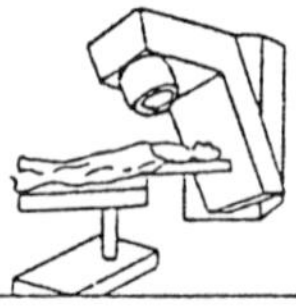
FOLLOW - UP - Regular check-ups & exams - EDP in archiving & reporting	RT-RECORDS/WORDPROCESSOR 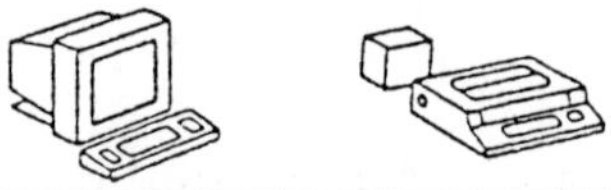
EVALUATION - Critical failures analysis over time & T.N.M.-compartment	MED. DATA ANALYSIS SYSTEM 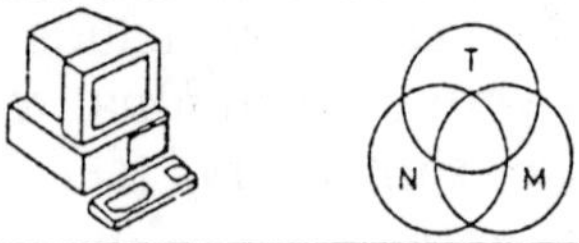

TABLE 4.6b High Precision Procedure and Equipment (Catharina Hospital, Eindhoven, The Netherlands) (after M.A. Crommelin, D. Bakker and W. Dries [4.127]).

IMMOBILIZATION Whole body fixation	FOAM MOULD 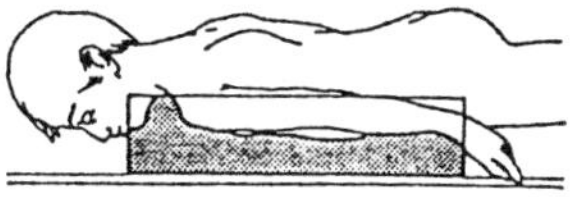
CT - SCANNING Multiple slices for patient model reconstruction and 3D - dose planning	RT-ADAPTED CT-SCANNER 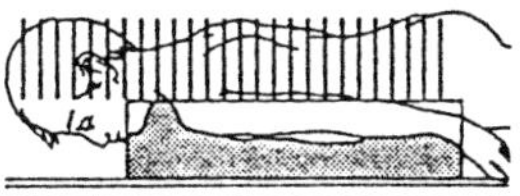
PATIENT/TARGET VOLUME MODELLING 3D - dose planning for photon and electron beams	3D - DOSE PLANNING AND VIEWING SYSTEM 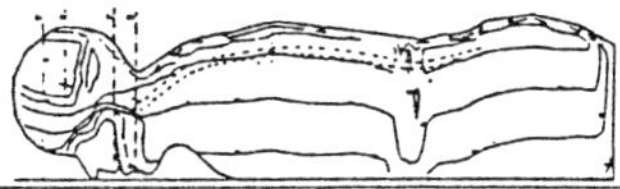
FRACTIONATION Fraction size & timing based on physical and radiobiological dose-planning optimization	CLINICAL RADIOBIOLOGICAL MODELS Linear Quadratic model $S = N / No = e^{-(\alpha + \beta d) D}$
PORTAL MARKING Marking of treatment portals on mould, cast or patient's skin for guidance of the subsequent daily set-up	DIGITAL OPTICAL SIMULATOR 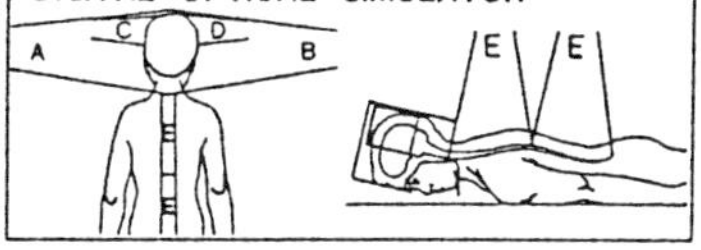
TREATMENT Radiotherapy, fully exploiting potentials of photon- and electron beams, including auto set-up / dynamic treatment	LINAC: PHOTONS & ELECTRONS 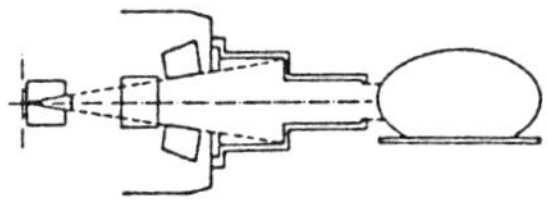
FOLLOW - UP - Regular check-ups & exams - EP in archiving / reporting of data & images	DATA AND IMAGES RECORDING AND RETRIEVAL 
EVALUATION Critical assessment of efficacy and precision of 3D - dose distribution overlying 3D - anatomy images	BIOMEDICAL / STATISTICAL ANALYSIS SYSTEM 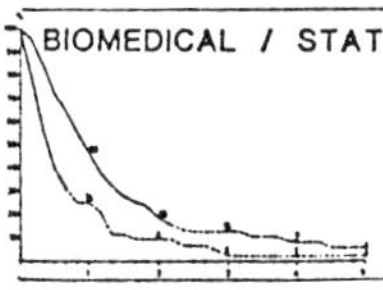 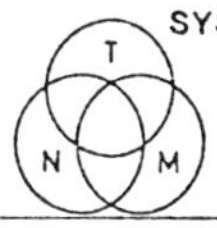

The modern simulation process is a simulation of the dose distribution based on CT data, from which a patient model can be constructed. The dose distribution is calculated with the assistance of refined beam models and displayed in arbitrarily selected planes within the field set-up. Assuming certain conditions, such as very accurately reproducible patient positioning and the ability to mark the chosen field position, X-ray imaging techniques are in principle no longer necessary. The X-ray simulator could, therefore, be replaced by a simple piece of equipment that can project the chosen light fields on the basis of the treatment prescription in digital form (Fig. 4.25). The functions that should be included in such a simulation process are: (1) Construct a patient model from serial transversal CT-slices, (2) Localize the tumor (target volume) and critical structures, (3) Provide dose distribution for the selected fields, (4) Generate digital verification film templates similar to the films taken with a conventional X-ray simulator, (5) Assist in shaping and tailoring of treatment fields, (6) Generate a template from which shielding blocks can be fabricated, (7) Mark the treatment fields on mould, cast or patient's skin.

The new 3D imaging modalities of CT, MRI, SPECT and PET, apart from their other advantages, have made it possible to better delineate tumors and normal tissues in patients. This has numerous benefits, as shown in Table 4.7. In the 1990s, it is

TABLE 4.7 Impact of Improved 3D Anatomical Information (after C.G. Orton [1.80]).

Information	Impact
External contour definition	Improved dosimetry
Precise delineation of normal tissues and organs	Higher tumor doses without increased morbidity
Precise delineation of primary tumor volume	Optimal field design reduction in geographical misses
Detection of regional spread	Avoidance of regional recurrences

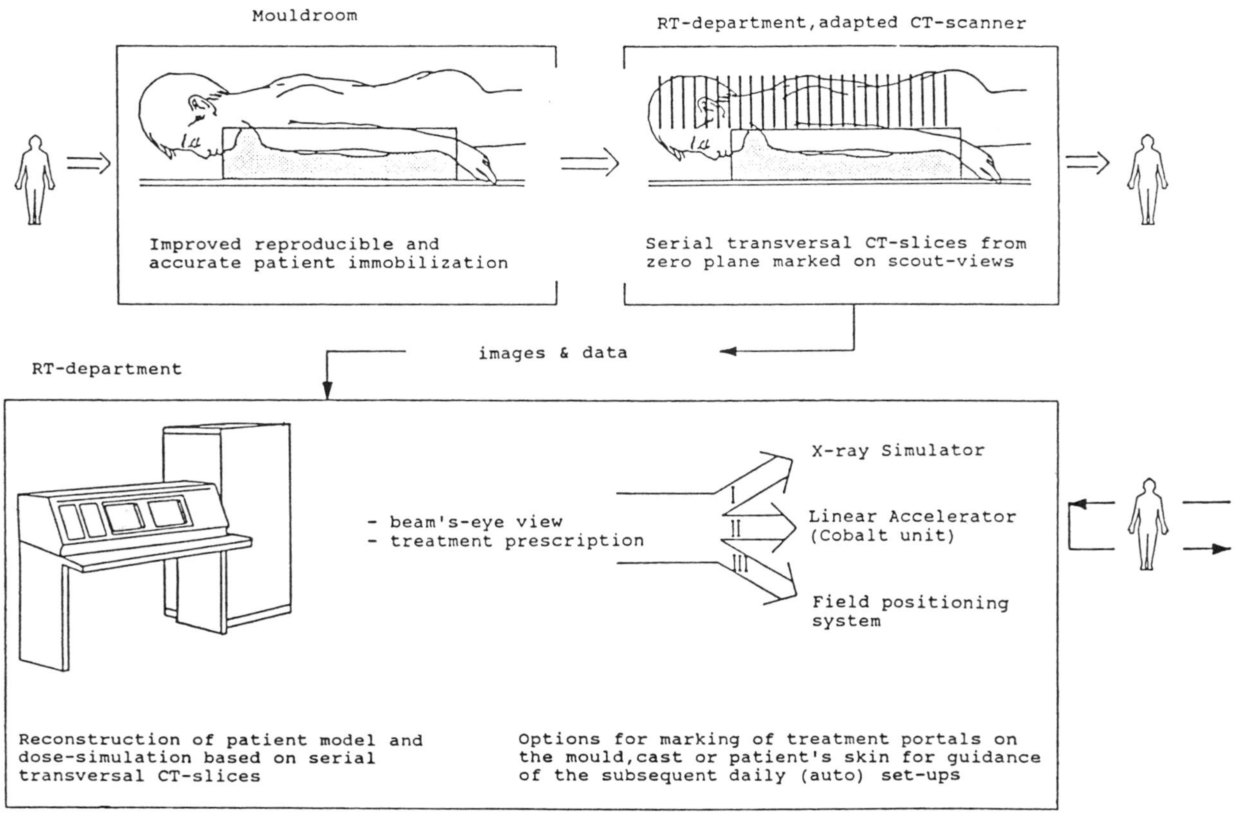

FIGURE 4.25 Outline of a digitized dose planning procedure as a part of an Integrated Radiation Oncology Process, (after D. Bakker et al. [4.125]).

certain that far more reliance will be placed on CT information and such techniques as CT simulation will become routine. This will result in more precise dosimetry and more rational selection of machine parameters and treatment modalities. With 3D external and internal contour data available in digital form, design and fabrication of tissue compensators using computerized milling machines or other such techniques will make their application commonplace in most facilities. With all this data available, it will be possible to attempt to optimize the treatment of each patient [1.80].

The reader interested in modern methods of radiotherapy planning is referred to textbooks [1.75 and 4.112]. In what follows, we shall be dealing with some instrumental aspects of radiation planning.

4.2.4.3 Commercial Planning Systems

Treatment Planning Systems (TPS) are usually incorporated in the overall commercially available computer systems for simulation, planning and management of radiation therapy (Fig. 4.11).

FIGURE 4.26 View of the Siemens Mevaplan treatment planning system.

TABLE 4.8 Parameters of Some Commercial Radiation Therapy Planning Systems

Parameter	Siemens Mevaplan	AECL Theraplan L
Central Unit	PDP 11/84-A	L SI 11/73
External memory	soft disc 1.6 MB, 8" hard disc 26 MB[1)]	soft disc 1.2 MB, 5.25" hard disc 75 MB[1)]
Techniques of therapy		
– stationary beams	+, max. 10/plan	+, no limitations
– movable beams	+, max. 10/plan	+, no limitations
– step-wise	+	+
– tangential fields	+	+
– arc	+	+
– with collimator rotation	+	+
– rotating with variable speed	+	+
– abaxial		+
Energy and type of radiation	Co-60, >1 MV photons, e–	0.25 to 40 MVphotons, e$^-$
Data for calculations:		
– shielding masks	+	+
– wedge filters	+	+
– shielding blocks	+	+
– scattering foils	+	+
– mantle shields	+	+
– smoothing filters	+	
– beam incident angle	+	+
– nonhomogenity of irradiated volumes	+	+
– collimator rotation	+	+
– scattered radiation	+	+
Input parameters:		
– depth dose and profile curves	water phantom or supplied from a digitizer	water phantom or supplied from a digitizer
– depth dose and profile curves with wedge filters	listed in tables	listed in tables
Calculation algorithms:		
– electrons	pencil beam method from tables	pencil beam method
– irregular fields	the Clarcson and Wrede method	?
– nonhomogenity tissue	path equivalent method	
Normalization of results with respect to:		
– maximum dose	+	+
– selected point	+	+
– maximum & minimum dose	+	+
– isodose value	+	+

1) optionally additional disc

Fundamental parameters of therapy planning systems will be discussed with the example of the AECL Medical — Theraplan and the Siemens Mevaplan (Fig. 4.26).

The central units of both systems operate together with arithmetic coprocessors and 32-bit matrix processors. They are also furnished with color graphic monitors with color resolution of 16 and 32 steps, and grey scale resolution of 256 and 32 steps for Mevaplan and Theraplan, respectively. The patient's anatomical images are converted from radiography in both systems by analog-to-digital digitizers which feature an active area of 51 x 51 cm^2. The systems include printers with printing speeds of 240 characters s^{-1} and 120 characters s^{-1} for Mevaplan and Theraplan, respectively, as well as multicolor x-y plotters.

The software for both systems include programs for: 1. external therapy, 2. internal therapy, 3. joint operation with a CT scanner, and 4. patient data and system data management.

The Mevaplan and Theraplan make it possible to perform internal therapy with the use of a large range of radioisotope applicators. Parameters of external radiation therapy systems such as Theraplan and Mevaplan are listed in Table 4.8.

An important parameter to compare the quality of various planning systems is the *speed of operation.* Notable progress in this respect has been achieved with the development of new generations of planning systems. In the mid-1970s, the calculation time of a single distribution by the EQTAR method was about 10 min, whereas it was reduced to less than 20 s in the late 1980s. In the latest planning systems, three-dimensional calculations (3D therapy) have been included as one of the options.

4.2.4.4 Fundamentals of Integrated Systems

External and internal accelerator control systems (Sec. 3.1.2.4) and radiotherapy planning systems (Sec. 4.2.4.3) are representative examples of computer facilities designed to operate only a fragment of the whole radiotherapy process. In the 1980s, the development of radiotherapy computer systems was characterized by a tendency to design *integrated systems* to embrace all stages of radiation treatment together with the patient's follow-up period (cf. Figs. 4.1 and 4.10). Such systems are referred to as *radiation management systems.* In addition to developing local systems intended for radiotherapy hospitals or departments (Secs. 4.2.4.2 and 4.2.4.3), attempts are now being

TABLE 4.9 Summary of Prototype Knowledge Based Systems for Cancer Therapy (after S. Hyödynmaa [4.88]).

System	Domain	Approaches & interests	Development environment
ONCOCIN	chemotherapy management	critical approach	DEC-20, XEROX AI-workstation
OPAL	knowledge entering		InterLisp
ONYX	treatment planning		
ROUNDSMAN	therapy sel. (breast c.)	case based reasoning	HP-9836 GLISP
BREASTCAN	chemotherapy selection	frames, production rules	PROLOG
NEWCHEM	cell kinetic & molecular knowledge	simulation & optimisation	Franz-Lisp PEARL, C
ISP	treatment sel. (prostate c.)	statistical approach	VAX/VMS FORTRAN 77
Kalet & Jacky, 1987	radiotherapy of H&N cancers	object oriented design	VAX/VMS, Common Lisp, CLOS
CARTERS	Diagn. & radiother. (lung c.)	critiquing approach	XEROX AI-workstation, LOOPS
ESTHER	therapy sel. (laryngeal c.)		PC, Smalltalk
SENEX	management of tr.protocols (breast c.)	data coll. & documentation	PC, Personal Consultant Plus
SKIPMAO	prescr. of treatment (Hodgkin's d.)	data management	PC, PROLOG
Giron et al. 1988	dg & therapy (ovarian c.)		PC, O+, PC+ Turbo Pascal
EXACT	chemotherapy management		PC, Personal Consultant Plus

made to devise large-scale nation-wide or regional *intelligent systems* (Secs. 4.2.4.6 and 4.2.4.7) to allow decision making, verification and correction also at the departmental level.

Characteristic of current cancer therapy is that there are more opinions than truths about the most suitable treatment. Since the etiology of cancer diseases is still partly unknown, the knowledge of these diseases is mostly *empirical.* In these circumstances, careful collection, recording and analysis of patient data is a necessity in monitoring the quality of care and developing better treatment methods. Because of the chronic character of many cancer diseases, the treatment cycle of a cancer patient is often several years, even decades. Although there is an endeavor at uniform therapy in various cancer diseases, the response of a patient to a given therapy varies requiring in some cases a more individual therapy. This, in turn, makes it difficult to obtain statistical evidence for the most suitable treatment. The long treatment cycle and individual therapy jointly put special emphasis on *patient management* [4.81 and 4.85].

Some of the decisions that are made during cancer treatment are complex requiring expertise from several medical specialities. Oncological knowledge derived from medical textbooks, reports of clinical trials, etc. is involved in each decision-making stage, and applied at the institutional level according to the local circumstances. In handling the problems of decision support and patient management in radiation oncology information technology has been shown to be benefitial. Espe-

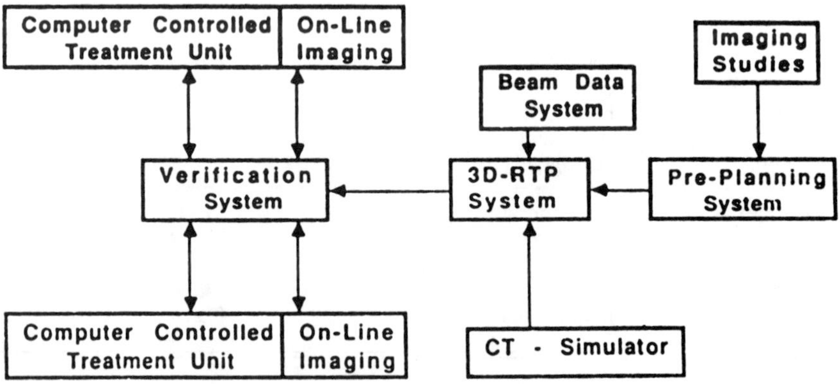

FIGURE 4.27 Block diagram of a computer 3D radiation therapy management (after J.A. Purdy [4.78]).

cially *knowledge based systems* (KBS) have been suggested as a solution [4.89, 4.92 and 4.129].

The summary of prototype KBS systems and their domain, approaches and development environments are given in Table 4.9 [4.88]. The application of these systems started with the famous ONCOCIN system and its offspring. ONCOCIN was originally made to assist the physicians in the management of patients enrolled in experimental cancer chemotherapy protocols.

Integrated knowledge based systems are now being developed predominantly at large radiation therapy centers in teaching hospitals with highly qualified staff. The costs of such systems are usually covered by national cancer institutes or national health care departments.

The requirements for radiation treatment management systems are presented in [4.81]. In what follows, examples of several integrated systems will be discussed in more detail.

4.2.4.5 3D Therapy Control Systems

As a result of the widespread use of computed tomography and the appearance of computers equipped with monitors to represent spatial configurations, several leading centers have developed *three-dimensional therapy planning and management systems.* A detailed bibliography is given in [4.82].

A block diagram of a 3D system commissioned by the National Cancer Institute (NCI) and designed by the Mallincrodt Institute of Radiology, Washington University School of Medicine [4.78] is shown in Fig. 4.27. In 1984, the NCI initiated a three-year project entitled "Evaluation of High Energy Photon External Beam Treatment Planning", and again in 1986 continued with another three-year project on electron treatment planning [4.78 and 4.82]. This work is still (1991) in progress as a 5 year collaborative project of the NCI [4.128].

It can be seen from Fig. 4.27 that the system incorporates a *pre-planning unit* which delivers information to the proper 3D system. The beam data and those from the CT simulator are also fed to the main 3D system operating with a monitor which displays in 3D the tumor and critical organ areas as well as beam contours from any arbitrary direction.

The present-day radiotherapist has at his or her disposal various diagnostic images, such as radiographs, CT, NMR, scinticamera scans or ultrasound pictures. These images differ in respect of format, projection, enlargement, resolution, etc. The aim of the pre-planning system is to assist the radiotherapist in evaluating this multiplicity of information by having it preliminarily processed, whereas the main 3D planning system utilizes the *method of virtual simulation* [4.73 and 4.83] and the "beam's eye view' (Fig. 4.20). The results of the planning are subsequently conveyed to a verification system connected to two multi-energy accelerators (4 and 18 MV photons and 6 to 20 MeV electrons) which must incorporate an option for asymmetrical movement of the collimation jaws (Fig. 3.43) and, in future applications, for a multi-leaf collimator (Fig. 3.49). An ambitious attempt to attain the goal of 3% or better accuracy in dose fractionation has been undertaken by the designers of the system [4.78]. The Collaborative Working group is committed to publish its results and any software produced by the group [4.128].

4.2.4.6 The IRTIS System

The *Integrated Radiation Therapy Information System* (IRTIS) has been designed for American hospitals at the New England Medical Center [4.84]. This system was developed from the observations that the volume of patient-related data used to be so large that there was a possibility of information overload. In response to this, a comprehensive system was designed which would easily provide information useful for the patient care and administrative operations as well as for the cooperating institutions.

The IRTIS architecture (Fig. 4.28) was based on a network of single microcomputers coupled to *work stations* operated by medical physicists, oncologists and administrative staff and also to the systems of the overall hospital information, radiology information RIS and other peripheral systems. The IRTIS can also operate with commercial radiotherapy management systems, such as Varian RMS-2000.

The system is based on five principal modules of routine programs. Two modules are intended for administrative applications. The first module is an *operations management module* and performs such functions as patient registration, treatment scheduling, patient throughput and cost analysis, time and calendar management, spread sheet analysis, materials and

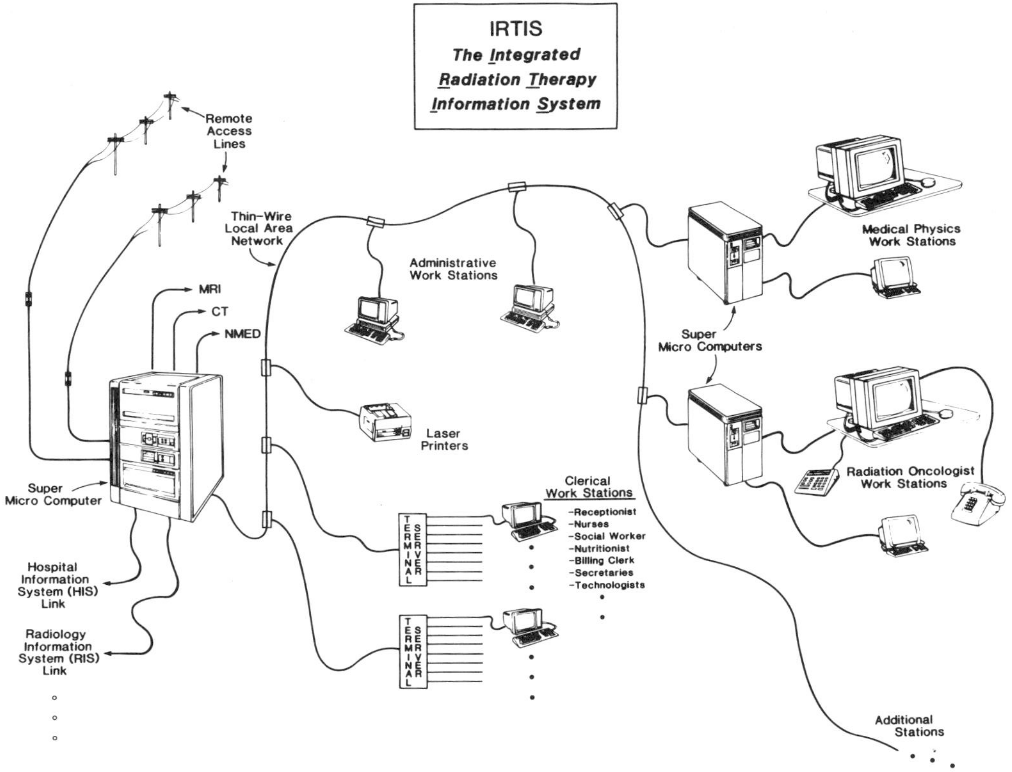

FIGURE 4.28 Block diagram of the Integrated Radiation Therapy Information Systems (IRTIS) developed at the New England Medical Center (Boston, USA), (after E.S. Sternick [4.84]).

equipment management, personnel and grants management, etc. The second module is designed for strategic planning, i.e. it concerns environmental analysis, budgeting and forecasting.

Another three modules are to be used for clinical applications. Module three covers *medical management,* that is medical records, quality assurance documentation, image archiving and treatment planning analysis. Module four deals with *clinical research* and provides analysis of patient data base and the patients survival. Finally, module five, called *continuing education module,* is used for literature search and retrieval and gives access to the extramural data.

The research leading to the development of the IRTIS and the main features of this project are discussed in detail in [4.84].

4.2.4.7 The CART and CARTES Information Systems

The five Nordic countries: Denmark, Finland, Iceland, Norway and Sweden — with populations of 0.2 to 8 million — have cooperated in clinical radiation physics since 1962. The main efforts have been to improve basic and clinical dosimetry. In the early 1980s, the representatives from these countries initiated work on a joint regional information project designated by the acronym CART *(Computer Aided Radiotherapy).* The principle of the information flow of the system is shown in Fig. 4.29. The CART system envisages a design of a data bus with separate channels for different kinds of information, coupled through some indirect systems to individual parts of oncological and radiotherapy machines.

After completing two preliminary stages, the main stage of the CART system was implemented in the years 1985-1987 at the expense of $ 4 M [4.85]. In addition to designing the information network model, parameter standards necessary for proper operation of the network were determined at that stage. At the same time, construction of some module prototypes was started in cooperation with specialized commercial firms. Details of the CART project are given in [4.85 through 4.87].

By using the CART standards, work on another intelligent project called *Computer Aided Radiation Therapy Expert System* (CARTES) was initiated [4.85, 4.88 and 4.89). The main idea of this system is computer aided radiotherapy decision making with the use of a large body of data representing the existing knowledge in the field. The system is able to make a critical

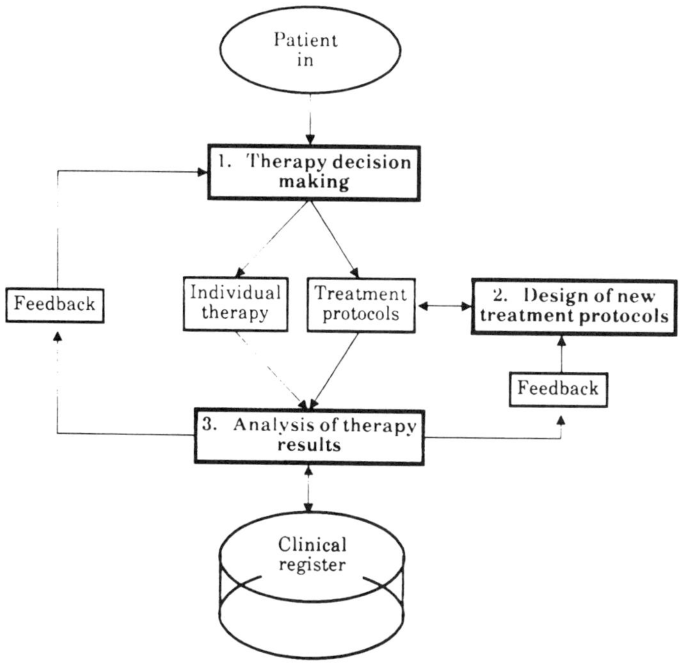

FIGURE 4.29 Model of decision making in the CARTES system (after S. Hyödynmaa et al. [4.89]).

evaluation of decisions in the light of the collected information. The main blocks in the system are: therapy decision making, design of new treatment protocols and the analysis of therapy results. Thanks to a feedback channel the final stage can affect the initial stage of decision making and the stage of designing new treatment protocols.

In the CART system, provision has been made for three separate modules for the implementation of the three stages of therapy [4.86]. The first module is intended for making *primary therapy decisions* together with classifying patients to various groups, specifying kinds of treatment and selecting standard or individual treatment protocols for the patient. The second module helps to design *new treatment protocols.* It contains a register of protocols used routinely and methods necessary to

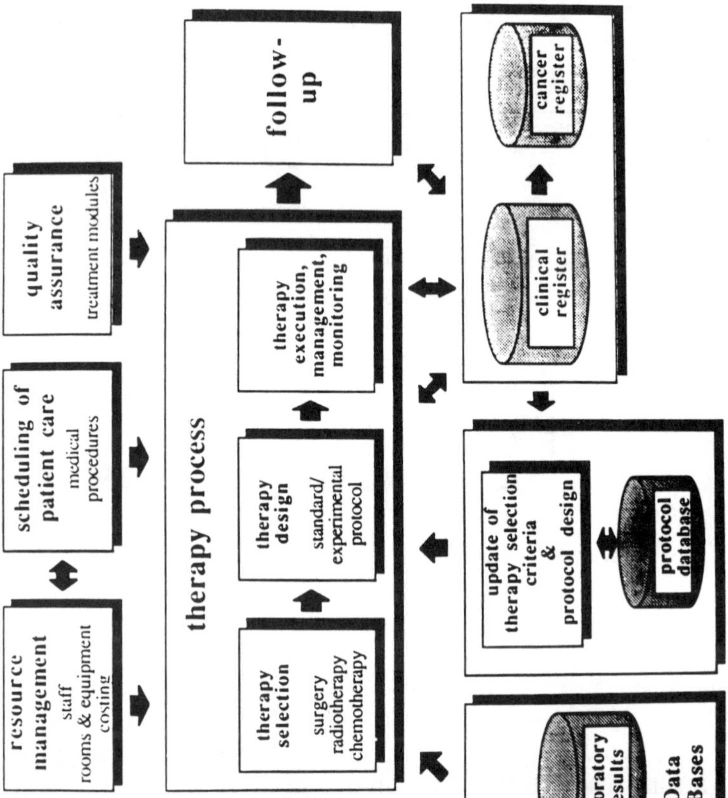

FIGURE 4.30 Stages in the radiotherapy process envisaged for cooperation with an intelligent information system (after S. Hyödynmaa et al. [4.86]).

design new ones on the basis of the current knowledge in this field. To this aim, oncological and statistical information contained in the module on the planning and implementation of clinical procedures is utilized. The third module is destined for the *statistical analysis of treatment results* based on the data in the clinical register on the treatment follow-up.

Thus far, a prototype of the first model has been developed and further work is under way [4.86]. In the next several years the CART and CARTES information systems are expected to find wider use at all stages of radiotherapy processes. These stages are presented in a block diagram in Fig. 4.30. The design of a computer work station in the CART system for clinical purposes is discussed in [4.88].

4.2.4.8 Trends in the Development of Artificial Intelligence Systems

Radiotherapy has become one of the first fields of clinical medicine to employ computers to assist clinical decision making on a large scale. Progress in computer technology has taken place at a faster rate than could ever have been predicted. A desktop computer can now command memory and disk storage size which five years ago would have been possible only on a minicomputer and twenty years ago only on a mainframe computer [4.130]. Radiotherapy has in consequence already undergone major changes and we are probably on the threshold of more extensive ones to come [4.131].

For the past two decades, computers have evolved from simple dose distribution calculations to present-day interactive systems which permit graphical 3D simulations [4.132 through 4.134]. These simulations can help to design sophisticated radiotherapy planning procedures in which external high-energy photon, electron or ion beams are combined with interstitial and intracavitory radioisotope sources. The rapid advance in the flexibility of modern accelerators and other radiotherapy apparatus, and, above all, the greater possibility of beam modification during treatment (Sec. 3.1.4.4) as well as the development of computer programs able to simulate anything the machinery can do constitute a real challenge to utilize these elements in a more effective radiation treatment. It is the new *intelligent or, artificial intelligence systems* that permit systematization of the present knowledge about the variety of the existing treatment plans and design of new treatment strategies. Without any claim to replace physicians, medical physicists or dosemetrists, artificial intelligence systems will soon form an effective tool to assist treatment planning teams in producing far more powerful and comprehensive plans. It seems axiomatic that the treatment plan should be shaped so as to fit or conform to the target area as closely as possible with minimal inclusion of adjacent or vital organs. This has recently been referred to as *conformal or conformation therapy* (Sec. 3.1.4.4). The multileaf collimator (Fig. 3.50) has opened up a new chapter of possibilities. Multileaf conformation radiotherapy with computer-assisted systems has already been used at several hospitals in Japan and applied to various sites in the body including the brain, maxillary sinus, esophagus, lung and pelvic organs [3.133].

One of the aims in developing such systems is to use elements of *picture archiving and communication systems* (PACS) in radiation therapy. PACS developments and tests have been dedicated mostly to the field of *diagnostic radiology,* with emphasis on the requirements of the diagnosticians as "human viewers". Today, a second generation PACS is appearing, which integrates image generation, display and analysis, storage, hardcopy generation and management, also implying integration with other information systems to support image management and to integrate the use of image and non-image information. The PACS concept covers the whole organization of the PACS clinic, and the new tools available will affect the ways of working with images on all application levels.

The requirements for radiotherapy PACS are different from the requirements for the pure diagnostic applications (Table 4.10). Concerning e.g. selective retrieval and response time, interactive access to all images for the active patient is required during planning, while radiotherapy procedures of today make it possible to use some simple type of prefetching of image data for the active patient.

The increasing use of diagnostic PACS in radiotherapy, apart from gaining experience and obtaining unification of both systems, is stimulated by the fast development of *optical computing and optical disks* [4.130].

The optical disk of 12 inches has a capacity of 2 GB up to 6.8 GB (access time 90 ms, cost 0.2 \$/MB). This device is employed as an external memory media for medical image very

TABLE 4.10 Difference Between Radiotherapy PACS and Diagnostic PACS (after K. Inamura [4.130].

	Diagnostic PACS	Radiotherapy PACS
Data volume	large	small
Time constant	short	long
Economical	must be	not necessarily
Image quality	mandatory	good contrast
3D display	not necessarily	necessary
Service area	large	small
Organizer		must have excellent staff

often nowadays. It can store up to 8,000 chest films in case of 1/5 data compression. Memory capacity of the magneto-optical disk is not so big, because its diameter is only 5.25 inch at present. However, this memory is rewritable and compact so that the volume of annual shipment increases very rapidly. Therefore, today's magneto-optical disk of 600 MB will have the capacity of 5 GB in 5 or 6 years. It means that a 5.25 inch magneto-optical disk could accommodate 1,500 sheets of chest film, 1/5 compressed [4.130]. For example, the *Radiation Oncology Data Base* by PACS for radiotherapy developed at the College of Biomedical Technology (Osaka University, Japan) is shown in Fig. 4.31. This experimental system employs a 600 MB magneto-optical disk for database media and/or for hand carried circulation media. One pair of the mirrored system is installed on the 2nd floor, and other PACS and ISAC *(Image Save And Carry)* systems on the 3rd floor. The mirrored systems are connected to each other by a public telephone line or hand carried magneto-optical disks. One of the above systems will be moved to Osaka University School of Medicine in the near future (1991) and clinical evaluation will be carried out by telephone line communication. Exchangeability or interchangeability of magneto-optical disks will be also tested after standard protocol and standard file manager of ISAC has been implemented on four systems shown in Fig. 4.31. Another feature of this system is to employ a personal computer, so that the cost could be lower.

An optimized design of radiotherapy PACS for future use is shown in Fig. 4.32. Here, again a magneto-optical disk and a public telephone line are employed, as in Fig. 4.31. But the PACS workstation is in the center of the system. The treatment planning system with 3-D processing, CT, MRI, and a simulator is connected to the PACS workstation together with an expert system supported by follow-up data as well as other clinical data.

A more detailed discussion of the above problems is beyond the scope of this book and the reader is referred to the overviews [4.90, 4.91]. The artificial intelligence systems used in radiotherapy, with special emphasis on the US systems, are reviewed in [4.92] provided with extensive bibliography. The main ideas behind the IRTIS, CART and CARTES intelligent systems have already been presented in Sec. 4.2.4.6 and 4.2.4.7. It should be noted, however, that a new system called NEO *(Netherlands Expertise Center for Oncology)* is currently being

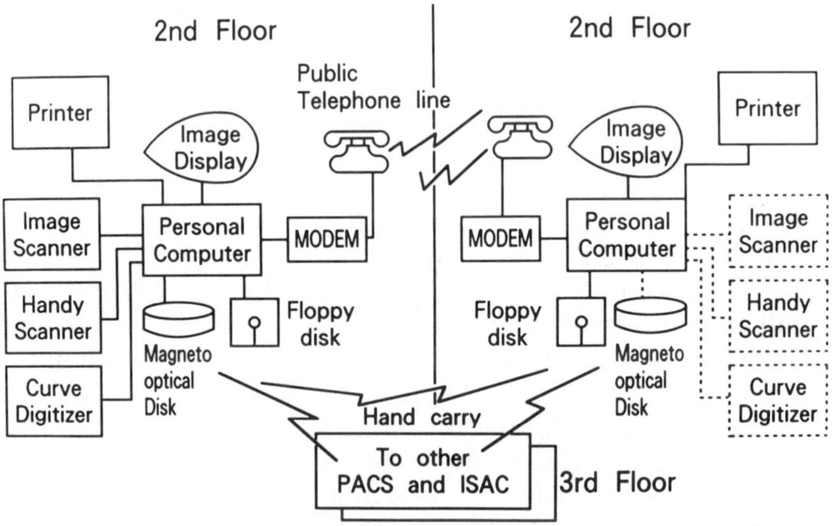

FIGURE 4.31 Radiation Oncology Database by PACS (College of Biomedical Technology of Osaka University) (after K. Inamura [4.130 and 4.135])

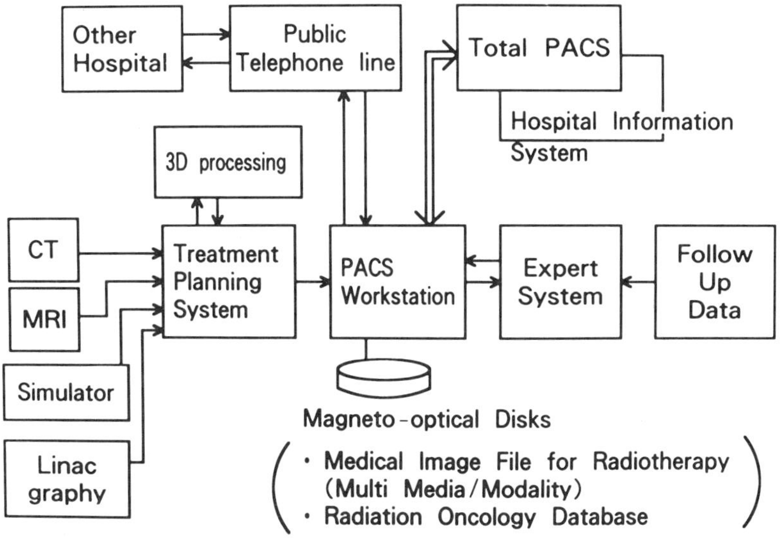

FIGURE 4.32 Optimized design of radiotherapy PACS (after K. Inamura [4.130]).

developed [4.93]. In the Netherlands, like in other countries, only a small fraction of patients diagnosed as having tumors can be treated at large university centers or specialized oncology institutions. In order to ensure a high and uniform standard of radiation therapy, eight consultative regional service centers have been set up in that country. They operate on a two-week schedule, which has been regarded only as a partial solution of the problem since not all cases and treatment decisions can be thoroughly discussed at such infrequent meetings and decisions are often belated. The NEO system should help to overcome these difficulties by ensuring that oncological centers and departments have an easy access to all the relevant information about the standard treatment protocols, research protocols, and to the total body of information on protocol modification, non-standard treatments and side-effects.

4.3 General Principles of Radiation Protection

The problems of the generation of stray radiation by radiotherapy accelerators and the requirements imposed on radiation protection of the patient have already been presented in Secs. 3.1.6.5 and 3.1.6.6. Although the radiation safety of the patient is of primary importance, the accelerator user is also responsible for eliminating any radiation hazard to the personnel and the population at large. In other words, general principles of radiological safety in force in a given country must be respected. These principles will be dealt with for the following categories of applications: 1. radiotherapy electron accelerators for routine treatment (Chapter 3) and those used in unconventional therapy (Chapter 5), 2. medical radioisotope production (Chapter 7) and 3. other biomedical uses. Our considerations will be mainly based on the US standards since it is in the USA alone that over 1300 electron linacs are installed, and, with a few hundred other machines, that country ranks first in the world in the development of particle acceleration techniques [4.99].

4.3.1 Types of Hazards and Permissible Dose Rates

In accordance with the considerations presented in Chapter 3, the accelerator's aim is to produce a well-defined radiotherapy field confined to a relatively small area which is penetrated by a high intensity primary beam. The concurrent secondary radiation (bremsstrahlung, neutrons, scattered electrons, etc.) may generate very high dose rates over large areas around the accelerator. Some of the secondary radiation is highly penetrating and if the primary beam energy is sufficiently high, some induced radioactivity may be produced.

The permissible dose rates result from the adopted international radiation protection regulations. For a long time, the maximum occupational mean dose per working week of 40 h used to be 1 mSv (100 mrem) and, consequently, the maximum mean dose rate was 25 mSv h^{-1} (2.5 mrem h^{-1}). For the general population, i.e. for persons who are not occupationally exposed to ionizing radiation and, therefore, not subjected to dosimetric control, the permissible dose rate was established to be 2.5 mSv h^{-1} (0.25 mrem h^{-1}).

The above levels were understood to be maximum values which should not be exceeded in spite of their legal permissibility. The actual exposures to the personnel at US accelerator facilities may be inferred from the SLAC Report [4.99] published in 1988. The average dose equivalent per person from measurable exposures was 0.48 rem per year in 1974, whereas in 1984 it was reduced to 0.16 rem (about 33% of the 1974 value).

In connection with the rapid development of electron radiotherapy accelerators, the US National Council on Radiation Protection and Measurements has published several reports and guidelines on radiological protection for accelerator facilities [4.94 through 4.96]. Although the problems of radiological protection for large accelerators, i.e. those for energies above 1 GeV, are treated in separate publications (cf. IAEA Reports [4.97 and 4.98]) the fundamental principles of radiation protection are common regardless the type of the accelerator. It should be noted that even a small machine operating at 3 MeV can produce radiation levels as dangerous as those from the highest energy accelerators [4.99].

In the early 1980s, a new concept, recommended by the ICRP, of ALARA, which stands for a dose "As Low As Reasonably Achievable" has become more widely popularized [4.99 through 4.101]. According to this new concept, the doses received by the personnel, or other persons, should be kept at a "reasonable" level. Although the annual occupational maximum radiation dose in the USA is 50 mSv (5 rem), the ALARA principle requires that in new or modernized accelerator facilities radiation protection should be designed in such way that the real exposure should not exceed 20% of the limiting value.

Hazards from electron accelerators. Electron accelerator radiation hazards depend on the energy range of the primary beam. For energies between 6 and 10 MeV, electrons lose energy only through ionization and bremsstrahlung (Sec. 3.1.4.2). Accelerators operating in this range must be equipped with shields to absorb only primary and secondary electrons and bremsstrahlung (X-rays).

Electron machines operating in the range of 6 to 50 Mev, in the *giant resonance region,* apart from primary radiation, can also generate neutrons (Sec. 3.1.6.5, Fig. 3.67). In this region, energy can be transferred to the nucleus by photons produced along the electron's track, which leads to a nucleus in an excited state which subsequently de-excites through *neutron emission.* Although bremsstrahlung must still be eliminated by shielding, the presence of neutrons requires the use of shielding materials rich in hydrogen. The neutron radiation is nearly isotropic, whereas the bremsstrahlung, which is produced during electron conversion, is highly directional (cf. Fig. 3.34).

For the energy range between 30 and 150 MeV neutrons are produced by the pseudodeuteron process. The cross section for this process is smaller than that for the giant resonance and the bremsstrahlung region so that the latter two phenomena decisively determine the kind of shielding used. In the region of very high energies, those above 150 MeV, when the threshold for pion production is exceeded (Sec. 5.2) neutrons can be generated whose energies are nearly equal to those of the primary particles. Because of the high penetrability of neutrons with such high energies, concrete shields thicker than 120 cm must usually be used. Of course, shielding against high-energy neutrons is not a problem for routine radiotherapy accelerators since the upper energy limit of such machines is usually not higher than 20 to 25 MeV.

Hazards from proton accelerators. In the low energy domain, of the order of a few MeV, the main interaction is that of elastic scattering. For most target materials, inelastic collisions can only be produced when proton energies are in excess of 6 to 8 MeV, however; for some materials inelastic reactions can occur at much lower energies. The range of protons with energies lower than 8 MeV is rather small: for example, it is less than 1 mm in solids and less than 1 m in air.

If the initiating protons have energy sufficient to penetrate the Coulomb barrier they can transfer energy to the nuclei of the target which then emits neutrons or other nuclear fragments. This energy transfer is placed in the group of inelastic processes. For energies up to about 100 MeV, the prevailing inelastic process is the isotropic neutron emission from the target nuclei. The evaporation neutrons have energies up to about 20 MeV, depending on the energies of primary particles. In view of the possible intensive neutron emission, radiological shields must then mainly be designed in such a way that the neutron component is attenuated.

Above 100 MeV (proton energies for radiotherapy, Sec. 5.1.2) a strong high energy neutron component exists primarily in the forward direction. These neutrons can have energies up to that of the primary proton beam. The shielding measurements for a 230 MeV proton beam and measurements of neutron dose equivalent has been analyzed in [4.138].

Residual (induced) radioactivity. When the beam energy in an electron accelerator exceeds the order of a dozen or so MeV

(Fig. 3.67), *residual radioactivity* appears. The isotope ^{22}Na is produced ($T_{1/2}$ = 2.6 yrs) in concrete, ^{22}Na and ^{26}Al ($T_{1/2}$ = 7.4 x 10^5 yrs) is produced in aluminum and ^{54}Mn ($T_{1/2}$ = 303 d) in steel. Among the construction materials with low susceptibility to activation with high-energy electrons are common concrete, lead (without antimonum added), aluminum, plastics and other organic materials.

The *activation threshold of water* is 15.67 MeV. Among the radioisotopes which can be produced, mention should be made, first of all, of ^{15}O, ^{11}C and ^{7}Be ($T_{1/2}$ = 123 s, 20.34 min and 53.6d respectively). Possible activation of water must be considered with respect to cooling water used to carrying away heat from the accelerator if the beam power is in excess of about 10 kW. This, therefore, does not concern common radiotherapy accelerators for routine applications. Their cooling system is supplied with water from the public water-pipe network. For high-power accelerators, however, a *double water circulation system* with a heat exchanger is usually employed with all of the parts of the primary circulation being installed inside the radiation shield.

Residual induced radioactivity may pose a serious problem for ion accelerators. A fraction of the accelerated beam, which cannot be extracted outside the machine, causes quite strong activation of the internal walls of the accelerating chamber (cyclotrons and proton synchrotrons).

Radiotoxic and toxic gases. Radioactivation of air occurs for energies of X-ray higher than 10.5 MeV, which corresponds to the threshold energy of the reaction ^{14}N (X, n)^{13}N. The reaction ^{16}O (X, n) has the threshold energy of 15.6 MeV. These reactions are of fundamental importance because nitrogen and oxygen make up 78.1 % and 21.2% of the composition of air. ^{13}N and ^{15}O, produced during activation, have half-lives of 10.1 min and 2.1 min, respectively. The isotope ^{38}Cl ($T_{1/2}$ = 37 min) generated during the reaction ^{40}Ar (X, np) ^{38}Cl is of lesser importance. Since the above radioisotopes are produced from an indirect reaction by bremsstrahlung, they can occur in greater concentrations only in electron accelerators with energy above about 50 MeV. High concentrations can be eliminated by ventilation or by being left to decay for isotopes with short half lives.

Among the toxic gases which may be produced during accelerator operation, when a high-energy beam passes through air, mention should be made of ozone O_3 and nitric oxides (NO, NO_2, NO_3, N_2O_4, N_2O_5). Ozone is particularly harmful because its acceptable concentration is a mere 0.1 ppm, whereas that for NO is 25 pmm. In addition, ozone is very active chemically and thus causes quick corrosion of most construction metals, with the exception of stainless steel, concrete, aluminum and ceramic casts.

Since ozone can be detected by smell at the level of 0.02 to 0.05 ppm, its measurements are seldom made. If the smell of ozone is found to prolong, its source must, of course, be detected and eliminated by ventilation and/or the personnel must be prevented from entering the facility.

Leakage microwave radiation. The assessment of health hazards both to the patients and to the personnel including operating as well as maintenance staff is usually based only on the measurements of the dosimetric parameters of ionizing radiation around a linear accelerator installation. The possible biological risk resulting from *microwave radiation* is usually ignored. In some cases, however, this may not be fully justified without an ade-

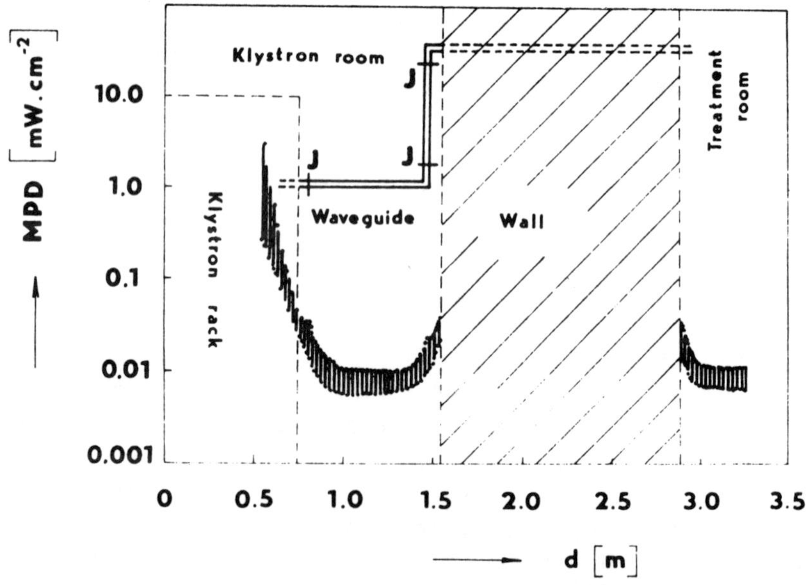

FIGURE 4.33 Microwave power density distribution along the waveguide connecting klystron rack with the Mevatron 77 accelerator structure (MPW — microwave power density, d — distance from the front panel of the klystron cabinet, J — joints of the waveguide system), (after J. Sobol and F. Pernicka [4.137]).

quate measurement of the microwave component of the existing electromagnetic field.

The microwave survey included measurement of power density around the klystron cabinet and along the transmission waveguide [4.137]. The values of power density measured at a distance of 2 cm from the surface of the waveguide, klystron rack and klystron amplifier (Mevatron 77, 18 MV X-rays and electron beams with nominal energies 7 to 18 MeV) lie in the range of 0.02 to 3 mW cm^{-2}, the highest reading corresponding to the klystron amplifier (Fig. 4.33). Farther from the measured surfaces power density falls rapidly so that a distance of about 0.5 m is at the level of 1W cm^{-2}, which is slightly above the sensitivity of the used monitor (RAHAM Model 4, General Microwave Corp., Farmingdale, N.Y.). Such microwave power density levels do not present any hazard to the patients under the treatment.

The American National Standards Institute guidelines on safety levels with respect to human exposure to radio-frequency electromagnetic fields, 300 kHz to 100 GHz, are based on a permissible specific energy absorption rate SAR of 0.4 W kg^{-1} averaged over the whole body; the permissible peak SAR averaged over any gram of tissue is 8 W kg^{-1}. Attention should be paid, however, during the maintenance and manipulation with klystron amplifier where microwave exposure might not be neglible [4.137].

4.3.2 Radiation Shielding of Radiotherapy Accelerators

Fundamental systems. Fig. 4.34 shows, by way of example, radiation shielding for a multi-energy radiotherapy 25 MV linear accelerator (Philips SL-25). The main shield is a *massive concrete shell* with walls 30 to 250 cm thick, the thickest walls facing the incident primary beam.

The shield as in Fig. 4.32 incorporates a typical *labyrinth* to extend the paths of scattered rays so that the dose rate at the entrance, equipped with a solid door, should not exceed the maximum permissible value.

Exposure levels. As an example, for a 40 hr working week the exposure of the personnel working in a continuous mode in the adjacent rooms should not be higher than 0.1 mSv (10 mrem). Since the hazard from stray radiation is usually specified as percent of the main beam fluence (Sec. 3.1.6.5) the weekly total dose received by the patient during treatment (e.g. 10^3 Sv = 10^5 rem) serves as a starting point from which all the subsequent calculations are made. Consequently, the fraction of the dose corresponding to the main beam radiation scattered on the walls, floor and ceiling (e.g. 25% of the main dose) is determined. The resulting thickness of the shielding is then differentiated with respect to the type of adjacent rooms, that is, whether personnel works there continuously or only sporadically. For rooms accessible to general public such as corridors, waiting rooms, etc. the thickness of the shielding and/or its efficiency must be increased.

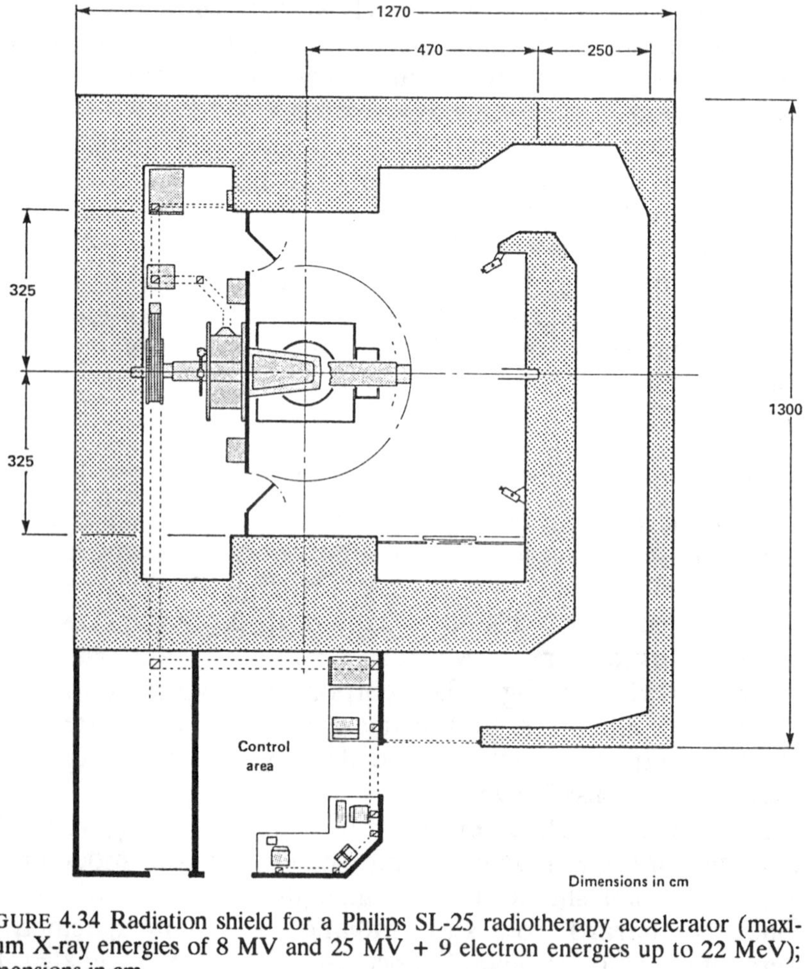

FIGURE 4.34 Radiation shield for a Philips SL-25 radiotherapy accelerator (maximum X-ray energies of 8 MV and 25 MV + 9 electron energies up to 22 MeV); dimensions in cm.

It follows from Fig. 4.34 that the maximum thickness of the shielding corresponds to the area of the primary beam interaction. It may be further drastically reduced if the accelerator is furnished with a *beam absorber.*

A detailed discussion of the shielding systems for radiotherapy accelerator facilities and the estimation of radiation hazards with particular attention given to the effect of neutron radiation is presented in [3.55]. Manufacturers have available standard designs of radiation shielding installations for various types of accelerators. In the last few years, computer codes have also been developed useful in the design of accelerator shielding [4.99] as well as those intended for calculating shielding against photon and neutron sources with energies less than 20 MeV. The *Radiation Shielding Information Center* (RSIC) at Oak Ridge National Laboratory (USA) maintains a large collection of computer codes useful for shielding accelerators. Quality Assurance measurements of shielding a 25 MeV radiotherapy accelerator are presented in detail in [3.59].

Shielding materials. For most accelerators, *ordinary concrete* (ρ = 2.4 g cm^{-3}) is the most commonly used material (2.4 x 10^3 kg m^{-3}). The class of normal concrete also includes *lighter concrete* with the density of about 2.2 g cm^{-3}. The advantage of ordinary concrete is its moderate cost, flexibility and self-supporting construction as well as high hydrogen content which is of special importance from the point of view of shielding against neutrons. Attenuation in ordinary concrete, steel-concrete, steel and lead of 30 MV X-rays is shown in Fig. 4.33.

In addition to ordinary concrete, *heavy concrete* with such fillers as limonite (brown iron ore) ilmenit (titaniferrous iron ore), barite, magnetite or steel can also be used. In this way, densities up to 6.5 g cm^{-3} can be obtained. However, heavy concrete is much more expensive and can cause some technological difficulties since, for example, it cannot be transported by pumps designed for ordinary concrete. The problem of cost is now being solved by using materials that are by-products of the metal industry. The paper [4.103] describes a mixture of sand (35% in volume), powder cement (15%) and steel-punchings (50%) in the form of 2 mm thick and 10 mm in diameter discs. The price of steel-punchings is about 20% of the price of steel with the same physical properties. When water is added to the mixture, a concrete mortar is obtained with a density of

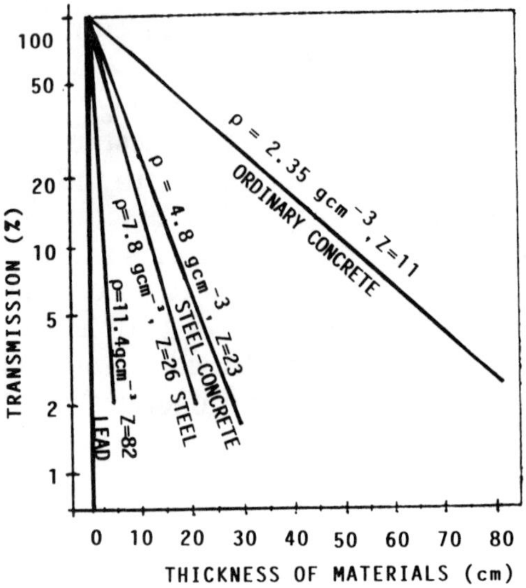

FIGURE 4.35 A 30 MV X-ray beam attenuated in concrete (2.35 g cm^3, Z=11), steel-concrete (4.8 g cm^{-3}, Z=23), steel (7.8 g cm^{-3}, Z=26), and lead (11.4 g cm^{-3}, Z=82), (after S. Koutroubas et al [4.103]).

about 4.7 g cm^{-3} and the attenuation characteristics shown in Fig. 4.35. This mortar is used to construct steel-concrete 40x40x50 cm^3 slabs.

From the comparison of the conventional and new shielding materials for a 30 MV accelerator (weekly work-load of 100 000 R and maximum dose rate of 100 mR behind the shielding), it follows that the total cost of shielding 150 cm thick and weighing 316 t constructed from ordinary concrete would be 9500 ECU (the price of concrete is 0.030 ECU kg^{-1}). If steel (0.408 ECU kg^{-1}) is used for shielding, the equivalent thickness and weight would be 34 cm and 168 t respectively for the total cost of 68,500 ECU, whereas steel-concrete (0.0112 ECU kg^{-1}) shielding would cost half of that, 21, 300 ECU, its thickness and weight being 59 and 190 t, respectively. The advantage of employing steel-concrete, especially when steel or lead shielding must be applied, apart from the reduction in relative costs, is that the thickness of the walls is only a third of that in the case of concrete. Hence more room can be saved up. Besides, the cost of shielding represents only about 2% of the price of the accelerator [4.103].

Due to their higher costs when compared with concrete, steel and lead are only employed when, because of the space requirements, shielding thickness must be reduced. For comparison purposes, the term the *Tenth-Value Layer* (TVL) is used in the English-language professional literature, instead of the *Half-Value Layer* (HVL) commonly accepted in radiology. For example, for 6 MV X-rays the TVL are 34 cm for ordinary concrete (2.35 g cm^{-3}), 23 cm for barite concrete (3.5 g cm^{-3}), 10.0 cm for steel (7.8 g cm^{-3}) and 5.5 cm for lead (11.3 g cm^{-3}). Another class of shielding materials are those used for attenuating neutrons. They are needed in the head (gantry) shielding as well as in radiotherapy rooms accommodating accelerators with energies in excess of about 10 MeV (Sec. 3.1.6.5). Originally, paraffin, due to its low lost, was used for this purpose. For

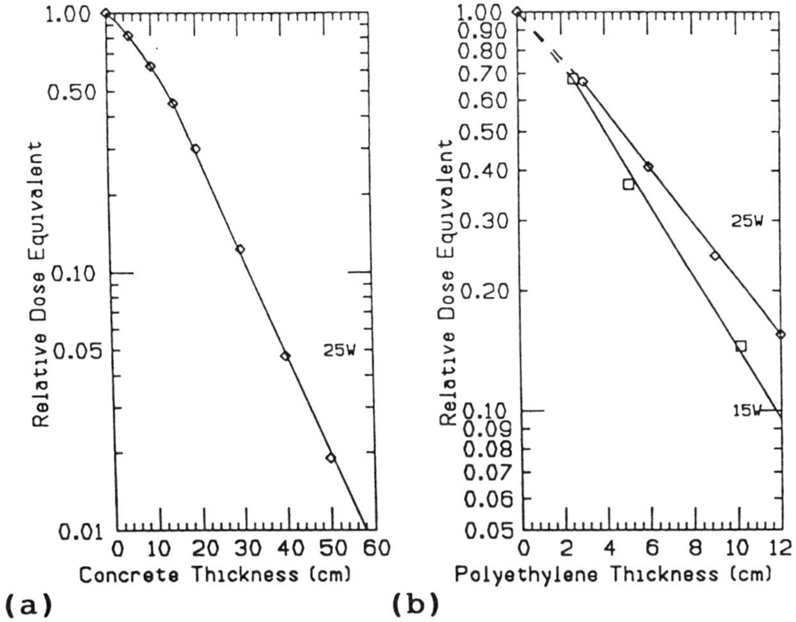

FIGURE 4.36 Neutron attenuation by: (a) — concrete and (b) — polyethylene (symbols 15 W and 25 W denote curves for photo neutrons produced by bombarding tungsten with 15 and 25 MeV primary beams) (after Mc Call and T.M. Jenkins [4.104]).

example, the entrance door, provided with additional steel shielding, would be covered with a layer of paraffin. The neutron flux generated during the accelerator operation is also attenuated in air, however, the TVL in the absence of other shielding materials is as large as 5 m [3.55]. The attenuation of accelerator neutrons by concrete and by polyethylene (ρ =1.05 gcm^{-3}), is shown in Figs. 4.36a and b. The degree of attenuation depends to a large measure on the neutron spectrum. Symbols 15 W and 25 W denote curves which correspond to photoneutrons generated in tungsten acted upon by 15 and 25 MeV electrons, respectively [4.104].

Effective shielding against a neutron beam can be achieved now-a-days by using commercial *composite materials* made from polyethylene with some boron added in the form of boric acid or borax. The main purpose of boron is its large thermal neutron cross-section. Commercial composites for neutron shielding in medical accelerators are manufactured by the US company Reactor Exp [F18].

4.3.3 Planning of Radiotherapy Centers

Machines and Treatment Rooms. Planning of a regional or national network of radiotherapy centers is usually started with determining the tumor structure and the morbidity rate (Sec. 1.1.2, Fig. 1.5) for the near future. This serves as a basis for rough estimation of the number of patients who would need radiation treatment. In some countries, national guidelines have been prepared to determine the number of patients per one accelerator. According to US 1986 Blue Book (Sec. 1.1.2) one treatment machine should be employed for every 250 new patients [1.61]. Therefore, a nation wide net of treatment machines can be planned. Details of planning and design for British radiotherapy are given in [4.105].

If a department or center is to be built from scratch, the design has to be based on the manufacturer's specifications which define such requirements as the minimum door and corridor clearances, determined by the transport of accelerator parts, floor load-carrying capacity, building power parameters (typical power requirements of 20 to 50 kVA), hoisting capacity of cranes, ventilation, air-conditioning, cooling water supply, etc.

The design proper is usually in the hands of a specialist architect, whose role is of particular importance when old accelerators are exchanged for new ones to be installed on the premises of the older generation accelerators. The task that the architect has before him is to design pleasing decor of the treatment and waiting rooms since patients who undergo treatment are usually in stress or even depression. At the same time, the rooms must meet all radiation safety requirements and help the personnel in performing their various functions. For example, the design of good lighting installation is an important factor, with the provision made for dimming the lighting when the patient is being positioned with laser systems. Ventilation is also an important consideration in view of the heat generated by the accelerator and the necessity to cools the machine. If an anesthetics are being used in the treatment room, a separate ventilation system should be provided so that they do not enter the general ventilation.

Considering the large mass of the radiation shielding, the accelerator facility is, as a rule, situated on the ground floor or partly underground. Top manufacturers usually have typical designs which match the machines they produce. The facility offered by the Siemens company has three radiotherapy units: one X-ray machine for deep therapy and two accelerators. The X-ray unit, together with a simulator, darkroom, CT scanner, a dose planning system, electronics laboratory and a physicists' room are located in one part of the facility. In another part, there is the first accelerator and the waiting room with a special recess for patients transported on beds. This part also includes workshop, physics laboratory and a mold room as well as store rooms, staff room and an auditorium.

The third part of the diagram incorporates the second accelerator and the examination rooms, doctors' rooms, secretary's office and reception. The total area of the facility is about 30x45 m^2.

Manpower. The staff in radiation oncology centers consists of (1) radiation oncologists, (2) medical physicists, (3) dosimetrists and (4) technicians.

The US Blue Book recommends four to five radiation oncologists for 1000 patients in addition to the oncologist-in-chief, and this ratio was nearly achieved in the U.S.A. in 1986 (4.54 to 5.01 [1.60]). The physics support, however, is lower than the recommended level of 2.5 physicists/100 patients in both hospital based and freestanding facilities (2.2 and 1.5). The Blue

TABLE 4.11 Availability of Dosimetrists and Technologists in the US Radiotherapy Facilities (after J.C.H. Chu et al [1.61]).

	1980			1986		
	Actual	Recommended	%	Actual	Recommended	%
No. Dosimetrists	405	1259	32	522	1484	35
No. Technologists	3097	4026	76	4251	4950	86

Book recommends one dosimetrist for every 300 patients, not including the support required for brachytherapy, and about 2.5 technologists for every megavoltage treatment machine, in addition to the requirement for simulation support. All types of US radiotherapy facilities show a lack of adequate dosimetrist support, indicating a national shortage in the availability of this personnel. The number of full-time equivalent dosimetrists per 100 new patients in 1986 ranges from 0.99 for freestanding to 1.56 for government facilities. This is less than 50% of the optimal level [1.60]. Consequently, a significant number of facilities are using technologists for dosimetry duties. Table 4.11 shows extremely serious deficit of dosimetrists and technologists with a clear immediate need for 950 dosimetrists and 700 technologists to achieve adequate staffing.

The shortage of qualified personnel is made more serious by the fact that the treatment capabilities of modern radiation oncology departments are becoming more complex and now include intraoperative radiation therapy, high-dose rate intracavitory, hyperthermia and three-dimensional treatment planning.

In Japan, there are about 300 active radiotherapy institutions. In about half of them an average of 50-100 patients per year is treated with a linear accelerator by only one radiation oncologist and 1 or 2 technicians [4.136].

CHAPTER 5

Accelerators for Nonconventional Therapy

5.1 Ion Beam Accelerators

5.1.1 Proton and Heavy Ion Therapy

Sec. 1.1.3 provided the main reasons for the development of *nonconventional therapies* with heavy charged particles, mostly protons, light and heavy ions and mesons, as well as neutrons produced in accelerators. Therapies of this kind are still at the experimental stage of development.

In 1946, R.R.Wilson was first to suggest that protons could be used in radiotherapy [5.1], and in the 1950s pioneering studies were initiated on using protons and helium ions in Berkeley and only protons in Uppsala. Until mid-1970s, radiotherapy with heavy-ion beams was confined to small benign and malignant tumors. It was only in the 1970s that the newly developed *technique of beam spreading* and *heavy ion range modulation* made it possible to irradiate large tumors with uniform (homogenous) fields. By 1981, about 4000 patients had been treated with proton and light and heavy ions with the use of accelerators initially designed for physics studies and later adapted for medical applications in nonmedical institutions. In the 1980s, investigations of this kind became more widespread, and by June 1989 [5.2] the total number of patients who had undergone therapy with heavy charged particles was as high as 11254, including those treated with meson beams (859), ion beams (2257) and protons (8138).

In the subsequent 2.5 years, i.e. by Jan. 1, 1992, these numbers increased in the following way [5.114]: the number of patients treated with heavy particles rose by 2582 to the total of 13836;

TABLE 5.1 Tumor-related Causes of Radioresistance (after L.J.Peters et al. [5.115])

Cause	Treatment strategies
Large number of tumor clonogens	Combined modalities Hyperfractionation
Intrinsic cellular radioresistance	High LET Chemical radiosensitizers
Poor reoxygenation	Hypoxic cell sensitizers High LET Hyperfractionation Hyperthermia
Poor redistribution	High LET Hyperthermia
Rapid regeneration	Accelerated fractionation

the number of patients treated with mesons increased by 132 to 991, and that treated with heavy ions by 230 to 2487, while the number of patients irradiated with protons was higher by 2220. The largest percent increase in the number of patients over the above period was thus observed in proton radiotherapy (27%): of the total of 10358 patients irradiated so far, more than half (5419) were treated at the Harvard Cyclotron Laboratory (USA).

The good results of the proton and ion therapy programs have enhanced the tendency to use nonconventional therapy in hospitals. This, in turn, has lead to the design of medical proton (Sec.5.1.2.3) and ion accelerators (Sec.5.1.3). The first dedicated proton accelerator was set in operation in the USA (Loma Linda University Medical Center, LLUMC, California). This machine was designed and built at Fermilab, where R.R. Wilson was director from 1967 to 1978.

Some types of tumors are *radioresistant.* If it can be predicted, for example, that the tumor is likely to be *resistant to conventional radiotherapy,* it might be possible to employ a different strategy which would give a higher probability of control. Some of these *treatment strategies* are shown in Table 5.1.

It can easily be seen that the above strategies envisage the use of high LET radiation such as heavy charged particles (Sec. 1.1.3, Fig. 1.15).

High-energy protons and heavier ions have similar physical properties. Since they have masses much greater than that of an electron, as they pass through material they are little scattered, and consequently the radiotherapy fields have much smaller penumbras. Their depth-dose distributions (Fig.5.1) are completely different from those of photons, electrons and even neutrons (cf. Fig.1.16). In accordance with the proton depth-dose curve shown in Fig.5.1, which is sometimes called the *Bragg curve*, the energy of the particle penetrating the tissue is partly deposited in the near-surface layers *(the plateau range)*, whereas a characteristic sharp *Bragg peak* appears near the proton or other ion range end. Therefore, there is a fall-off of the dose beyond the peak, which is, of course, of big advantage in radiotherapy. Heavier ions ranging from carbon up to argon, produce a high ionization density, known as high LET (Fig. 1.15). The so called *Relative Biological Efficiency* (RBE) increases and the *Oxygen Enhancement Ratio* (OER) decreases

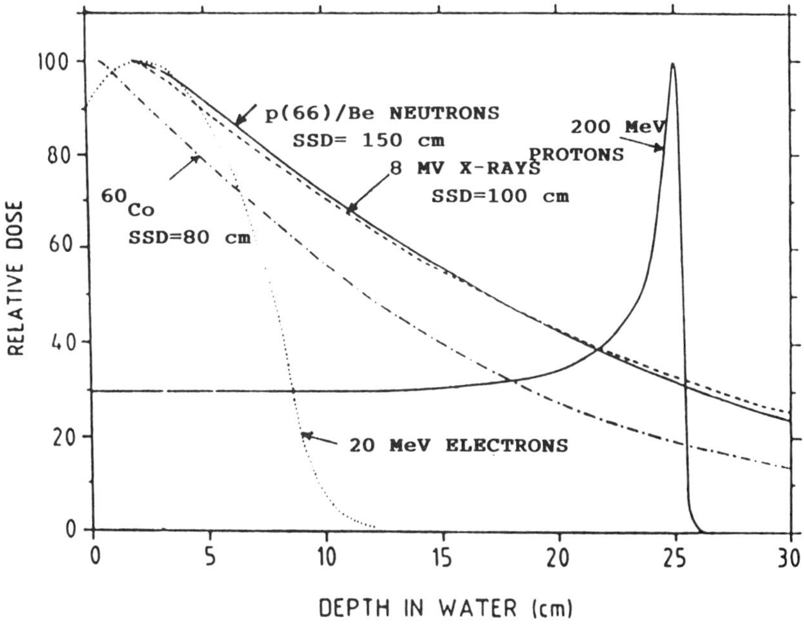

FIGURE 5.1 Depth-dose distribution curves for a monoenergetic 200 MeV protons, ^{60}Co photons, 8 MV bremsstrahlung and p(66)/Be neutrons in a water phantom with a 0.3 cm^3 ionization chamber (after D.T.L.Jones [5.5])

[1.34]. This latter effect is a serious limiting factor in radiotherapy of large tumors which are poorly vascularized and whose cells are less oxygenated and therefore less sensitive to radiations. These two factors which magnify the effect of the Bragg peak are shown in Fig. 1.15 which presents typical variations of RBE and OER and characteristic LET bands for several charged particle beams. Hence a modification of quality of the radiation received by the patient becomes possible.

Since the maximum width of the Bragg peak for protons and other heavy ions (Fig.5.1) is only a few millimeters, only small targets can be uniformly irradiated. The actual width of tumors encountered in heavy ion radiotherapy is often much larger, and, therefore, the particle range must be appropriately matched by passively or actively changing energy of ions (Fig.5.2a). Individual peaks become superimposed in such a way that the approximately uniform region of the dose extends over the whole target volume. By additionally correcting the beam intensity, the RBE of individual peaks can be equalized so

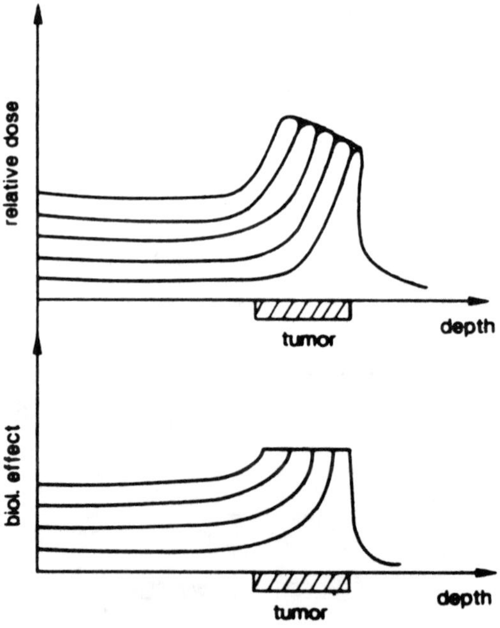

FIGURE 5.2 Modification of the depth-dose distribution from heavy charged particles: by changing beam energy, and by modifying beam intensity.

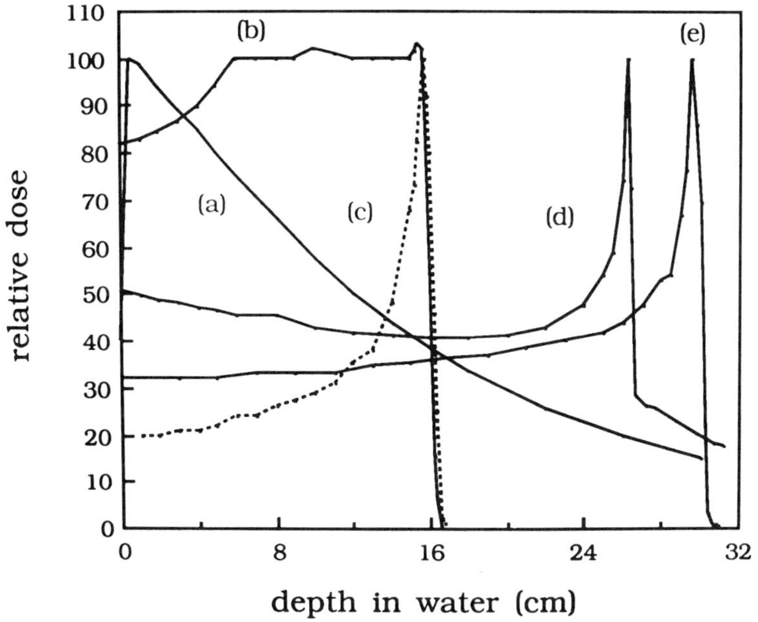

FIGURE 5.3 Representative central axis depth-dose distributions: (a) — for a 10x10 cm^2 ^{60}Co field (SSD=100cm), (b) — a spread out proton beam to give a uniform dose (to ±3%) over a depth of 11.8 cm, measured at the 90% isodose level, (c) — a 160 MeV proton beam, (d) — a 669 MeV/nucleon neon ion beam, (e) — unmodified helium ion beam of total energy 934 MeV (after J.M. Sisterson [5.1]).

that an identical biological effect is obtained over the entire tumor volume (Fig.5.2). This method is referred to as the *range or peak modification or modulation technique.* It may also be called an *energy modulation method.*

Practical applications of the proton modification technique are shown in Fig.5.3. Curve c describes a 160 MeV monoenergetic proton beam, whereas curve b represents a modified beam which provides a uniform dose (3%) over a depth of 11.8 cm. Curves d and e represent distributions for 669 MeV/nucleon neon and 934 MeV (total energy) helium ion beams, respectively.

In proton therapy, the upper energy used is usually 250 MeV, which corresponds to the range of about 38 cm in water. The minimum energy is close to 70 MeV, giving the range of about 4 cm.

As for heavy ions, the present-day accelerators provide the possibility of accelerating almost all ions, uranium included. From the clinical point of view, the most interesting and relevant group of ions are light ions ,up to and including argon (Z=18, A=40) [5.6,5.7]. Their range-energy curves are shown in Fig.5.4. To obtain a 30 cm range in tissue with ^{28}Si ions the energy required is approximately 800 MeV/nucleon.

The main advantage of the proton and ion radiation therapy is that the depth distributions can be made much more accurate and they better conform to clinical criteria than those from conventional radiations (Fig.5.3). In addition, treatment times for small intracranial lesions and eye tumors can be reduced to about one week as compared with six to eight weeks in conventional therapy; larger tumors are fractionated like those treated with photons. The interest shown by the medical profession in the new kinds of radiotherapy is best exemplified by the fact that an international group has been formed, called *Proton Therapy Co-Operative Group* (PTCOG), which publishes a newsletter [5.8] and arranges bi-annual meetings.

As for routine electron and photon radiotherapy (Secs. 3.1.1.2 and 4.1.6) there exists a *Protocol for Heavy Charged-Particle Beam Dosimetry* [5.42] written by Task Group 20 of the American Association of Physicists in Medicine (AAPM). This protocol provides general recommendations and prescriptions for dose determination and measurements for both proton and heavy charged-particle beams. In Europe, the working group of ECHED (European Clinical Heavy Particle Dosimetry Group) has also prepared a Code of Practice for Clinical Proton Dosimetry, confined specifically to the determination of absorbed dose for proton beams only [5.147]. It is recommended that calibrations be carried out in water at the center of the spread-out-Bragg-peak and that dose distributions be measured in a water phantom (Sec. 4.1.3). It is estimated that the error in the calibration will be less than ±5% (1 S.D.) in all cases.

The world's largest facility for proton therapy, which carries out about 50% of all irradiations, is the Harvard Cyclotron Laboratory which operates a 160 MeV synchrotron. By June 1988, 4300 patients had been treated in collaboration with the Massachusetts General Hospital (MGH) and Massachusetts Eye and Ear Infirmary (MEEI). The therapy was implemented in three programs. In the neurosurgery program, 2549 patients

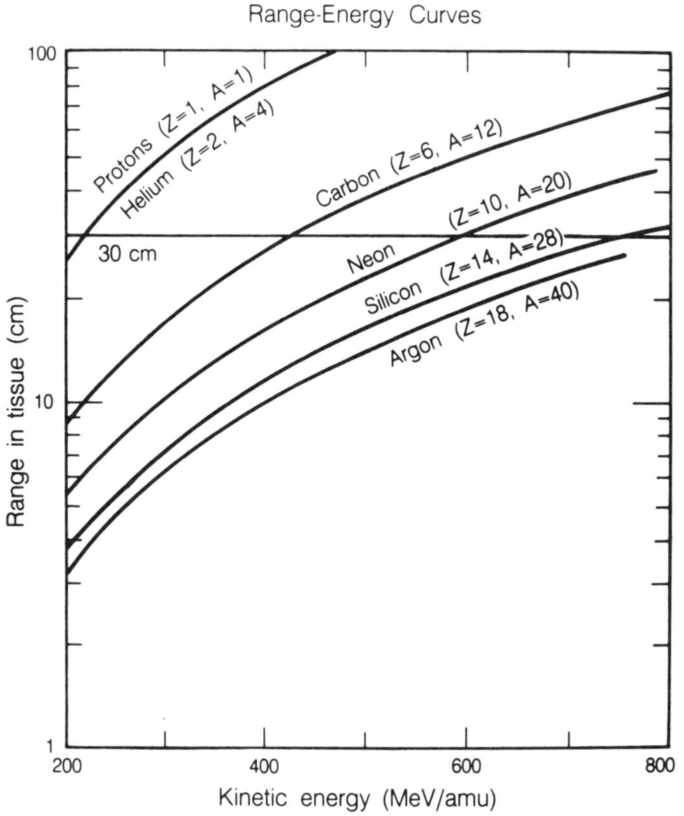

FIGURE 5.4 Light ion ranges in therapeutic applications vs. kinetic unit energy (after E.L.Alpen and R.A.Gough [5.7]).

were treated, of which the majority (1143) were those with arteriovenous malformations. Two other large groups of patients were those with acromegaly (576) and pituitary adenoma (501). The eye program included 1242 patients with uveal melanoma, and fractionated radiation therapy was applied to 540 patients, of which 139 were treated for prostate tumors. The results of the above programs were at least as good or better than those achievable with other radiation modalities. For example, for 1006 patients treated for uveal melanoma the 5-year survival rate was 78.53.6% [5.1].

At present, two major applications of proton beams are encountered in clinical practice. These are, firstly, the use of relatively low energy protons: $E_p \leq 80$ MeV or higher energy degraded beams, specifically for the treatment of the ocular tumors using field sizes smaller than 10 cm^2; and secondly, the treatment of large or deep-seated lesions using higher energy beams ($E_p \geq 150$ MeV) with small or large fields.

5.1.2 Proton Accelerators

5.1.2.1 Chronology of Development

Table 5.2 is a list of accelerators used so far in proton therapy. For this purpose, *synchrocyclotrons* have been mostly employed (Sec.2.6.3, Table 2.5). The installations in Moscow and Tsukuba (Japan), however, have been using proton synchrotrons (Sec.2.7, Table 2.6).

The first generation of accelerators used in proton radiotherapy was designed for physics studies. Therefore, they had to be specially adapted to proton therapy, which often required quite complex technical procedures. For example, systems for energy degradation had to be used. With increase in interest in this new modality, some centers had their accelerator facilities completely restructured by devoting a large part of the facility to radiation therapy. The best example is the Swedish Gustaf Werner Institute in Uppsala, which has greatly contributed to the development of proton therapy (Fig.2.39) [5.9]. Recently (1991), the 200 MeV synchrotron facility at Orsay near Paris (France), initially used for physics research studies since 1958, has been adapted exclusively to proton therapy and is called Centre de Protontherapie d'Orsay (CPO).

Accelerators of the second generation employed for proton therapy are those added to medical cyclotrons originally intended for the production of radionuclides and neutron therapy (Sec.5.3.4). The proton channel recently set in operation at the Clatterbridge Hospital in England, dedicated to the treatment of uveal melanoma, is an example of such a design. A similar scheme is being implemented now at the National Accelerator Centre in Cape Town , Republic of South Africa (cf. Fig.5.27), the proton energy used being 200 MeV.

The above design cannot be regarded, however, as a universally adopted solution since, according to the US experience,

TABLE 5.2 Accelerators for Proton Therapy, status 1992.

Institution	Accelerator	Energy [MeV]	Year of operation	No. of patients at Jan 1, 1992
Donner Lab., Berkeley USA	Synchrocyclotron	340	1954-57	30
Werner Inst. Uppsala, Sweden	Synchrocyclotron	85	1957-76	73[4)]
Harvard Cycl. Lab. USA	Synchrocyclotron	160	1961	5419
Dubna, USSR	Synchrocyclotron	90-200	1964-74	84
Inst. Teoret. i Eksp. Fiziki., Moscow, USSR	Synchrotron	70-200	1969	2200
LIJF, Leningrad, USSR	Synchrocyclotron	70-1000	1975	719[1)]
Nat. Inst. Nucl. Phys. Chiba, Japan	Synchrocyclotron	70	1979	74[2)]
PARMS, Tsukuba, Japan	Synchrotron	250	1983	242
SIN[5)], Villingen, Switzerland	Cyclotron isochron.	70-590	1985	1150
Werner Inst., Uppsala, Sweden			1988	23[4)]
Clatterbridge Hospital, Bebington, England	Cyclotron isochron.[3)]	62	1990	189
LLUMC, Loma Linda, USA	Synchrotron	70-250	1990	76
Louvain-la-Neuve[7)], Belgium	Cyclotron	90	1991	8
Nice[8)], France	Cyclotron	65	1991	45[6)]
CPO Orsay[9)], France	Synchrocyclotron	70-200	1991	13
NAC[10)], South Africa	Cyclotron	200	1993	–

1) June 1991; 2) May 1991; 3) Set in operation in June, 1990; 4) Treatments with a 72 MeV stationary (fixed) beam resumed after reconstruction; 5) At present Paul Scherrer Institut (PSI); 6) December 1991; 7) Universite Catholique de Louvain, Louvain-la-Neuve, Belgium; 8) Centre Antoine Lacassagne-Cyclotron biomedical, Nice, France; 9) Center for Protontherapy (CPO) — Orsay Synchrocyclotron Cedex; 10) National Accelerator Centre at Faure, Republic of South Africa.

a single facility like one at Harvard can serve a population of about 20 million. Thus a total of 10 machines intended solely for proton therapy would be necessary for the whole population of the USA [5.10]. For this reason, in the last few years several designs of dedicated accelerators have been developed to ensure that the proton therapy will eventually become a routine method of treatment. Among such designs, to be later discussed at length in Sec.5.1.2.3, is a proton synchrotron designed by the Fermi Laboratory (Batavia, USA) and already installed at Loma Linda University Medical Center.

5.1.2.2 Beam Transport and Proton Gantries

In the overwhelming majority of proton therapy facilities the accelerator is physically separated from radiotherapy rooms. Therefore, a *special system to transport the beam* from an accelerator to the room where the patient is located has to be used. Ion-optical systems employed for this purpose are presented in [5.11]. The simplest method of proton or heavy ion beam delivery is to use a *horizontal immobile channel* (Fig.5.21). The patient is irradiated in a lying or seated position, which is only good for irradiating lesions in the head since the tumor volume localization can then be relatively easily reproduced. For the reasons presented in more detail in Sec. 3.1.3 irradiation of tumors localized at other parts of the body can be more readily implemented with the patient in supine or prone position. Therefore, various designs have been initiated to develop *proton gantries* with full 360° rotation, which provide treatment fields achievable in electron accelerators for routine therapy. Details of these studies are discussed in [5.12 and 5.14].

The magnetic rigidity (Sec.2.4.1) of a proton beam accelerated to the energy of 250 MeV, that of the upper limit of the range used, is Br = 2.4 Tm. By using conventional magnets at room temperature to bend the proton beam an induction of about 1.5 T can be achieved, the radius of the curvature being approx. 1.6 m. For 50 MeV protons, the radius of curvature will be 0.7 m. Thus the beam bend systems in proton therapy facilities must be twice as large as those used in proton generators in neutron therapy (Sec.5.3.4).

Fig.5.5 is a schematic representation of a magnetic proton bend system, called the *Corkscrew Gantry* [5.14]. The proton beam coming from the left is bent by about 90° by two small

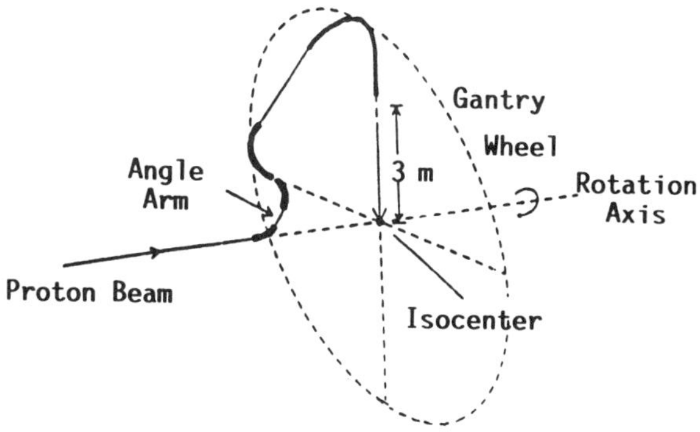

FIGURE 5.5 The "Corkscrew" proton beam deflection system developed at Harvard (after M.S.Z. Rabin et al.[5.14])

dipoles. When the beam enters the rotation plane, in the plane at right angles to the primary proton direction, it reaches two consecutive larger dipoles with the bend angles of about 135° each. The dipoles bend the beam in the plane of gantry rotation until it points towards the isocenter.

The Corkscrew system minimizes the need for floor space and also reduces shielding requirements to 1400 metric tons of concrete. In other designs, shielding may be as high as 1900 t of concrete [5.12]. The system shown in Fig.5.5 has been employed in the dedicated accelerator installed at Loma Linda (Sec. 5.1.2.3).

A characteristic feature of the newly designed systems for proton therapy is that *fixed stationary beams* are synchronously combined with *rotating gantries*. Fig.5.6 is a plan view of such a system utilizing a Corkscrew Gantry designed at the Harvard Cyclotron Laboratory [5.14]. In a three-level building (Fig.5.6b) the accelerator will be located at the lowest level. The accelerated beam will be delivered to the middle level (Fig.5.6a) and directed either to two fixed-beam channels or to the gantry. The first fixed-beam channel would be used for eye treatments with 70 MeV protons, whereas the other horizontal fixed-beam channel would be used for brain tumor irradiations and in a multi-field therapy. The shielding has been designed in such a way that the radiation dose outside treatment rooms is less than 5 mrem h^{-1} for the 20 nA proton beam intensity at the gantry [5.14].

Passive Beam Modification Systems. Heavy charged particles beam spot sizes have usually 3-4 mm, which when they stop in water increase to 1-2 cm. The volume in which they stop is the product of the beam area times the range. This is about 40 cm x 1 cm^2 = 40 cm^3 without a scattering system. The 1 cm^3 represents the volume where the maximum dose deposition occurs ($\pm 10\%$ of the maximum). Without additional devices to modify the beam the volume in which particles are stopped would be small. The size of this volume, characteristic for a given ion used and often referred to as a *voxel,* is dependent on the depth of treatment, multiple scattering and ion range (straggling). For heavy ions, such as neon, the voxel size is about one order of magnitude smaller than that for protons. Matching the beam stopping volume to that of the target volume requires not only spreading the beam laterally but also proper adjusting of its energy so that adequate penetration depth is obtained (Fig.5.2). Shaping of the lateral field, like that of the electron beam (Sec.3.1.4), can be made by using a system of scattering foils (Fig.3.37) or by magnetic scanning (Fig.3.41). On the other hand, adjusting the range is made with a variable absorber or by varying the energy of the beam extracted from the accelerator. Precise shaping of the target volume requires that the lateral and transverse beam control systems be independent of each other.

The main technique of obtaining large treatment fields is that of *scattering foils and ridge filters.* Scattering on lead foils spreads the beam in an extended Gaussian shape, but does not ensure that the field is sufficiently uniform. Good uniformity (flatness) can be achieved with the aid of cylindrical blocking rings as depicted in Fig.5.7a. Circular fields of up to 20 to 30 cm in diameter can thus be obtained. The final outline of the field is tailored with proper collimators to the outline of the irradiated volume.

With the use of a system of scattering foils and blocking rings as in Fig.5.7a the beam utilization is, unfortunately, low, not exceeding 10 to 25%. The rest of the beam is absorbed by the blocking rings or forms an edge field outside the flat area and, therefore, must be collimated out. This high beam efficiency loss reduces the available dose rate, and, in addition, increases neutron background in the treatment room.

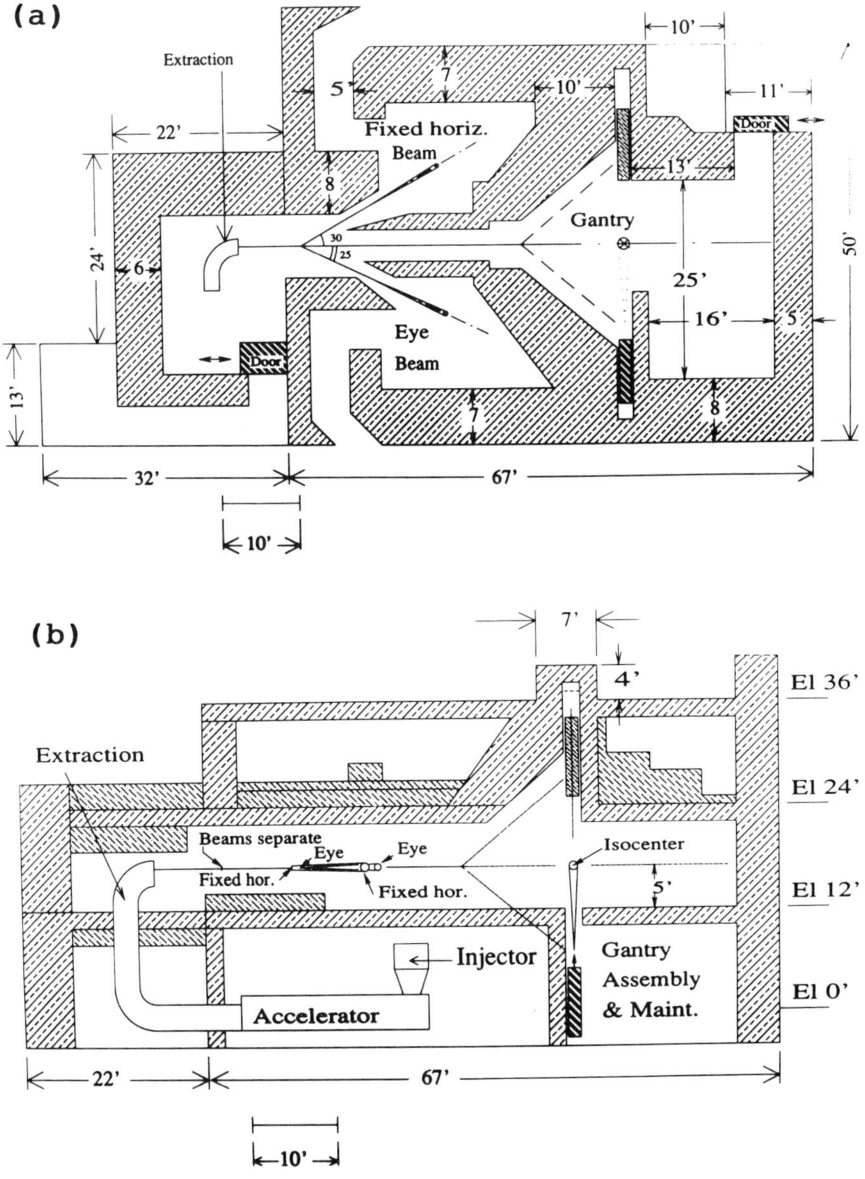

FIGURE 5.6 Plan view of the design of a three-story proton therapy facility: (a) — horizontal section of the middle floor, (b) — vertical section of the building (after M.S.Z.Rabin et al.[5.14])

Range modulation requires, as mentioned earlier, *ridge filters* made from a material of non-uniform thickness placed between the beam accelerator exit and the patient. Particles which pass through the thinner regions of the filter lose less energy compared with those which must traverse thicker regions. Thus the deeper tissue layers become uniformly irradiated (Figs. 5.2 and 5.3, curve c). Ridge filters are usually fabricated from brass with a properly grooved surface. Lucite propellers with fan-shaped blades are also used. Filters of this kind are rotated at a constant speed so that all areas of the treatment field "see" the full energy modulation range. The depths of the grooves are usually adjusted to the desired depth modulation, and, therefore, a proper set of filters is required to cover the whole range of thicknesses irradiated.

The modality shown in Fig. 5.7a used for scattering protons and helium ions is referred to as a *passive system* since it uses passive mechanical elements. Its great advantage, however, is simplicity, accuracy and reliability. Its disadvantage is that the

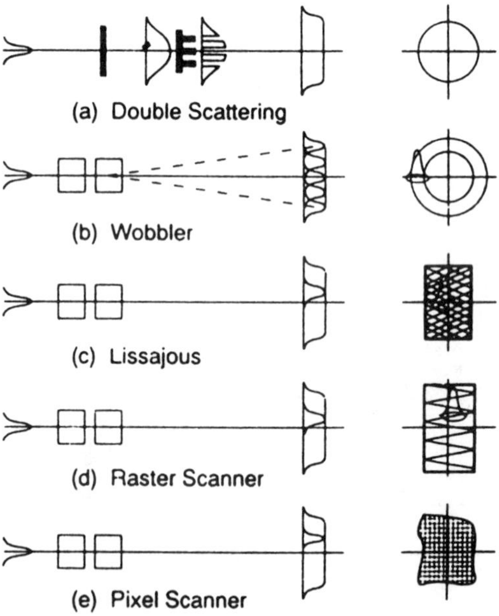

FIGURE 5.7 Proton beam modification (spreading) systems: (a) — double scattering foil, (b) — wobbler, (c) — Lissajou scanning, (d) — raster scanning, (e) — pixel scanning (after J.R.Alonso [5.15])

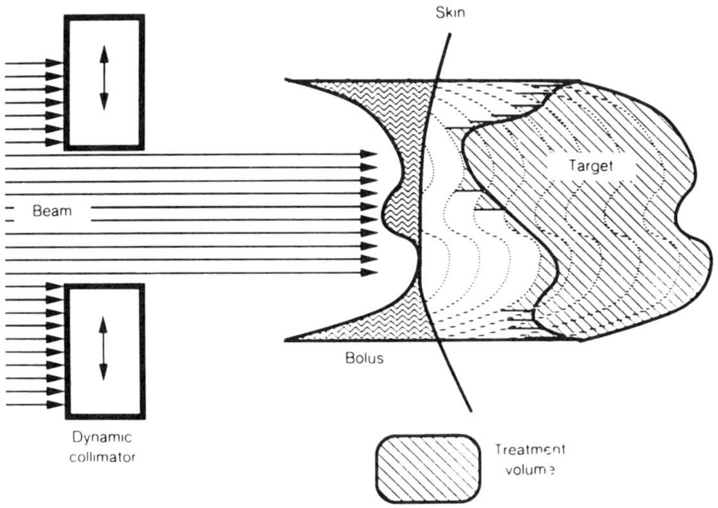

FIGURE 5.8 A system for minimizing normal tissue irradiation in 3D modality with a uniform beam spreading system, a multi-leaf collimator and an individual compensator on the patient skin (after J.R.Alonso (5.14]).

treatment field must be of uniform modulation depth. There is also no correlation here between the beam lateral position (x,y) and depth modulation (z), all the elements of the treatment volume are subjected to identical range modulation. Therefore, in order to irradiate an irregular target volume by placing it in a cylindrical volume a relatively large portion of a normal tissue has to be also treated. As shown in Fig. 5.8 an *active collimator*, can be employed to irradiate an irregularly shaped field. An *individual compensator* placed on the patient's skin can additionally adjust the deepest range of the beam to the deepest layer of the target volume. However, the maximum thickness of the target still determines the necessary energy modulation range over the whole treatment volume, and by irradiating the thinner parts of the target, normal tissues also become also irradiated.

Magnetic Beam Modifying Systems. The above mentioned shortcomings of the passive system can be overcome by employing *magnetic control of the beam* in the x-y plane in conjunction with an independent system of depth modulation control. The fundamentals of scanning systems have been described in Sec.3.1.4.3 and the systems used for deflecting proton and heavy ion beams have been presented schematically in Fig.5.7b through e. In the *wobbler system* (Fig.5.7b), the deflecting magnets are driven by a sinusoidal current 90° out of phase so that the beam travels in a circular path on the treated field. The radius of the circle can be controlled by changing the amplitude of the sinusoidal supply current. A *Lissajous scanner* (Fig.5.7c), well known in the oscilloscope techniques, is another type of the wobbler. The deflecting magnets are driven by see-saw waves of independent frequencies. In a *raster scanner* (Fig.5.7b) the beam is deflected by a fast single see-saw sweep of the current and a second magnet is also driven by much slower see-saw waves which ensure slow linear scanning movement. The system shown in Fig.5.7e, called *pixel or spot scanning,* moves the beam from pixel to pixel,i.e. from one spot to the next. The beam can be maintained at a given pixel until the full predetermined dose is achieved, or it can be moved several times so that each pixel receives a partial dose during one scan and then all the partial doses become added to give a total dose (cf. Fig. 3.42).

An overview of the therapeutic options of the various scanning systems is given in [5.15]. The common trend in the development of these systems is to adjust them to the implementation of 3D conformal therapy (Sec.4.2.4.5). In this therapy, the treatment volume is made to closely conform to the target volume through variable modulation of the width of the spread-out Bragg peak across the target volume (Fig. 5.18). Magnetic scanning methods offer the ability to truly optimize the dose localization within the target volume, and they also provide a flexible and rapidly reprogrammed beam delivery system. Magnetic scanning can easily be implemented in the beam optics of an isocentric beam transport systems, permitting the design of the gantry to be reduced to dimensions similar to those used in conventional therapy [5.116].

Fig. 5.9 shows such a system developed by the Swiss Paul Scherrer Institut (formerly SIN, Villigen). The target volume is scanned with a 250 MeV proton *pencil beam* slice by slice, like in computed tomography, by magnetic deflection, concomitant range modulation and patient translation. In the first stage of the project, the beam used is horizontal and stationary (fixed).

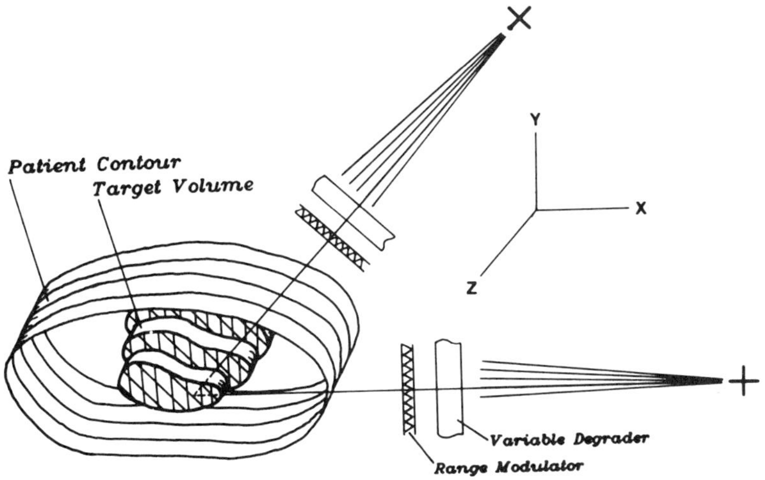

FIGURE 5.9 A system for a 250 MeV pencil proton beam scanning and modulation, developed at the Paul Scherrer Institut, Villigen (after M.Blattmann et al. [5.16])

In order to adjust the irradiated volume to the irregular and complex target volume a large number of pixels, or voxels, are necessary. If the target volume is, for example, 1000 cm^3, or 1 liter, the number of voxels is about 10 000 for a 5 mm distance between successive pixels. Since the treatment time should be limited to several minutes only, the irradiation times per voxel must be about 15 ms. Between two successive spots the beam is turned off and on by a fast switching magnet. Another similar magnet ensures that the beam travels to the desired coordinates. Both the magnet response times and those of the beam monitoring system should be of the order of 0.1 ms to achieve dose uniformity better than 1%.

In the system shown in Fig.5.9, energy or range modulation is carried out by injecting small 5 mm thick polyethylene plates whose purpose is to reduce the beam range. The *variable beam degrader* shown schematically in Fig. 5.9 is intended for degrading the accelerator output energy down to the treatment energy of 250 MeV.

A similar system has been developed at the National Institute of Radiological Sciences (NIRS) in Japan for 70 MeV protons.

The dimensions of each voxel, 1 cm square, result from the relatively low energy of the protons used. In this way, various irregular target volumes can be covered slice by slice [5.17].

5.1.2.3 Hospital Proton Accelerators

Basic Requirements. A hospital accelerator intended for proton therapy should deliver beams of accelerated protons with the maximum energy of at least 250 MeV. What is particularly important is that this energy should be easily controlled over a wide range of values because of the need to have the depth distribution quickly modified. For the dose rate of several Gy min^{-1} to be produced in the volume of 1000 cm^3 the beam intensity should be about 10^{10} p s^{-1}. At the present stage of development of proton acceleration techniques these requirements are easily met since protons are found to be the most frequently accelerated particles second only to electrons.

However, other requirements that allow proton accelerators to be transformed into medical machines are far more difficult to satisfy. There is, first of all, the problem of bulk, weight and size of the shielding, or in other words, the problems of costs. The main aim of the proton therapy, which is a prerequisite of its widespread use, is to have treatment unit costs approximate those of conventional photon or electron therapies.

At the world's largest facility for proton therapy at Harvard (USA) the 160 MeV synchrocyclotron, equipped with a 583 ton magnet, delivers a 20 nA beam, typical for medical purposes. This makes it possible to irradiate a target area 30 cm in diameter in about 2 min, which meets standard therapeutic demands.

Some Prospective Designs. Protons can be accelerated to the energy of 250 MeV in several types of accelerators. An rf linac, with a standard linear gradient of 5 MeV m^{-1}, would have to be about 50 m long, and thus would require a large floor space. On the other hand, isochronous cyclotrons with the above energies would have to be equipped with very heavy magnets weighing several hundred tons and made from magnetic steel, if operated at room temperature. This involves high costs and, therefore, their use seems to be rather problematic. Greater hopes are placed in the application of sychrocyclotrons which were, in fact, the first generation of proton accelerators used in therapy.

The main difficulty in the construction of a modern sychrocyclotron for hospital use is that, like in isochronous cyclotrons, a large and heavy electromagnet has to be employed. In order to overcome this difficulty a *synchrocyclotron equipped with superconducting magnets* has been designed. As shown in Fig.2.23, superconducting magnets allow their mass to be reduced by a factor of about 17 as compared with conventional magnets. Fig.5.10 is a layout of a synchrocyclotron facility designed at the National Superconducting Laboratory, Michigan State University (USA) [5.18]. The synchrocyclotron proper, weighing 52 t would be installed, together with a beam transport system and a gantry, in such a way that it could rotate as one unit, which would greatly simplify the construction of the gantry. The energy could be controlled by a graphite shifter in the form of a wedge. A simple transport system could be

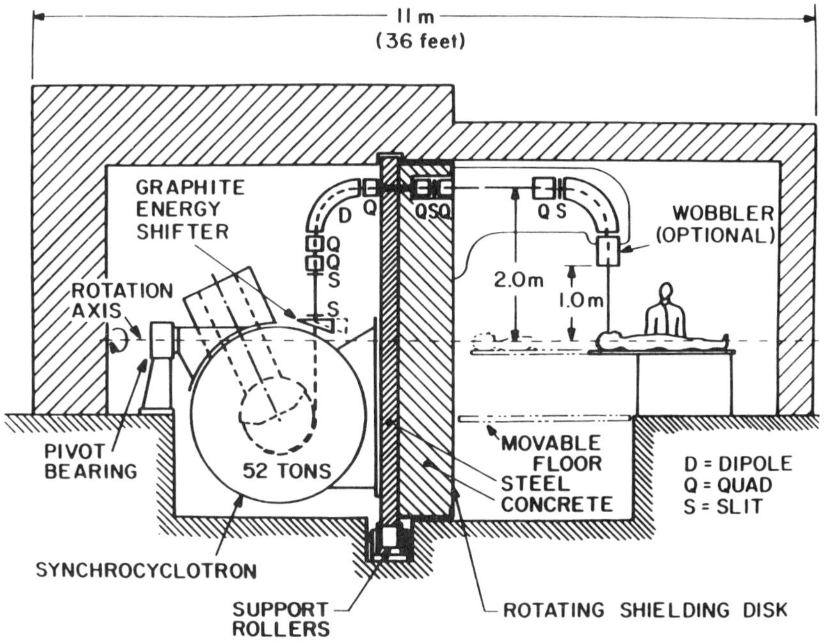

FIGURE 5.10 Stand-alone radiotherapy facility using 250 MeV proton synchrocyclotron with superconducting magnets, developed at the National Superconducting Cyclotron Laboratory, Michigan, USA (after H.G.Blosser [5.18])

provided with an optional wobbler. A 70 MeV version of the machine intended for eye therapy would feature superconducting magnets weighing 15 t. The whole facility, set up with a horizontal stationary (fixed) beam and a shielding, would require 6.5 x 9.8 m^2 floor space [5.19].

At the Harvard proton therapy facility, a new small proton accelerator is being developed which will deliver 250 MeV and 20 nA beams [5.10, 5.19]. Its weight will be a mere 5 t but it will require 50 tons of shielding. It will fit into a 9.15 x 9.15 m^2 room, excluding treatment and control rooms. The supply power will be of the order of 250 kW and each treatment will require about 10 kWh of energy. It is estimated [5.10] that the commercial version would costs two million US dollars, which is comparable with the cost of a large electron radiotherapy accelerator.

Another US conceptual design of a radiotherapy proton accelerator, making use of the experience of the Argonne National Laboratory, also envisages a construction of a *synchrotron intended, however, for negative hydrogen ions* H^- [5.20,5.21]. This method is also used in isochronous cyclotrons (Secs.2.6.2 and 7.1). Hydrogen ions here are accelerated as negative ions which are subjected to charge exchange to produce extraction at the end of the acceleration process. In this way accelerated protons are obtained with maximum energies of up to 250 MeV and a beam current of 60 nA at a repetition rate of 5 Hz, the diameter of the electromagnet system being 7 m. Hydrogen H^- ions would be injected to the synchrotron from a preaccelerator, in this case a tandem 1.5 MeV Van de Graaff machine (Sec.2.2.2.1). With the financial support from the National Cancer Institute, ACCTEK Associates [F18] has started the construction of the prototype synchrotron magnets, raster scanning system and rotating beam delivery system. Considerable care has been paid to the commercial side of the project and to the transfer of experience from research laboratories to industrial practice [5.22].

The University of California at Davis is developing plans to build a proton therapy facility as a part of its Cancer Center in Sacramento. The *UC Davis Proton Facility* will be installed in a new building to be erected adjacent to the present Cancer Center building. UC Davis and Lawrence Berkeley Laboratory (LBL) are developing an agreement to advance this project; LBL will design the accelerator, beam transport and beam delivery systems, will oversee their construction, and will be responsible for commissioning and initial operation of the Proton Facility [5.123].

The CYCLONE 230 cyclotron proposed by IBA [F23] (Belgium) is an isochronous, high field, non-superconducting cyclotron able to accelerate protons to 230 MeV (Fig. 5.11a). The basic question in designing the magnet for this cyclotron is whether to use superconducting coils or not [5.120]. In both solutions the field levels are rather similar: 3.09 T on the hills (Sec. 2.6.2, Fig. 2.33), 0.985 T in valleys, 2.165 T average field at extraction radius, 1.74 T average field at the center.

Such field levels are mostly dictated by the requirement of obtaining adequate vertical focusing without having to use excessive spiraling of the sectors. The advantages of a superconducting system are: the reduction of iron for the return magnetic flux, the reduction of the electrical power and, possibly, reduction in the total investment costs. The disadvantage is the mechanical complexity of a split-coil cryostat, with very large forces involved, the potential problems associated with the liquid helium supply and the very long thermal time constants; for such a large mass, warm-up or cool-down times are in excess of one week.

Using the *deep-valleys concept* pioneered by IBA with the CYCLONE 30 series, it is possible to design a high-field, non-superconducting 185 t (weight of iron) magnet for a 230 MeV cyclotron (Fig. 5.11a) using not more than 185 kW of electrical power. The cyclotron can be opened for maintenance or repair, closed, pumped down and restarted for operation in less than two hours, excluding the repair time.

As a part of this proton therapy facility IBA has developed the concept of *compact gantry* (Fig. 5.11b). The beam is scanned in three dimensions. Range scanning is performed by a variable thickness rapidly rotating cylindrical absorber. The beam pulsing is synchronized with the absorber rotation so as to deliver a dose only in the specified range. Axial scanning is achieved by a sweep magnet located upstream of the final 90° magnet. Due to the optical features of the system the beam reaching the patient is swept parallel with a maximum amplitude of 40 cm. The effective source to isocenter distance is therefore infinite. The axial scanning speed is controlled by a dose integrator allowing an automatic correction of small beam intensity variations.

(a)

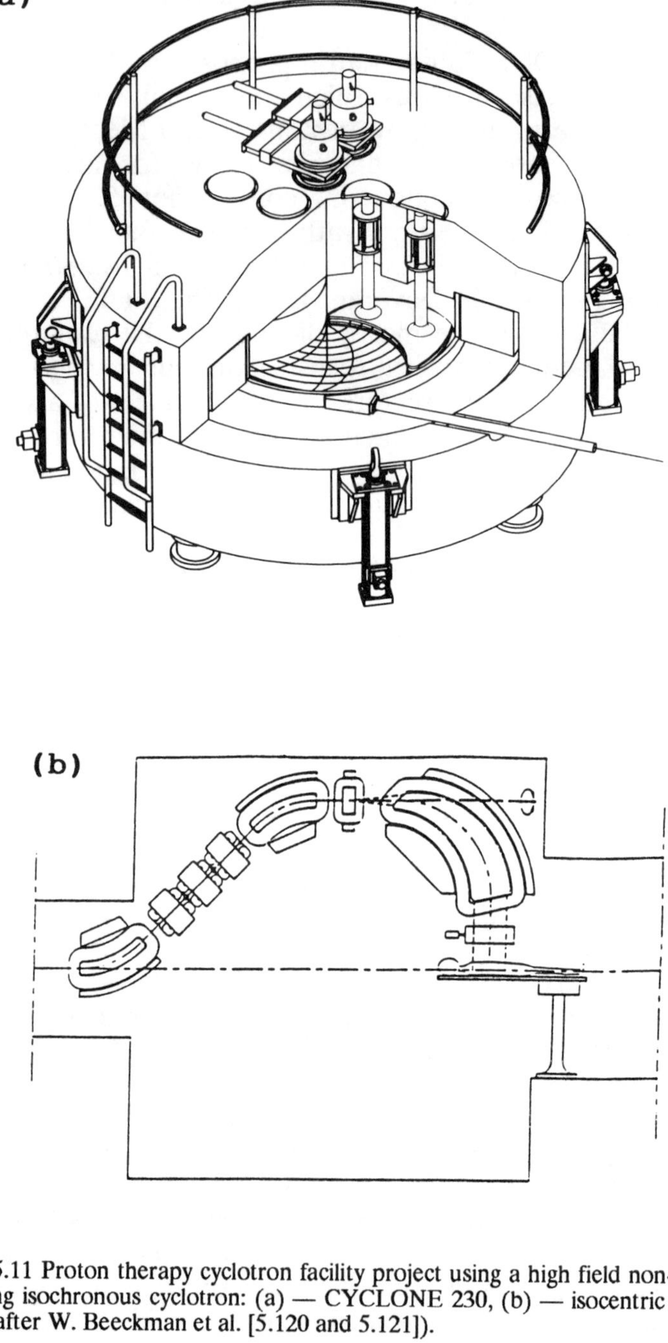

(b)

FIGURE 5.11 Proton therapy cyclotron facility project using a high field non-superconducting isochronous cyclotron: (a) — CYCLONE 230, (b) — isocentric gantry scheme (after W. Beeckman et al. [5.120 and 5.121]).

Transverse beam scanning is obtained by a slow rotation of the final 90° magnet around the axis of the incoming beam. By combining this motion with a slight gantry motion, a parallel beam translation, or even an arc therapy is also possible. The 5 m diameter of the gantry permits its installation in therapy rooms of a much more reasonable size.

Ion Beam Application (IBA, Belgium) has obtained a governmental loan worth over $ 4 million to finance part of the development of this proton therapy system and to begin immediate construction of a prototype installation to be operational by mid-1994 [5.122].

In Japan, proton therapy is being realized at two facilities: at the National Institute of Radiological Sciences (NIRS), Chiba, and at the Particle Radiation Medical Sciences Center (PARMS), University of Tsukuba. At the NIRS, protons are accelerated to an energy of 70 MeV with a range of 36 mm, and the PARMS uses 500 MeV protons from the High Energy Physics Laboratory (KEK). For therapy, proton energies are degraded in a graphite absorber. The dose rate exceeds 1 Gy min^{-1} [5.23]. Promising clinical results obtained with the above machines have led to the design of a dedicated synchrotron with a maximum energy of 230 MeV (range in water of 33 cm) and beam intensity of 20 nA [5.117 and 5.118]. Like in the US project Argonne National Laboratory (ANL) + ACCTEK, it will consist of a multicusp H^- ion source, a tandem electrostatic accelerator (NEC 9SD-2 Pelletron) as an injector, a slow-cycling synchrotron and beam lines to two treatment rooms. The synchrotron will accelerate 5 MeV protons up to max. 230 MeV. The treatment rooms will be equipped with fixed isocentric beams. One will have a horizontal beam and two vertical beams, from upper and lower directions. The other will have two vertical beams but no horizontal beam [5.118]. The circumference of the synchrotron magnet system will be approx. 35 m. The NIRS envisages development of a medical heavy ion accelerator intended for helium to argon ions (Sec. 5.1.3.3).

Dedicated Proton Synchrotron. In 1986, the US Loma Linda University Medical Center entrusted the Fermi Laboratory (Fermilab) with the design and construction of an accelerator for the world's first dedicated proton therapy medical facility. At the present time, i.e. in 1992, the development of the facility is nearing completion [5.26 through 5.28, 5.153].

A schematic of the Loma Linda University Proton Therapy Facility is shown in Fig. 5.12 and the accelerators shown in more detail in Fig. 2.42 (Sec. 2.7). The facility, in a large hospital located on the Loma Linda University campus in Southern California, is divided into three areas: the accelerator, the beam transport lines, and the treatment rooms. The accelerator is a *zero gradient synchrotron* capable of accelerating protons to final energies between 70 and 250 MeV. The proton beam is initially accelerated to 2 MeV by a radio frequency quadrupole (RFQ) and subsequently injected into the ring after a 180° bending magnet. Shortly after the ring is filled, the beam is accelerated to its final energy by an rf cavity, located diametrically opposite to the point of injection. After achieving final energy, the beam is resonantly extracted at the first 1/2 integer betatron frequency. This frequency, often called the *tune of the accelerator,* is regulated by quadrupole magnets in the accelerator. During resonant extraction, the portion of the beam crossing the electrostatic septum is deflected into a Lambertson extraction magnet which in turn deflects the beam out of the synchrotron into the beam transport lines, The pulsed beam is extracted with a typical repetition rate of 2.0 s with a 0.4-s spill duration (Fig 2.41b).

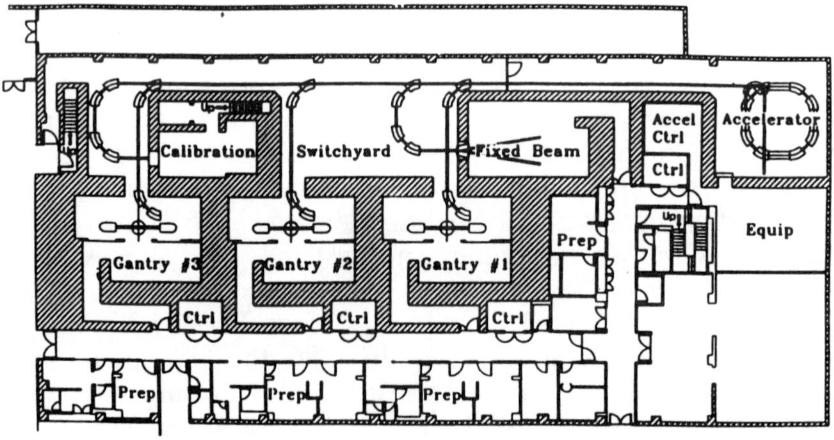

FIGURE 5.12 The Loma Linda University Proton therapy facility (after G. Coutracon et al. [5.153])

The beam transport lines guide the beam into any one of four treatment rooms. As seen from Fig. 5.12, three of the rooms contain isocentric gantries. In all gantries, there is about 3.4 m space between the last bending magnet and the gantry isocenter. This space contains the beam delivery system referred to here as the *nozzle.* It contains dosimetry instrumentation and beam spreading devices to monitor and control the dose to the patient. The beam entering the nozzle is approximately 1-cm FWHM with a divergence typically less than several milliradians. It is therefore necessary to spread the beam into a larger area with good field uniformity. Initially, field sizes up to 20-cm diameter will be required with field uniformity better than ±3%. The dose should also be as uniform as possible for target depths ranging from 1 to 15 cm [5.153].

At present (1992), preliminary studies have been completed on the horizontal (fixed) beam line. It delivers 155 MeV and 200 MeV protons through a double foil scattering system to produce a field size of 15 x 15 cm^2 with a uniformity better than ±3%. Uniform depth dose profiles are achieved through the use of a rotating variable thickness polycarbonate modulator wheel. At 200 Mev, treatment depths up to 21.1 cm are achievable, whereas at 155 MeV, a maximum depth of 13.1 cm can be obtained. Dose uniformity of better than ±2% is observed with modulations from 3 to 10 cm. The penumbra (90% - 50%) on the lateral dose profiles with the 155 MeV beam varies from 0.2 cm at 1 cm penetration to 0.4 cm for 12 cm penetration. At 200 MeV the penumbra varies from 0.2 cm at 1 cm penetration and to 0.6 cm at 20 cm penetration. Dose rates over 100 cGy min^{-1} have been achieved, which results in treatment times under 2 minutes [5.27].

Since January 1991, improvements have been made in the clinical capabilities of the accelerator and beam transport system. The beam can now be switched between treatment rooms at different energies, in less than five minutes without beam tuning procedures. In addition, a 250 MeV beam has been delivered successively to both treatment rooms; patients requiring this energy are now being treated in the gantry room with a beam having an available range of 31.5 cm. Soon patients requiring 250 MeV beam will be treated in the horizontal beam room also.

Treatments began in March 1991 using the head and neck (horizontal) beam line and in June 1991 using the world's first proton gantry beam line. The facility is serving 18 to 20 patients per day [5.28].

CPO — Synchrocyclotron. It was decided some years ago that the Orsay Synchrocyclotron (OS) located about 20 km from Paris and originally designed for physics research would be shut down at the end of 1989. For many years, the possibility of using this machine for medical application had been under consideration [5.124], the 200 MeV proton energy and the relatively high current output making it suitable for treatment of deep seated tumors. Although built in 1958, the machine was in good repair and substantially upgraded in 1977 leading to a breakdown times of less than 5% of the operational time.

The OS has been structured like a public hospital and the Centre de Protonthèrapie d'Orsay (CPO), was established in April 1990 [5.125 and 5.126]. In full operation, at the end of 1992, the permanent personnel will comprise an ophthalmologist, 2 radiation oncologists, 2 physicists, a nurse, a technologist, and 3 members of administration staff. Eight or nine cyclotron operators will run the machine. Budget figures will be in the range of 5 to 10 million French francs for the complete treatment room equipment plus 6 to 10 million francs yearly for CPO operations, mainly personnel salaries.

The maximum available current of $2 \mu A$ leads to very stable dose rates and fluxes of about 10 Gy min^{-1} (73 MeV) or 5 Gy min^{-1} (200 MeV protons). Two rooms are planned. The first one opened in the Fall 1991 is devoted to small fields (less than 4 cm in diameter), such as those used in eye and brain treatments. Large and deep-seated lesion irradiations will be performed in the second room. An isocentric gantry construction can bring up the number of patients treated per year from approximately 100 without the gantry to 300 with the gantry [5.126].

Multiregional cooperation. The high total cost of a new proton facility partly explains why there are so few installations of this kind dedicated to cancer treatment in the world. On the other hand, relatively small number of patients eligible for low energy proton therapy (or for neutron therapy) in Europe calls for multiregional use of these installations, if more than a hundred patients per annum are to be treated. Therefore, in France, a group has been set-up around the Centre Antoine-Lacassagne (CAL) in Nice. Eight major radiotherapy departments in the South-Eastern France have already joined the group. In its present form the group represents almost 10,000 cancer cases irradiated per annum and covers 5 regions (Auvergne, Corse, Languedoc-Roussillon, Provence-Alpes-Cote d'Azur, Rhone-Alpes). This represents one fourth of the French population and constitutes a sort of recruitment warrant. Another group has been set-up at CAL facility to prepare treatment protocols and logistics for patient recruitment both for proton and neutron therapy. The structure remains open and new members, in and out of France, are expected to join soon. A communication network is under preparation [5.127].

The NAPT. The National Association for Proton Therapy (NAPT) was founded in 1990 to promote the therapeutic benefits of proton therapy in the U.S. and abroad [5.128]. The mission of the NAPT is to educate and inform the medical community, the U.S. Congress, and the general public about the opportunity to control the spread of tumors through the clinical application of proton therapy. The NAPT has proposed to the Congress a government-supported *Proton Therapy Technology Transfer Demonstration Program* to help finance a limited number of qualified medical centers in the design, research and development of proton beam treatment facilities to increase the availability of this breakthrough technology nationwide. This legislative initiative in the form of a congressional bill calls for $ 150 million to be appropriated over a five year period. It is the goal of the NAPT to effectively spearhead this legislative effort resulting in the passage of a Proton Demonstration Act in the 1991 Congress [5.128]. The NAPT also works closely with the National Cancer Institute (NCI) and supports NCI's objectives for establishing the need for 5 state-of-the-art hospital-based dedicated proton research and treatment facilities in the U.S. In its recent report NCI addresses the potential advantage of proton beam therapy and concludes that this technology has resulted in greater tumor control rate, lesser morbidity, and no increase in marginal failures.

The NAPT has also taken the initiative to establish funding in 1991 for an independent Economic Analysis Study of Proton Therapy to determine the cost benefits of this new treatment modality when compared with the other forms of treatment such as conventional radiation therapy, surgery, and chemotherapy. This study will establish uniform criteria for selecting future proton beam sites within the medical community.

5.1.2.4 Other Research and Development Studies.

Proton Computed Tomography. The increased interest in accelerated protons used in clinical studies is best evidenced by the attempts at designing and constructing a *proton computed tomography system* [5.29, 5.30]. It is believed that using a high energy proton beam will provide a tool to obtain image resolution comparable with that achievable with conventional X-ray CT and a greatly reduced patient radiation dose. According to the results of investigations carried out at the Japanese National Laboratory for High Energy Physics (KEK) [5.30], by using

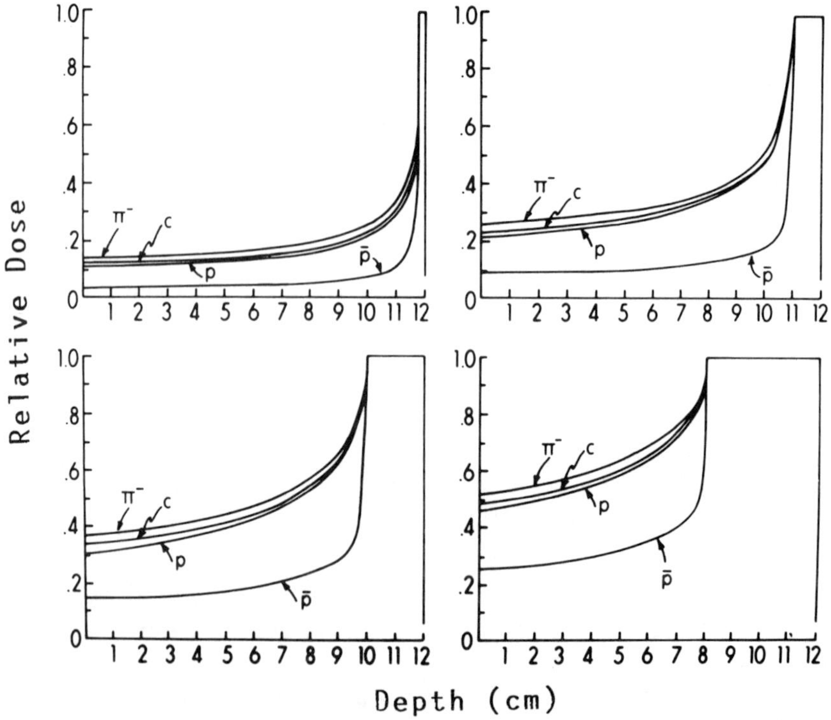

FIGURE 5.13 Depth-dose distributions from antiprotons $\bar{p}$, π^- pions, p protons and C ions for a tumor of variable thickness located at a depth of 12 cm (after T.E.Kalogeropoulos and R.Murator [5.32])

250 MeV protons and a pulsed beam, tomograms featuring spatial resolution of 1.7 mm were obtained in 8 min and the absorbed dose was about 1 cGy. The corresponding values for conventional X-ray CT were: 0.8 mm and 1.4 cGy. Work on the design of proton computed tomographs are under way.

Applications of antiprotons. Thirty five years after the discovery of *antiprotons,* $\bar{p}$, studies are being carried out on the application of these particles in therapy and biomedical investigations. Today, beams of accelerated antiprotons are available from three centers involved in research in high-energy physics: Brookhaven, Fermilab and CERN in Europe. It is estimated that at present, the cost of producing 10^8 antiprotons is roughly 1 US dollar, and as many as 10^{15} $\bar{p}$ can be accumulated annually [5.31, 5.32]. It is expected that with present-day technology transportable "bottles" can be constructed in which some 10^{12} antiprotons could be stored over months or years.

Studies carried out so far on the applicability of antiprotons in medicine have centered on *antiprotonic stereography and therapy* [5.129]. Fig.5.13 shows computer simulated depth dose distributions for artificially projected tumors located at a depth of 12 cm and of variable thickness. For comparison, curves obtained in proton, pion and C ion treatments are also presented. Curves for antiprotons decrease much more rapidly beyond the distal edge of the tumor, which corresponds to the depth of 12 cm. And what is more, antiprotons deposit less energy in normal tissues outside the tumor.

Details of the antiproton techniques have been presented at the RAND Workshop [5.31] and summarized in [5.32]. It is estimated that 10^9 to 10^{10} antiprotons will be needed to treat 1 cm^3 when all the biological factors have been taken into consideration.

5.1.3 Heavy Ion Accelerators

5.1.3.1 Clinical Requirements and Chronology of Development

The reasons which were behind the attempts made to develop heavy ion radiation therapy have already been set out in Sec.1.1.3, whereas the clinical fundamentals have been discussed in Sec. 5.1.2.1.

As of July 1, 1992, the number of patients treated with heavy ion beams was only 2487 [1.78], that is, several times lower than that of patients treated with protons or neutrons the reason being that the number of accelerators that can be utilized for heavy ion therapy is much smaller. Practically, all patients who had undergone heavy ion therapy were treated at one facility, i.e. Lawrence Berkeley Laboratory (USA) (Sec.5.1.3.2). Compared with the technique of electron and proton acceleration, methods of heavy ion, or heavy charged particle acceleration are still at an early stage of development mostly due to technical difficulties, but also because of relatively higher costs.

Comparing protons with the heavier ions like oxygen or neon the dose profiles are in many cases identical and the major reason to use heavier ions than protons are the radiobiological advantages of these ions. With increasing energy deposition the linear energy transfer LET (see Sec. 1.1.3, Fig. 1.15) to the cellular system increases [5.36]. This yields a significant change in the radiobiological efficiency (RBE). For the lighter ions like protons, RBE is nearly constant over the whole range of penetration and the small increase in RBE at the end of the particle range is not sufficient to produce a clinically relevant change. For He-ions a much larger change in RBE was observed. But the clinically relevant contribution of the high LET fraction is too small to produce a therapeutic effect. This situation is changed for the heavier ions between carbon and neon, where larger fractions of the particle range can be correlated with higher RBE-values. These elevated RBE-values affect different cellular reactions such as the repair capacity, the oxygen effect and the response to fractionated radiation exposure (Table 5.1).

Since the Oxygen Enhancement Ratio (OER), as expressed by the ratio of the biological effect for normal and anoxic cells, is largely dependent on RBE, heavy ions, as well as neutrons, are characterized by an ability to overcome the radioprotective barrier exhibited by apoxic cells. This is depicted in Fig.5.14 which shows the dependence of the *oxygen gain factor* (reciprocal of OER) for various therapy modalities, ion, neutron and photon beams. The values given in Fig.5.14 were determined as a function of a biologically effective dose adjusted to the peak to plateau ratios for a hypothetical tumor volume at a depth of 10 cm extending to the further depth of 4 cm [5.35]. The oxygen enhancement ratio is a measure of how far the protective action due to tissue anoxia can be suppressed by a given radiation modality.

A detailed discussion of other aspects of heavy ion therapy is beyond the scope of this book, and, therefore, the reader is referred to specialized literature. For example, the fundamentals of this therapy have been discussed in [1.34, 5.36 and 5.37]. Some recent results, together with the chronology of development, have been summarized in [1.33, 5.35 and 5.38]. Report on the results of helium ion treatment of uveal melanoma is given in [5.39], whereas [5.40] describes helium ion treatment of tumors adjacent to the critical central nervous system. A comparison of treatment planning for carcinoma of the esophagus with helium ions with that using photons is given in [5.41]. The AAPM Report No.16, similar to that on conventional therapy beam dosimetry described in Sec.4.1.4, provides guidelines for heavy charged-particle therapy beam dosimetry [5.42].

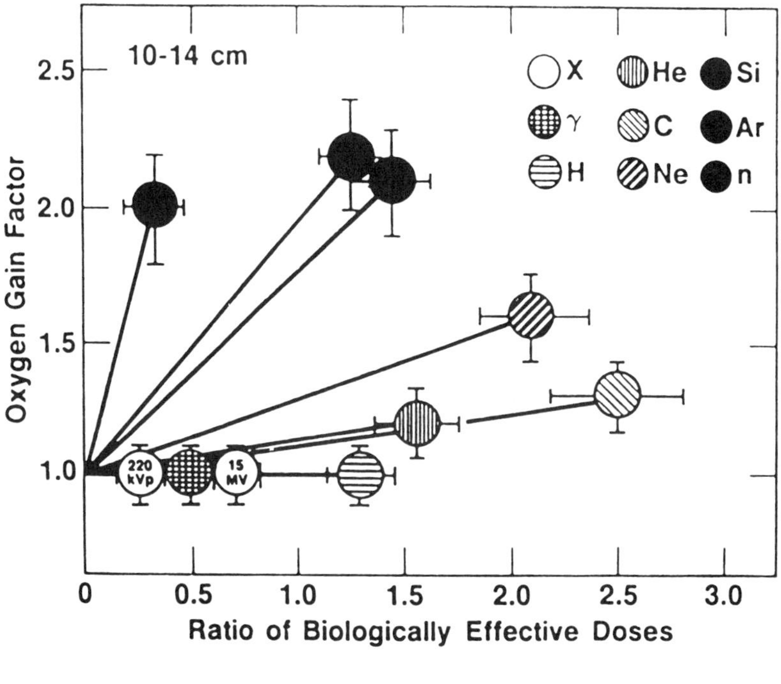

FIGURE 5.14 Dependence of the Oxygen Gain Factor (reciprocal of OER) and the relative biological doses on the type of radiation (after E.L.Alpen [5.35]).

Types of Ions. A list of ions of the greatest interest to radiotherapy is given in Fig.5.15. These are ions from Z=1 (protons) to Z=10 (neon). In some applications, silicon (Z=14,A=28) and argon (Z=18,A=40) are also being taken into consideration. Apart from *stable ions,* depicted by black squares in Fig.5.15, there is also considerable interest in the production of accelerated *beams of radionuclides* [5.36, 5.43]. In Fig.5.15 they are shown as white squares. Beams of this kind can be produced by bombarding a conversion target with an accelerated beam of stable parent nuclides. As a result, secondary beams of radionuclides are produced whose half-lives are of the order of several seconds to several hours. For example, ^{11}C and ^{19}Ne are obtained by the bombardment of a beryllium target with stable ^{12}C and ^{20}Ne beams [5.36]. Radionuclide beams are of particular diagnostic importance since they enable precise localization of the Bragg peak deposition. It is because nuclides of the same element, stable or radioactive, are absorbed in tissue in the same way. Accelerated beams of the

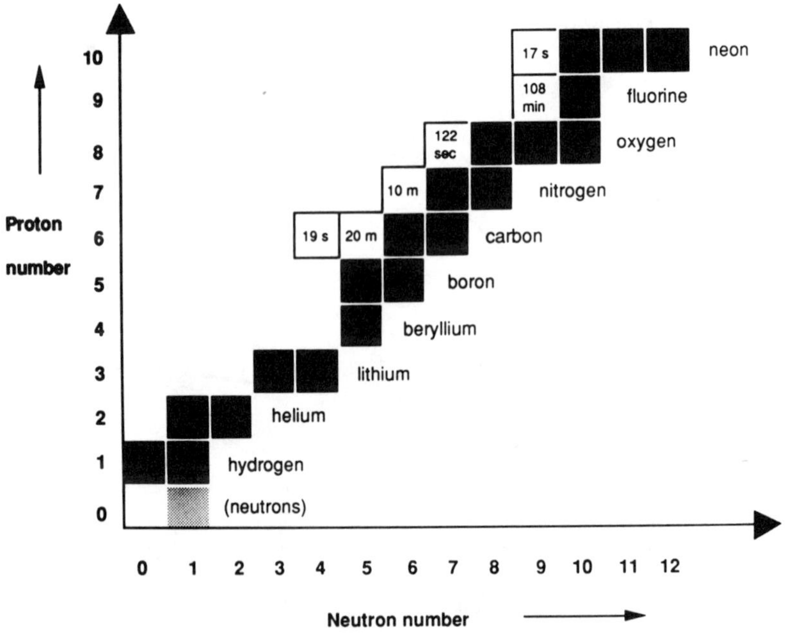

FIGURE 5.15 Portion of the chart of the nuclides accelerated as ions for radiation therapy purposes (after R.A.Gough [5.6]).

positron emitting nuclides may subsequently be detected with a PET camera. This is a simulation (conformation) technique of a kind, which helps to localize the beam stopping point (the Bragg peak) [5.36]. In addition, radionuclide beams per se can be utilized in radiation therapy as they represent a noninvasive method of introducing radionuclides into a pre-selected region of the body.

Energy Ranges and Beam Intensities. The ranges of light ions of interest to radiotherapy have been already presented in Fig. 5.4. The full penetration range of these ions should usually lie between 4 and 30 cm, although in some considerations the upper limit is taken to be 20 cm [5.44]. By choosing the penetration range the required unit energy of ions can be determined. For the range of 4 to 30 cm, the range of unit energies required has already been given in Fig.2.41 as the region of ion therapy. The maximum value is usually taken as 400 to 500 MeV/nucleon. Carbon ions accelerated to an energy of 500 MeV/nucleon have the range of 38 cm in water, whereas the respective values for oxygen and neon ions are 29 and 23 cm.

It should be noted that the above are *energy unit values,* that is they are expressed as per single nucleon of an ion. *Total ion energies* must be, of course, respectively greater. For instance, a neon ion accelerated to an energy of 470 MeV/nucleon (20 cm range) has a total energy of 9.4 GeV. Similarly, the total energy of an argon ion accelerated to the energy of 630 MeV/nucleon is 25 GeV.

Beam intensities necessary to obtain the desired therapeutic effect are quite varied. In the LIBRA project [5.6] (Sec.5.1.3.3) it was assumed that a dose of 6 Gy (600 rad) should be delivered to the volume of 1 l (200 cm^2 x 5 cm) in 5 min. Therefore, the intensities of individual ion beams should vary from 1 x 10^{10} for protons to 8 x 10^7 for argon. For the dose of this magnitude, the ^{16}O ion beam intensity should be 3.5 x 10^8 ions per second.

Other specific requirements that must be met by accelerators used in heavy ion radiotherapy involve a high degree of flexibility of operation. Provision must also be made to quickly change from one type of ions to another or to change their energy, and to have the possibility of conversion from radioactive to stable ions. As a result, specific technological requirements are placed on the design of the accelerator, especially on the flexibility of its control system.

5.1.3.2 The BEVALAC Facility (Berkeley, USA)

The world's, as yet, only facility devoted to heavy ion therapy is located at Lawrence Berkeley Laboratory (USA) [5.45,5.47]. Originally, the well-known 184" synchrotron, set in operation in the 1950s and dismantled in 1988, provided a source of ions. The medical program at the 184" Synchrocyclotron has enjoyed a record of great accomplishments and longevity. Since its inception nearly 2000 patients have received treatments, primarily with helium ions at 225 Mev/nucleon, for pituitary disorders, malignant tumors of various kind, and arterio-venous malformations in the brain [5.130]. In its latter years the accelerator was devoted exclusively to the medical program, and as such compiled a truly spectacular performance record, with reliability figures generally greater than 99%. In its last five years, not more than a handful of treatments were delayed or postponed due to machine failures. With a total staff of 3 FTE's, and with its excellent performance record, this accelerator has provided solid evidence to reinforce the case that ion-beam technology is sufficiently well-developed to be transferred to a hospital environment for routine therapy treatments.

At LBL in 1954, a 6 GeV proton synchrotron, called the *Bevatron* (coined from 1 Billion eV = 1 GeV), was set in operation. When its research potential was exhausted in the 1960s, the machine was modified to accelerate heavy ions. This new facility, started in 1974, was developed by coupling Bevatron with a heavy ion linear accelerator called *SuperHILAC*. The difficulties involved in the conversion of proton to heavy ion synchrotrons have been discussed in detail in [2.12, p.362]. At the same time, a specialized *Biomedical Facility* was set up at Berkeley. Its plan view is given in Fig. 5.16. Beam time at the Bevalac has been programmed in such a way that one-third was devoted to biomedical studies and two-thirds to nuclear science research. The biomedical program consists of radiation therapy, radiobiology as well as basic radiation physics and chemistry research.

Biomedical studies have available a full range of ions from 150 MeV protons to ^{235}U ions with energies of 400 and 600 MeV/nucleon,their ranges in water being 2.3 and 4.3 cm, respectively. In radiotherapy, the most frequently used ions are those

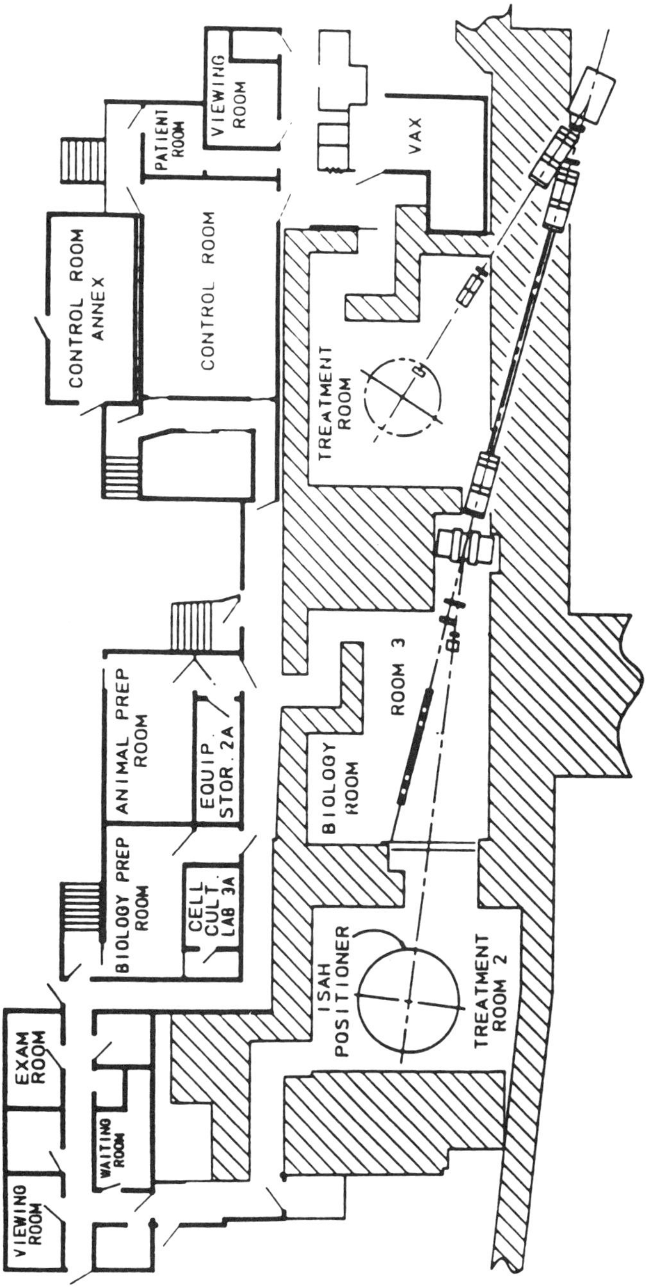

FIGURE 5.16 Plan view of the Biomedical Facility at the Bevalac at the Lawrence Berkeley Laboratory (USA) (after W.T. Chu [5.47]).

of ^{20}Ne accelerated to energies of 425 to 670 MeV/nucleon, which corresponds to ranges of 13.9 and 31 cm.

Radiotherapy rooms occupy nearly half of the total floor space of the facility. They include a shielded treatment room,a waiting room, treatment planning rooms,etc. The beam is extracted horizontally and the patients are irradiated in the seated or lying position. The *radiosurgery station,* called a *medical cave,* is described in Sec.5.1.3.4 (Fig.5.21). Another shielded room, adjacent to the treatment room, is intended for biological irradiations, i.e. experiments on animals, bacteria, tissue cultures, etc. There is also a third room with a shielding where the beam scanning system is installed.

Initially, a system with two scattering foils, represented schematically in Fig. 5.7a, was used for beam modification. One of its drawbacks, however, was that quite a lot of material had to be put in the beam path. For example, in order to produce a uniform (flat) neon ion field 20 cm in diameter for the energy of 670 MeV/nucleon the first lead foil was 0.90 cm thick, the second brass foil was 0.32 cm thick, and, in addition, a spiral stepwise filter and a water absorbed had to be used. In this arrangement, the particle useful range was reduced and the beam became inferior in quality due to its fragmentation.

The disadvantages of the scattering system were eliminated by developing a *wobbler* schematically shown in Fig.5.17. It consists of two identical dipole magnets in series, of which one deflects the beam vertically, and the other horizontally. The effective length of the dipoles is 0.7 m, each being driven by a maximum 0.4 T induction field. Both dipoles are powered from a transformer that provides a two-phase current, 90° out of phase. As a result, a rotating field is produced, which wobbles along the axis of the transport channel. The wobbler operation is also shown in Fig.5.17. When three approximately Gaussian-shaped beams are superimposed on each other a field 30 cm in diameter can be obtained, its nonuniformity being less than 5% of the mean dose.

The scattering and wobbling methods are suitable for delivering a uniform dose to a large target, if the thickness along the beam direction is constant across the entire field *(fixed modulation of the spread-out Bragg peak).* Usually, the target thickness varies across the field, and the fixed modulation method invariably deposits unwanted Bragg peak dose in the tissue out-

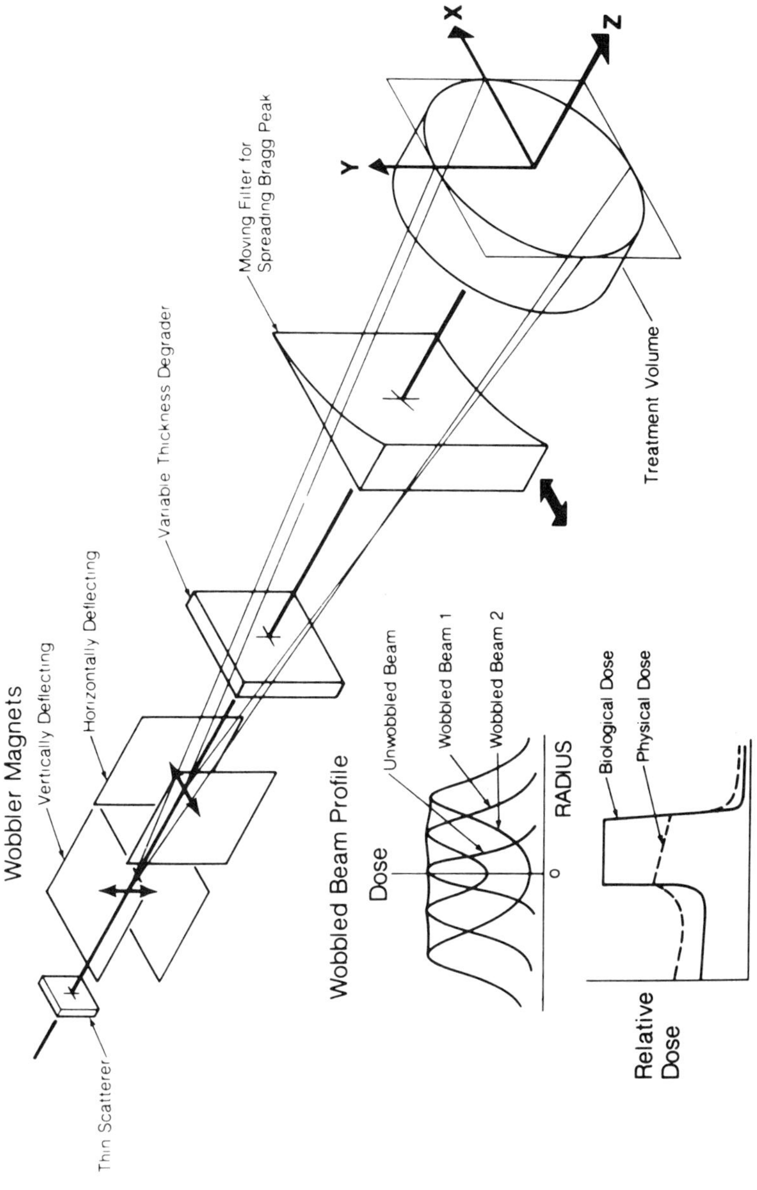

FIGURE 5.17 Principle of heavy ion beam wobbler operation at the Biomedical Facility at the Bevalac (after W.T.Chu [5.47]).

side of the target volume. To reduce such an unwanted dose, it is necessary to vary the width of the spread-out Bragg peak across the field according to the local target width *(variable modulation)* [5.134].

In order to irradiate the irregularly shaped tumor, the target volume is divided into slices of equal particle range (Fig. 5.18, cf also Fig. 5.9) by variation of the beam energy. The resulting irregularly shaped slices are then exposed to the beam. The treatment starts at the distal part of the tumor. The proton beam spreads to ±15 mm at the end of the range. In contrast, the ion beam is localized to about ±1.5 mm laterally at the end of the range and 5 mm in depth. This allows the tumor volume

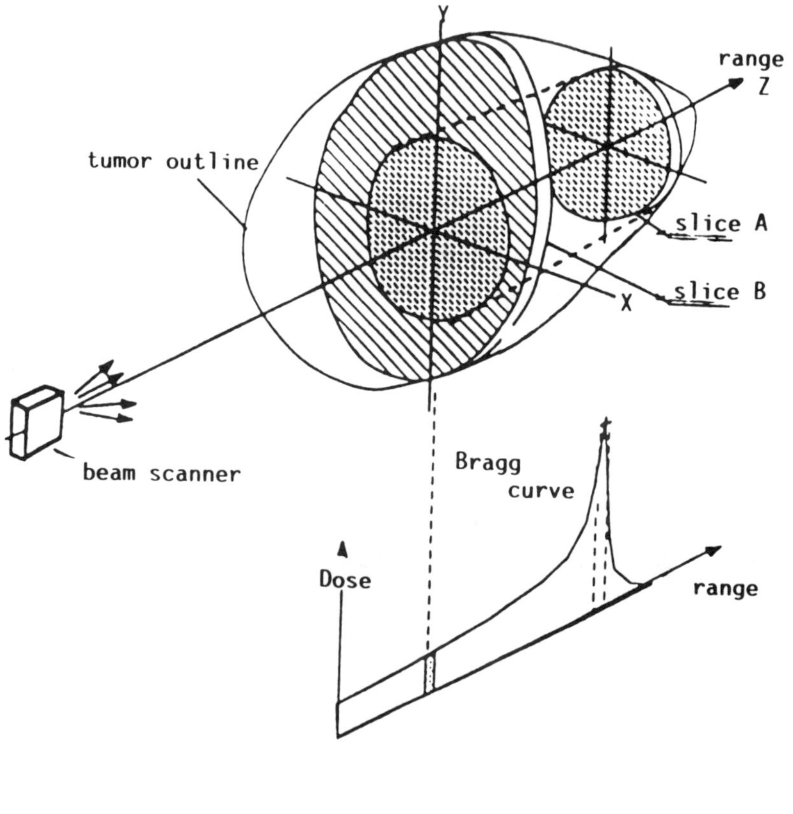

FIGURE 5.18 Three-dimensional conformal ion therapy (after F.J.M. Farley [5.131]).

to be treated by scanning in three dimensions (Fig. 5.18). By adjusting the range, the Bragg peak is placed successively at various depths. Each depth slice is scanned over the tumor cross-section at the depth; that is, when distal slice A is treated, the center of central slice B will receive some dose (plateau dose). This must compensated for when slice B is treated, by giving more dose at the edges than at the centre. Therefore, in general, each slice requires a carefully computed non-uniform dose. Scanning in depth is achieved by varying the particle energy. For example, to sweep from 5 to 20 cm in depth in tissue the energy per nucleon must vary from 140-340 MeV (carbon beam) or 170-420 MeV (oxygen beam).

It is impractical to adapt a scattering or wobbling method to perform the variable modulation [5.132]. At LBL they have developed a raster scanner with a variable scan speed, a multi-leaf collimator, and a large-area fine-resolution dose detector to achieve an optimum 3D conformal therapy using high-ion beams. This makes it possible to produce large field of an useful area of up to 30x20 cm^2 with 450 MeV/nucleon neon ion beams and 215 MeV/nucleon helium ion beams. Dose distribution measured on films shows uniformity of $\pm 3\%$ over the area larger than 30x30 cm^2. The dose uniformity obtainable through the raster scanner is very insensitive to the beam tuning [5.132].

Particularly for beam particles heavier than protons, a magnetic beam spreading system avoids the degradation of beam quality caused by nuclear reactions in the scattering material. In the LBL system, the field is fully scanned once every beam pulse of 1 s duration by moving the beam spot fast (20-40 Hz) in one dimension and slowly in the orthogonal dimension. Like in the previous design (Fig. 5.17), two magnetic dipoles produce two mutually perpendicular fields. Downstream of the second magnet the Bragg peak is spread by a ridge filter, an assembly of parallel metal ridges, and the residual range is adjusted with a variable water column or with metal range-shifter [5.133]. The scan speed can be controlled with a dynamic range of 20. The multi-leaf collimator (cf. Fig. 5.8), which consists of two opposing jaws, each with twenty 9 mm thick leaves, automatically varies the shape and size of the aperture, up to maximum aperture of 20x20 cm^2. In constructing a 3600-element ionization chamber with 0.5 spatial resolution, the main difficulty was the crowded condition of bringing out 3600 signal leads from the collecting area of the size 30x30 cm^2. By placing the signal leads in four different planes, the spacing between the leads are made as large as 0.2 mm.

Depending on the field size between 20 and 60 scans are needed to deliver a 2 Gy dose. This corresponds to treatment times between 2 and 4 min. In May 1991, the first patient was treated using a scanned beam [5.132].

In addition to stable ions, radionuclide beams, like those emitting positrons, i.e. ^{11}C, ^{13}N, ^{15}O, ^{17}F and ^{19}Ne, can also be generated in the Bevatron Biomedical Facility. Such beams are used in therapy planning and other diagnostic procedures. The conversion of a stable beam into a radioactive one occurs in a beryllium target. For ^{19}Ne ions the yield is 0.3% net. It means that one nucleus delivered to the patients results from three hundred ^{20}Ne nuclei extracted from the synchrotron.

A Quality Assurance program has been implemented at the LBL to insure the radiation quality delivered during patient treatments. This system of QA will improve the performance of the entire system, decrease system down-time and improve patient throughput [5.134].

The biomedical facility BEVALAC is well equipped with various devices intended for radiotherapy and radiobiological studies. The research capabilities of this facility are described in [5.43, 5.46].

5.1.3.3 Medical Accelerator Designs

Intensive work is under way on four projects for heavy ion therapy. The first one is the Medical Heavy Ion Therapy Project at the Gesellschaft fuer Schwerionenforschung Facility at Darmstadt (Germany) in collaboration with medical centers in Heidelberg. At GSI, a new heavy ion sychrotron, called SIS *(Schwerionen-Sychrotron)* is now being set in operation. It accelerates light ions to the maximum energy of 2 GeV/nucleon with the beam intensity of 10^8 to 10^9 ions s^{-1}. The project [5.48] envisages addition of a special medical laboratory to use accelerated particles for a few minutes for patient treatment in between running physics experiments. Thus a facility would be built, similar to that of LBL at Berkeley at a moderate cost of 20 million DM (about $ 12 million). By comparison, the cost of the dedicated proton therapy facility at Loma Linda (Sec.5.1.2.3) would be about 50 million US dollars, whereas that of the dedicated heavy ion therapy center in Chiba (Japan) is estimated at 300 million US dollars [5.48]. The German project GSI is planned for 5 years.

However, up to now (1992) the government has not yet made a final decision on the project. At present, in collaboration with EULIMA, radiobiological problems as well as problems concerning construction of an advanced beam delivery system are being studied [5.135 and 5.136].

The other three projects involve dedicated facilities, in which a heavy ion accelerator would be employed exclusively for radiotherapy or similar purposes. These are:

1. The US Project LIBRA *(Light Ion Biomedical Research Accelerator)*,
2. The Japanese Project HIMAC *(Heavy Ion Medical Accelerator in Chiba)*, and
3. The European Project EULIMA *(European Light Ion Medical Accelerator)*.

Some basic parameters of the above projects are listed in Table 5.3. The Japanese project has the best chance of implementation due to the already guaranteed financial support: it is to go into operation in 1994. Now we shall discuss these projects in some more detail.

TABLE 5.3 Designs of Medical Ion Accelerator for Radiotherapy (after P.Mandrillon et al. [5.44])

Project	Accelerator MeV/ nucleon	E_{max}	Particles	No. of rooms	Total cost M ECU	Cost per room M ECU
EULIMA Europe (199?)	cyclotron (7)	400	p,α,C,O,Ne	3	21	7
LOMA-LINDA USA (1991)	synchrotron (6)	250	p	5	33	6,5
LIBRA USA (199?)	synchrotron (16)	500	p,α,C,O,Ne	3	54	18
HIMAC, NIRS Japan (1994)	synchrotron (35)	600	p,α,C,O,Ne, A,Si	4+6	205	20,5

The LIBRA Project. It was originated at the Lawrence Berkeley Laboratory [5.6,5.7]. The Merritt Peralta Medical Center (MPMC) in Oakland is also interested in this project. The ions preaccelerated in a Radio Frequency Quadrupole linac (RFQ) to the energy of about 200 keV/nucleon, will be injected into an 8 m orbit of a heavy ion synchrotron operating with a variable repetition rate of 4 to 5 Hz. With the variable extraction from 70 to 500 MeV/nucleon, the maximum beam intensities will be 6×10^9 and 6×10^8 ions s^{-1} for ^{12}C and ^{20}Ne, respectively. At the initial construction stage, the beam will be delivered to two treatment rooms, but there will be opportunities for expanding the number of rooms at a later date. A treatment room for intraoperational applications is also envisaged. In 1986, the total cost of the project was estimated at 50-55 million US dollars [5.6]. However, there is no information available on whether this project has been started.

The HIMAC project. It was designed by the National Institute of Radiological Sciences (NIRS) in Chiba, in which Japan's first cyclotron system for proton therapy started operation. Details of the project were presented and discussed at an International Workshop in March of 1986 [5.34]. HIMAC will probably be the world's first *dedicated light ion therapy facility* [5.49]. The construction is planned to be completed in 1993 and regular treatment will be started in 1994 [5.137].

In HIMAC, helium to argon ions will be accelerated to the variable unit energy range from 100 to 800 MeV/nucleon. The dose rates over the target volume of 1000 cm^3 will be 5 and 2 Gy min^{-1} for helium and carbon ions, respectively, whereas for silicon and argon the corresponding value will be 0.5 Gy min^{-1}. As for the extracted fluxes (beam intensities), these will be 1.2×10^{10} and 2.7×10^7 ions s^{-1} for ^4He and ^{40}Ar, respectively. The maximum field diameter will be 22 cm. A cross sectional bird's-eye view of HIMAC is shown in Fig.5.19. After preacceleration by an RFQ linac and an Alvarez linac, ions are injected into the synchrotron proper consisting of two separated-function superimposed rings. As a result, the beam intensity and the repetition rate are increased. Both rings have identical construction, their diameter being 41 m and the circumference about 130 m.

Each beam, extracted from the upper or lower ring, will be dedicated to the horizontal or vertical transport system, respectively. The horizontal system will deliver horizontal beams to

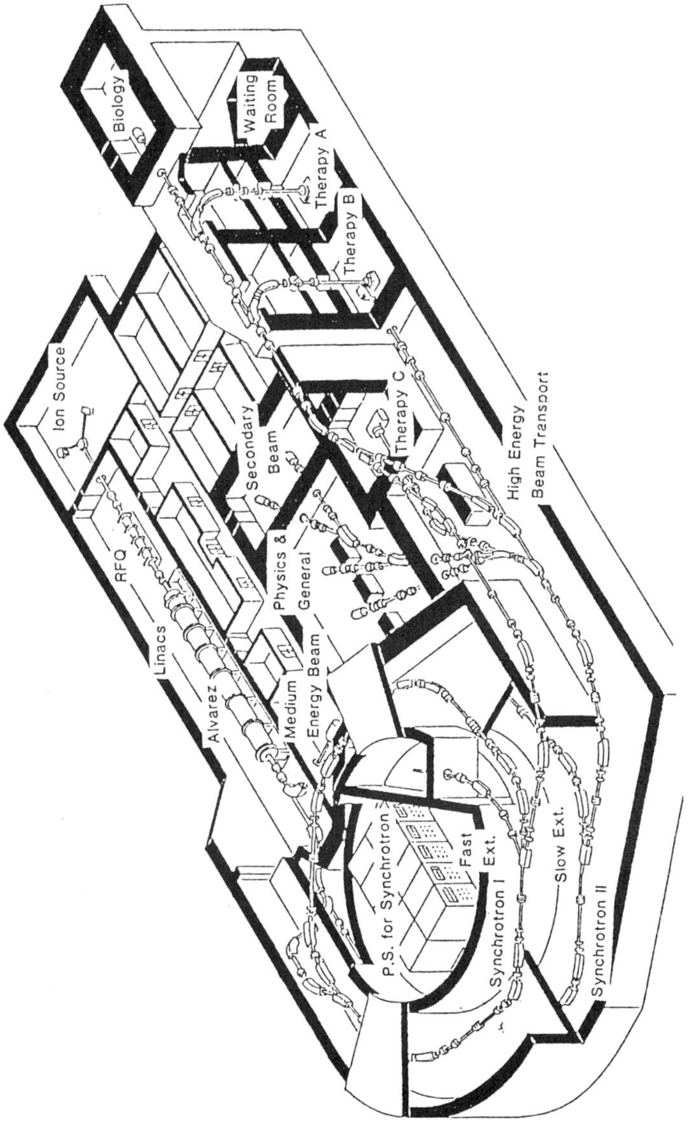

FIGURE 5.19 A cross-sectional view of the HIMAC (Heavy Ion Medical Accelerator) facility (Japan) (after K.Kawasaki et al.[5.49]).

two treatment rooms B and C (Fig.5.19) as well as to rooms intended for general applications and to a secondary beam irradiation room. The vertical channels will deliver vertical beams to treatment rooms A and B, and horizontal beams to a biological irradiation room. Thus, room A could be used for patient irradiations with a vertical beam, whereas room B could be employed for both vertical and horizontal irradiations. However, room C is intended for treatments with only horizontal beams. The general radiation laboratory is dedicated to oncology applications as well as to experimental irradiations in physics and biology. The laboratory utilizing secondary beams is intended for diagnostic and therapeutic uses of *radioactive ion beams.* The biological laboratory will carry out radiobiological investigations on small and medium size animals, on tissue-cultured cells, and is also adapted for the use of small amounts of radioactive materials.

Due to the use of two *parallel synchrotron rings,* the HIMAC project features high degree of flexibility and reliability of operation. In the long run, the lower ring could be used as a storage ring for radioactive beams to be used in therapy in the horizontal or vertical modality. This would also enable accelerated stable ions to be stored in the rings for subsequent patient treatment for a very short time with a high intensity single "shot" of the accumulated beam. The idea of two-stage *cascade acceleration* together with a stripper for ions heavier than argon is under consideration. Studies on the use of magnetic scanning combined with multileaf collimators for 3D treatment are also in progress [5.49].

The EULIMA project. At the European Organization for Research and Treatment of Cancer (EORTC) meeting held in Nice in 1985 the idea of a *EUropean high energy LIght Ion Medical Accelerator* (EULIMA) was put forward.

This installation should allow the treatment of 1000 patients per year [2.22] with ions ranging from carbon to neon and unit energies up to 400 MeV/nucleon, corresponding to a penetration of about 20 cm for oxygen in biological tissue. The Commission of the European Communities and the Conseil General des Alpes Maritimes, France, have supported a feasibility study of this complex [5.50].

Medical requirements are as follows:
(1) fully stripped ions from ^{12}C to ^{20}Ne, (2) energy range for

irradiation: 140-400 MeV/nucleon, (3) irradiation dose 10^{10} to 10^{11} ions, (4) number of energy slices: 10, (5) instantaneous rate: 10^9 to 10^{10} ions s^{-1}, (6) delivery time for the total dose: 200 s [5.138].

Two options are available for the accelerator: (a) a machine with a continuous and simultaneous process of injection, acceleration and ejection, i.e. of the cyclotron type; (b) a machine in which these three phases are separated and follow each other in time, leading to a cyclic mode of operation of synchrotron type.

The cyclotron concept is based on a separated sector cyclotron with a single cylindrical superconducting excitation coil, contributing as much as 50% of the necessary average magnetic field of about 3 T [2.22].

It was considered appropriate at the beginning of the feasibility study to explore the possibilities of operating EULIMA in either of the two regimes, as a *stand alone dedicated facility*, or as a *booster cyclotron* for an already existing machine with an active therapy programme, such as that in Nice, Louvain-le-Neuve and Liverpool (Table 5.2). In the former case, due to its excellent reliability and high intensity, an *Electron Cyclotron Resonance* (ECR) ion source may be used to produce high charge state ions that are matched to the first accelerating orbits in EULIMA by a suitable axial injection system.

The cyclotron concept can lead to a compact cost-effective (Table 5.3) and overall technically feasible design. The cyclotron has the best possible time structure of the beam suitable for the realisation of a beam scanning technique with three dimensional conformation (Fig. 5.18). But the superconducting technology involved in such a machine is sophisticated and needs a very thorough design study. In addition it requires a heavy mechanical, handling and cryogenic infrastructure, which reduces the advantage of its compactness. The major drawback of these accelerator types is the fixed beam energy and the unavoidable energy degrader in the extracted beam line.

The possibility of varying the energy from one cycle to the next is probably the strongest argument in favour of the compact synchrotron. A synchrotron accepts further modifications and improvements, in particular the possibility to add, at a later stage, beam manipulations such as storage, cooling and stretching. These techniques would also prepare the machine for the use of radioactive beams [5.139].

In the framework of the EULIMA feasibility study, two machines of the synchrotron type are proposed [5.138]: (1) separated function machine, Fig. 5.20 (SFM, pure bending magnets, focusing and defocusing quadrupoles) of about 60 m in circumference, and (2) a combined function machine (CFM) with a circumference of about 50 m. In this case, the focusing is obtained by adding a quadrupole component, alternatively focusing and defocusing, to the dipole component of the bending magnets. Basically, a machine cycling from 0.1 Hz to 1 Hz is envisaged. The accelerator consists of an injector (injection energy 5 MeV/nucleon) and a synchrotron. The injector consists of an ECR source, as in the cyclotron option, and a standard RFQ and linac available from industry. The complete

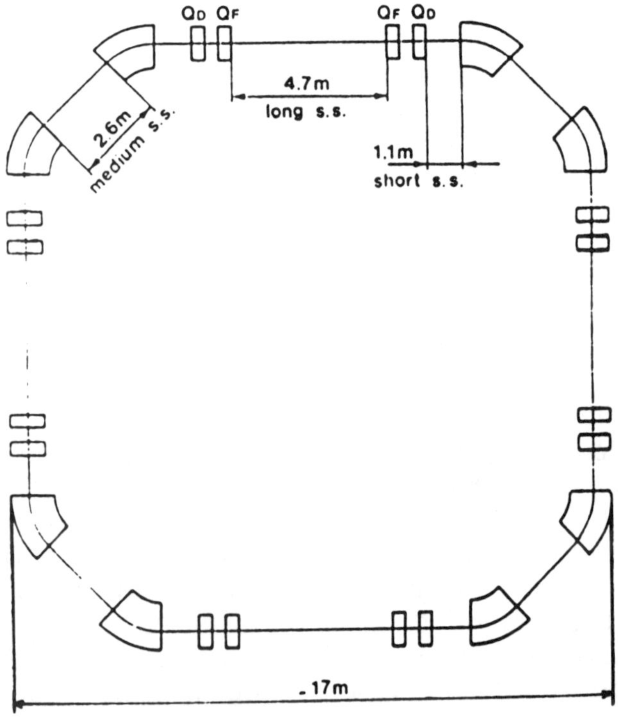

FIGURE 5.20 An arrangement of EULIMA synchrotron lattice of SFN type (after P. Mandrillon [2.21]).

TABLE 5.4 Cyclotron versus Synchrotron in Light Ion Therapy (after P.Mandrillon [2.21])

	Cyclotron	Synchrotron
Energy variation	degrader	machine adjustable
Beam intensity	high (CW mode)	low (pulsed mode)
Injection energy	0.01-0.1 MeV/nucleon	1-5 MeV/nucleon
Typical diameter		
— ions	8 m	18 m
— protons	3 m	6 m
Operation simple PLC control	computer control	
Beam delivery system	raster and pixel	raster

design, study, construction and running-in time is estimated as 3.5 years, at a cost of 25 million Swiss francs for SFM + storage option and 23 million Swiss francs for CFM.

If the scope of the irradiation program can be limited to well defined operations with stable ions, then the recommended machine is the CFM without options. The SFM with a storage option remains recommended in the perspective of radioactive ion applications [5.138].

The relative merits of any of these solutions, the cyclotron or synchrotron are quite difficult to evaluate at present. The ion energy required by the physicians is high for a cyclotron and somewhat low for a synchrotron. The cyclotron needs a large magnet which requires a superconducting coil. On the other hand, with a small synchrotron it is hard to reach the desired intensity due to space charge limitation at injection. Some important features that may influence the final choice of the facility are presented in Table 5.4 [2.21].

5.1.3.4 Systems for Ion (Charged-Particle) Radiosurgery

The requirements imposed on accelerator systems using electron and photon beams for *radiosurgery,* and in particular those for stereotactic brain surgery have been listed in Sec.3.4.3. At the early stage of investigating radiotherapeutic capabilities of proton and light ion beams it was already found that these beams are particularly well suited to radiosurgery, mainly because of their Bragg ionization peaks (Figs. 5.1 and 5.3). Special systems for *stereoencephalotomy* have been developed both in the USA (Berkeley and Harvard) and in Sweden (Uppsala) [5.51, 5.52]. Some US studies were carried out in the Donner Pavilion and Donner Laboratory program which aimed at the diagnosis and radiosurgical treatment of life-threatening intracranial vascular disorders that affect approximately one million Americans [5.53]. An overview of studies on radiosurgery is provided in [5.54] which states,among other things, that heavy-ion radiosurgery avoids the morbidity and mortality associated with extensive neurosurgical procedures and is superior to alternative methods, such as cryosurgery, electrothermal surgery and radiosurgery with photon beams.

Fig. 5.21 shows a system of *ion radiostereosurgery,* developed at Berkeley [5.53, 5.54], for the treatment of the diseases of the central nervous system. It was used in combination with the synchrocyclotron that accelerated helium ions to an energy of 230 MeV/nucleon, which corresponded to the water range of

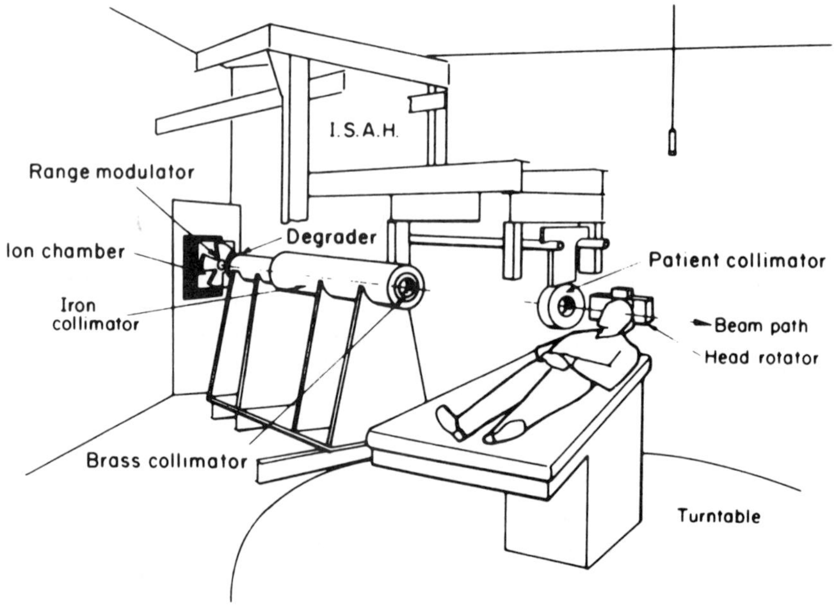

FIGURE 5.21 A system for radiosurgery with helium ions — I.S.A.H. (Isocentric Stereotactic Apparatus), (after J.T.Lyman et al. [5.53 and 5.55])

32cm. Since the synchrocyclotron energy could not be varied, it had to be degraded to 145 MeV/nucleon to obtain the maximum desired range. This was made by using a two-component absorber consisting of a 15.3 cm thick polyethylene layer and 0.34 cm copper foil. The modulation of the range, i.e. the modification of the depth-dose distribution in the Bragg peak, was obtained by means of a range modulator which had a form of an acrylic propeller placed in front of the energy degrader. The ion beam then passed through a brass collimator and a patient collimator to the patient located on a turntable. Total equivalent doses of 45 Gy, 35 Gy or 25 Gy could be delivered to treatment volumes ranging from 120 to 25 000 mm^3. The dose depended on a number of factors, such as size, shape and location of the lesion. For most patients the treatment volume ranged from 1500 to 5000 mm^3. The treatment cycle often included irradiation with the use of three to five fields.

The system depicted in Fig.5.21 was intended for eliminating shunting vessels of the arteriovenous malformation and for protecting against rebleeding with reduced morbidity and mortality [5.53]. It now operates in conjunction with the Bevalac synchrotron (Fig. 5.16, Treatment Room 2).

One of the most recent systems for radiosurgery, manufactured by Product Genesis [F24], is shown in Fig.5.22. It was developed at the Massachussetts General Hospital in Boston (USA) in collaboration with the Harvard Cyclotron Laboratory and is intended for the therapy of lesions of the craniocervical axis. To establish the treatment procedure, stereotactic target volume coordinates are defined from computed tomography, magnetic resonance imaging and angiographic images. After the patient's stereotactic head ring has been fitted to the positioning system, the patient is supported in a reclining position and then rotated about the vertical and horizontal axes to have treatment fields properly selected. Each rotational axis passes through the stereotactic target, so that an appropriate beam path is matched to the treatment volume. Optimum beam angles and the individually designed collimators and compensators are determined with the aid of a 3D treatment planning system. The resulting dose nonuniformity in the irradiated volume does not exceed $\pm 5\%$ of the prescribed value. Using this technique, fields of arbitrary shape of up to 10 cm in diameter can be treated.

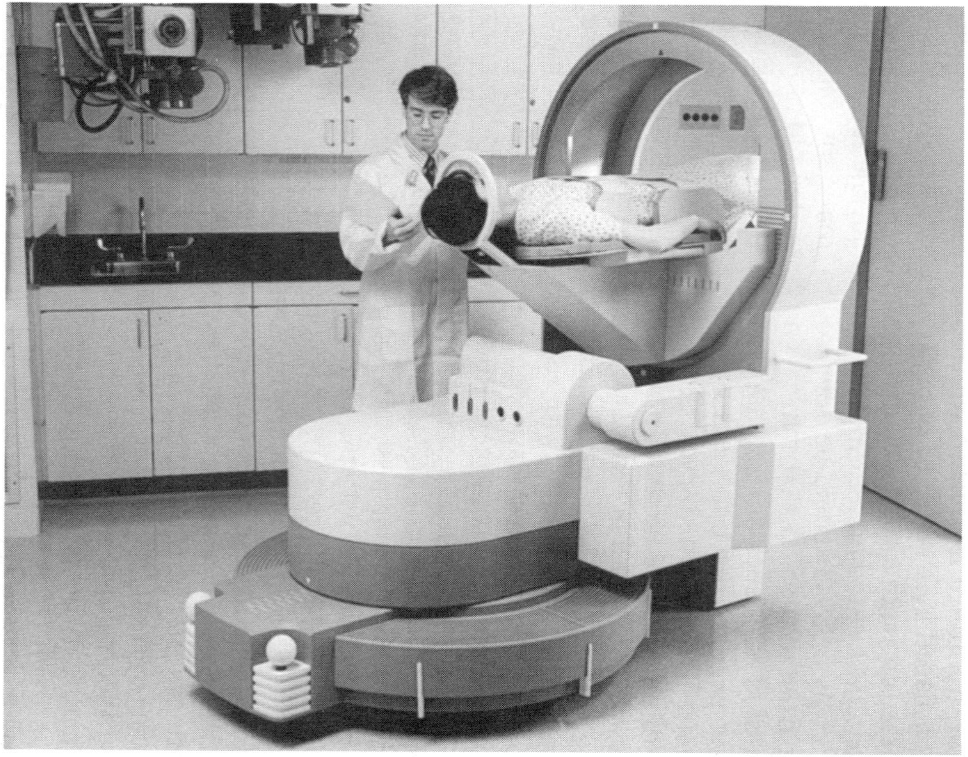

FIGURE 5.22 A system for radiosurgery with a proton beam, developed by the Massachussetts General Hospital and Harvard Cyclotron Laboratory, and manufactured by Product Genesis Inc. [5.56]

The transportable positioning system (Fig. 5.22) is moved into the treatment room and is engaged into indexing pods in the treatment room floor, positioned relative to the proton accelerator. The system is initialized and checked to verify positional accuracy. The patient is installed in the five axis positioning system; the system is aligned to place the target at the isocenter of the system, X-ray images are made to verify correct placement, and treatment is made. The apparatus has a fixed coordinate system in respect to the radiation source and moves the entire patient for target positioning and portal selection. This reduces both the risk of patient injury and the time to set up each exposure, and thus the entire treatment. Closed-loop positioning accuracy at the target location is ±0.25 mm or less for patient up to 109 kg (240 lbs). Positioning resolution for any load is 0.001 cm.

5.2 Meson Generators

5.2.1 Chronology of Development

Negative *pi-mesons (pions),* like other high-energy charged-particles, feature a characteristic depth-dose distribution curve (Sec.5.1.1, cf also Fig.1.16). Since the pion rest mass is 139 MeV, a large fraction of this energy is absorbed in the region of mutual absorption in the form of short-range secondary charged-particles which cause a high density of ionization. As for their LET, pions, together with such ions as He and C for which RBE is less than 2, are characterized by medium LET values. A comparison of radiobiological properties of pions with those of other charged-particles can be found in [5.57].

Although the idea of using pions in radiation therapy is about 30 years old (P.H.Fowler and D.H.Perkin,1961), and the first artificial pions were obtained from a 184" synchrocyclotron as early as 1948, the clinical use of pions was only possible when accelerators became available which produced beams with intensities of the order of 10^8 pions s^{-1}; meson intensities obtainable in the early machines were as low as 10^5 s^{-1}. Higher intensities were achieved in the 1970s with the advent of the high intensity medium energy accelerators called *meson factories.* Typically, a meson factory accelerates protons over the range between 500 and 1000 MeV with beam intensities of the order of several hundred μA. To produce pions with the rest mass of

139 MeV in p-p collisions, the kinetic energy of the initiating proton should be at least 290 MeV. Of particular importance is the high intensity of the beam because only one pion is generated in 10^6 interactions in a carbon target for the proton energy of 700 MeV.

Among the best known meson factories are LAMPF (Los Alamos Meson Physics Facility, now Clinton Anderson Meson Facility, USA), TRIUMF (Tri-University Meson Facility, Canada), and SIN (Schweizerisches Institut fuer Nuklearforschung, now: Paul Scherrer Institute, Switzerland). The LAMPF is equipped with an rf 800 MeV proton linear accelerator (1 mA beam intensity), whereas TRIUMF and SIN use large isochronous cyclotrons with maximum energies of 520 and 588 MeV, respectively. The construction of the above facilities is described in [2.12]. In the early 1970s, in all the facilities, experimental work was started on meson therapy.

5.2.2 Radiotherapy Meson Channels

LAMPF. The meson therapy channel was put into service in 1974. Its construction is described in [5.58] and the meson beam properties are given in [5.59]. Until November 1981, a total of 227 patients were treated there and then the channel was shut down. The evaluation of medical results together with the analysis of a long-term follow-up is presented in [5.60]. Among the consequences of these studies was a design of a compact hospital ion accelerator, called PIGMI *(Pion Generator for Medical Irradiations)* [5.61]. The design envisaged the construction of an rf linear proton accelerator with a maximum energy of 650 MeV and beam mean intensity of 100 A. Its length would be 134 m, so that it would be installed under a parking lot of the hospital facility.

TRIUMF. Fig.5.23 is a schematic of the pion radiotherapy beamline at TRIUMF. Pions produced by conversion of accelerated protons in a 10 cm thick beryllium target are subsequently delivered to the treatment room using a channel with nine electromagnets (two dipoles, five quadrupoles and two sextupoles). The system produces a beam of the energy up to 210 MeV, which corresponds to the range of 32 cm in water. To minimize pion loss by decay and scatter, the beam travels in vacuo and the channel length is 7.5 m [1.18]. The first beam in the biomedical channel was obtained in 1975, but the first ir-

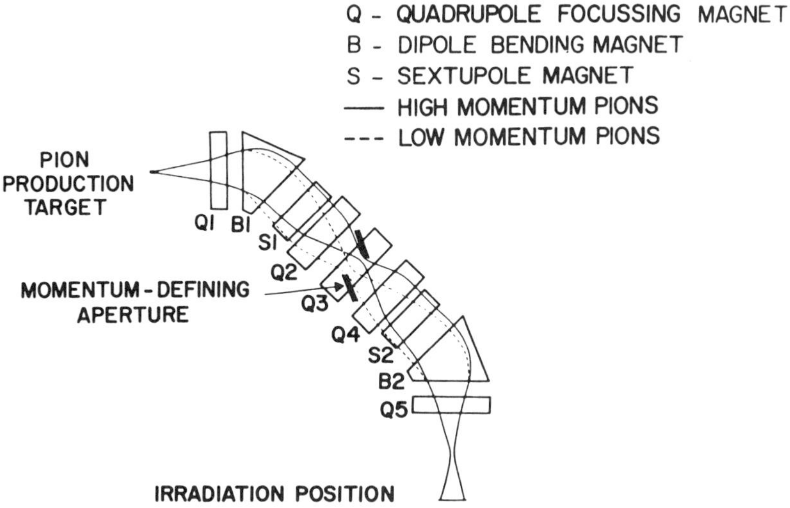

FIGURE 5.23 Schematic of the meson radiotherapy beam at TRIUMF (after H.Schumacher et al.[5.68])

radiations of deep-seated tumors were started in 1983. The channel shown in Fig.5.23 delivers a meson beam to a 5 x 7 m^2 treatment room shielded with a 2 m thick concrete ceiling. The patient can be irradiated in a 3D modality. For this purpose, the beam is scanned by a magnetic system [5.62], its range being modulated in discrete steps by a device called *a range shifter* [5.63] which is a Plexiglas disc of variable thickness driven by a stepping motor which presents disc sectors of various thicknesses to the path of the pion beam. The modulation range is 14 cm in steps of 0.5 cm.

When the initiating proton beam intensity is elevated to 200 μA the TRIUMF radiotherapy channel provides a dose of 2.5 Gy to a 500 cm^3 volume during a 25 min treatment.

The TRIUMF is located at about 9 km away from the A.Maxwell Evans Clinic (AMEC), the main provincial cancer treatment center in the British Columbia. Meson therapy program is being implemented in blocks of 6 to 8 weeks, totalling 22 to 26 weeks per year [5.65]. The average treatment time is 40 min. Details of the treatment of 109 patients with brain and pelvis cancer are reported and discussed in [5.64].

TRIUMF aspires to develop a KAON facility [5.140]. The pion therapy program will be affected by the KAON project because the pion flux will be drastically reduced when the energy of the cyclotron is lowered to 450 MeV for kaon operation. However, prior to this, a sufficient beam is expected to be delivered to complete the present randomized trials. A continuing program at KAON would be made possible by the installation of a backward negative pion channel for the desired pion momenta ($p < 210$ MeV/c). Space has been requested for pion therapy facilities at KAON. Summary of particle beam therapy programs at TRIUMF is presented in Table 5.5.

SIN (PSI). A channel, based on the concept developed at Stanford using 60 individual beams [5.66], called PIOTRON,

TABLE 5.5 Summary of Particle Beam Therapy Programs at TRIUMF (1990), (after G.B. Goodman et al. [5.64]).

1. IMMEDIATE (i.e. CURRENT OR WITHIN NEXT 5-6 YEARS)

A) PIMESONS (PIONS)

Clinical Program

Program used site restricted tumor models. Brain-glioblastoma multiforme (G.M.) - highly radioresistant tumor. Pelvis-relatively radioresistant tumor.

Preliminary dose-escalation studies to tissue tolerance levels determined the dose for use in comparative clinical trials.

Comparative (Pions vs Photons), controlled, randomised trials are operative for brain (1988) & in planning stages for carcinoma prostate.

Basic Program

"In vivo" animal experiments to evaluate the combination of pions and the biological response modifier Sizofilan (SPG) are ongoing.

B) PROTONS

Conceptual designs and first order cost estimates.

— For appropriate energy proton beam for treatment of eye melanoma (70 MeV) and neurosurgical application (200 MeV).

— For associated beam alignment and patient immobilization systems (nearing completion).

2. LONG TERM (i.e. 7 YEARS OR LONGER)

? New pion treatment facilities located closer to KAON.

? Proton beam facility for more general use in other cancer sites.

?? Antiproton imaging and/or therapy.

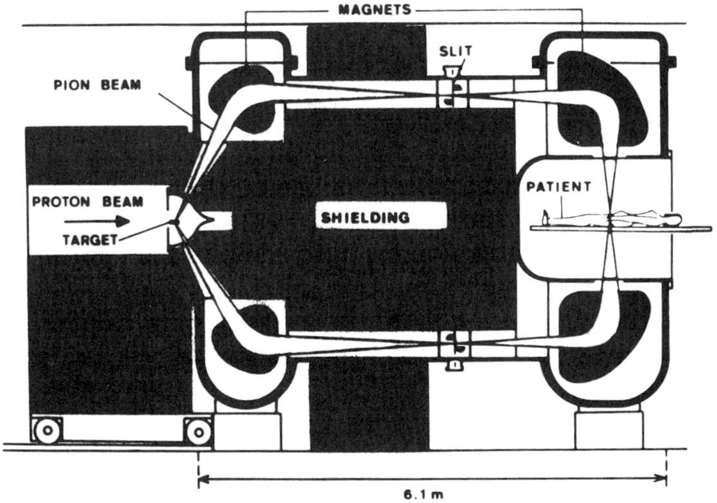

FIGURE 5.24 Schematic of the "Piotron" system for meson therapy (after H.Schumacher et al. [5.68])

was developed at SIN, now called the Paul Scherrer Institute (PSI). Its schematic is shown in Fig.5.24 [5.67, 5.68]. Pions produced by the bombardment of a 7 cm thick beryllium target with 590 MeV protons enter superconducting toroidal magnets which deflect the beam by 60°. The beam is separated into 60 individual rays through separate magnetic channels. The magnets also focus pions which enter an identical output magnet through a system of slits and then are again focused in 60 individual channels at the isocenter in the patient's body. Thus single beams converging at isocenter seem to perform a role of a quasi-rotational therapy. The total rate of particles delivered to the patient is about 10^9 s^{-1} for the proton beam current of 16 μA, the range of the pions being about 31 cm in water.

Irradiations in the PIOTRON are carried out by means of a compensation system consisting of a water jacket and a plastic insert. Dynamic scanning can achieve 3D treatment due to a magnetic beam scanning system [5.67]. Preliminary clinical results obtained with the dynamic spot scanning method in the years 1982-1984 and compared with the US results are presented in [5.69]. Typically, the dosage administered in 20 treatments over 5 weeks was 3000 to 4000 cGy. These studies, continued till the end of 1988, totalled 406 patients. A more detailed discussion of the results obtained was presented at the EORTIC Symposium [5.70].

According to the most recent overview (as of July 1, 1992 [1.78]) of patients treated with charged-particles, the total number of patients irradiated with pions is 1011 : 230 in Los Alamos, 283 at TRIUMF and 498 patients at PSI. There is no information on the construction of a dedicated pion therapy facility.

5.3 Accelerator Neutron Generators

5.3.1 Neutron Therapy

The number of cancer cases which, because of their *radiation resistance,* cannot be treated by conventional therapy is estimated to be about one-third of all patients referred [5.73]. What is meant by radiation resistance is the feature of the tumor which requires that the dose needed to sterilize the tumor leads to unacceptable complications in the adjacent normal tissues. Neutrons have LET and RBE values which are much higher than the respective values for conventional radiations, i.e. high-energy electrons and X-rays. Therefore, the use of accelerator-produced neutrons for radiotherapy has been known for quite a long time (Sec.1.1.3). The fundamentals of the present-day neutron therapy are given in [5.72, 5.73], and some radiobiological aspects are discussed in more detail in [5.74, 5.75], whereas the physical aspects are treated in [1.37].

Opinions on the efficacy of neutron therapy vary [5.76 through 5.83]. By 1987, the total number of patients irradiated with neutrons in about 20 facilities the world over was roughly 10 000 [1.38]. From the analysis of the results of therapy on the above 10 000 cases of inoperative radioresistant tumors published in [5.38] it was found that control rates vary from 71% for salivary gland tumors and melanoma to 48% for bladder cancers and 33% for recto-sigmoid cancers. In spite of these quite diverse results, they are probably the best that are achievable at present [5.84]. Similarly, the results of an eight-year study on the long-term survival of patients who received neutron therapy for advanced prostate cancer indicate that this modality helps to achieve higher rates of survival [5.82]. Opinions are also voiced that, thanks to their high penetrability and RBE, neutrons are predisposed to use in about 20% of all cases now being referred to photon treatment. However, the question of turning neutron therapy into a routine treatment modality is still open, which is evidenced by the lack of any final conclusions from the many-year long US neutron therapy program [5.83] (Sec.5.3.3). From the point of view of apparatus availability, the above transformation is quite realistic.

5.3.2 Accelerator Techniques of Neutron Production

Radiotherapeutic requirements. For neutron therapy to become a routine method it must enjoy parameters such as penetration, the SSD distance (typically about 100 cm) and dose rates similar to those obtained with other conventional techniques, e.g. photon and electron modalities. It is generally assumed that irradiation times should not be in excess of 5 min because of the patient's comfort. Typically, the dosages that the patient receives in a single treatment usually range from 100 to 200 cGy. Consequently, the neutron source should deliver 20 to 40 cGy min^{-1} in the treatment volume. For the neutron beam to have the penetration power similar to that of, for example, ^{60}Co gamma radiation, the mean spectrum energy should be close to about 15 MeV. It is, of course, desirable that both the dose rates and penetration be higher than the above values.

A detailed overview of accelerator techniques of generating beams of accelerated neutrons may be found in [2.16, p.161]. Here, we shall be concerned with some selected methods which have been or shall be used for radiation therapy purposes.

Deuteron methods. High-energy neutron beams, or, otherwise, fast neutrons can be produced by bombarding thin and thick targets of light elements with charged-particles from an accelerator. The energy spectrum of neutrons thus obtained has a rather wide range, from zero to a certain maximum value. The spectrum mean energy and the neutron beam intensities depend both on the energy of the initiating particles and on the type of nuclear reaction and the target used. Generally, the higher the energy of the initiating particles the higher the intensity and penetration power of the neutron beam generated.

The deuteron-tritium (D-T) reaction, ^{3}H (d,n) ^{4}He, makes an exception to the above rule. Fig. 5.25 gives neutron yields per 1 μA beam current for various types of reactions and target thicknesses vs. energies of the bombarding ions E_k. As seen from Fig. 5.25, relatively high neutron yields can be obtained from the D-T reaction even for very small deuteron energies over the range from 100 to 300 keV. Consequently, intensive neutron beams can be produced in low-cost high-voltage machines, such as cascade accelerators. The D-T reaction can be written as

$$D + T \rightarrow {}^4He + n + 17.6\,MeV \tag{5.1}$$

The energy of the neutron thus produced will be

$$E_n = 17.6 \frac{m_{He}}{m_{He} + m_n} = 14.1\,MeV \tag{5.2}$$

where m_{He} is helium nuclear mass and m_n is the neutron mass.

The D-T reaction may thus be regarded as a source of fast monoenergetic neutrons with energies of about 14 MeV, whose energy spectrum is not continuous. The D-T reaction used to be employed in neutron radiotherapy devices, but, due to reasons to be discussed later in Sec.5.3.3, it does not find use in the more recent generations of therapeutic neutron systems. In contrast to these modern facilities, 14 MeV D-T generators can be used in routine clinical work at a lower cost. As regards the dose distribution, 14 MeV neutrons correspond roughly to 60-Co gamma rays, but due to radiobiological findings it is more essential to spare adjacent normal tissue in neutron therapy than in photon therapy [5.149].

At present, in medical applications it is mainly isochronous cyclotrons which produce accelerated deuterons initiating neutron reactions (Sec. 2.6.2). It follows from Fig. 5.25 that high-intensity neutron beams can be obtained by bombarding a beryllium target with cyclotron produced deuterons. The dose rate at 100 cm from the target is

$$D = 2.12 \times 10^{-4}\, E_k\, 2.97 \qquad [\text{rads min}^{-1}\, \mu A^{-1}] \tag{5.3}$$

Hence, even when small cyclotrons are employed to accelerate 10 MeV deuterons with an extracted beam intensity of 100 μA, neutron beams can be generated which will give a dose rate of about 20 cGy min^{-1} (20 rad min^{-1}). The simple relation-

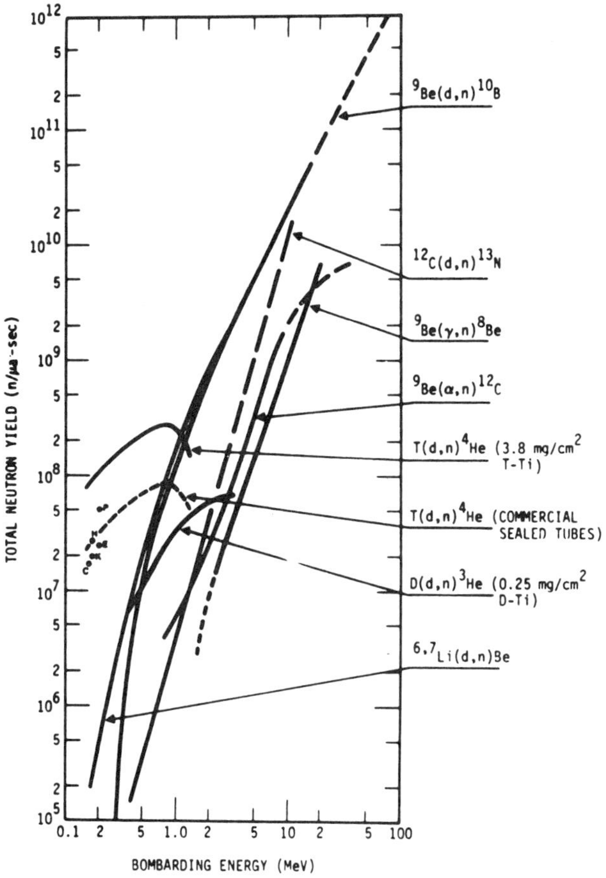

FIGURE 5.25 Dependence of neutron yields on the bombarding particle energy E_k and various target thicknesses per 1 μA of the initiating particle intensity (after J.A.Stokes et al.[5.85])

ship between neutron spectral mean energy, E_n, and the energy of deuterons is given by

$$E_n = 0.42\,E_d \tag{5.4}$$

which considerably limits the generation of high-energy neutrons. For example, even when a relatively energetic 50 MeV deuterons are used, the neutron beam obtained has a mean energy of only 20 MeV. It is clear that to accelerate deuterons to such high energies a large cyclotron is needed.

Proton Reactions. The present-day isochronous cyclotrons employed as typical sources of initiating particles in neutron therapy systems make it possible to accelerate protons to energies twice as high as those achievable in the same machines for deuterons (Sec. 2.6.2). Studies on the feasibility of generating neutron beams from protons on beryllium [5.86, 5.87] have revealed that protons have an edge on deuterons. Fig. 5.26 gives neutron spectra generated by 35 MeV protons and deuterons. The spectrum initiated by protons is more useful since it has a larger high-energy component. The mean energy of neutrons generated by protons for the beryllium reaction is

$$E_n = 0.52\,E_p - 1.5\text{ MeV } (\pm\, 1\text{ MeV}) \qquad (5.5)$$

Thus neutron-generated protons feature better penetration, higher intensity and higher energies. In addition, there is a better skin sparing in this modality as compared with deuteron-generated neutron beams [5.88].

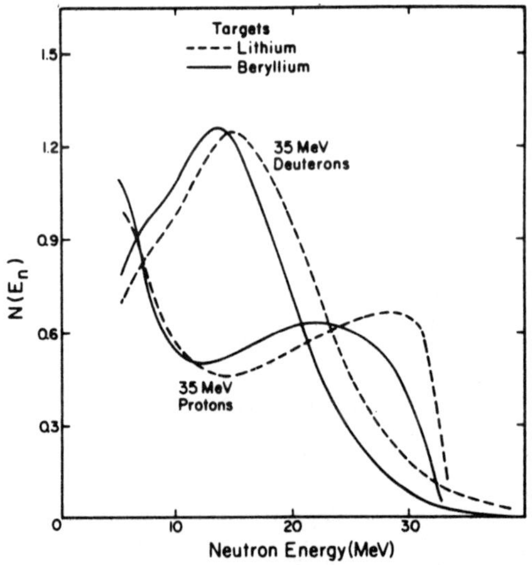

FIGURE 5.26 Neutron energy spectra (0° emission angle) generated by 35 MeV protons or deuterons (after D.K. Bewley et al. [5.87])

Due to the above properties, proton reactions are now commonly used for the production of neutrons in radiotherapy facilities (Table 5.7). The proton maximum energies are about 66 MeV. In this case, neutron beams can be generated whose mean energy is as high as about 30 MeV. Some facilities which had previously used deuterons to produce neutrons now have their systems modified and generate neutrons by means of protons [5.89]. There exists also a tendency, shown at some research centers, to upgrade proton energies to the already mentioned level of 66 MeV [5.84].

The existence of a large low-energy component in the proton-generated neutron spectrum is clinically disadvantageous since it lowers beam penetration and increases RBE near the surface layers. This low-energy component can be reduced by two methods. In the first one, like in that employed for photon beams, filters can be used (Sec.3.1.4.3). They are usually made from polyethylene or other hydrogen-rich materials. The effect of various filters on the quality of a neutron beam is described in [5.90].

The other method uses thin targets in which the thickness of the beryllium layer is smaller than the range of protons which lose only a fraction of their energy in the beryllium target. The use of thin targets to eliminate the low-energy component is more economically sound than that of using thick targets combined with filtration [5.87].

5.3.3 Chronology of Development of Accelerator Systems

Among 18 accelerator installations for neutron therapy in operation in 1985, five were equipped with *D-T generators* [5.91]. They were located in Western Europe (Amsterdam, Glasgow, Hamburg, Heidelberg and Manchester). The other 12 facilities employed cyclotrons, and one installation used an rf linear machine (cf Table 1.5).

In the 1970s several companies (Haefely, AEG/RDI, The Cyclotron Corp.) developed radiotherapeutic versions of D-T neutron generators, whose advantages were: monoenergetic beams, a compact design of the unit which could be easily installed in the gantry, and a relatively low cost. Unfortunately, the low dose rates, ranging from 5 to 20 cGy min^{-1}, necessitated long irradiation times. In addition, the service life of neutron lamps was not longer than 300 to 500 h, their emission yield quickly decreasing with use. Due to the very principle of their operation it was impossible to obtain neutrons with energies higher than about 14 MeV. That is why the production of D-T generators has been discontinued.

In the USA, more intensive work on neutron therapy was started as early as the 1970s using cyclotrons and rf linacs intended initially for physics research. The preliminary results proved to be quite promising, so much so that the National Cancer Institute initiated a 10-year research project with the aim of designing clinical facilities and implementing clinical trials as the third stage of the project. The list of US neutron therapy installations is given in Table 5.6. In the years 1971-1989, 7 facilities covered by the NCI project were provided with funds totalling \$69.5 million. The grants varied from \$1.4 million for the MANTA facility to \$16.7 million dollars for the MDAH facility [5.83].

The presently implemented third stage of the clinical project includes studies on the carcinomas of the brain, neck, prostate and lungs with special emphasis placed on radioresistant can-

TABLE 5.6 U.S. Neutron Beams, 1972-89 (after S. Zink et al. [5.83])

Facilities	Start of patients treatments	Particle energy on targets, MeV	Average energy of neutrons, MeV
TAMVEC	1972	50 (d Be)	20
MDAH	1985	42 (p Be)	19
University of Washington[a)]	1973	22 (d Be)	8.8
	1984	50 (p Be)	22
MANTA	1974	35 (d Be)	14
GLANTA	1976	25 (d Be)	10
	1982	42 (p Be)	21
Fermilab	1977	66 (p Be)	25
University of Chicago	1981	8 (d Be)	8
UCLA	1986	46 (p Be)	18

TAMVEC, Texas A&M variable energy cyclotron, College Station, Texas; MDAH, M.D. Anderson Hospital, Houston, Texas; MANTA, Mid-Atlantic Neutron Therapy Alliance, Washington, D.C.; GLANTA, Great Lakes Neutron Therapy Alliance, Cleveland, Ohio; Fermilab, Fermi National Accelerator Laboratory, Batavia, Illinois; UCLA, University of California, VA Wadsworth Medical Center, Los Angeles, California.

cers. In the latest contract with the NCI, 1698 patients had been treated through April 1989. The final analysis of the results is planned after less than 4 years.

In order to standardize dosimetry procedures, the first US protocol for neutron therapy was drafted in 1980 [5.23], and soon after that a European protocol was also prepared [5.94]. The differences between the two protocols were later eliminated in a subsequent document [5.95] which was a joint US and European declaration. Intensive studies have also been carried out on neutron dosimetry [5.96, 5.97]. One of the problems is that the neutron beam is accompanied with gamma radiation produced in the target by initiating particles as well as with that resulting from neutron interaction with the medium and the irradiated object. Gamma contamination is usually a small fraction (less than 15%) of the total dose and is usually negligible since its RBE is lower by about two-thirds than that for neutrons. The full dosimetry pattern, however, requires that the gamma component should be determined.

A distinctive feature of the present-day neutron therapy facilities equipped exclusively with cyclotrons is that therapy is being combined with the production of radionuclides, especially those with short half-lives (Chapter 7). Recently, installations have been set up, in which, in addition to production of radionuclides and neutron therapy, proton therapy can also be implemented. The aim of such programs is not only to extend the scope of therapeutic modalities and to combine the various types of treatment but also to improve beam utilization and the cost-effectiveness of therapy as the treatment costs are still quite high. According the UCLA estimate [5.98], the cost of a neutron therapy facility is $9 million US (including the machine cost of $ 6 million), whereas the cost of a comparable photon therapy facility is $2.75 million (including the machine cost of $ 1.25 million). The authors predict that when all cost factors have been considered, the cost per neutron treatment will be about 4 to 6 times that of a conventional radiotherapy treatment. Other authors contend that these costs could be comparable. For this to happen, the equipment provided for neutron therapy facilities should be mass produced. Recently, Scanditronix (Sweden) [F14] has been specializing in this field. The Scanditronix NT60 system comprises the 60 MeV isochronous cyclotron and the isocentric neutron gantry designed for an SSD of 150 cm. The gantry can be supplied both for ±110° rotation without the need for movable parts in the floor or for ±190° rotation with a retractable floor system. The isocentric height is 120 cm. Field sizes of up to 32.5x30.0 cm^2 are available. The total dose rate (neutron plus photon) is 50 cGy min^{-1} for 35 μA proton beam current on the beryllium target. The photon dose is less than 13.5% of the total dose in air.

The first facility to be equipped with cyclotron based clinical neutron therapy system is the University of Washington Medical Center in Seattle (USA) [5.99, 5.100]. It was built at the cost of $ 11.2 million (including machine cost of $ 4.2 million) and put into operation in 1984. Patients are treated with an average of 2 to 2.2 treatment fields per session. A typical two-field treatment takes 30 min, including the therapy run of 6 to 12 min depending on the beam intensity and dosage prescribed [5.101]. The facility is provided with two treatment rooms: in one room irradiation is carried out with a rotating gantry, and in the other a stationary (fixed) beam is used. A five-year experience in the operation of this facility, which in addition produces radionuclides, is presented in [5.100].

At present (1991), the status of clinical high energy neutron therapy in the United States is not as favourable as it was earlier [5.141]. There are, however, three facilities actively treating patients and a new facility which should initiate treatment in 1992. The continued utilization of these facilities will be strongly influenced by the economics of reimbursement by the insurance companies and governmental bodies. Any future facilities will most likely incorporate the superconducting isocentric cyclotrons as the source of charged particles. Assuming the equipment costs are comparable or less than those of the existing cyclotron — isocentric gantry systems, the lower capital costs would have a very favourable impact on the economics of neutron therapy [5.141].

The cyclotron for the Neutron Therapy Facility at Harper Hospital (Gershenson Radiation Oncology Center, Detroit, USA) represents the first application of superconducting technology to a medical cyclotron (Sec. 5.1.2.3, Fig. 5.10). The cyclotron was delivered from the National Superconducting Cyclotron Laboratory (Michigan State University) [5.142]) to the Harper Hospital in early July 1990 and the first neutron beam was obtained in late August. The approximate total dose rate at the isocenter is 0.60 Gy min^{-1} for the unflattened beam

TABLE 5.7 Parameters of Some High-energy Neutron Therapy Facilities (initiating proton energy E_p = 60 MeV) (after D.T.L. Jones [5.104]); results are for 10x10 cm^2 field in a water phantom at quoted SSD unless otherwise stated.

Institute	MRC	UCL	MINTF	NAC
Accelerator	Cyclotron	Cyclotron	Linac	Cyclotron
E_p (MeV)	62	65	66	66
Be target thickness (mm) (MeV)	17.45 36	17.0 35	22.1 49	19.6 40
Beam direction	Isocentric	Vertical	Horizontal	Isocentric
SSD (cm)	150	162.5	190	150
Collimator	Variable	Inserts	Inserts	Variable
Depth. of dose max.	1.6 (g cm^{-2})	1.6	1.6	1.7
Entrance dose (% of max.)	30-40	51	40 (6x8 cm^2)	33
Depth. of 50% dose (cm)	16.2	17.6	16.1 (TE liquid)	16.9
Flattening filter	Yes	No	No	Yes
Hardening filter (cm CH_2)	-	2	-	2.5
Total dose rate at dose max. (mGy min^{-1} μA^{-1})	16.3	19.6*	19.8*	15.1
Component γ at dose max.	2-3 (%)	3.1 (CH_2)	2-4 (air)	5.2
Penumbra (20-80%) at 10 cm (cm)	1.85	1.7	2.0	1.31

* Normalised to SSD = 150.0 cm (SSD = Source-to-surface distance)

MRC:MRC Cyclotron Unit, Clatterbridge Hospital, Merseyside, UK.
UCL:Universite Catholique de Louvain, Louvain-la Neuve, Belgium.
MINTF:Midwest Institute for Neutron Therapy at Fermilab, Batavia, Illinois, USA.
NAC:National Accelerator Centre, Faure, South Africa.

with $10 \mu A$ of deuteron beam (50 MeV) incident on the beryllium target. On 12 September 1991 the first patient treatment took place [5.143].

In the United States the reduced interest in high energy neutrons has been replaced by a revived interest in neutron-boron capture therapy [5.141], see Sec. 5.3.5.

Another desired feature in the development of neutron therapy installations is that they become equivalent to the similar accelerator facilities for conventional photon or electron therapy with respect to treatment and service. Table 5.7 lists some data on several neutron therapy facilities. In what follows we shall describe, by way of example, the design and construction of the most recent installation at Faure (R.of South Africa) set in operation in 1988.

5.3.4 Medical Facilities for Neutron Therapy

In the Republic of South Africa, a neutron therapy facility was built as part of the National Accelerator Centre at Faure near Cape Town. Started in 1988, it is the world's most recent installation. The layout of the facility (Fig.5.27) shows two injector cyclotrons SPC1 and SPC2 which supply the main cyclotron SSC intended for accelerating protons to energies ranging from 27 to 200 MeV, the most common energy being 66 MeV and maximum intensity 100 μA. The beam can be delivered either to physics research channels (lines A, F and G) or to a biomedical laboratory. The latter one (the bottom part of Fig.5.27) includes a treatment room for neutron therapy TC and two rooms for proton therapy: TL with a horizontal stationary (fixed) line and TR for future proton treatment isocentric gantry or vertical line. The accelerated beam can also be delivered to a medical isotope production unit.

The design of the neutron therapy gantry used at NAC is shown in Fig.5.28. Neutrons are generated by a 66 MeV proton beam which impinges on a 19.6 mm beryllium target, in which protons lose 40 MeV of their energy. The proton beam is delivered to the target by a system of two 70° and 160° dipole bending magnets which includes a quadrupole triplet, a scanner and a stray beam detector. Below the target are placed, in order, filters to "harden" the spectrum and flatten the field, an X-ray tube for localization purposes, transmission ionization

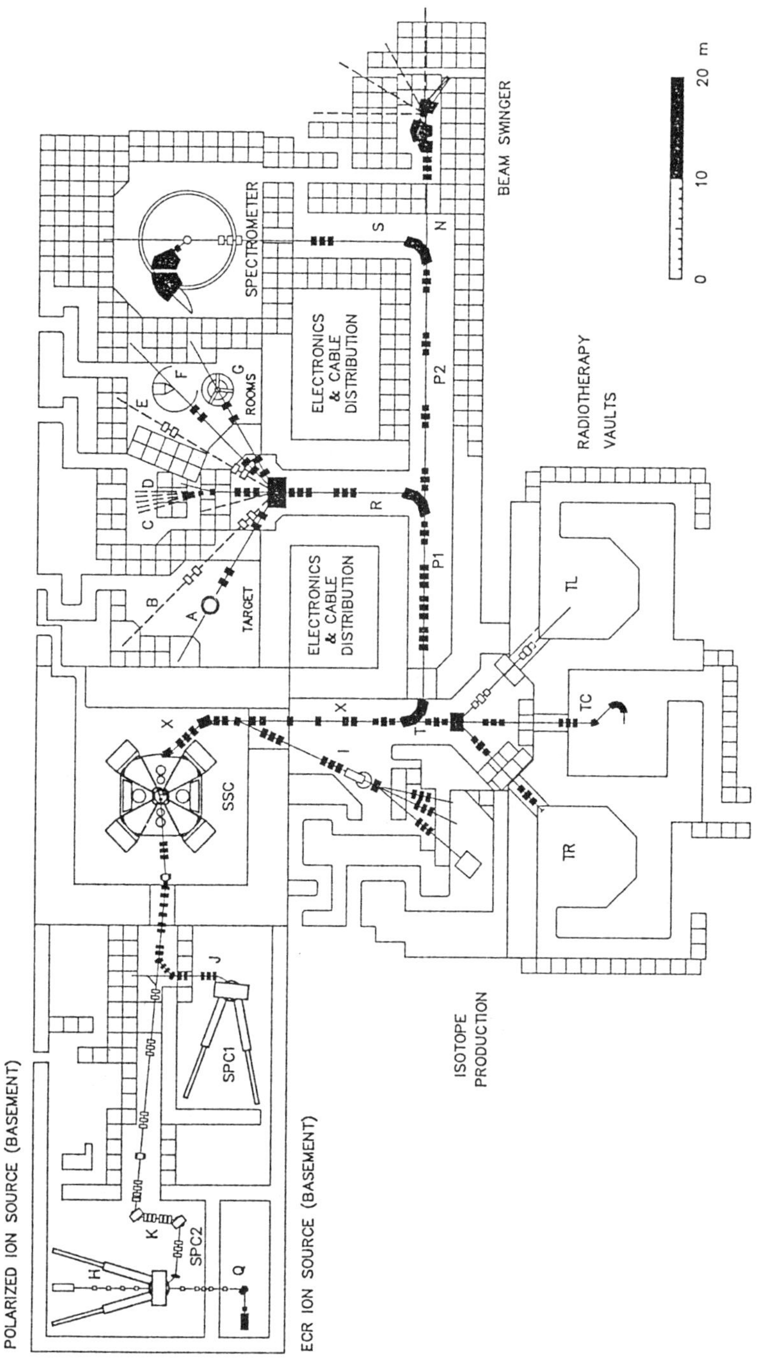

FIGURE 5.27 Layout of the medical facility of the National Accelerator Centre (NAC) at Faure (South Africa) (after D.Reitman [5.102])

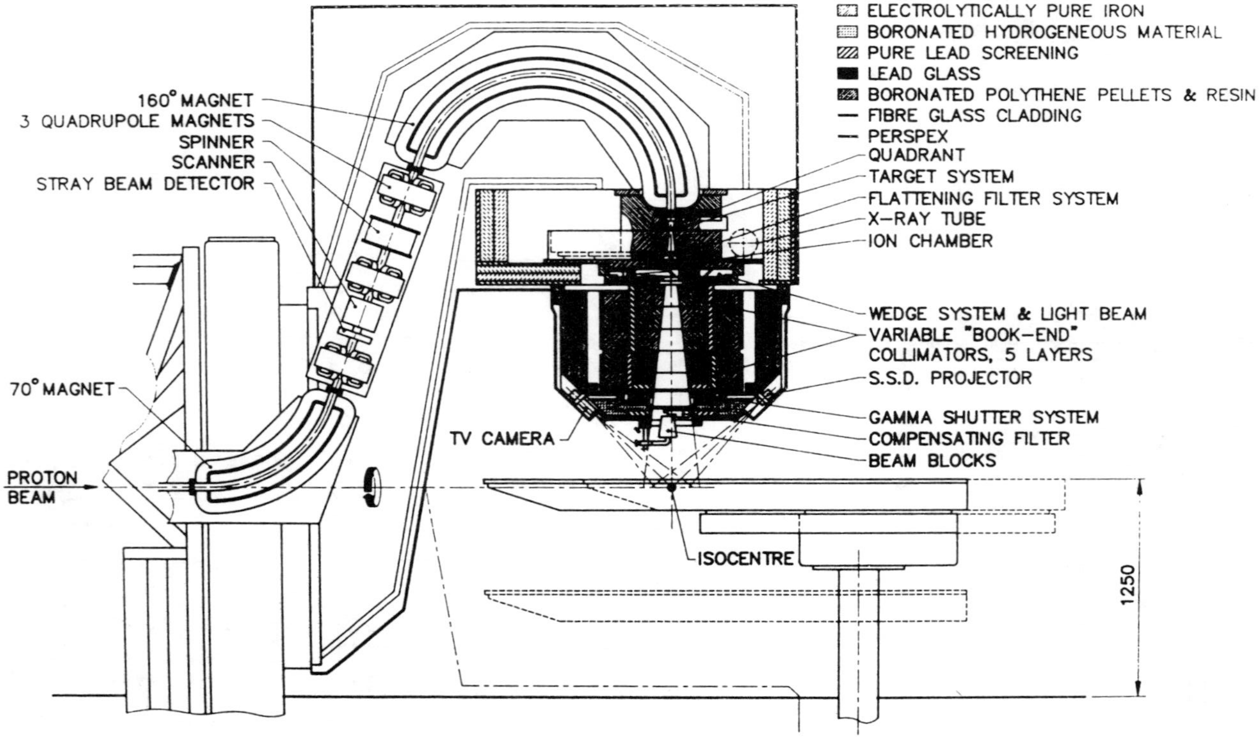

FIGURE 5.28 Schematic view of the neutron therapy gantry at NAC at Faure (South Africa) (after D.T.L.Jones [5.103])

chambers, a wedge filter head and a collimator made from five steel blocks and boronated polyethylene. A 2.5 cm layer of polyethylene constitutes a hardening filter, whereas two 8 mm thick circular steel filters are used for flattening the field: one filter for fields with sides less than 16 cm long and both filters for larger fields.

Wedge filters made from tungsten (25°, 35° and 45°) mounted on a turret are used for modifying the beam. The above mentioned collimator provides a tapered aperture, its total thickness being 63 cm. The collimator has a 2 cm lead outer facing which absorbs gamma rays produced directly or those generated by activation of its inner components. The collimator can be rotated through the full 360° angle and the distance from its end to the axis is 35 cm. In the absence of the beam, gamma rays produced in activation are attenuated by a lead block close to the target and by another block made of lead glass placed at the collimator end. Like in accelerators for routine therapy, beam blocks, compensators and other devices can be mounted around the collimator. The head also includes a light beam localiser, an SSD projector and TV cameras.

In order to rotate the gantry through 360° the floor segment under the gantry is removed during rotation. The automated patient treatment couch has a removable table top and the patient can be set up outside the treatment room and then transferred to the treatment position.

Fig.5.29 shows a build-up curve from a neutron beam for a 10x10 cm^2 field determined with a A 150 TE-plastic ionization chamber in a PMMA phantom. For comparison, depth curves for ^{60}Co and 8 MV X-rays are also given. Some fundamental parameters of the NAC facility, compared with those of other neutron therapy installations are listed in Table 5.7.

The NAC facility devotes 15% of the beam time to radiotherapy and 27% to the production of medical radionuclides [5.102]. The facility also cooperates with two large cancer hospitals located at a distance such that it is possible to reach the facility in about 20 min. They serve the Cape Town area with the population of 1.8 million and admit about 5 000 new patients per year who receive treatment on an out-patient basis. In addition, a 30-bed hospital ward located at the NAC is intended for hospitalized patients. Typically, the treatment beam time ranges from 2 to 3 min, whereas the patient positioning takes 10 to

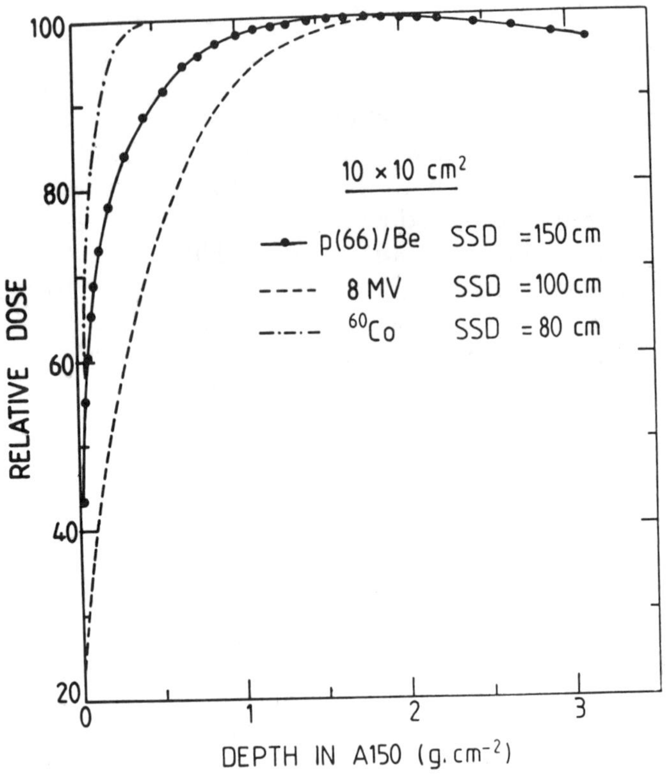

FIGURE 5.29 Build-up curve for a p(66)/Be neutron beam (SSD=80 cm) and an 8 MV beam (SSD=100 cm) (after D.T.L.Jones [5.103])

30 min during which the beam is delivered to the radioisotope production unit, the switching-over time being about 20 s. A detailed description of the operational parameters of the NAC neutron facility is given in [5.104].

The collimator enables continuous aperture control over the field range from 5x5 cm^2 to 30x30 cm^2 at SSD = 150 cm. The gantry has a total weight of 50 t and can be rotated through ±185°. For the full rotation the measured isocenter positional tolerance is 2.25 mm.

TABLE 5.8 Summary of Patient Treatment Statistics for NAC Faure Neutron Therapy Unit (after D.T.L. Jones (private communication).

Period	Scheduled treatments	Treatment days	Completed Treatments	Completed Fields	Rescheduled treatments[1)]	Patients started
1988	12	7	12 (100%)	18	–	2
1989	880	113	767 (87.2%)	1738	113 (12.8%)	72
1990	1261	130	1196 (94.8%)	2944	65 (5.2%)	101
1991 Feb	89	12	85 (95.5%)	253	4 (4.5%)	13
Mar	127	11	127 (100%)	257	–	9
Apr	90	13	87 (96.7%)	206	3 (3.3%)	10
May	148	14	148 (100%)	379	–	11
June	94	12	93 (98.9%)	270	1 (1.1%)	6
July	62	11	56 (90.3%)	168	6 (9.7%)	7
Aug	139	13	139 (100%)	333	–	13
Sept	155	12	148 (95.5%)	394	7 (4.5%)	13
Oct	224	15	224 (100%)	519	–	17
Nov	186	12	186 (100%)	416	–	21
Dec	114	9	114 (100%)	262	–	–
Total 1991	1428	134	1407 (98.5%)	3457	21 (1.5%)	120
Total	3581	384	3382 (94.4%)	8155	199 (5.6%)	295

Average to date (19 Dec 1991)

Treatments per day	8.8	Fields per day	21.2
Fields per treatment	2.4	Time per field	11.7[2)] min
Time per day	299[2)]		

[1)] Due to machine-related problems only. Patient-related rescheduling is not included.
[2)] Since June 1990.

The daily maximum patient throughput is estimated at 15-20. The main types of tumors treated at present are salivary glands, soft tissue sarcomas, as well as breast, head and neck carcinomas. During the nine months of 1989, that is after the accelerator was started, the total of 599 fractions were given 1311 treatment fields. Summary of patient treatment statistics as of 19 December 1991 is presented in Table 5.8.
Table 5.9 shows typical patient treatment file.

TABLE 5.9. Typical Patient Treatment File (National Accelerator Centre at Faure, Rep. of South Africa) [5.148]

SURNAME, INITIALS	
SEX:-	m
DATE OF BIRTH:-	17-7-1977
HOSPITAL NUMBER:-	7777 9999
RADIOTHERAPY NUMBER:-	91/0099
N.A.C. NUMBER:-	999/91

TREATMENT PLAN NUMBER	1	VERSION NUMBER	1
MODIFIED DATE	91-2-11		
FIELDS PER SESSION	3	TOTAL NO. OF SESSIONS	12
TOTAL PRESCRIBED DOSE (Gys)	23.4720	TOTAL NUMBER OF TREATMENTS	
DOSE PER USE OF THIS		TO DATE	7
PLAN (Gys)	1.9560		
DOSE TO DATE (Gys)	15.0130		
TUMOUR LOCATION:-	TONGUE		
AUTHOR:-	M A LE ROUX		
VERIFIED:-	Plan Entry		

NOTE:- NECK SUPPORT NO. 1, BITE-BLOCK MASK

FIELD NUMBER	1		
DOSE THIS FIELD (Gys)	0.6400	MONITOR UNITS	1.3980
COMPENSATOR	0	WEDGE FILTER	2
FLATTENING FILTER (1=12, 2=30)	1		
COLLIMATOR Y (cm)	7.0000	COLLIMATOR X (cm)	7.8000
COLLIMATOR ROT.(DEGREE)	90.0000	GANTRY ROTATION (DEGREE)	270.0000
COUCH TOP ROT. (DEGREE)	0.0000	COUCH FLOOR ROT. (DEGREE)	0.0000
BOLUS	N	BEAM BLOCK	N
PRONGS UP/DN STREAM	DS	COUCH OR CHAIR (0 OR 1)	0
SAD(O) OR SSD (NE O) (cm)	150.0000		
FIELD NUMBER	2		
DOSE THIS FIELD (Gys)	0.7110	MONITOR UNITS	0.7840
COMPENSATOR	0	WEDGE FILTER	0
FLATTENING FILTER (1=12, 2=30)	1		
COLLIMATOR Y (cm)	7.0000	COLLIMATOR X (cm)	6.5000
COLLIMATOR ROT. (DEGREE)	270.0000	GANTRY ROTATION (DEGREE)	0.0000
COUCH TOP ROT. (DEGREE)	0.0000	COUCH FLOOR ROT. (DEGREE)	0.0000
BOLUS	N	BEAM BLOCK	N
PRONGS UP/DN STREAM	DS	COUCH OR CHAIR (0 OR 1)	0
SAD(O) OR SSD (NE O) (cm)	150.0000		
FIELD NUMBER	3		

DOSE THIS FIELD (Gys)	0.6050	MONITOR UNITS	1.3200
COMPENSATOR	0	WEDGE FILTER	0
FLATTENING FILTER (1=12, 2=30)	1		
COLLIMATOR Y (cm)	7.0000	COLLIMATOR X (cm)	7.8000
COLLIMATOR ROT. (DEGREE)	270.0000	GANTRY ROTATION (DEGREE)	90.0000
COUCH TOP ROT. (DEGREE)	0.0000	COUCH FLOOR ROT. (DEGREE)	0.0000
BOLUS	N	BEAM BLOCK	N
PRONGS UP/DN STREAM	DS	COUCH OR CHAIR (0 OR 1)	0
SAD(O) OR SSD (NE O) (cm)	150.0000		
	LOG RECORD		
TREATMENT START	91-2-26	15:47:16	
START FIELD	0	START FIELD DOSE	0
END FIELD	0	END FIELD DOSE	0
FULL DOSE FOR END FIELD	0		
TOTAL DOSE THIS SESSION	1962		
FIELD COMPLETED THIS SESSION	1 2 3		

Installations with rf linacs. The present-day dedicated installations for neutron therapy are equipped only with isochronous cyclotrons. One such facility, in which protons are accelerated by an rf linac, is *Fermilab's Neutron Therapy Facility* (USA) set in operation in 1976. Table 5.7 gives the system's parameters. It is a typical *parasitic installation* which operates in collaboration with a large rf linac delivering about 1.2 x 10^{13} protons per pulse at 15 Hz repetition rate, whereas the high energy physics program implemented at the Fermilab requires repetition rates not higher than 0.5 Hz. Until 1985, that is until the US National Cancer Institute stopped funding the NTF activities, a total of 1300 patients had been treated and the clinical results were very good. About 500 patients have been treated since the grant was terminated [5.84]. The number of side effects resulting from neutron therapy has been found to be smaller than that at other facilities [5.105]. These favourable results have led to the study on the design of a medical proton linac for neutron therapy and radioisotope production [5.105, 5.106]. Table 5.10 lists the proton intensities expected from the different accelerator types. An linear accelerator would have a length of 24 m [5.105] or less [5.106]. The predicted advantages would be: power requirements of 200 kW, which are lower by a factor of three than those required to run a comparable cyclotron, and the dose rate of about 2 Gy min^{-1} at SSD = 190 cm, which exceeds the maximum dose rate obtained in the present-day cyclotrons.

TABLE 5.10 Expected Accelerator Intensities (after F.T.Cole [5.144])

Accelerator	Pulse rate, s^{-1}	Intensity, proton s^{-1}
Synchrotron	0.5 - 1.0	$(0.5 - 1)\ 10^{11}$
Synchrocyclotron	1000	$5x\ 10^{12}$
Cyclotron	Continuous (MHz rf structure)	$5x\ 10^{13}$
Linear accelerator	60	$5x\ 10^{14}$

5.3.5 Designs of Boron Neutron Capture Therapy

The idea of a neutron therapy modality employing neutron capture by boron atoms was first put forward by G.I.Locher as early as 1936. Boron is a light element which in human tissue is present in small quantities. It includes the isotope ^{10}B which constitutes about 20% of natural boron and has a very large cross-section (about 4000 barns) for thermal neutrons. For comparison, the geometrical cross-section of the nucleus is about 0.2 barn. Typically, a tumor can be loaded with 30 ppm by weight of ^{10}B and 5-10 times less in the neighbouring healthy tissue. As a result of *neutron capture,* an alpha particle is emitted and ^{7}Li nucleus undergoes recoil. The total energy released in this reaction is 2.8 MeV. Depending on the size of individual cells, the ranges of these heavy particles vary between 9 and 5 μm, the LET value being quite high (150 keV μm^{-1}).

The *boron neutron capture therapy* (BNCT) modality consists in saturating the tumor tissue with ^{10}B atoms incorporated into an appropriate pharmaceutical, and then irradiating the tumor with thermal and epithermal neutrons. If the ^{10}B concentration in tissue is assumed to be 10 to 100 ppmw, irradiation requires 10^{12} to 10^{13} thermal neutrons per cm^2, which corresponds to a dose of about 8 Gy. Details of the BNCT may be found in [5.111, 5.112]. Elements other than boron have been suggested as possible agents for neutron capture therapy [5.150]; one of them is gadolinium, a contrast medium used in magnetic resonance imaging (MRI) in the form of gadopentetate meglumine. The ^{157}Gd which comprises 15.6% in abundance has a thermal neutron cross section 66 times larger than that of ^{10}B. In gadolinium, neutron capture reactions, prompt gamma-rays with an energy spectrum of up to 7 MeV, X-rays, high energy conversion electrons and high LET Auger electrons are released.

The boron neutron capture therapy could not be used in the past because no satisfactory specific pharmaceuticals nor proper neutron sources were available. At present, work in both these fields has greatly intensified [5.112]. Although it is usually nuclear reactors that are employed as neutron sources (or ^{252}Cf implants), because they are less useful clinically, other accelerator sources have also been designed, mostly low energy

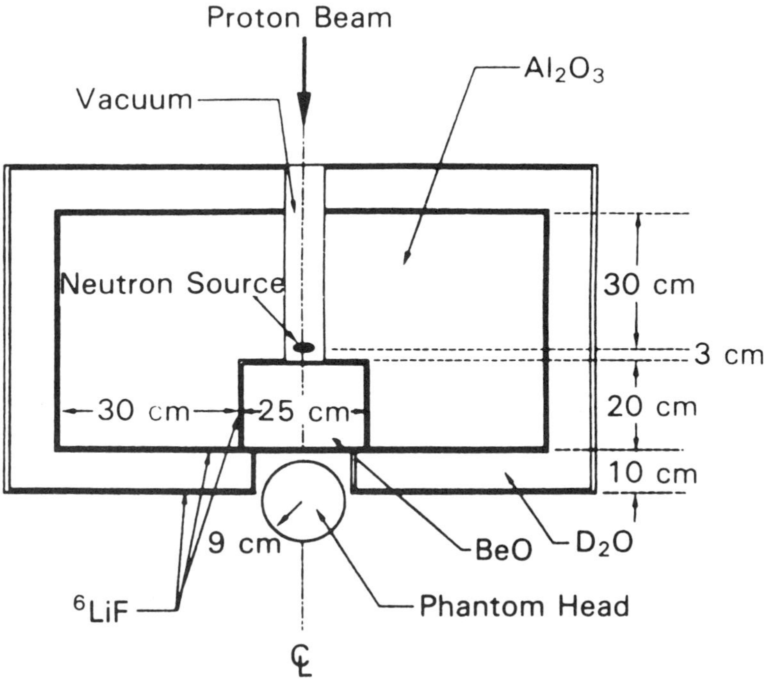

FIGURE 5.30 The configuration of the design of the moderator assembly for the boron neutron capture therapy, developed at the Ohio State University (after C.K.Ch.Wang [5.108])

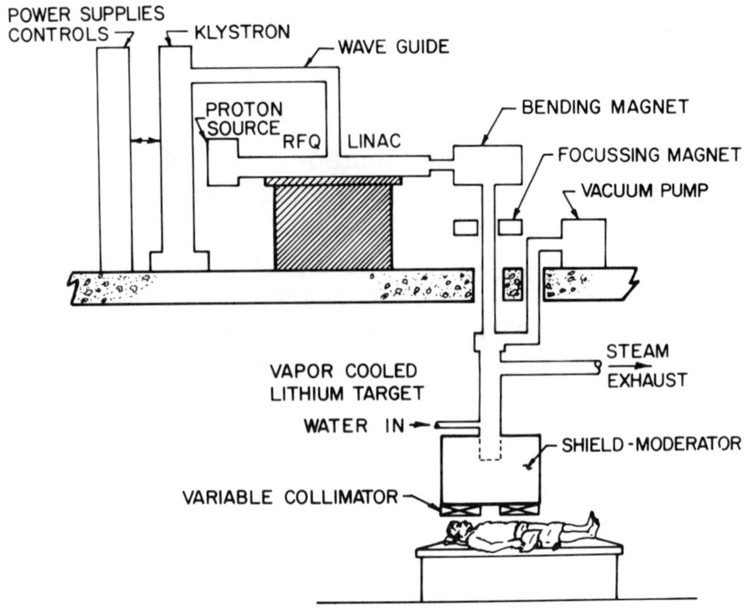

FIGURE 5.31 Schematic of RFQ linac BNCT treatment facility (Los Alamos National Laboratory), (after T.P. Wangler et al. [5.113])

(2 to 3 MeV) proton machines. An overview of the suggested designs is given in [5.109].

Fig.5.30 is a schematic of a moderator assembly for the BNCT [5.107, 5.108]. A 2.5 MeV proton beam with intensity of 10 mA or higher generates protons in the ^{7}Li (p,n) ^{7}Be. Fast neutrons thus produced are then thermalized (degraded in energy to between 1 eV and 10 keV) in a cylindrical moderator, 25 cm in diameter and 20 cm in height, made of beryllium oxide (BeO). The moderator, in turn, is surrounded by a 30 cm thick Al_2O_3 reflector. In order to reduce contamination of the thermal neutron beam, an additional 0.01 g cm^2 layer of ^{6}Li is also used. An additional 10 cm thick layer of heavy water, which serves as a neutron shield to minimize the irradiation of the whole body, is placed around the alumina reflector. For the proton beam intensity of 10 mA, 8.0 x 10^{12} neutrons per second will be generated, corresponding to the dose of 23.4 cGy min^{-1}. This would require 1.5 h for a single treatment dose of 20 Gy.

Treatment times could only be reduced if beam intensities were increased. In the design of a linac with rf quadrupoles recently developed at Los Alamos proton beam intensities up to 100 mA are envisaged, with treatment times reduced to about 11 min [5.113].

Fig. 5.31 shows a schematic of the overall RFQ linac BNCT treatment facility, including the accelerator system, a vapor-cooled liquid lithium target, the neutron moderator assembly (Fig. 5.30), and a patient undergoing treatment.

Apart from the feasibility of using linacs in radiotherapy, the possibility of generating thermal and epithermal neutrons obtained from a cyclotron have also been investigated. Details are provided in [5.110]. The production of keV neutrons by spallation reactions induced by accelerated protons in heavy element targets has been studied as part of a Swedish-Swiss collaboration [5.145 and 5.146]. The aim is the design of a *keV neutron* source that could be rotated relative to the patient. Special attention has to be paid to the cooling and maintenance of the spallation target, where many kilowatts of heat will be produced. In the possible solution (Fig.5.32) the horizontal 72 MeV proton beam is delivered to a magnetic field on the axis, where it is deflected to hit the ring-shaped neutron production target. In normal operation, the magnetic field is rotated about the axis about once per minute, so that the point from which neutrons are

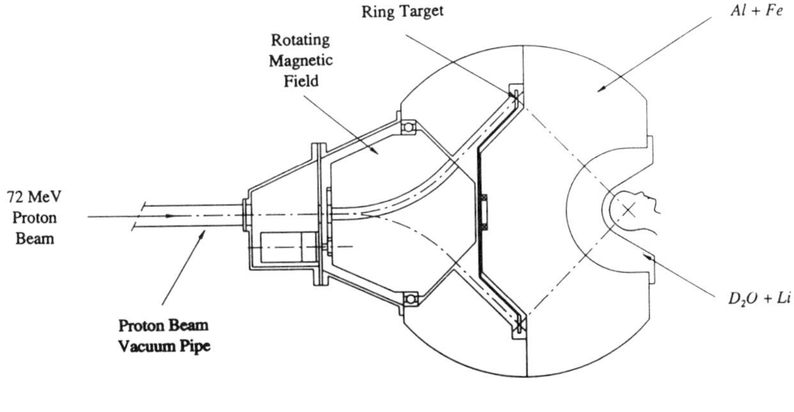

FIGURE 5.32 Layout of a possible design of spallation source for keV neutron BNCI for brain tumor treatment (after J.F. Crowford et al. [5.151])

emitted orbits the patient's head (Fig. 5.32). This is done largely to distribute the skin dose and improve the neutronics inside the phantom, but also to facilitate cooling the production target. A preliminary design for the magnet, including the current and cooling water supplies, has been worked out at PSI [5.152 and 5.153].

The ring target is buried in a semispherical moderator: D_2O+Li layer 20 cm thick, which in turn is surrounded by a 90 cm layer of Al/Fe, 50:50 by atom. The moderator materials extend as far as the plane tangential to the downstream side of the phantom. The ring source has a radius of 47.9 cm, and is centred 47.9 cm from the centre of the phantom. Since 10^{17} initial neutrons can be produced in 1000 seconds by an 80 μA proton beam, a dose of 30 Gy can be delivered to a deep-seated tumor in a total of about two hours, fractionated as required. This facility is comparable, in terms of advantage depth and dose rate, with reactor facilities.

CHAPTER 6

Synchrotron Radiation Sources and Their Uses

6.1 Properties of Synchrotron Radiation. Fundamentals of Generation

The source set-up. Fig.6.1 is a schematic view of a typical *dedicated synchrotron radiation source* (SRS), that is, one which is exclusively intended for generating synchrotron radiation for industrial and research uses. Within the *storage ring* electrons are accelerated to 2 GeV (99.999997% of the speed of light) and magnetically constrained to a quasi-circular orbit for many hours; at this energy the electrons make three million orbits per second.

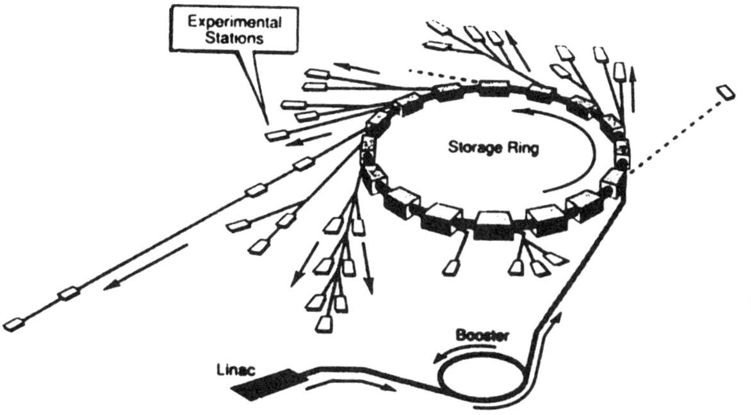

FIGURE 6.1 Fundamental set-up of a synchrotron radiation source (SRS) (Daresbury Laboratory, England)

A high-energy beam, usually made up of several tens of bunches with well-defined positions, travels in an orbit in a storage ring. The beam intensity usually ranges from several tens of mA to about 1 A. The beam is confined to an approximately circular orbit by approximately ten or up to several tens of magnetic dipoles (Sec.2.4.1).

In order to obtain maximum energy ranging from several hundred MeV to several GeV, electrons are pre-accelerated in an rf linear accelerator and then are delivered to an *electron synchrotron* which serves as a *booster.* After attaining nominal energy they are injected in the *storage ring orbit.* The *beam lines* are arranged tangentially to the storage ring to accept the fan of radiation emitted. The number of beam lines which use the synchrotron light in dedicated sources usually ranges from several to about 60.

In addition to the bending dipoles, the magnetic structure of the ring is also made up of quadrupole and sextupoles (not shown in Fig.6.1) which focus the beam horizontally and vertically.

Synchrotron radiation emission. A charged particle that travels along a curved trajectory (Fig.6.2) with relativistic velocities emits strong electromagnetic radiation. This radiation was indirectly discovered by J.P.Blewett in 1945 in his studies with a 100 MeV betatron, but it was first observed experimentally in 1947 in a 70 MeV electron synchrotron (Schenectady NY, USA)

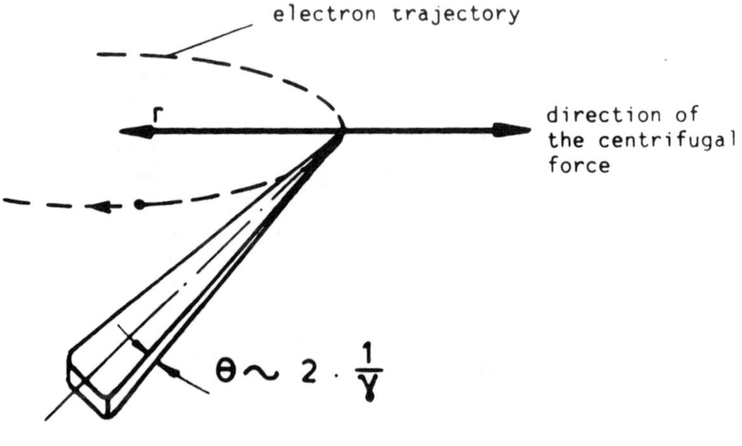

FIGURE 6.2 Synchrotron radiation emission

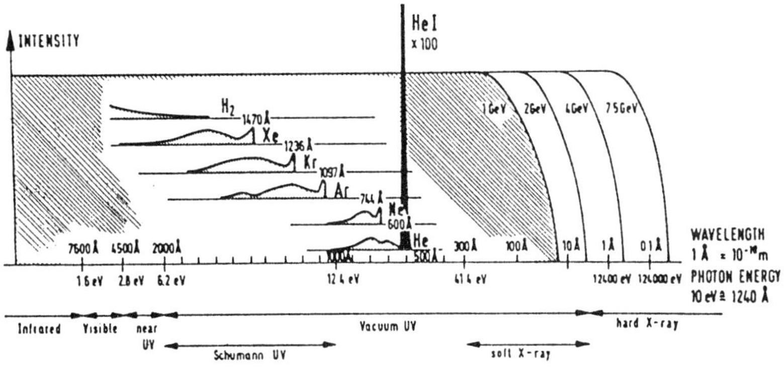

FIGURE 6.3 Schematic comparison of the spectral emission of synchrotron radiation with other sources of photon radiation vs. electron energies 1-7.5 GeV, (after E.E.Koch (1000 Å = 100 nm = 0.1 μm = 12.399 eV))

from which it took its name. It encompasses a wide continuous spectral range from the infrared to hard X-rays (Fig.6.3). For comparison, Fig. 6.3 shows spectral curves for emission radiation from conventional discharge lamps.

Synchrotron radiation (SR) was initially regarded as *undesirable and parasitic.* It led to great losses in the rf energy utilized for accelerating electrons in large synchrotrons and storage rings. In the largest machines these losses during operation may be as high as several to several tens of megawatts of rf energy ! It was, however, soon found that SR may serve as an excellent research tool, and, at a later date, as a technique used in practical applications. As early as the 1960s, a 180 MeV synchrotron was employed as a source in atomic spectroscopy (NBS, Washington, USA).

Relationships and fundamental properties. The *total power of SR* emitted by a relativistic particle travelling in an orbit of the radius r is given by

$$P = \frac{2}{3}\frac{q^2c}{3r^2}\frac{E_k{}^4}{(m_0c^2)^4}, \qquad 6.1)$$

where q is the particle charge (e in the case of an electron), r is the radius of the orbit, E_k is the kinetic energy of the particle and

m_0c^2 is the rest energy of the particle. It follows from Eq.(6.1) that the emitted power is proportional, among other things, to the ratio of $q^2m_0^{-4}$. Thus, in order to obtain intensive SR, particles with large charge-to-mass ratios, such as electrons or positrons, should be used.

It follows from Eq.(6.1) that the power emitted is proportional to E_k^4 and $1/r^2$. That is why the designers of large electron synchrotrons, especially large electron storage rings, tend to maximally increase the radius r, which for rings with energies of the order of several GeV becomes as large as several kilometers. As mentioned earlier, SR in this case is parasitic and undesirable, and, therefore, the tendency is to make it as small as possible. On the other hand, in practical applications this tendency is opposite: in order to *maximize the SR power* generated, the electron beam is made to move in a small orbit whose radius is of the order of a few meters (Table 6.1). In this case what is maximized is the energy loss, which expressed per one revolution, is given by a practical relationship

$$\Delta E_k = 88.5 \frac{E_k{}^4}{r}, \tag{6.2}$$

where ΔEk is electron energy loss in keV per revolution, E_k is the kinetic energy in GeV, and r is the radius of the orbit in m.

Thus the *total power emitted in an electron synchrotron or in a storage ring* of the magnetic induction B, in which electrons are accelerated to an energy Ek and the beam current I, is given by

$$P_t = 26.6\, E_k{}^3\, BI, \tag{6.3}$$

where P_t is the total power in kW, E_k is the energy in GeV, B is the induction in T, and I is the beam current in A.

The SR spectrum can be characterized by the parameter referred to as *critical wave-length* λ_c, or the corresponding *critical energy* E_c, which divides the emitted spectrum in two parts of equivalent total energies (Fig.6.4); λ_c is determined by

$$\lambda_c = 5.59 \frac{r}{E_k{}^3} = \frac{18.64}{B\, E_k{}^2} \tag{6.4}$$

where λ_c is the critical wave-length in Å, r is the radius of the orbit in meters, E_k is the energy in GeV, and B is the induction in T.

In the SR technique, described in detail in many publications [6.1 through 6.4], the angstrom unit Å = 0.1 nm is still used. The photon energy E_{ph} is given by

$$E_{ph} = \frac{12.40}{\lambda} \tag{6.5}$$

where E_{ph} is expressed in eV and λ in Å.

The critical SR photon energy corresponding to the critical wave-length λ in Eq. (6.4) can be expressed as

$$E_c = 2218 \frac{E_k{}^3}{r} = 2.96 \cdot 10^{-7} \frac{\gamma^3}{r} \tag{6.6}$$

where E_c is the critical photon energy in eV, E_k is the kinetic electron energy in GeV, r is the radius of the orbit in meters, γ is the relative electron energy in terms of the rest energy 0.51 MeV (e.g. for the kinetic electron energy of 511 MeV, $\gamma = 100$, and for 5.1 GeV, $\gamma = 1000$).

Synchrotron radiation is characterized by a *very narrow emission angle* (Fig.6.2)

$$\theta \approx \frac{1}{\gamma} = \frac{m_0c^2}{E_k} \tag{6.7}$$

where γ is the electron relative energy, m_0c^2 is the electron rest energy, and E_k is the kinetic energy.

Cavity resonators, which compensate for the energy lost in the emission of SR, are mounted on the storage ring. They are powered with rf energy of frequencies between 50 and 500 Mz. Like in electron synchrotrons, electrons in storage rings also travel in orbits (Fig.6.2) forming *bunches* whose lengths are matched to the frequency of the accelerating voltage used to supply resonators. As a result, SR is *pulse-like in nature.* The *electron orbital period* must be equal to the integral multiple of the periods of the resonance cavities. This multiple is referred to as *the ring harmonic number* and usually ranges from 100 to 500. Thus a larger number of bunches *(multi-bunch mode)* can

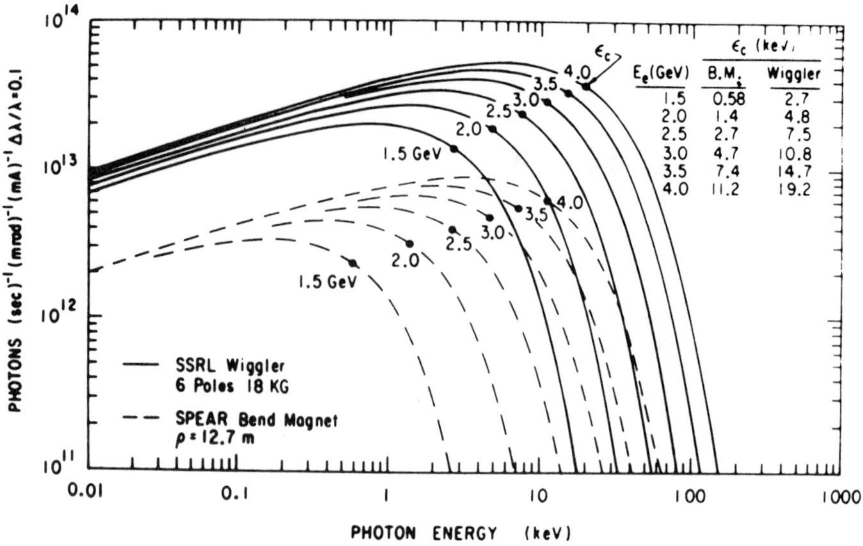

FIGURE 6.4 Synchrotron radiation (SR) spectra obtained from electron-positron storage ring (1.5 - 4.0 GeV): dashed lines represent spectra from bending magnets, and solid lines represent spectra obtained from wiggler magnets (after A.Bienenstock)

orbit the ring, which makes possible the control of the time structure of SR. The length of a single bunch is sometimes determined by the time of the radiation flash in a given beam line (Fig.6.1) and is typically 50 ps to 1 ns. For example, for 160 200 ps bunches distributed along the SRS ring orbit (Daresbury, England) synchrotron radiation is emitted as 200 ps pulses at the repetition rate of 2 ns. When, however, a single electron bunch travels in the orbit *(single-bunch mode),* the flash radiation repetition is 320 ns. The respective values for the DORIS ring in single-bunch mode are 150 ps and 960 ns.

The SR time structure, described above, opens up wide possibilities of SR application in the study of dynamic processes in live cells as well as in stroboscopic methods.

The synchrotron radiation has the following fundamental properties:

1. The wide *continuous spectrum* (Fig.6.4) from the infrared to hard X-rays has no equivalent in any other technical source of electromagnetic radiation. The maximum photon energy,

depending on the parameters of the presently used rings, is 150 kV (HASYLAB, Hamburg, Germany). It should be noted that the spectrum from discharge lamps ends with the He resonance line at 21.2 eV, i.e. 58.4 nm (584 Å). There is a gap towards the shorter wave-lengths up to the onset of the X-ray region (1000 eV), in which no radiation of suitable intensity has been generated so far owing to the lack of appropriate techniques.

2. High *emission power,* ranging from 1 kW to several MW for storage ring systems. With respect of intensity, SR exceeds that achievable with such sources as X-ray tubes by several orders of magnitude. Over the X-ray spectrum, SR sources exceed the power of the largest X-ray tubes with a rotating anode by 2-3 orders of magnitude for the K-line and by 5-6 orders of magnitude for the continuous spectrum (cf Fig. 6.23). By using booster magnets, the SR power can be increased by another two or three orders of magnitude.

3. A high degree of *beam collimation,* ranging from approximately 1 mrad to 0.1 mrad.

4. Total *linear polarization* in the orbital plane.

5. High *beam brightness* (Figs.6.5 and 6.6).

6. The useful *SR photon energy range,* obtainable with storage rings and circular accelerators, is usually $\Delta E \approx 4E_c$, critical energy E_c being given by Eq.(6.6). An optimum frequency band for a given application can be extracted from the above range by means of a monochromator.

7. As each single bunch of electrons traveling along the synchrotron or storage ring orbits is a source of radiation, pulses as short as 50 ps can be obtained, which is particularly useful in a variety of spectroscopic uses.

8. An electron beam emitting SR revolves in a *high vacuum* 10^{-7} to 10^{-8} Pa (10^{-9} to 10^{-10} Torr), which ensures very pure environment required in many studies.

Emittance and brightness. Charged particles accelerated in accelerators show some energy scatter. In most accelerators the acceleration process occurs in the gaps, and is localized. During this process, particles are often focused by means of strong focusing magnetic quadrupoles (Fig.2.24) also localizing in nature. Thus the beam has a spatial structure, pulsating along the accelerating orbit. In other words, the accelerated beam is a spatial structure which moves, for example, in circular accelerators, so that its central part coincides with the so-called *stable orbit,* whereas single particles move in separate orbits which are at various distances from the stable orbit. The aim of the designers is to make the separate orbits lie as close as possible to the stable orbit. From this point of view the beam can be described by the concept referred to as *emittance* which can be expressed in terms of the product of the *beam diameter and the angular divergence* usually given in mm·mrad.

In the SR techniques, the intensity of the beam incident upon a sample is described with the concept of *brightness.* From the experimenter's point of view, beam size and its divergence are also important (Fig.6.5). The smaller the lateral dimension of the elliptical cross-section of the beam and the divergence, the higher the brightness. Practically, the SR source brightness is determined by the beam focusing system. Since the brightness also depends on the spatial parameters of the source, on the bunch of electrons orbiting in the storage ring, in order to achieve high brightness the dimensions of the bunch should be kept as small as possible. Therefore, *strong localized beam focusing* is specially important in designing SR sources.

In characterizing the properties of SR sources, in addition to brightness, another concept has been recently employed. This concept is referred to as *spectral brilliance* [6.1, 6.2] expressed in terms of photons per second, per mm^2 of the source area, per

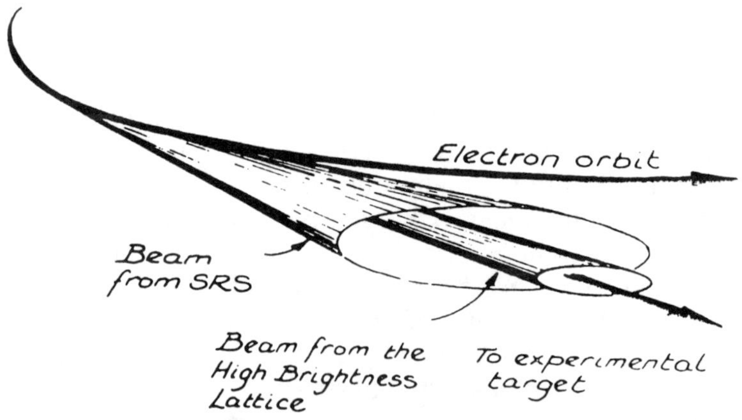

FIGURE 6.5 SR beam brightness

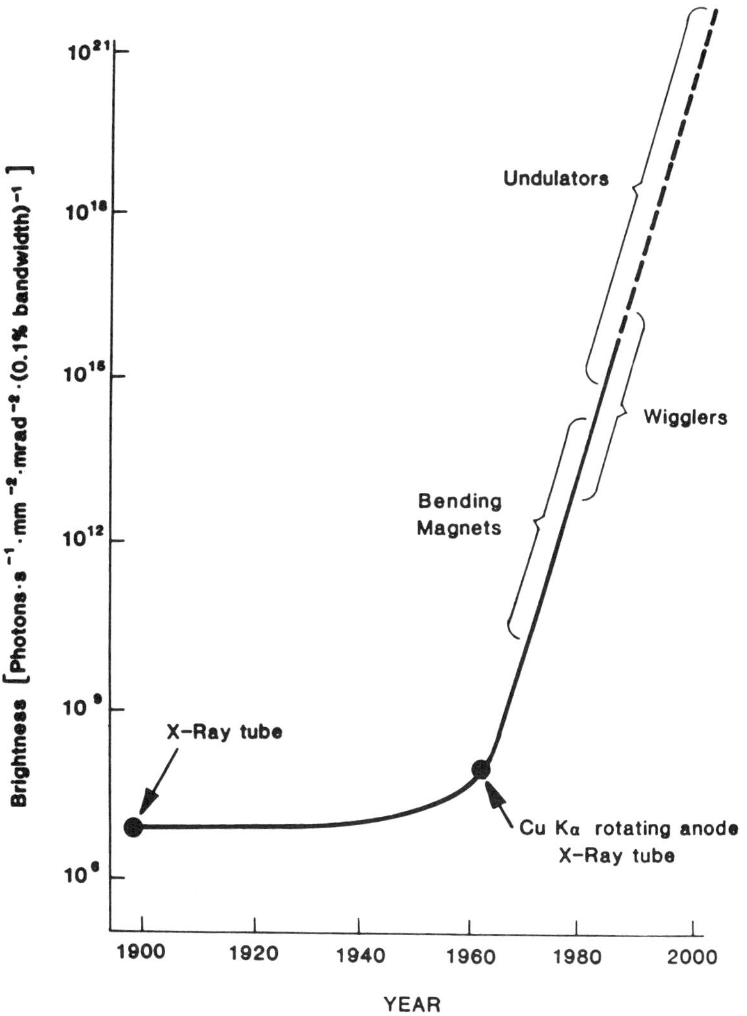

FIGURE 6.6 Chronological development of X-ray source spectral brilliance (after H.Winick [6.3])

mrad^2 of the source angular divergence and per 0.1% of the emitted bandwidth ($\Delta\lambda/\lambda$ or $\Delta E/E$). Thus

$$\text{spectral brilliance} \rightarrow \frac{\text{photons}}{\text{s mm}^2\ \text{mrad}^2\ (0.1\%\Delta E/E)} \tag{6.8}$$

By integrating over the source area the earlier mentioned spectral brightness is obtained

$$\text{spectral brightness} \rightarrow \frac{\text{photons}}{\text{s mrad}^2\,(0.1\%\Delta E/E)} \qquad (6.9)$$

By integrating brightness over all emission angles we obtain

$$\text{spectral flux} \rightarrow \frac{\text{photons}}{\text{s}\,(0.1\%\Delta E/E)} \qquad (6.10)$$

What is particularly important in the studies with a focused beam, in which the sample image is projected onto a detector, is the achievable spectral brilliance.

Fig.6.6 shows the chronological development of spectral brilliance obtained from X-ray sources. Until the early 1960, conventional X-ray tubes were the only sources to provide a moderate degree of brilliance. It was only when tubes with a stationary anode were replaced with those with a rotating anticathode that the brilliance increased tenfold. The true breakthrough was caused by the development of SR sources, which resulted in an increase in spectral brilliance by several up to ten odd orders of magnitude (Fig.6.6).

Physical fundamentals of SR generation can be found in the major textbook [6.4] and in [6.5, 6.6]. Physical aspects of SR sources and the limitations of their potentialities are discussed in [6.7, 6.8].

6.2 Ancillary Magnets

Synchrotron radiation emission occurs when particles move in circular orbits (Fig.6.2). Both the energy loss ΔE_k per one revolution in an orbit with diameter r, and the critical energy E_c, in accordance with Eqs. (6.2) and (6.6), are proportional to the reciprocal of the radius r. Therefore, initially the SR beams used to be extracted in the region of the bending magnets where the curvature of the orbit is the largest.

An essential step in the direction of developing practical SR sources was taken with the design of ancillary magnets, referred to as *wigglers or undulators.* These magnets permit generation of

SR with parameters unobtainable in conventional bending magnets by causing local bending of electron beam trajectories. In addition, wigglers and undulators can be installed at the rectilinear segment of the orbit, where research apparatus or work stations can be easily placed. Because of these advantages, several wigglers and undulators are usually installed in the present-day SR sources.

Wigglers. Fig.6.7a illustrates the operating principle of a *wiggler.* It is a *linear màgnetic structure,* whose length of a single period is Δ_0, and the induction B is maximum. The number of periods n usually varies between a few and about ten. In this structure, electrons oscillate at a relatively small radius thus generating intensive SR. Since radiations from each oscillation are added together, the resultant intensity increases considerably. On the other hand, as the critical wave-length λ_c is inversely proportional to the induction B (Eq.6.4), the *whole spectrum is*

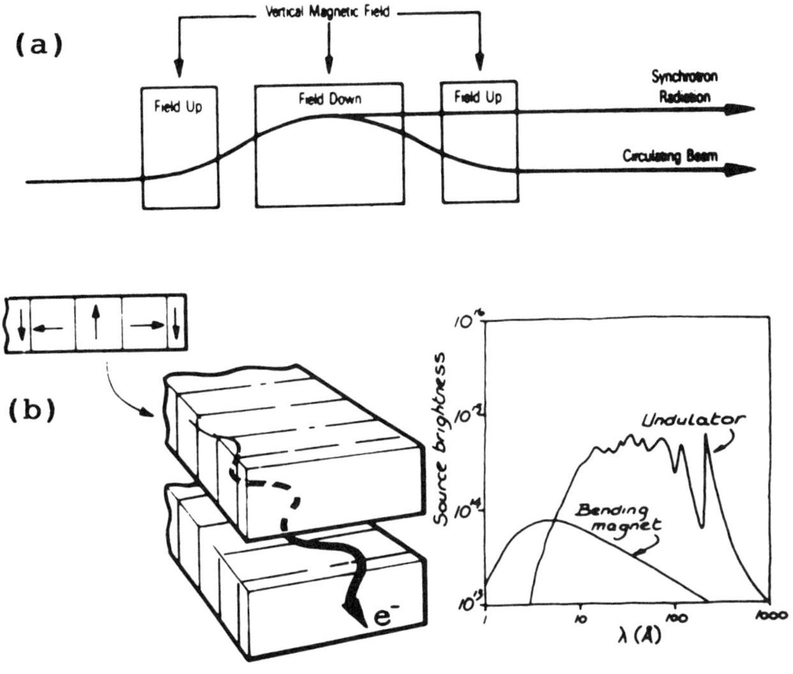

FIGURE 6.7 Ancillary magnets: (a) operating principle of a wiggler magnet, (b) construction of an undulator, (c) SR spectrum (SRS Daresbury)

shifted towards shorter wave-lengths. This is because induction in the wiggler can be much higher than that in the bending magnets.

The effect of the wiggler on the SR spectrum is shown in Fig.6.4. Dashed curves represent spectra generated by a storage ring SPEAR with electron energies ranging from 1.5 to 4.0 GeV at 100-300 mA. The induction of the bending magnets is less than 0.6 T. The bending radius is 12.7 m, and thus the SR critical energies vary from 0.58 to 11.1 keV. Solid curves represent the corresponding spectra obtained with a 7-pole wiggler with induction of up to 2.0 T (Fig.6.4). The critical energies vary from 2.7 to 19.2 keV: for example, for 3.0 GeV electrons the critical energy of the bending-magnet radiation is 4.7 keV, whereas that with the use of the wiggler increases to 10.8 keV.

In the latest designs wigglers with superconducting windings attain induction between 4 and 6 T. Synchrotron radiation spectra are obtained in a storage ring with a superconducting wiggler which correspond to seemingly higher electron energies without any changes in the maximum electron beam energy E_k.

Undulators. These are devices which incorporate *multiperiodical magnets* and correspond, in respect of construction, to wigglers (Fig.6.7a). The length of the *magnetic period,* compared with a wiggler, is much shorter (up to several cm), whereas the number of periods, n, increases up to 100. The induction is moderate and does not exceed 1 T. Currently, undulators employ more often permanent magnets (Fig.6.7b), denoted by REC *(Rare Earth Cobalts),* since they are distinguished by relatively low cost and do not require a power supply.

The operation of an undulator differs from that of a wiggler in that the radiation emitted during oscillations undergoes *interference.* As a result, quasimonoenergetic lines appear in the spectrum and thus *monoenergetic radiation* of relatively high intensity can be obtained at work stations if suitable *monochromators* are used. Therefore, undulators provide the maximum available spectral brilliance (Fig. 6.6). Fig. 6.7c is an example of such a spectrum compared with a normal emission from bending magnets (Synchrotron ring SRS, Daresbury, England) [6.11, 6.12].

Details of the operation and construction of wigglers and undulators are given in [6.9, 6.10].

6.3 Systems of Synchrotron Radiation Sources

The first *SR laboratories* were installed at electron synchrotrons and they made use of the *parasitic radiation.* In 1976, nine such laboratories were in operation. The mean electron beam currents usually ranged from 10 to 30 mA, their energies varying from 0.5 GeV (Bonn II) to 12 GeV (Cornell). The critical SR energies were in the 0.16 to 38 keV range, respectively. Another similar group of SR sources was made up of facilities set up at storage rings intended for high-energy physics research implemented with the method of colliding beams. The synchrotron radiation, like that in the first group of laboratories, was also parasitic. Sources of both types have pioneered the way in the development of applied SR and are still being used quite intensively. Fig. 6.8, which is a layout of one of the largest and oldest SR facilities in Europe, called HASYLAB, is an example of such a source. The abbreviation HASYLAB denotes Hamburger Synchrotronstrahlungslabor, Germany. Initially, a small laboratory was set up to make use of the parasitic radiation from a DESY (Deutsches Electronensynchrotron) 7.5 GeV electron synchrotron of 31.7 m radius. As early as 1972, the European Molecular Biology Laboratory (EMBL) located in Heidelberg, Germany, established an SR facility for biological studies at the above accelerator. When storage rings DORIS (Fig.6.8) designed for high-energy physics research went into operation in 1974, new prospects of generating SR opened up in view of a much greater beam intensity than that afforded by a synchrotron. In the storage rings with the equivalent radius of 12.12 m, colliding beams of electrons and positrons were accelerated to the initial energy of 3.5 GeV. Another facility was set up in the vicinity of the DORIS rings (Fig. 6.8). Now the rings have been redesigned, and, as a result, the maximum electron energy has been increased to 5.7 GeV, which corresponds to the SR critical energy E_c = 33.92 keV. Radiation is delivered to work stations located in a 50 x 30 m^2 experimental hall. As of 1989, HASYLAB used one beamline for the spectral range between 2 and 6 eV, four beamlines for the 5-50 eV range and another four beamlines for the 20-1500 eV range. In addition 20 beamlines are being prepared with higher energy X-ray up to max. 150 keV, of which three are intended for studies carried out by the EMBL.

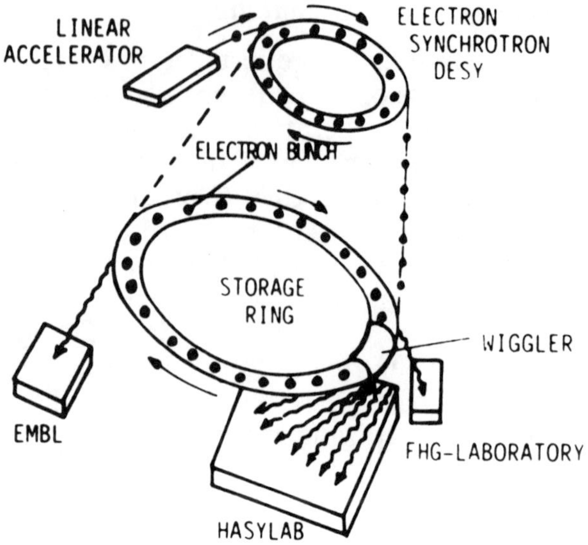

FIGURE 6.8 Layout of a synchrotron radiation facility HASYLAB (Hamburger Synchrotronstrahlungslabor, Germany)

Around the world, in 1989 there are laboratories in 12 countries engaged in the operation, construction or planning of electron storage rings, ranging from a few hundred MeV to the above 10 GeV, as sources of synchrotron radiation for basic and applied research. More than 6000 scientists now use radiation from 27 operational rings in a wide range of studies in biology, chemistry and physics and their numerous subfields. About 10 more rings intended for research are in construction or design [6.3].

Some basic parameters of these machines are listed in Table 6.1. Most of the SR sources in operation are those of the *dedicated type,* intended exclusively for generating SR for research and industrial applications. As an example of the SR source of this kind Fig. 6.9 provides a schematic of the facility installed at Brookhaven. Because of the difficulty in designing a system which would be able to generate SR over a wide spectral range shown in Fig. 6.3, two storage rings have been employed in the NSLS (National Synchrotron Light Source, Brookhaven, USA). Electrons preaccelerated in an rf linac and a 750 MeV auxilliary accelerator are either injected to a smaller storage ring of the same energy, which emits *vacuum ultraviolet,* or fed to a larger ring where they are again accelerated to an energy of 2.5 GeV. It is the latter ring that emits X-rays. For this radiation to be emitted (E_c > 1000 eV) electrons must be accelerated to an energy at least 2 to 3 GeV. The beam intensity in the first ultraviolet ring is as high as 1 A, whereas that in the 2.5 GeV X-ray ring is 250 mA and can be elevated to 500 mA. The facility has 50 work stations mostly installed round the 2.5 GeV ring.

TABLE 6.1 Electron Storage Rings (final target), (after T. Tomimasu [6.13]).

Machine and location	E(GeV)	R(m)	I(mA)	λ_p (Å)
TANTALUS I (Stoughton, USA)	0.24	0.64	150	108
SURF II (Gaithersburg, USA)	0.28	0.83	100	89
NIJI-I (Tsukuba, Japan)	0.27	0.70	340	83
INS-SOR II (Tanashi, Japan)	0.4	1.1	450	40
SIBERIA-I (Moscow, USSR)	0.45	1.0	(300)	26
ACO (Orsay, France)	0.55	1.11	300	16
MAX (Lund, Sweden)	0.55	1.2	60	17
COSY-II (Berlin, FRG)[a]	0.63	0.44	4.1	
VEPP-2M (Novosibirsk, USSR)	0.67	2.0	100	15
NSLSI (Brookhaven, USA)	0.7	1.9	(1000)	13
IMS-UVSOR (Okazaki, Japan)	0.75	2.2	300	12.2
NTT-I (Atsugi, Japan)[a]	0.8	2.22	10.2	
NTT-II (Atsugi, Japan)[a]	0.55	0.52	7.4	
TERAS (Tsukuba, Japan)	0.8	2.0	250	9.2
BESSY (Berlin, FRG)	0.8	1.83	500	8.4
SUPER-ACO (Orsay, France)	0.8	1.7	7.8	
HESYRL (Hofei, China)[a]	0.8	2.22	10.1	
ALLADIN (Stoughton, USA)	1.0	2.08	(500)	5.0
SPRC (Stanford, USA)[a]	1.0	2.0	(1000)	4.7
ADONE (Frascati, Italy)	1.5	5.0	100	3.36
DCI (Orsay, France)	1.8	3.82	400	1.55
SRS (Daresbury, UK)	2.0	5.55	(1000)	1.64
VEPP-3 (Novosibirsk, USSR)	2.2	6.15	(500)	1.39
PHOTON FACTORY (Tsukuba, Japan)	2.5	8.66	(500)	1.30
NSLS II (Brookhaven, USA)	2.5	6.93	(500)	1.05
SPEAR (Stanford, USA)	4.0	12.7	60	0.46
DORIS (Hamburg, FRG)	5.0	12.12	500	0.23

[a] Under construction

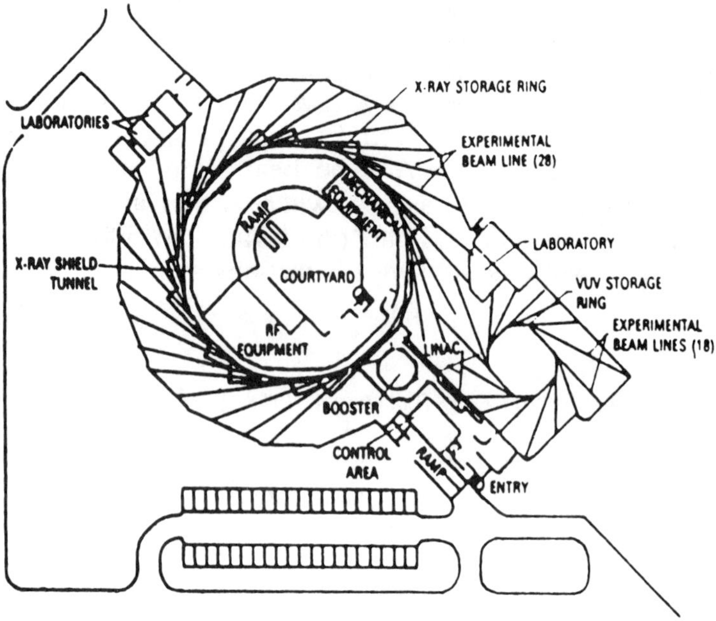

FIGURE 6.9 Layout of the dedicated SR source at the Brookhaven National Laboratory (National Synchrotron Light Source, USA)

The current costs of dedicated SR sources of small and medium dimensions are between ten and several tens of million US dollars.

Two large SR facilities for 6 and 7 GeV are at present under construction. The first one is being built by the European Science Foundation at Grenoble (France) at a cost of 2.6 billion FF, about $ 470 million (price as of 1987)[6.14]. The ring circumference will be 844 m. The operation of the whole facility, called the European Synchrotron Radiation Facility (ESRF), will be started in 1992, and a regular service for external users will start in mid-1994 [6.15]. A similar SR source is under construction at the Argonne National Laboratory (USA). It has a form of a ring 1000 m in circumference, in which a positron beam of about 100 mA intensity and 15 h beam life time will be stored. Instead of electrons, positrons have been chosen in this design because the stored beam will then not trap positive ions from the residual gas in the vacuum chamber. Equipped with 60 work stations the facility will be put in operation in 1992 at a cost of 375 million US dollars [6.16].

Apart from the tendency to construct new *national sources of synchrotron radiation,* the fact that some older SR sources employing storage rings are now being adapted for use as dedicated sources means that investigations with SR have recently become increasingly popular. The US SPEAR storage ring, after completion of a high-energy physics research project in 1990, has become a fully dedicated SR source, being now used, among other things, in intensive work on intravenous subtraction angiography (Sec. 6.5) [6.17].

6.4 Free Electron Laser Sources

The first system of a nonconventional source of electromagnetic radiation, referred to as a *free electron laser* (FEL) was invented by John Madey and put in practice in 1971 at Stanford. A schematic of an FEL design is shown in Fig.6.10. Electrons preaccelerated in an rf linear accelerator, microtron or a storage ring are delivered to a wiggler or undulator (see Fig. 6.6) together with a laser beam of the correct phase. The laser beam propagating parallel to the electrons decelerates the electrons slightly, and thus energy is transferred, which, in turn, increases light intensity. It is obvious that the principle of operation of the FEL is highly unconventional. The lasing action is not based on the stimulated emission from an atomic or molecular system, it is rather a beam of high-energy electrons, which oscillates in the wiggler and emits its energy in the magnetic field in a way characteristic of synchrotron radiation (Sec.6.2). In this case, emis-

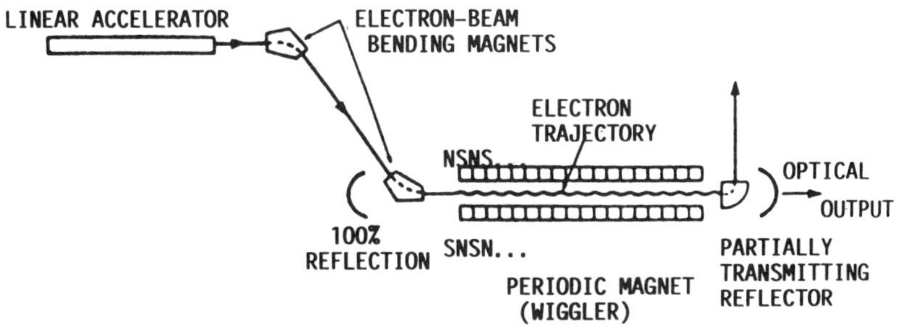

FIGURE 6.10 Schematic of a free electron laser

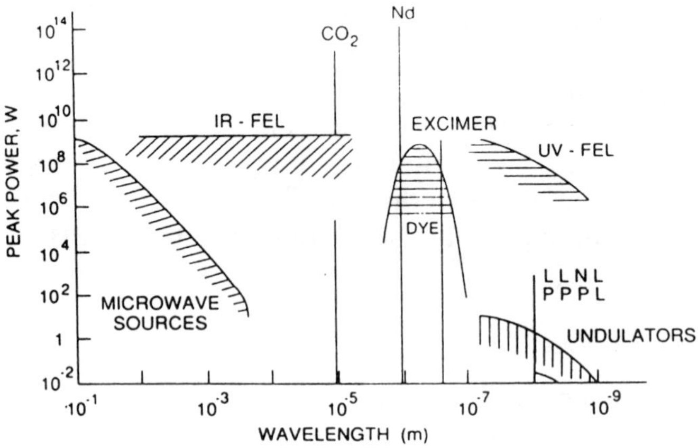

FIGURE 6.11 Comparison of the spectral range of FEL radiation with other sources of electromagnetic radiation (after C. Pellegrini [6.18]).

sion is identical with the spontaneous emission from stimulated atoms and molecules.

It can be seen from Fig.6.10 that the wiggler structure is surrounded by two optical mirrors, of which the left one reflects the light totally and the right one is semi-transparent. Therefore, light can be bounced through the structure many times. The system of mirrors ensures that the light pulses are in coincidence with the electron bunches. Similarly, electrons can be used many times: many FEL systems operate together with storage rings.

The *FEL wave-length* can be easily changed over a large region, from centimeter to nanometer (Fig. 6.11), [6.18], by simply changing the beam energy. Since the radiation pulse structure reflects that of the electron beam, it is also easy to produce pulse lengths ranging from almost CW, using an electrostatic accelerator, to picosecond, using an rf linac. In addition, for a high energy electron beam, it is easy to obtain large peak power, on the order of GW to TW (10^9 to 10^{12} W). The FEL has thus become a conceptual and practical alternative to other radiation sources such as microwave tubes and lasers, and can extend the operational range of both these systems. This point is illustrated in Fig. 6.11, which compares the FELs and other radiation sources, microwave tubes, lasers, undulator radiation from high brightness synchrotron radiation storage rings, and from plasma based X-ray lasers at Livermore and Princeton [6.18].

TABLE 6.2 Summary of the FEL Applications (after C. Pellegrini [6.18])

FEL Applications		
Condensed Matter	Surface Science	
	–Catalysis	IR
	–Adsorption	IR
	–Selective excitation of surface layer or adsorbed molecule	IR
	Semiconductors	
	–High Tc SC bandgap	IR
	–Carrier dynamics	IR
	–Electron gas dynamics at metal-insulator junction	IR
	Superconductors	IR
	Magnetic Properties	IR
Chemistry	Molecular Vibrational Excitation	IR
	Reaction Dynamics	IR
	Photochemistry	IR
	Molecular Cluster, van der Waals - molecule	IR
	Electronic excitations	UV
	Raman Spectroscopy	UV
	Crossed Photon-Molecular Beams	UV
Biology	Microscopy, Holography, Cellular Dynamics	1 - 3 nm
	DNA sequencing	0.1 - 1 nm
Medicine	Surgery	3 μm
	Phototherapy-Selective absorption in pigmented elements	0.7 - 1 μm

One can see from Fig. 6.11 that there are two main regions of interest for FEL applications, where they can be far superior to other sources: one is the IR (InfraRed) and millimeter to centimeter region; the second is in the short wavelength region, below 100 nm. A summary of some of the applications is given in Table 6.2 [6.18].

Theoretical fundamentals of the FEL are given in [6.19, 6.20], and the advances in the generation of radiation with FEL sources is discussed in [6.18]. Most of the FEL systems in operation or under construction are intended for operation at IR wavelengths (1 to 500 μm). Their properties are discussed at length in [6.21].

The FEL systems are becoming increasingly popular as prospective unconventional electromagnetic radiation sources [6.22]. However, practical applications of the FEL in medicine and biology as routine sources seem to be, as yet, quite a way off, since the FEL is still at the experimental stage of development. Potential applications in experimental *laser surgery and laser phototherapy* (in the 1-10 μm range) are being considered. *Photodynamic tumor therapy,* most likely in the visible or, perhaps, in the near infrared, is also being suggested [6.23]. Examples of prospective uses in biology in the nucleic acid synthesis and in biopolymer spectroscopy are given in [6.21]. The region around 1 to 5 nanometers would be particularly useful for X-ray microscopy and holography. *X-ray microscopy* of biological samples will produce images resolving cellular substructure in the natural state, without dehydration or staining, and on a picosecond time scale, before the hydrodynamic expansion of the sample becomes important. If one can produce even shorter wavelengths, below 1 nm, with the required intensity and spectral characteristics, one might be able to sequence DNA pairs by direct imaging [6.18].

6.5 Subtraction Angiography

Premature disease of the coronary arteries is the leading health problem in the industrialized countries. For example, in the USA the number of patients with symptoms of a coronary disease is estimated at three to four million. More than 1.3 million heart attacks occur each year, half of which are fatal [6.24, 6.25]. So far, no reliable method of diagnosing coronary diseases has been found. *Conventional angiography,* radiological examination of blood vessels, consists of injecting an *iodine-containing contrast* of high concentration (30 to 50%) in the vicinity of the region to be examined. A relatively large amount of contrast is injected by means of a catheter from a peripheral artery, usually in the thigh. Then a series of fast radiographs is made, in which, due to the strong absorption of X-rays by iodine, the image of the blood vessels filled with the contrast can easily be seen against the weaker background of bones and tissues.

With modern X-ray devices this method gives excellent images [6.26 and 6.27], but has a certain risk for the patients with respect to morbidity (1.2 to 2.2%; 0.5% severe) and mortality (0.07 to 0.23%). This risk, as well as the high cost of the examination (up to $ 3000), limit the use of angiography to patients suspected of having a far-advanced disease [6.24].

That is why efforts are being made to develop a method without entering the arterial system by a catheter. This *non-invasive* method could be used to investigate out-patients with mild symptoms of coronary heart disease, to monitor disease progression routinely, to perform follows-ups after surgery, to study effects of therapeutic interventions, etc.

The use of SR in *digital subtraction angiography* will probably help to overcome the above difficulties [6.25, 6.28]. A small quantity (bolus) of commercially available iodine-containing contrast (370 mg/ml) is injected in 1 s intravenously, which eliminates any hazard from using a catheter. After reaching the coronary vessels, the contrast agent is diluted and its concentration would be about 10 mg/ml, corresponding to a mass density of 1 mg·cm^{-2} in a coronary artery of 1 mm diameter, therefore, visualization by means of a conventional X-ray technique becomes impossible. The image, however, can be obtained using two intensive SR beams with energies of E_1 and E_2 (Fig. 6.12) bracketing the K-edge of iodine (33.17 keV). The images for both beams are taken alternately within the same, strictly defined, period of time. The visualization of the coronary arteries filled with iodine is only possible because the absorption coefficient near the iodine K-edge is characterized by a 6-fold change, whereas that for the soft tissue or bone is much smaller and changes monotonically (Fig. 6.12). Hence, the logarithmic subtraction of such image pairs taken at different energies enhances the iodine contrast by eliminating the effect of soft tissues and bones. For an *energy separation* of the two mono-

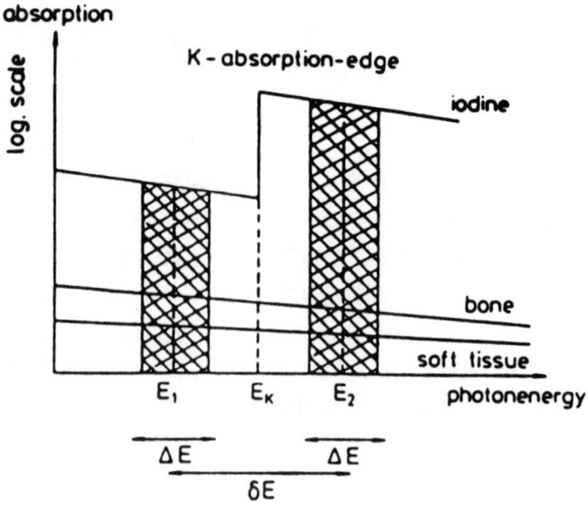

FIGURE 6.12 Absorption coefficient vs. photon energy

chromatic energies of 300 eV the sensitivity to iodine is 10000 times higher than that to soft tissue [6.26].

The K-edge of iodine is at 33.17 keV. Photons of this energy in soft tissue have an absorption length of only 2.1 cm. The energy separation δE (Fig. 6.12) must be smaller than 300 eV, otherwise after substraction, bone and soft tissue give higher contrast than the smallest artery. This means that the bandwidth ΔE must be less than 300 eV, too.

This method requires monochromatic X-rays with very high intensity. Therefore, the development of a system for the *energy substraction mode (dichromography)* only started when these X-rays were available at synchrotron radiation facilities. Since the early 1980s intensive work has been going on the application of SR to digital subtraction angiography at Stanford (USA) [6.25, 6.30, 6.31], Novosybirsk (USSR) [6.32], Hamburg (Germany) [6.29, 6.33, 6.34] and Tsukuba (Japan) [6.35]. An overview of partial results of this work was made at the Symposium at Frascati (Italy) [6.36]. In 1989 the Stanford system was transferred to Brookhaven (USA). Up to now (1992), human studies were only performed with the systems at Stanford [6.30], Brookhaven [6.31] and in Hamburg [6.26].

The quality of the subtracted image depends on the *energy separation*, δE, of the two beams, i.e. E_1 and E_2, which can usually be as high as 300 eV. The *energy spread*, that is the widths of the analyzing beams, are of the order of several tens of eV, for example, they are 20 eV and 55 eV for the Stanford and HASYLAB (Hamburg) facilities, respectively.

The requirements which have to be met by an SR system for digital subtraction angiography can be formulated in the following way [6.33]: 1. Arteries down to 1 mm in diameter should be visualized, which means that the matrix size should be 256x256 pixels with 0.5 mm pitch, 2. Intravenous injection of a 10 ml iodine-containing contrast agent into a brachial vein within 1 second results in an iodine concentration of about 10mg ml^{-1} in the coronary arteries corresponding to 1 mg ml^{-2} surface concentration. Consequently, arteries of 1 mm diameter should give a 3%-signal in the subtracted image. If the required signal-to-noise ratio is 3, noise has to be suppressed to about 1%. 3. The two images processed by the digital subtraction technique must be taken within the same heart cycle at a *phase of minimum motion of the heart* (usually within 200 ms) in order to avoid motion artefacts and strong image distortion.

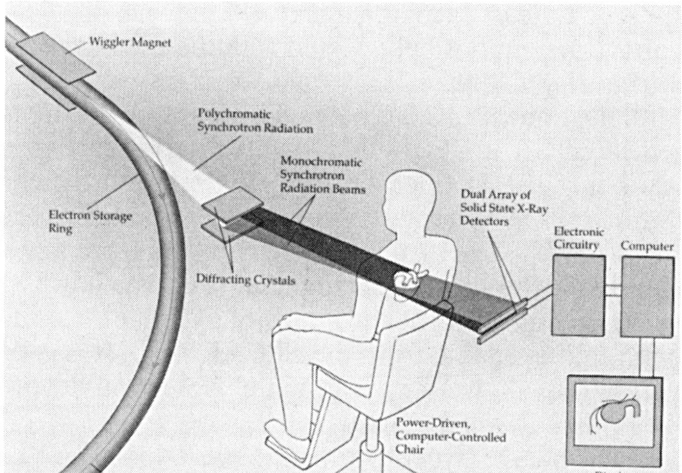

FIGURE 6.13 A system for transvenous coronary angiography using crossed X-ray beams, developed at Stanford. The polychromatic radiation is monochromatized by diffracting elements that provide two monochromatic beams with energy closely bracketing the K-edge of iodine; the beams cross at the position of the subject's heart, diverge and then impinge the dual detectors (after E. Rubenstein [6.30])

The 1% noise requirement leads to a minimum flux of 1.4 x 10^4 detected photons per pixel at each energy. If the measurement is made within 200 ms in a soft tissue 20 cm thick, the flux intensity on the patient's surface must be about 5 x 10^{10} photons mm^{-2} s^{-1}. At existing SR centers this intensity can be obtained only in wiggler-beam lines (Sec.6.2).

Fig.6.13 is a schematic of a *system for transvenous coronary angiography* developed at Stanford [6.25, 6.30]. Two initially convergent monochromatic beams of band widths of about 15-25 eV and the energy separation of 300 eV, cross at the position of the patient's heart. Then they diverge and are incident on a double linear monitor. Each beam, which is 125 mm wide in the horizontal plane, impinges on an array of 300 pixel elements of the lithium-drifted silicon detector [6.30].

The efficiency of the detector for 33 keV photons is about 70%, whereas that for the transmitted third harmonic (99 keV) is as low as 5%. The dynamic range of the electronic system is 40 000 : 1 for the readout time less than 2 ms. The pixel resolution is 0.5 x 0.5 mm^2. In order to obtain an image of the whole heart, the *patient is scanned vertically* at a speed of 12 cm s^{-1}. The exposure time of each line image is 4 ms.

The above system has been tested with the use of the SPEAR synchrotron ring [6.30, 6.31]. Fig.6.14 shows an angiogram made with the energy of 3.3 GeV and beam intensity of 50 mA.

A dedicated SR angiographic facility shown in Fig. 6.13 is currently being set in operation at the National Synchrotron Light Source (NSLS) (Fig. 6.9) at the Brookhaven National Laboratory [6.25]. The system will be supplied from a new wiggler equipped with superconducting magnets, which will make it possible to increase the beam intensity by a factor of 5 compared with that available from the SPEAR ring. It is estimated that by operating 40 weeks per year the NSLS will devote 25% of the beamtime to angiography.

In the Hamburger Synchrotronstrahlunslabor HASYLAB at DESY, the system NIKOS II (Nicht-invasive Koronarangiographie mit Synchrotronstrahlung) consists of six main parts: the 20-pole wiggler HARWI, a two-beam monochromator, a safety system, a fast scanning device, a fast low-noise two-line detector, and a computer system [6.26].

In the *two-beam monochromator* two Ge (111) crystals are used for filtering the quasi-monochromatic beams out of the

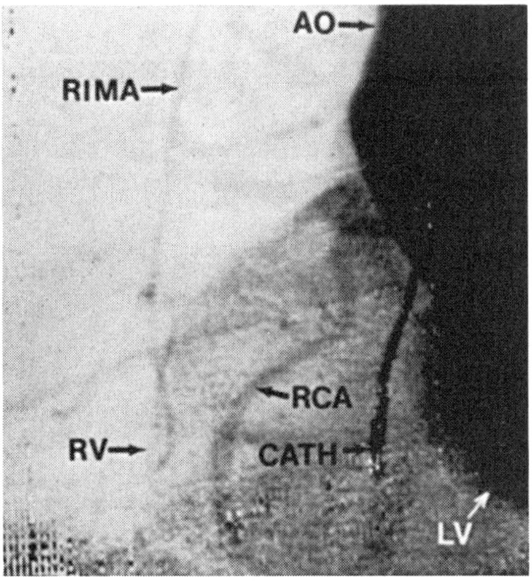

FIGURE 6.14 Transvenous synchrotron radiation coronary angiogram (left anterior oblique, 30°, with cranial angulation) of a human subject in April, 1987, showing that the right coronary artery (RCA) can be visualized through the right ventricle after the contrast agent has entered the left side of circulation (after A. Rubenstein et al. [6.30])

white synchrotron radiation beam. The bandwidth of the beams is 55 eV with a separation of δE = 350 eV. The crystals vertically split the incident synchrotron radiation beam into two parts. Each crystal is delivering one of two monochromatic beams, 10 cm x 0.7 mm. A flux of maximum 0.13 x 10^{11} photons/(mm^2 s) was measured in the monochromatic beams for a current of 44 mA in the storage ring DORIS (5.2 GeV, 2 e^- bunches). After investigation of 16 anesthetized dogs [6.59] with the system NIKOS II in May 1990, the first three human subjects were studied. Thirty ml of contrast medium were injected within 2 s into the vena cava superior. The scan velocity was 9 cm s^{-1} corresponding to 5.6 ms per line. The resulting images have a size of 10.0 cm (horizontal) x 12.3 cm (vertical). The pixel size in the images is 0.5 x 0.5 mm^2. The radiation dose per scan (two energy images) was 25 mGy on the average. Up to now the images have had bad quality and are not suited for routine investigations in a clinical environment. The main technical reasons are: (1) the intensity in the monochromatic beams is insufficient — a factor 8 is missing; (2) the fast beam fluctuations decrease the quality dramatically; (3) the detection efficiency and the dynamic range of the detector are too low. These problems must be overcome before routine medical work can start [6.26].

In addition to the improvement of the existing detection systems in various facilities [6.60 through 6.62] studies on new and interesting methods are also being carried out. The choice of iodine as a contrast agent is poor, since iodine has a K-edge at 33 keV, which is too low for good transmission of the monochromatic X-rays through the human body. It is proposed that the diagnostic quality can be improved by changing from iodine to a higher Z contrast agent, such as *gadolinium or ytterbium* [6.63] and by replacing the presently used Si (Li) multichannel detector with a kinestatic charge detector using krypton gas [6.60]. Since gadolinium has its K-edge at 50 keV and ytterbium at 61 keV, the use of these contrast agents can improve the X-ray transmission through the patient by more than one order of magnitude. At the same time, the patient radiation dose from the procedure will be reduced by at least a factor of three. Gadolinium-DTPA has already been approved by the US FDA as a contrast agent for MRI.

It seems that dichromography, with its techniques and systems fully developed, could become the most important application of synchrotron radiation in medical diagnosis.

6.6 Synchrotron Radiation Microscopy and Microanalysis

Microscopy of biomedical specimens with the use of synchrotron radiation became more popular in the early 1980s. Compared with conventional microscopic techniques, optical and electron microscopy, SR microscopy has many advantages. It is distinguished by better resolution than optical microscopy. Less energy is absorbed in the specimen than in electron microscopy, and therefore the specimen is less exposed to structural damage. It is easier to observe natural as well as wet and thick specimens. Since the image is produced as a result of differences in absorption characteristics of its elements, in SR or soft X-ray microscopy the specimen need not be stained.

Soft X-rays with wave-lengths from 0.5 to 10 nm (5 to 100 Å) are absorbed primarily due to photoabsorption. According to the Lambert-Beer law

$$I = I_0 \exp(-\alpha d) = I_0 \exp(-\alpha/\rho\, d\rho) \qquad (6.11)$$

where I_0 is the intensity of the primary nonabsorbed beam, I is the intensity of the beam after passing the material with density ρ, thickness d and the linear absorption coefficient α. Fig. 6.15 shows the spectral response of the *linear absorption coefficient* for several materials with high and low absorption of X-rays. The reciprocal of α determines the radiation penetration capability, which in the above spectral range is usually greater than 1 μm for organic materials, whereas for metals with high atomic numbers it is typically smaller than 0.1 μm. The bottom curve in Fig. 6.15, denoted by PMMA, represents *polymethylmethacrylate (organic glass)* used in the fabrication of *radiation-sensitive resists*

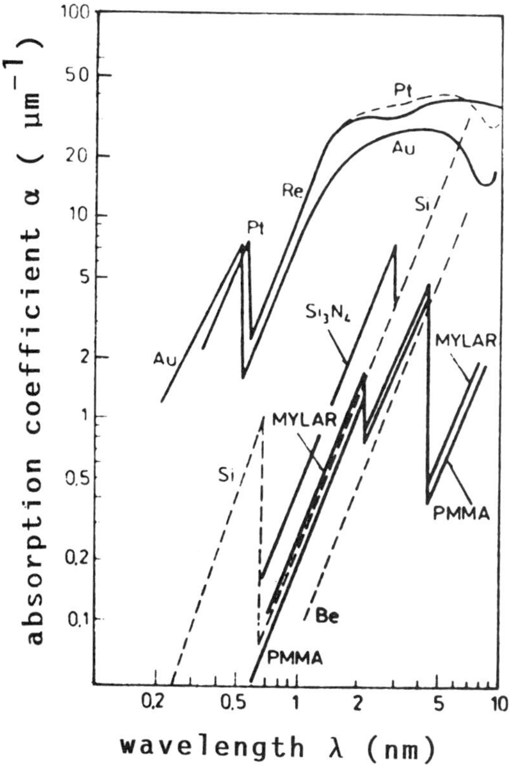

FIGURE 6.15 Spectral response of the linear absorption coefficient of soft X-rays for some materials with high and low absorption: PMMA is a curve for polymethylmethacrylate (organic glass) used in the fabrication of positive photoresists (after W.Gudat [6.37])

utilized most frequently in microscopy. Curves are also shown for Mylar and beryllium, typical materials used for windows in soft X-ray generators.

Among the techniques employed in *soft X-ray microscopy, contact microscopy* is the most prevalent [6.38]. Another technique using SR and scanning systems is that of *transmission*

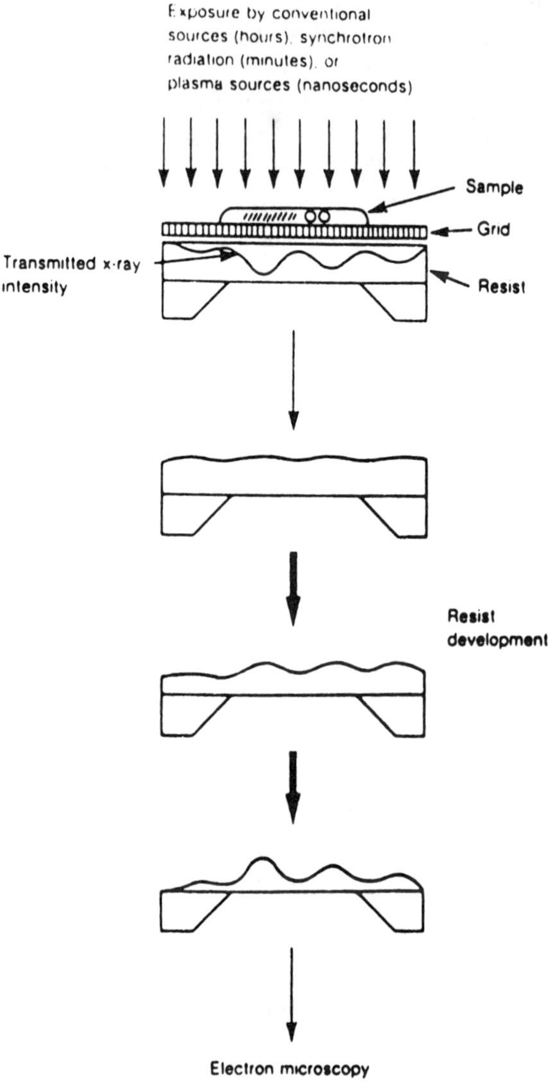

FIGURE 6.16 The principle of producing contact radiograms

microscopy, which, in addition to *structural analysis,* also makes it possible to perform *chemical analysis and mapping.* Holographic methods are less often used. Readers interested in SR microscopy are referred to the fundamental textbook [6.4, 6.5] and papers [6.39, 6.40].

Contact microscopy. The principle of producing radiograms with the *contact method* is presented in Fig. 6.16. A thin photoresist foil which is an organic polymer, usually PMMA, is placed on a polished silicon substrate. The sample to be investigated is, in turn, placed directly on the foil and then exposed to soft X-rays emitted from a storage ring. The photoresist is positive, that is, areas exposed to radiation are subsequently etched away. As a result, a relief-like replica of the sample is obtained, which can then be observed under a conventional electron microscope. In this case, the replica surface is additionally coated with a thin metal layer to avoid space charge.

The structure of the sample is replicated due to the differences in soft X-ray absorption, which is a function of the thickness of the investigated layer, the type of the material and the energy of the radiation used (Eq. 6.11). To obtain *adequate contrast and good penetration* the following expression is usually adopted

$$\mathrm{d}\alpha \sim 1 \qquad (6.12)$$

where α is the linear absorption coefficient in a sample with thickness d. Generally, the curves in Fig. 6.15 — except for the region near the absorption edges — show that the absorption coefficient decreases with the wave-length of X-rays. At a given wave-length, coefficients for various elements may differ even by a factor of 100. In the 1-2 nm (10-20 Å) wave-length range the absorption coefficients for most materials vary from 0.2 to 20 μm^{-1}. Therefore, the wave-length is chosen depending on the structure of the object investigated and the required contrast. In general, the longer the wave-length the better the contrast and the lower the penetration.

In addition, the above method permits *chemical analysis* of the object investigated. For this purpose, two pictures are taken bracketing the absorption edge (see Sec.6.5). When investigating biological samples the region between the K-edge for oxygen (λ = 2.3 nm) and carbon (λ = 4.5 nm) is used. For wave-lengths longer than λ = 4.5 nm (45 Å) all organic materials become

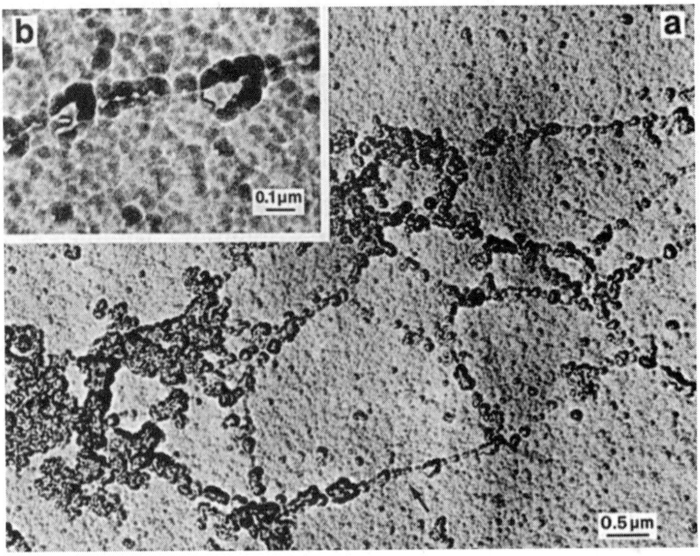

FIGURE 6.17 Radiogram of a human chromosome sample obtained by SR contact microscopy: (a) — complete image, (b) — enlarged portion (after K. Shinohara et al. [6.41])

very transparent so that heavy components contained in them can be observed. The full range of the soft X-rays utilized in contact microscopy encompasses wave-lengths between 0.5 and 10 nm (5 to 100 Å).

Fig. 6.17 shows, by way of example, the potentialities of contact microscopy. It represents an X-ray image of a stretched portion of a *human chromosome* obtained with the use of the PMMA resist exposed to 2.98 nm monochromatic undulator radiation. After etching, the radiograms were observed under a transmission electron microscope [6.41]. It follows from the enlarged portion of the image (Fig. 6.17b) that fibers with diameters of the order of some ten nm can be observed.

Scanning transmission microscopy. Fluorescence analysis. The principle of operation of a scanning microscope using synchrotron radiation is shown in Fig. 6.18 (Daresbury Laboratory). A soft X-ray beam from an undulator and a monochromator is incident on a small pin hole aperture (diameter 100 m). Then the beam strikes a Fresnel zone plate which focuses it further down to less than 100 nm in diameter. The specimen to be examined is placed in a scanning system which can be scanned in the x-y plane in 20 nm steps. The movement in the z-direction is also possible so that the position of the specimen can be matched with that of the beam focus. All the movements of the scanning system are computer controlled.

The microscope shown in Fig. 6.18 operates in the *transmission mode.* The X-rays transmitted by the specimen are counted in a gas-filled continuous flow proportional counter which registers the transmission component. The image is displayed on a black-and-white or color TV monitor. Intensities corresponding to various pixels are represented as grey levels, or in false colors.

Among the particularly attractive features of soft X-ray or SR microscopy is that of its suitability in the biological analysis of unstained or hydrated samples. By using X-rays with the wavelength below the K-edge for oxygen, (2.3 nm), relatively high transparency for water is obtained, which makes it possible to image organic constituents of cells. For example, at a wavelength of 3.2 nm the absorption coefficient in water is 0.22 μm^{-1},

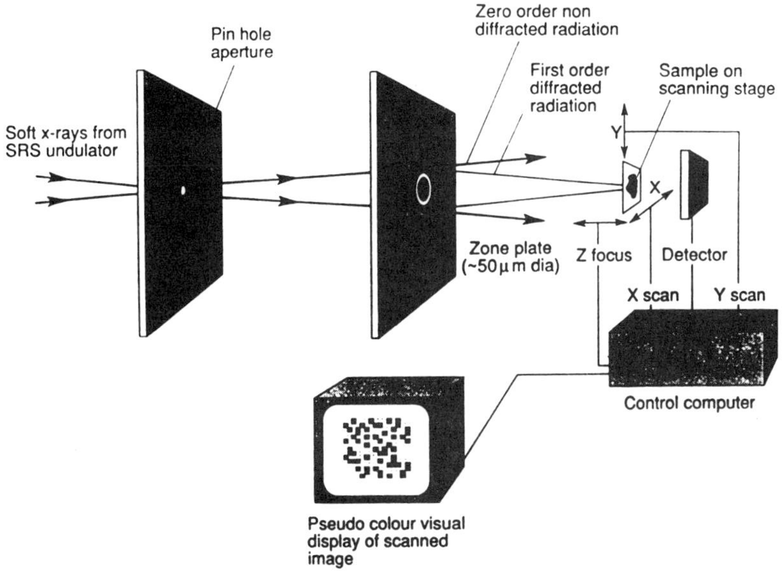

FIGURE 6.18 Schematic of a SR scanning transmission microscope (Daresbury Laboratory)

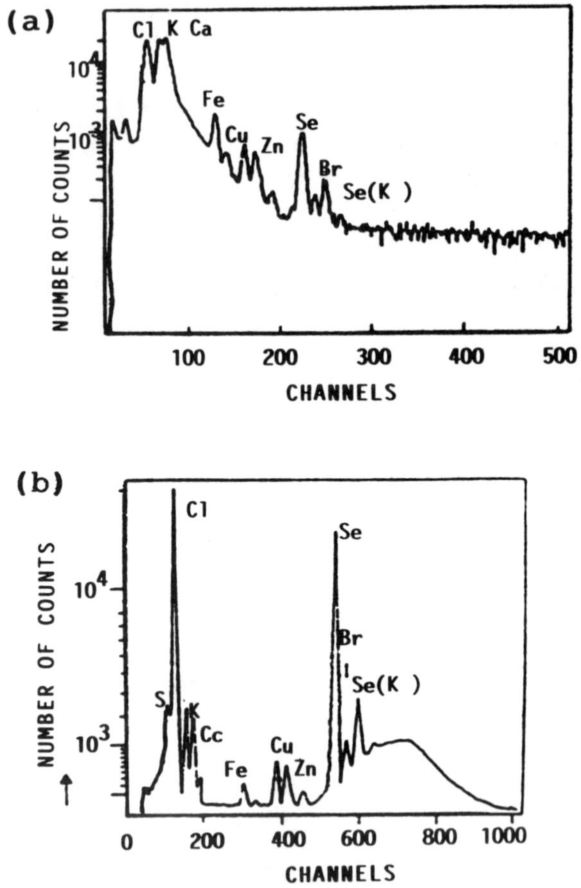

FIGURE 6.19 X-ray spectra of blood serum: (a) — excited with 2.5 MeV protons (PIXE), (b) — excited with SR of 15 keV (after W.J.M.Lenglet et al.[6.45])

whereas that in proteins or lipids is 1.9 μm^{-1} [6.46]. Therefore, the soft X-ray microscope appears well suited to imaging whole, wet, unstained and cultured cells.

SR microanalysis. In addition to structural studies, *mapping* of the composition of individual elements in specimens can also be performed with the *X-ray fluorescence (XRF) spectrometry* which is among the fundamental methods of investigating biological and medical specimens [6.51]. It consists in *inducing charac-*

teristic X-ray radiation by means of synchrotron radiation. This has several advantages over the conventional method of using X-rays from an X-ray tube or radiation from radioisotope sources [6.42 through 6.46]. Due to the high brilliance of synchrotron radiation, unobtainable from X-ray tubes (Sec. 6.2 and Fig. 6.23), better sensitivity and spatial resolution can be achieved. For example, for biological specimens and low Z elements, quantification limits are less than 1 mg kg^{-1}, and for high Z elements they are in the 1-10 mg kg^{-1} range. These values are true for 1 min irradiations [6.45]. Moreover, since SR is distinguished by strong polarization, the contributions from the Compton and Rayleigh scattering in the signal are greatly reduced. For comparable sensitivities, the energy deposited in the sample excited with SR is smaller than that released by high-energy particles, most frequently protons (PIXE, Sec. 9.2.3). Chemical analysis with the use of SR can be carried out either in air or in helium. By controlling the X-ray energies below and above the K-edge, individual elements can be excited selectively.

Fig. 6.19 shows, for comparison, the potentialities of the fluorescent analysis with SR and the PIXE spectroscopy of the emission spectra of a blood serum sample [6.45]. The spectrum in Fig. 6.19a was excited with 2.5 MeV protons, whereas that in Fig. 6.19b was excited with 15 keV synchrotron radiation, both spectra being quite different from each other. Employing monochromatic SR excitation, the background can be significantly reduced and the signal-to-background ratio S/B for a

TABLE 6.3 Comparison of XRF Microprobe Analysis Methods (after B.M. Gordon et al. [6.64])

Microanalysis method	MDL [ppm]	Resolution	Wet samples	Analysis at atmospheric pressure	Beam damage	Selective elemental excitation	Depth profiling	Chemical speciation	Crystal structure
Electron μ-probe	100	1	No	No	High	No	No	No	No
Proton μ-probe	0.5	0.5	Hard	Hard	Medium	No	Yes	No	No
Synchrotron X-ray μ-probe	0.05	5	Yes	Yes	Low	Yes	No	Yes	Yes

1 ppm sample in a low-Z matrix could be about 10-100 times higher than that with PIXE [6.65]. Moreover, micro-SRXRF *(micro-SR X-ray fluorescence)* gives about 10^{-3} less heat load to the sample for the same *minimum detection limit* (MDL).

Synchrotron radiation microprobes. The use of *SR to induce X-ray emission* (SRIXE or μ-SRXRF) for analytical purposes allows the measurement of trace elements at levels not detectable with traditional microanalytical techniques (Table 6.3).

That is why we have recently witnessed increased interest in these techniques [6.64 through 6.67]. Like in PIXE, where *proton microprobes* (Sec. 9.1) have been developed, for this method *SR microprobes* have been constructed. At present, about 10 such systems are either in operation or under development, the parameters of some of them being listed in [6.67 and 6.72].

The synchrotron radiation is either *collimated or focused* using *shaped mirrors or crystals* to produce a 5-50 μm beam spot. Using an optical microscope the sample is positioned in the beam by (x, y, z, θ) motor-driven stages, because the *beam can-*

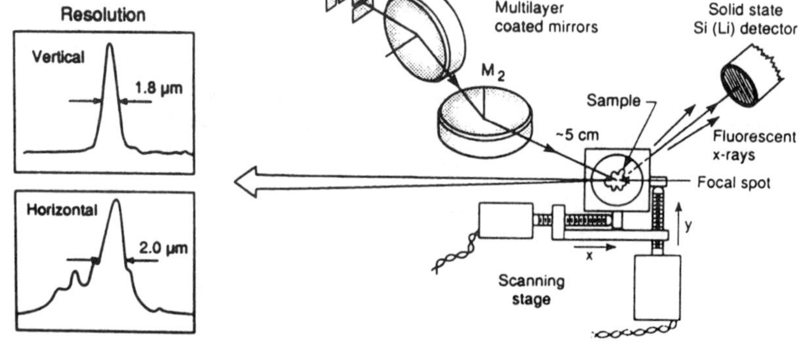

FIGURE 6.20 The National Synchrotron Light Source (Brookhaven National Laboratory, USA) hard X-rays microprobe.

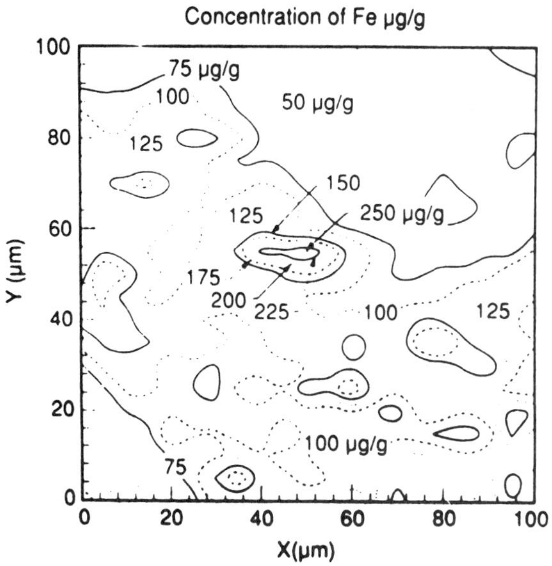

FIGURE 6.21 NSLS X-ray microprobe scan of a rat-kidney tissue section

not be moved electrostatically or magnetically, as can charged particle beams. The fluorescence X-rays and the scattered beam are detected with an *energy-dispersive spectrometer* and/or a *wavelength-dispersive spectrometer.* A major advantage of the X-ray microprobe is that the specimen does not have to be kept in vacuum or subjected to special contrast-enhancing preparation.

In order to minimize the MDL for a particular element a monochromatic beam with an energy just above the absorption edge of the element is required.

The *hard X-ray microprobe* (6-15 keV) has been used to achieve 2 μm spatial resolution (Fig. 6.20) along with femtogram elemental resolution at Brookhaven National Laboratory's National Synchrotron Light Source (Fig. 6.9) [6.68].

The heart of the microprobe is a pair of W/C multilayer mirrors (where reflectivities are enhanced by causing multiple reflections to add in phase) used at grazing incidence in Kirkpatrick-Baez geometry. They focus the beam of SR to a small, intense spot size much more effectively than would a pinhole collimator. The mirrors, coated with W/C multilayers act as broadband filters. The bandwidth (about 1 keV centered around 10 keV for these mirrors) combines good coverage with rejection of scattered background. The mirror that intercepts a projected area of 0.25 mm^2 with 50% efficiency and focused down to a spot diameter of 2 μm would pass $3x10^4$ more photons in a given time than would a 2 μm-diameter pinhole. By scanning a sample through the focused beam, the spatial distribution of many elements can be measured with excellent sensitivity (femtogram quantities of the elements K through Z can be detected with counting times in the order of 30 s).

Fig. 6.21 shows the potentialities of this microanalysis. An X-ray scan of a rat-kidney section shows the spatial concentration of iron. The contour interval is 25 parts per million (μg/g) and the peak Fe concentration is 250 ppm.

The NSLS Microprobe Facility is described in [6.64}. For biological samples sensitivities in thin samples at 10x10 μm^2 the spatial resolution is on the order of 1-10 fg.

Computed microtomography and tomography. Computed tomography (CT) is a widely used diagnostic method of obtaining cross sectional views of a patient on the basis of the acquisition of a large number of measurements of the attenuation coefficients. The radiation source for CT is most frequently an X-ray tube and the achievable resolution power is, at present, on the order of a few tenth of a millimeter.

The use of SR instead of an X-ray tube in CT offers several advantages. By taking two images above and below the X-ray absorption edge it is possible to determine a spatial distribution of a specific element. The much higher intensity of SR permits real-time imaging of moving samples, especially those which can withstand rapid rotation. As the SR photons are highly monochromatic, beam hardening artifacts will not occur. The contribution from scattered photons, i.e. from the background, is considerably reduced since the beam has a divergence of only several mrad and the transmitted radiation can be collimated so that the detector is reached by very little scattered radiation. Radiation damage to the sample can be limited to a large extent by tuning the beam energy to give the maximum signal-to-noise ratio.

Several studies have been embarked upon *SR microtomographic systems* [6.47 through 6.50] intended for imaging small objects, e.g. biopsy samples. Theoretically, systems of this kind should be useful in the analysis of hydrated samples of 100 μm^2 size with a resolution element size of 1 μm^2, ensuring sensitivity of about 5 ppm for elements from phosphorous to bromine [6.47]. In microtomographic systems already developed, the real linear resolution power approaches 1 μm [6.49,6.50].

At present, a monochromatic CT system called *multiple-energy CT* (MECT) is being developed at the National Synchrotron Light Source (NSLS, Fig. 6.9) at Brookhaven National Laboratory (BNL) for imaging the human head and neck [6.68 and 6.69]. The CT geometry will be one of a rotating seated subject in a fixed horizontal fan-shape beam (Fig. 6.22). The horizontal opening of the beam is 5 mrad, which provides a beam width of 20 cm at the patient, 40 m from source.

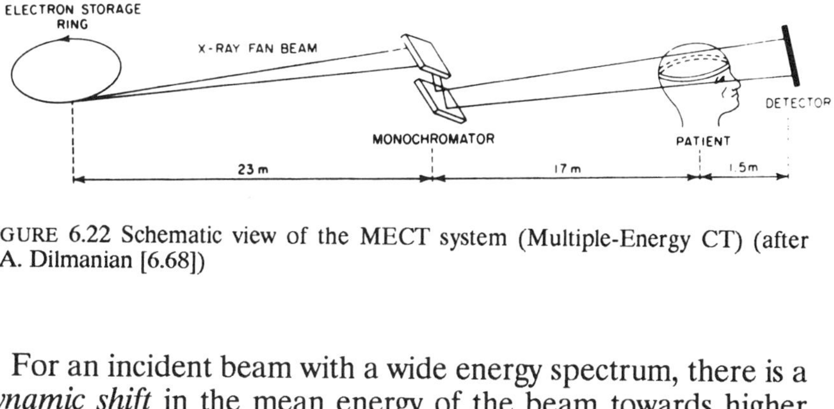

FIGURE 6.22 Schematic view of the MECT system (Multiple-Energy CT) (after F.A. Dilmanian [6.68])

For an incident beam with a wide energy spectrum, there is a *dynamic shift* in the mean energy of the beam towards higher energies as it penetrates the subject. This *spectral shift, called beam hardening,* degrades the contrast resolution of the reconstructed CT image. It is also detrimental to image sensitivity and quantification precision, and may produce image artifacts. Although beam hardening effects can be corrected to some extent by comparing the effect in the subject with that in a phantom, or by using dual-energy methods, in principle no correction can restore the image quality to that achievable with monochromatic X-rays. The typical energy bandwidth of a CT monochromator such as the one described in [6.68] is 0.1%, which is 10^2-10^3 times narrower than that of conventional CT X-ray beams (Fig. 6.23).

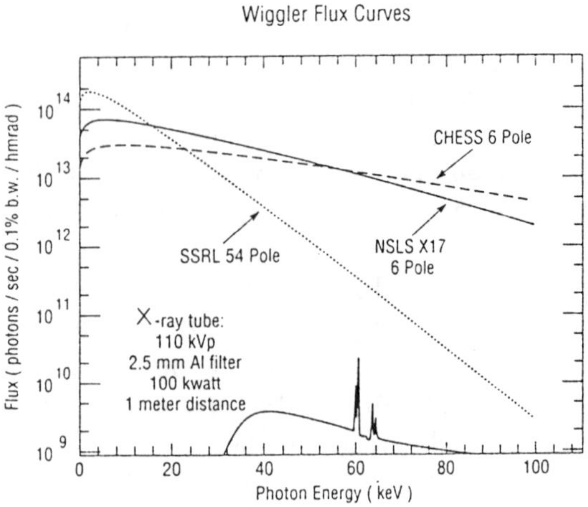

FIGURE 6.23 Comparison between the X-ray flux from the three wigglers in the US (given for maximum ring current), and X-ray flux from an X-ray tube; CHESS — Cornell High Energy Synchrotron Source, NSLS — National Synchrotron Light Source, Fig. 6.9 (Brookhaven National Laboratory), SSRL — Stanford Synchrotron Radiation Laboratory, (after F.A. Dilmanian [6.68])

The advantages of a synchrotron source of X-rays over a rotating-anode tube are the following: (1) *natural beam collimation as a divergent fan-shape beam:* this beam geometry is ideal for optimal beam utilization of the monochromator crystal, and ideal in the CT configuration itself. (2) *very high X-ray flux:* wigglers can provide monochromatic beam intensities several orders of magnitude higher than those from an X-ray tube (Fig. 6.23). (3) *small source effective spot size:* the image broadening caused by the source spot size is smaller than that of the conventional CT because the ratio of source-to-larger distance/subject-to-detector distance is much larger (the source spot sizes are comparable). (4) nearly *parallel beam*: the horizontal opening angle of the beam is typically a few milliradian; this allows positioning of the detector far behind the subject to eliminate the patient-to-detector scatter almost completely. (5) *beam energy stability:* compared to the fluctuations in the energy spectrum of beams from X-ray tubes caused by high voltage ripples, the fluctuations in synchrotron-produced X-ray spectra are usually much smaller; they can also be more easily corrected.

The MECT configuration (Fig. 6.22) is that of a fixed, horizontal fan-shape beam with a subject seated in a rotating chair. Two quantitative CT methods will be used: (a) *K-edge substraction* (KES) of intravenously administered iodine (or heavier element) to image brain tumors, large blood vessels of the lower head and neck, and arteriovenous malformations; and (b) *dual photon absorptiometry* (DPA) to obtain two brain CT images that map the low-Z elements and the intermediate-Z elements separately. This system is expected to provide 0.5 mm spatial resolution, horizontally, with unprecedented image constant and accuracy of quantification [6.68].

DPA is a method currently used for in vivo bone mineral content measurement in planar imaging and for bone mineral density measurements in CT. DPA is based on the difference in the energy dependence of the ratio of photoelectric to Compton cross sections for the two groups of elements. DPA CT of the human head will provide two separate images of the brain: an image mainly of the low-Z brain elements (i.e. H, C, N, O and Na), and one mainly of the intermediate-Z elements (i.e. P, S, Cl, K, Ca and Fe). Each of the two DPA CT images will be much more sensitive to changes in the concentration of any of its elemental constituents than are conventional CT images. In particular, the intermediate-Z image will be useful to detect neurologically significant lesions in the brain and its vasculature.

At the NSLS, the X17 superconducting wiggler and its beamline were commissioned in 1990. The flux from X17 (Fig. 6.23) is well beyond that required for the MECT project. At 33 keV, the beam must be attenuated by about a factor of 50, and at 80 keV it must be attenuated by about a factor of 5, to keep the total skin dose to the patient below 5 rad for two 5 s scans at different energies. The operating energy range extends from 33 keV (K-edge of iodine) to 100 keV (for DPA imaging).

Table 6.4 presents MECT system specifications [6.68]. A critical system component in providing a 0.5 mm in-plane resolution will be the *patient chair.* This requires that the axis of rotation at the height of the slice be fixed (i.e. wobble-free) with a precision of about ± 0.3 mm. The apparatus should allow translation of the patient about two orthogonal axes perpendicular to the fixed vertical axis of rotation. After measuring a single slice at two energies, the chair will be raised or lowered to the height of the next slice.

TABLE 6.4 MECT System Specification (after F.A. Dilmanian [6.68])

In plane spatial resolution	0.5 mm FWHM
Slice thickness	1 - 5 mm
Scan time	5 s first energy 5 s energy change 5 s second energy
Scan range	360°
Beam energy range	33 - 100 keV
Minimum angular sampling	0.14°
Minimum time sampling	2 ms
Maximum number of unattenuated 40 keV X rays/2 ms	2×10^6

The beamline in which the MECT system is being installed is the central branch (CX17B2) of the NSLS X17 beamline. This branch currently provides two convergent fan beams for transvenous coronary angiography (Sec. 6.5). A Medical Research Facility has been constructed around X17B2 beamline that will accommodate both the Angiography and the MECT project and will allow for future medical programs. The 140 m^2 (1500 ft^2) facility offers an out-patient care unit, including separate rooms for reception, fluoroscopy, imaging, and data acquisition and control. The facility is accessed from an elevator suitable for stretcher transportation. The safety system currently in place for the Angiography Project will be extended to accommodate the MECT project [6.68].

6.7 X-ray Absorption Spectroscopy and Other Techniques

EXAFS and XANES methods. X-ray absorption spectroscopy (XAS) can provide information on the coordinate geometry and electronic structure of a selected absorbing atom; this information is derived from analysis of two different regions of the XAS spectrum. The region near the onset of absorption consists of electronic transitions from core state into empty valence atomic or molecular orbitals, and also into continuum levels.

Both methods, XANES *(X-ray Absorption Near Edge)* and EXAFS *(Extended X-ray Absorption Fine Structure)* involves investigating the X-ray absorption spectrum above the absorption edge (Fig.6.24). In the range of up to about 1000 eV above each absorption edge, X-ray absorption coeffients vary within 10% or more. The range of these variations is referred to as EXAFS, whereas the 100 eV range adjacent to the edge is called XANES.

The investigation of the *fine structure of X-ray absorption spectrum above the absorption edge* (EXAFS) has become one of the widely used techniques of studying *multi-atomic complexes.* According to a generally adopted model, this structure or, in other words, *small modulations of the absorption coefficient,* is caused by scattering of D.Broglie waves of photoelectrons ejected from the K or L — orbit of the element in the sample on the neighboring atoms, followed by the interference of scattered waves with those transmitted. Since the range of photoelectrons is less than a few Å, information can thus be obtained about the nearest neighbors of a given atom, i.e. bond lengths and coordination numbers can be determined. Fig. 6.24 shows an X-ray absorption spectrum of a Cu, Zn containing protein at the Cu and Zn K-edge. In this way metalproteins, whose

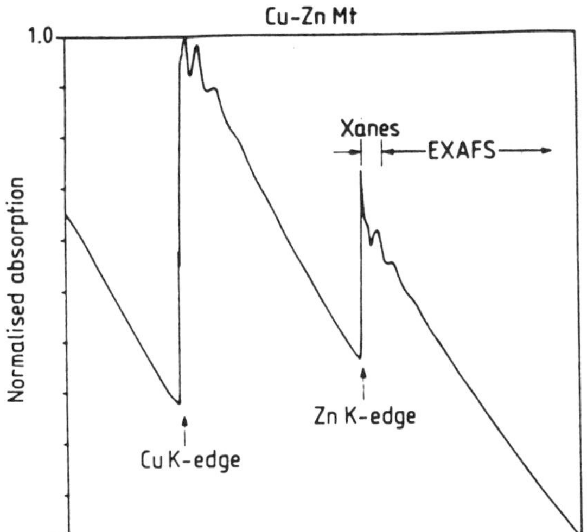

FIGURE 6.24 X-ray absorption spectrum of a protein containing Cu and Zn atoms at the K-edge (after S.S.Hasnain and C.D.Garner [6.52])

molecules usually contain 1000 to 5000 atoms with several metal atoms can be investigated.

Absorption coefficient modulation in the EXAFS method is small and decreases with the increase in the distance from the absorption edge. What is of special importance in this case is the high beam current, characteristic of synchrotron radiation (Fig. 6.23), since it enables the measurement range to be extended and the time of the analysis to be shortened down to a few seconds. Due to these advantages, EXAFS has become one of the basic methods to investigate metalproteins [6.53]. Details of the above methods can be found in [6.54, 6.55].

Other methods. Methods of investigating fine structures of absorption spectra constitute only a small fragment of a large set of techniques implemented with the use of synchrotron radiation for studying structures of biomedical samples. A more comprehensive overview of such methods is beyond the scope of this book. The reader is referred to materials of conferences held in Bristol and Frascati and published in [6.56, 6.57]. The fourth conference, next in the series will, be held at Tsukuba (Japan) [6.70] in 1992. A valuable source of information can be obtained from proceedings from national and international conferences devoted to SR equipment, for example, from the Sixth American Conference on this subject [6.58] or the 4th International Conference on Synchrotron Radiation Instrumentation [6.71].

CHAPTER 7

Production of Medical Radioisotopes

7.1 Accelerator Systems

7.1.1 Types of Accelerators

Present-day *medical radioisotopes* are produced in nuclear reactions (Sec.7.2), induced most frequently by protons or deuterons accelerated to energies between a few and several hundred MeV. The inducing beam intensities range from several tens to several hundred μA. These requirements can be satisfied by two kinds of accelerators used, rf linear ion accelerators and isochronous cyclotrons.

An example of an rf linear accelerator utilized for the production of medically useful radionuclides is afforded by the BLIP facility (Brookhaven Linac Isotope Producer), in which a 200 MeV proton accelerator serving as an injector to a large proton AGS machine was also employed "on the side" for the production of radioisotopes. Other designs of smaller linacs have also been developed [7.1]. For example, AccSys Technology [F20] has developed an 11 MeV positron isotope production linac for emission tomography (Sec.7.3) [7.2]. However, these systems are not widespread.

The present-day production of medical radionuclides — except for rf linear accelerators and some van de Graaff machines — is, above all, the domain of *isochronous cyclotrons* (Sec.2.6.2). Commercial cyclotrons for medical applications are now manufactured, first of all, by five companies: Swedish Scanditronix [F14], Japanese-French Sumitomo+CGRMeV [F25, F26], an American Company Computer Technology and Imaging (CTI) [F27] and Belgian Ion Beam Applications SA [F23]. Operating parameters of some of the facilities for the produc-

TABLE 7.1 Parameters of Some Isochronous Cyclotrons Used in Medical Radioisotope Production

Producer	Type	Protons energy MeV	Protons ext. intensity μA	Deuterons ditto MeV	Deuterons ditto μA
Scanditronix	MC17	17.2	65	8.6	65
Scanditronix	MC 40-II	10-40	100	5-20	100
Scanditronix	MC 60[1)]	15-60	50	8-30	50
CTI	RDS 114 (H^-)	11	50	–	–
Sumitomo + CGRMeV	370	10	100	10	100
Sumitomo + CGRMeV	750 PV	25-70	100	12.5-35	100
Ebco+TRIUMF[2)]	TR30(H^-)[2)]	15-30	500	–	–
Japan Steel Works	BC 105	10	60	5	60
Japan Steel Works	BC 1710	17	50	10	50
Ion Beam Applications	CYCLONE30	30[3)]	500		
Ion Beam Applications	CYCLONE10	10[4)]		5	
Ion Beam Applications	CYCLONE70	70[5)]	2000		

1) MC 60 PF type with nonvariable 62.5 MeV energy; in addition to ^{123}I production, adapted to neutron and proton therapy (carcinoma of the eyeballs).
2) Built by Ebco Technologies Ltd. [F32] with the technical and design assistance of TRIUMF, for NORDION International [7.39, 7.40]
3) H^- cyclotron
4) H^- self-shielding cyclotron for positron emitter production [7.41, 7.42].
5) Future H^- design

tion of radioisotopes manufactured by the above companies are listed in Table 7.1.

7.1.2 Energy Ranges and Beam Intensities

Cyclotron produced radionuclides for medical use are classifiable as *single photon emitters* (^{201}Tl, ^{123}I, ^{67}Ga) used for planar and tomographic imaging (SPECT — *Single Photon Emission Computed Tomography*), and as positron emitters (^{18}F, ^{15}O, ^{13}N, ^{11}C) as applied for tomographic organ imaging (PET — *Positron Emission Tomography*). Present success and future survival of clinical nuclear medicine critically depends on the availability of cyclotron produced radioisotopes [7.45].

Fig.7.1 shows the *energy ranges of protons* used in the production of medically useful radionuclides and in proton, and indirectly, neutron beams utilized in radiation therapy. In Fig. 7.1

energies necessary to induce some cyclotron-produced nuclides are also indicated. A group of positron-emitting radionuclides used in positron emission tomography (PET) can be distinguished here. Such isotopes can be produced with inducing energies on the order of a few to about ten MeV in the so-called *PET cyclotrons* which are small in size and fully shielded so that they can be installed directly in the ward or hospital without the need to build a costly shelter (vault). The construction of PET cyclotrons is discussed in greater detail in Sec. 7.1.3.

Energetically, another group of cyclotrons can be singled out, those with energies of 30 to 70 MeV. They are intended for the production of radionuclides at specialized manufacturers' (commercial cyclotrons) or for direct use in hospitals. *Commercial cyclotrons* are usually heavy, but they feature high current beams and energies variable over a relatively wide range. The 750PV cyclotron manufactured by the Sumimoto Heavy Industries (Japan) is a good example of machines of this type. It has proton energy variable over the 25-70 MeV range and the maxi-

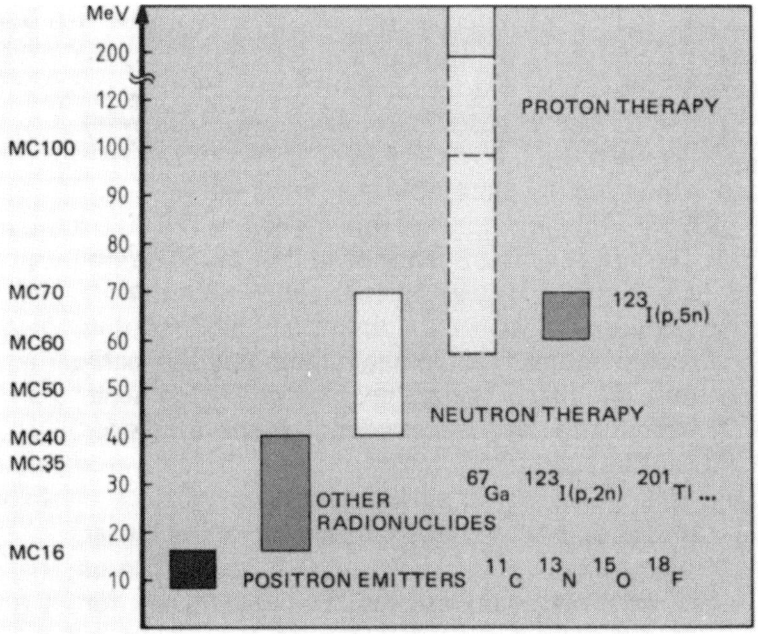

FIGURE 7.1 Energy ranges for the production of medical radionuclides and for neutron and proton therapies.

mum intensity of the internal beam of 250 μA. For installation, it requires a vault (shelter) with minimum inside dimensions of 7 x 8 x 5.5 m^3 and the thickness of the shielding walls of 4 m [7.3].

It follows from Fig.7.1 that the energy range of the above group of cyclotrons corresponds to that of protons used in the conversion to neutrons to be later utilized in neutron therapy (Sec. 5.3). For this reason these cyclotrons, when installed as medical cyclotrons, operate alternately producing either a neutron beam or medical radionuclides. According to the considerations in Sec. 5.1.1. the energy of 70 MeV corresponds to the lower limit of proton therapy, whereas the upper limit is as high as 250 MeV. Therefore, clinical facilities equipped with large isochronous cyclotrons are now being built, in which neutron and proton therapy combined with the production of radionuclides can be implemented. An example of such a facility is shown in Fig.5.24.

An essential parameter which determines the isotope yield and is important from the point of view of the manufacturer is the *beam current* (Sec. 7.1.4). In the 1970s, when only conventional isochronous cyclotrons were produced, the internal beam current was even as high as 100-300 μA, however extracted beams, more useful in production, had currents up to 50-100 μA. It was only when the fabrication of H^- commercial cyclotrons had been started (Sec. 2.6.2, Fig. 2.35b) that a solution was found to the above problem. In H^- cyclotrons, *negative hydrogen ions* are accelerated to the final energy (Fig.7.2) and then undergo *stripping* at a very thin carbon foil (0.0025 cm). As a result, the negative ion is transformed — by detachment of its electrons — into a positive proton. This method of beam extraction makes it possible to obtain yields of 90-98% of the internal beam intensity. On the other hand, in conventional isochronous cyclotrons the *external beam yield* is a mere 50 to 70%. By having the *beam extraction yield* elevated not only was the external beam current increased, but the activation of the internal walls of cyclotron parts was minimized since the losses inside the machine were reduced (Sec. 7.1.4). That is why the present-day cyclotrons used as producers of medical radionuclides are predominantly of the H^- type [7.4, 7.5]. In a variable 15-30 MeV H^- TR30 cyclotron (Ebco Technologies Ltd [F32]) intended for commercial production of radioisotopes, the internal beam intensity is 500 μA and two extraction probes

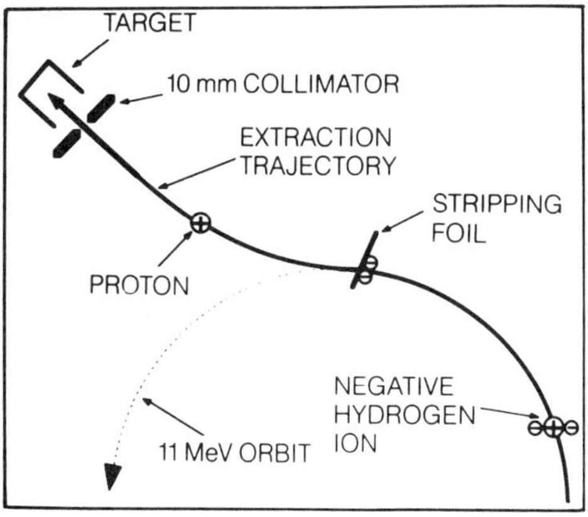

FIGURE 7.2 Schematic diagram of H^- ion beam extraction in an isochronous cyclotron

allow simultaneous extraction of two beams, each with an intensity of up to 250 μA [7.6, 7.40].

In view of the various requirements imposed on both clinical and commercial cyclotrons with regard to energy and beam intensity, manufacturers are making attempts at producing machines in several versions adapted to specific purposes. For example, Scanditronix (Sweden) produces at present an MC series with 6 types of compact cyclotrons for the use in radioisotope production and for other applications. The smallest of the series is the fixed energy MC16F machine delivering 17 MeV protons and 8.5 MeV deuterons with the external beam intensities of 50 μA. This cyclotron is designed for the production of positron-emitting radionuclides. Its total mass, together with *radiological shielding,* is 45 t and the dose rate around the machine is less than 2.5 mR h^{-1} for the beam intensity on the target of 50 μA. The most powerful of the series is the MC60 cyclotron for 15-60 MeV protons and 8-30 MeV deuterons with the external beam intensity of 50 μA. It is equipped with 160 cm diameter magnets weighing 120 t, the power supply being 120 kW. The MC60 machine can efficiently produce high purity ^{123}I by the (p,5n) reaction as well as nuclides requiring lower energies.

Twelve industrial cyclotrons in the Unites States operate almost continuously churning out radionuclides, the majority of which are compounded into *radiopharmaceuticals for nuclear medicine* [7.43]. These are long-lived machines, often operating for over twenty years. For example, CS-30 (Cyclotron Corporation) cyclotron capable of accelerating protons to energies of 26.5 MeV [7.44] was installed in 1972 at Mount Sinai Medical Center (Miami Beach, USA) and since that time it has been used for radiopharmaceutical research and development for nuclear medicine, for routine radiopharmaceutical deliveries, and for commercial sales of *radiochemicals.* The Mount Sinai cyclotron, which is configured to allow for both *internal and external irradiations,* is operated with protons only and has been used for the preparation of a number of radionuclides used in nuclear medicine (Table 7.2). There are seven available external beam lines with six operational lines and seven targets. The preparation of products for *in-house purposes* include those for research and development, both internal and external irradiations, and those for clinical delivery, both single photon and positron emitting radionuclides. The latter ones are primarily for use with positron emission tomography (PET). Internal irradiations for distribution of products outside the facility are performed to supply bulk *radiochemical grade materials* to commercial radiopharmaceutical companies. Such irradiations help provide revenue to cover some of the overall operational expenses of the facility [7.44].

TABLE 7.2 Some Radionuclides Routinely Produced at Mount Sinai for Nuclear Medicine Applications (after T.E. Bothe [7.44])

Internal irradiation	External irradiation
$^{68}Zn(p, 2n)^{67}Ga$	$^{14}N(p, \alpha)^{11}C$
$^{112}Cd(p, 2n)^{111}In$	$^{16}O(p, \alpha)^{13}N$
$^{203}Tl(p, 3n)^{201}Pb \rightarrow {}^{201}Tl$	$^{16}O(p, pn)^{15}O$
	$^{18}O(p, n)^{18}F$
	$^{124}Te(p, 2n)^{123}I$
	$^{82}Kr(p, 2n)^{81}Rb \rightarrow {}^{81m}Kr$

7.1.3 Systems for Production of Positron-emitting Radionuclides

Due to the *short half-lives of nuclides used in positron emission tomography* (PET), the systems in which they are produced are usually located in hospitals, although sometimes, when obtained outside medical clinics, they have to be transported from places a few kilometers away (Sec.7.4).

The RDS 112 accelerator, shown in Fig. 7.3a and 7.3b, is a clinical cyclotron manufactured by the US company CTI [F 27]. Suitable for producing short-lived, positron-emitting isotopes, it is compact and has a *self-shielded construction,* that is it does not require a *special vault,* since it consists of two stationary parts and four mobile parts (Fig. 7.3b) which can be moved hydraulically to allow access to the system. The self-shielding around the targets is made up of an inner layer of lead, and the remaining parts are fabricated from a mixture of polyethylene, portland cement and boron carbide. The reactions that take place in the targets during bombardment with a proton beam result in the emission of fast neutrons and hard 4.4 MeV gammas. Therefore, the 20 cm thick lead layer is necessary to absorb the gamma rays and also — through inelastic collisions — to reduce the neutron energy below 2 MeV. The lead shielding is then surrounded by a 100 cm layer of the above mentioned mixture of polyethylene, portland cement and boron carbide.

The *external dose,* that close to the external surface of the shield for a 1 curie ^{18}F production run in 110 min, is shown in Fig. 7.4 for proton energies of 8, 10.4 and 17 MeV. For the energy of 17 MeV and the external dose less than 0.5 mrem for a 1 Ci ^{18}F production the shielding should be 160 cm thick; the respective thicknesses for 11 MeV and 8 MeV should be 95 and 85 cm. On the basis of the above considerations in the production of a desired amount of positron-emitting radionuclides, the maximum energy of protons accelerated in the RDS 112 cyclotron has been chosen to be 11 MeV, and the shielding thickness 100 cm. Table 7.3 lists radioisotope yields for the RDS 112 accelerator, equipped with four beamlines to which independent targets can be inserted. Targets can be irradiated alternately, that is, two targets can be irradiated with two beams. Thus twice the amount of the same radionuclide, or two different isotopes, e.g. ^{15}O and ^{18}F can be produced at the same time.

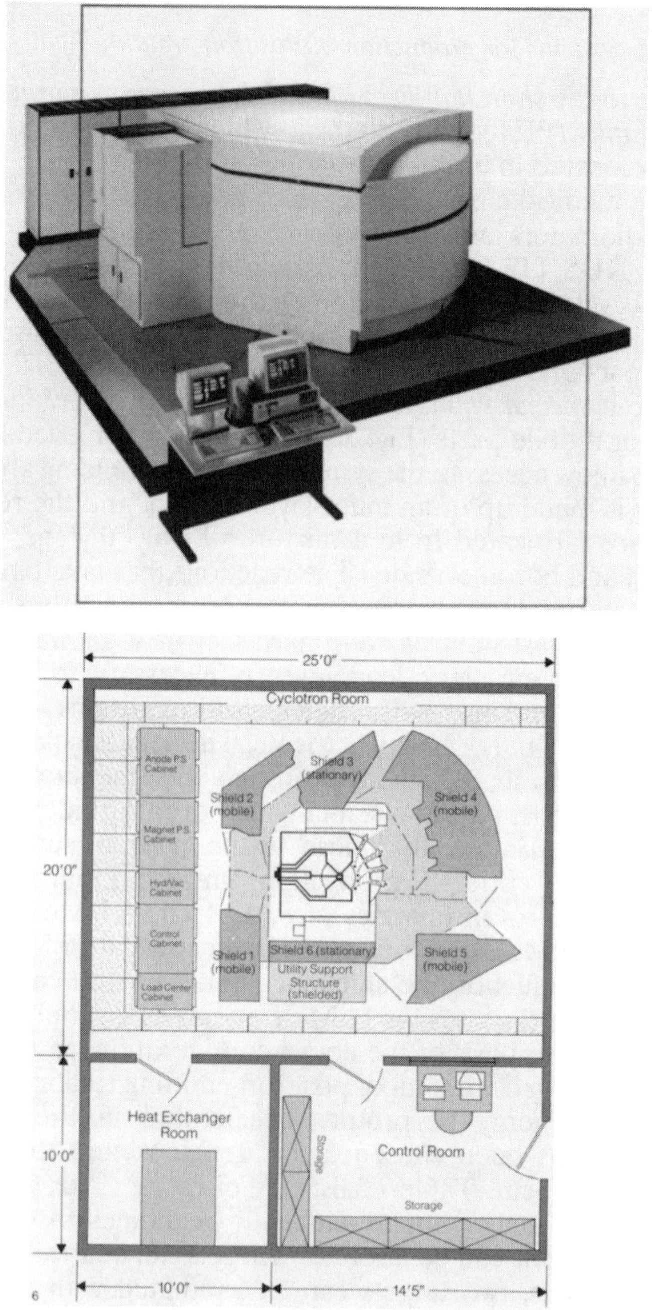

FIGURE 7.3 The RDS 112 clinical cyclotron, mfd. by CTI (Computer Technology and Imaging) intended for production of positron-emitting radioisotopes: (a) — general view, (b) — cross sectional view

TABLE 7.3 Typical Yields of Positron-emitting Radionuclides Obtained with 11 MeV protons (RDS 112 cyclotron)

Nuclide	Beam current μA	Irradiation time min	EOB activity[1)] mCi	Saturation activity mCi
^{11}C	30	40	1297	1746
^{13}N	20	20	308	410
^{15}O	20	10	750	726
^{15}O	20	10	19.0 mCi min^{-1}	–
^{18}F	20	30	410	2400
^{18}F	20	113	1170	

1) At the end of bombardment (EOB)

The RDS 112 cyclotron has a *beam extraction unit* like one shown in Fig. 7.2. H^- ions undergo stripping on 5-25 μm thick *carbon foils.* The foils are mounted on a *carousel feeder* which ensures prompt replacement, the life-time of a single foil at 50 μA being about 100 h. A 10 mm *tantalum collimator* is located 4 cm from the target. The intensity of a collimated beam may be as high as 100 μA, whereas the beams used to irradiate separate targets have maximum intensities of 50 μA.

The weight of the cyclotron itself, including targets, is 20 t, and that of the shielding and the electronic system is 18 t. The floor loading in the cyclotron room is 11.4 t m^{-2}, and in the electronics room it is 0.85 t m^{-2}. The cyclotron requires 38 m^2 floor space and another 4.5 m^2 for the remotely controlled heat exchanger. Input power is 100 kVA; 7.5 kW of heat is dissipated by air and 60 kW by water cooling.

Apart from the RDS 112 cyclotron, the CTI manufactures another system, model 217, for the energy of 17 MeV designed also for the fully automated production of positron-emitting radionuclides. Two other larger systems for 30 and 45 MeV are intended for isotope production and research in the field of physics and radiochemistry as well as for neutron activation analysis.

Details of the equipment necessary for the production of positron-emitting isotopes are discussed in Sec. 7.3.

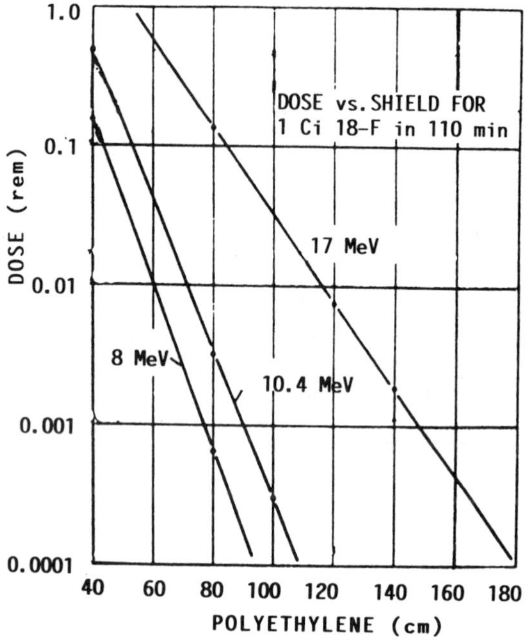

FIGURE 7.4 Dose vs. polyethylene shield thickness for 1 Ci ^{18}F production run in 110 min (RDS 112 cyclotron)

Significant progress in the construction of clinical cyclotrons for the production of PET radioisotopes will probably involve the introduction of systems equipped with superconducting magnets. Oxford Instruments Ltd [F28] and Amersham International [F29] have already designed such systems which have no internal yoke in the magnetic circuit [7.5]. In the version for the production of PET radioisotopes, the proton energy was established at 12 MeV with the external beam current of 100 μA. Protons would be extracted from an orbit of a mere 200 mm radius, which would enable the accelerator to be housed in a shielding with 1 m external diameter. The weight of the cyclotron is expected to be about 1250 kg and liquid helium used for cooling would be refilled every three months [7.5].

7.1.4 Systems for Commercial Production

^{67}Ga, ^{111}In, ^{201}Tl, ^{123}I can all be made in *multicurie quantities and at pharmaceutical purity* using 25-30 MeV proton bombardment for reactions detailed in Table 7.4. For production rates competitive with commercial manufacture, reactions (1), (2) and (3) require beam currents of 150-200 μA. Such currents are available with the internal beam of positive ion cyclotrons but not with external beams, and they have been achieved with external beam from negative ion machines. Reaction (5) performed using a gas target, is restricted to external beams with beam current limited by target window cooling to 50-100 μA. Reaction (4) can be carried out using either internal or external targets, but the radioisotopic purity of ^{123}I produced is much better using an external target where the length of beam path within the target can be more accurately controlled. Reaction (6) is included because ^{123}I is being commercially produced using such a large cyclotron, but the high proton energy is beyond the capability of any available noncommercial machines. Reaction (5), however, produces pure ^{123}I at competitive rates.

In commercial production of a large variety of medical radionuclides featuring high degree of radioactive purity, *large isochronous cyclotrons* with proton maximum energies from 30 to 40 MeV and beam intensities between 150 and 300 μA are usually employed. Cyclotrons of this kind have their energy *variable* over a wide range, e.g. between 11 and 42 MeV, to adapt it to some specific purposes. The machine installed at TRIUMF (Canada) operates in the above energy range for 100 h per week at a beam current of up to 200 μA and with an average beam delivery of 10 mAh per week. During 90% of the

TABLE 7.4 Radionuclides Commercially Produced with Proton Beams (after J.F. Lamb [7.43])

Reaction		Energy (MeV)
(1)	$^{68}Zn(p, 2n)^{67}Ga$	12-28
(2)	$^{112}Cd(p, 2n)^{111}In$	12-28
(3)	$^{203}Tl(p, 3n)^{201}Pb \rightarrow {}^{201}Tl$	17-28
(4)	$^{124}Tep, 2n)^{123}I$	20-25
(5)	$^{124}Xe(p, 2n)^{123}Cs \rightarrow {}^{122}Xe$	
	$^{124}Xe(p, pn)^{123}Xe \rightarrow {}^{123}I$	20-25
(6)	$^{127}I(p,5n)^{123}Xe \rightarrow {}^{123}I$	45-68

time it is used for commercial radioisotope production (Atomic Energy of Canada, Ltd.) and during 10% for the production of short-lived positron emitters sent via a fast pneumatic capsule transfer system to the positron emission tomograph at a hospital located 2.5 km from the cyclotron [7.4], (Sec.7.3.2).

In view of the above energy range and the high beam intensity required, cyclotrons employed in commercial production are provided with 50-100 t magnets supplied with 100 to 150 kW. Their supply units operate in the 20-30 MHz frequency range and require additional 100 kW power supply.

The 750PV commercial cyclotron designed and constructed by two companies Sumitomo and CGRMeV [F25, F26] accelerates protons over the variable 25 -70 MeV energy range. The upper energy limit was elevated beyond the usual range (see Table 7.1 and 7.4), i.e.to 70 MeV to enable the production of high purity ^{123}I [7.3]. The maximum intensity of the external beam depends on the final energy and for 25-40 MeV protons is as high as 100 μA, whereas for 70 MeV protons it is only 65 μA. In addition to protons, deuterons, ^{3}He ions and alpha particles can also be accelerated to energies of 12.5 to 35 MeV, 20 to 90 MeV and 25 to 70 MeV, respectively. The weight of the 750 PV cyclotron electromagnet is 115 t, and its pole diameter is 1.8 m. For installation, the cyclotron vault minimum inside dimensions are 7 x 8 x 5.5 m^3; the power supply room has minimum dimensions of 8 x 8 x 3 m^3 and the control room is 4 x 4 x 2.5 m^3. The total power consumption is 350 kW and cooling water consumption is 1,500 l min^{-1}. Cyclotrons of this type are employed at commercial facilities such as MediPhysics (USA), and Nihon MediPhysics (Japan) [7.3].

Details concerning installation and operation of H^- cyclotrons for commercial use in medical radioisotope production can be found in [7.4], whereas the construction of an H^- machine is discussed in [7.6, 7.39 and 7.42].

The development of present-day cyclotrons is reviewed at tri-annual international conferences. The 11th such conference was held in Tokyo (Japan) in 1986, and the latest one in Berlin in 1989. The proceedings from these conferences are presented in [7.7, 7.8]. Cyclotrons are also covered by a large number of seminars, of which mention can be made of one held in 1988 by the Svedberg Laboratory [7.9]. Symposia on the medical applications of cyclotrons are arranged in Europe also by the

Turku Medical Cyclotron Center (Finland), the latest having been held in 1989 [7.10]. The next (6th) Symposium was held in June 1992 [7.46].

7.2 Fundamentals of Medical Radioisotope Production

The main causes of the rapid development of medical radioisotope production in cyclotrons and other accelerators have already been discussed in Sec.1.4 (Table 1.9), and the list of isotopes produced is given in Figs. 1.21 and 1.22. Now we shall be concerned with the fundamental methods of radioisotope production, whereas the growth in the production of positron emitters will be covered in Sec. 7.3 and commercial production in Sec. 7.4.

Range of applications. The present-day medicine has available a wide range of diagnostic methods designed for studying the function of various body organs by following the kinetics of radiolabeled tracers. The first step in the elaboration of a diagnostic method is to use *biochemical identification* to determine the *metabolic pathway* which affects the disease process in question. Then a compound or radiolabeled substrate which would be suitable for investigating that metabolic pathway has to be identified. If this compound is labeled with such radionuclides as ^{3}H, ^{14}C or other *pure beta emitters,* their tissue distributions must be determined by *invasive biopsy or autoradiography;* if with ^{99}Tc, ^{123}I or other gamma emitters distributions in tissues can be quantitatively imaged with *conventional scinticameras;* finally, if labeled with ^{11}C, ^{13}N, ^{15}O, ^{18}F or other positron emitters, distributions in tissues can be quantitatively determined in transverse sections by *positron computed tomography* (PCT).

Both in the first and third case *genuine biological compounds are labeled,* whereas in the second and third cases *non-invasive localization in vivo* is possible.

Out of the wide range of cyclotron-produced medical radionuclides the following radioisotopes have recently found widespread use in thyroid and heart scanning as well as in tumor localization: ^{123}I (half-life 13.3 h), ^{201}Tl (half-life 73.5 h), ^{67}Ga (half-life 78.26 h) and ^{111}In (half-life 2.8 h). Iodine-123 is also used in the synthesis of various radiopharmaceuticals for kidney and brain studies and in the localization of the carcinomas of the lungs, breast, etc. *Radiopharmaceuticals, i.e. labeled compounds for medical use,* must meet several strictly defined pharmacological requirements, i.e. radiochemical, radionuclide and chemical purity. Table 7.5 is a list of some products derived from the three main nuclides and their uses. The paper [7.45] reviews the modern application of cyclotron produced radionuclides in clinical nuclear medicine for cardiology (myocardial scintigraphy), oncology (tumor scintigraphy), endocrinology (thyroid scintigraphy), dermatology (malignant melanoma), neurology (brain tumors), AIDS, nephrourology and for PET-imaging in neuropsychiatric diseases.

Parameters of nuclear reactions. Radionuclides can usually be produced in several different reactions induced by a beam of accelerated particles. The *excitation function* for a given reaction, i.e. for a given nuclide and type of charged particles, is frequently presented in the form of a relationship between the cross section and the energy of the inducing particles. Fig. 7.5 shows such a function for the ^{14}N (p,α) ^{11}C reaction used for production of carbon-11. The dashed curve has been drawn from the experimental data reported by various authors. Significant amounts of carbon-11 can be obtained with proton energy as low as about 6 MeV, but the maximum cross section is about 250 mb at a proton energy of 7.5 MeV. The energy range is relatively wide since large values of the cross section are found at an energy as high as 22 MeV.

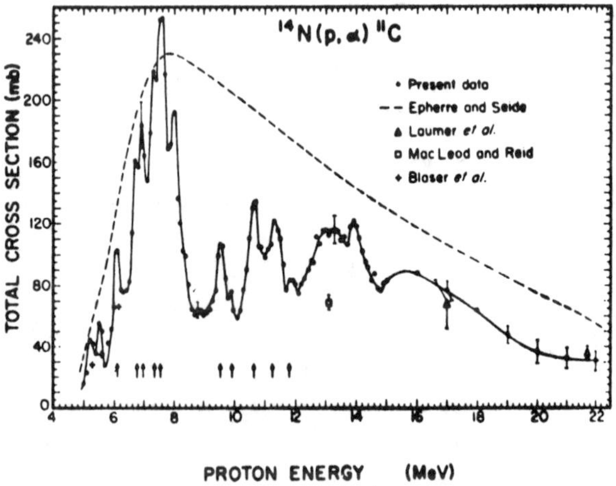

FIGURE 7.5 Total cross section for the $^{14}N(p,\alpha)^{11}C$ reaction as a function of proton energy (after W.W. Jacobs et al. [7.47])

TABLE 7.5 Products Derived from the Three Main Nuclides Produced by the Amersham Cyclotrons (after [7.38]).

Isotope	Product	Main use
gallium-67	gallium citrate	For imaging of inflammation and soft tissue neoplasm
indium-111	indium chloride	For protein labeling and a preparation for cell labeling
	indium oxine	For leucocyte and platelet labeling
	indium-calcium DTPA	For cerebrospinal fluid studies
	indium-bleomycin	For non-specific abdominal, thoracic and pelvic tumour localisation
thallium-201	thallous chloride	For myocardial perfusion studies in ischaemia

Parameters of reactions used by various researchers for the production of cyclotron radionuclides based on experimental data are listed in [7.11]. In addition to the type of reaction, the paper also lists the target materials, the degree of isotope enrichment, the energy of the bombarding particles, and the production yield expressed either in terms of mCi per 1 μA of the beam or per Ah of the beam current. For example, for the above mentioned ^{11}C (half-life 20.4 min), as HCN, the target material was 94.5% N_2 and 5.5% H_2 in the reaction $^{14}N(p,\alpha)^{11}C$. For the bombarding proton energy of 18 MeV the production yield was 140 mCi μA^{-1}. In the production of short-lived isotopes another parameter is the so-called *saturation activity* which is achieved after at least 4-5 times the half-life of the isotope being produced. The experimentally measured yield at the end of the bombardment is denoted by the symbol EOB *(end of bombardment)*. Tables in [7.11] provide a very good basis for the selection of a suitable reaction and range of bombarding particles.

Protons, which are most often used as inducing particles, exceed deuterons and alpha particles in respect to range. As a result, for protons the yield is higher and the beam power released in the target is reduced. This plays a decisive role in the large-scale production of radioisotopes, where reliability of the isotope producing system is of prime importance. In the case of protons,the reactions used are those of the (p,x) type over the 10-70 MeV energy range.

Other factors that have an essential effect on the isotope production process are the physical and chemical properties of the target material and those of the target containment system material. In view of the intensive *heat emission due to energy losses* in the target and in the target containment system material some physical properties can undergo change and the chemical reactivity among the various parts of the system may increase. These changes may be reflected in the target thickness and, consequently, in the reduction in the isotope yield. They may also lead to some variations in the *target radionuclide-yield ratios,* which affect radionuclide purity levels, and may also cause unwanted chemical reactions. Therefore, proper design of the target system must take into consideration all the accelerator parameters.

All the procedures of the isolation, purification and formulation into a desired radiopharmaceutical are carried out using *standard chemical methods.* What is often employed is a combination of precipitation, co-precipitation, solvent extraction, ion-exchange, electrolysis and chromatographic techniques. *Radiochemical processing* of the target often requires that time sequences of some procedures be carefully controlled because of the time-dependent changes in the composition of the radionuclide mixture. This is a particularly critical factor in the optimization of the production yield and purity in indirectly obtained nuclides, such as ^{123}I from ^{124}Xe [7.48, 7.49], and ^{201}Tl from ^{201}Pb or ^{201}Bi. The selection and sequence of radiochemical processing is strongly dependent on the type of the nuclear reaction and on the physical and chemical characteristics of the target and its containment system, as well as on the chemical properties and reactivity of the radioisotopes produced. Extremely useful and relatively simple in implementation are *physical methods,* such as vacuum distillation, vacuum deposition and phase changes.

The *radiochemical purity of a radiopharmaceutical* is one of the cardinal factors which determine its efficacy since it may affect the biological distribution and specificity of the drug. This is why methods employed in medical radioisotope production should make it possible to obtain a radionuclide in its *purest possible form,* especially without stable nuclides or those acting like *carrier or carriers.* Carriers or diluents can originate from various sources during irradiation, where stable nuclides are produced either indirectly or during the decay of the induced radioactivity. They can also be present in the target material or, in trace amounts, result from low-efficiency radiochemical separations, and, finally, can be found in reagents used. From the point of view of the synthesis of radioactive substances, nuclides of *high specific activity* (usually in terms of Ci g^{-1} or mCi mg^{-1}) or of *high concentration* (mCi ml^{-1}) should preferably be employed. High specific *activities* may be obtained by using *isotopically enriched target materials* and highly-purified reagents, and by maximizing production yields during all stages of the production process.

In what follows, methods of the production of ^{123}I will be presented to exemplify the problems involved in the production of the most popular medical radioisotopes.

Production of ^{123}I. The considerable interest taken in the production of ^{123}I results from the fact that ^{131}I obtained in reactors has become the most widely medical radioisotope in the world. However, ^{123}I has two advantages over ^{131}I. First, the patient radioactive burden is smaller by a factor of 50 to 100 (Table 7.6), and second, if used in scanning, the quality of images obtained is better. In addition, three-dimensional images can be made by *single photon emission computed tomography* (SPECT) in metabolic studies, which cannot be implemented with ^{131}I.

TABLE 7.6 Fundamental Parameters of ^{123}I and ^{131}I Radionuclides (^{124}I and ^{125}I, as production contaminants, are also included).

Nuclide	Half-life [h]	Gamma energy [keV]	Thyroid dose [mrad Ci^{-1}]	Total body dose [mrad Ci^{-1}]
^{123}I	13.3	159	20	0.07
^{131}I	193	660, 780	1200	0.2
^{124}I	101	511, 1690	1200	2.0
^{125}I	1440	35, 27	1200	2.0

Iodine-123 has a half-life of 13.3 h and it emits 159 keV (8%) gamma rays; its decay scheme, compared with that of iodine-131, is shown in Fig. 7.6. ^{123}I decay energy matches the detection sensitivity characteristics of the present-day gamma cameras very well. Due to the well-known chemical properties of iodine a wide range of ^{123}I-labeled compounds can easily be produced for many applications. As sodium iodide it is employed in thyroid studies,whereas incorporated in hippuran it finds use in renal investigations. ^{123}I-labeled amides and diamides are used in brain perfusion scintigraphy, whereas several agents labeled with that nuclide are intended for myocardial studies [7.17, 7.18].

Production and application of ^{123}I have been described in many papers [7.13 through 7.19, 7.48]. The production methods can be classified into *direct and indirect* techniques. In the *direct method*, ^{123}I is produced by bombarding an enriched (over 95%) tellurium target in the form of tellurium oxide with a proton beam according to the reaction

$$^{124}\text{Te (p,2n)}\ ^{123}\text{I} \qquad (7.1)$$

which can be produced with protons with energies on the order of ten MeV, i.e. accelerated in small cyclotrons designed for the production of positron-emitting nuclides. Unfortunately, in addition to ^{123}I, other iodine isotopes of ^{124}I and ^{126}I are produced,

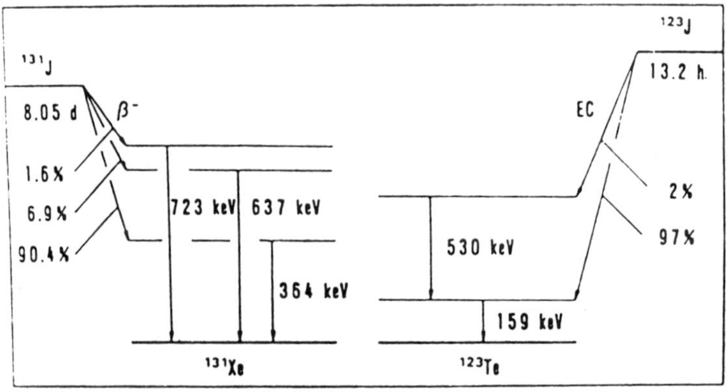

FIGURE 7.6. Comparison of physical properties between ^{131}I and ^{123}I (after V. Bechtold et al. [7.19]).

totaling a few percent [7.16]. High-energy photons emitted by those nuclides impair the quality of scintigraphic images and increase the patient absorbed dose (Table 7.6).

In one of the *indirect methods,* ^{123}I is obtained from the decay of its precursor ^{123}Xe (half-life 2.1 h). By bombarding a target of elemental iodine or potassium iodide with 50-70 MeV protons (Crocker Nuclear Laboratory and Brookhaven National Laboratory) ^{123}Xe is produced by the reaction

$$^{127}I\ (p,5n)\ ^{123}Xe \rightarrow\ ^{123}I \tag{7.2}$$

The contamination of the final product with ^{125}I is 0.1-0.4%

To produce high purity ^{123}I (radionuclide purity 99.8%) two indirect methods have been developed, in which the initial nuclide is ^{124}Xe (AECL)

$$^{124}Xe\ (p,2n)\ ^{123}Cs \rightarrow\ ^{123}Xe \rightarrow\ ^{123}I \tag{7.3}$$

$$^{124}Xe\ (p,pn)\ ^{123}Xe \rightarrow\ ^{123}I \tag{7.4}$$

In this method, enriched ^{124}Xe gas targets are irradiated with 75-100 μA currents and the proton beam energy may be about 42 MeV or less. Therefore, those methods can be implemented with commercial cyclotrons (Table 7.1). Impurity levels are as follows: ^{124}I and ^{126}I less than 5×10^{-5}%, ^{125}I less than 1×10^{-3}% and ^{121}Te less than 5×10^{-3}%.

In the method employing reaction (7.3), the intermediate nuclide ^{123}Cs decays according to the scheme

$$^{123}Cs \xrightarrow{5.9\ \text{min}}\ ^{123}Xe \xrightarrow{2.08\ \text{h}}\ ^{123}I \tag{7.5}$$

Unlike the reaction of the same (p,2n) type with ^{124}Te (Table 7.6), no contaminant ^{124}I is produced in reaction (7.5).

In order to obtain high purity ^{123}I with small contamination of ^{125}I, i.e. below 5×10^{-5}, the initial highly enriched ^{124}Xe (99.8%) must be used (Karlruhe [7.19]). Using a 30 MeV proton beam of 50 μA intensity (1.5 kW beam power), the activity produced in reaction (7.3) may vary between 300 and 500 mCi h^{-1}. In other words, the daily yield could then be 2-3 Ci. The advantage of this method is that the nuclide obtained features high degree

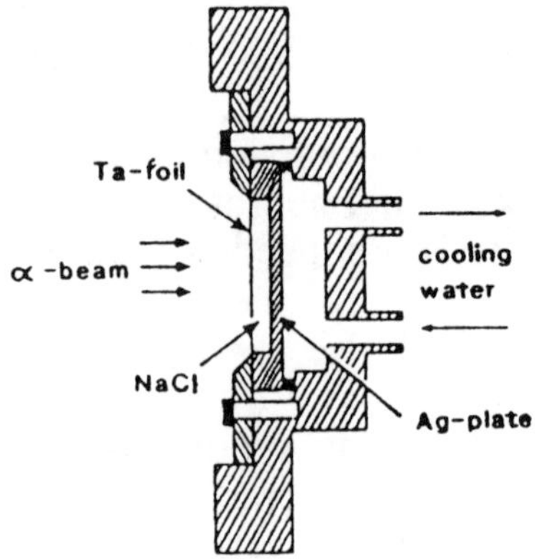

FIGURE 7.7 Solid sodium chloride system for the production of ^{38}K (after C.Vandecastede and K.Strijckmans [7.21])

of purity and the production costs are reduced by employing a relatively small cyclotron [7.19].

Construction of production targets. Targets are an essential component in the radioisotope production system [7.20 through 7.26]. The target usually consists of: (1) a *window* separating the cyclotron vacuum from the target material, (2) the *target material* (solid, liquid or gas), (3) the *holder and an electrically isolated beam absorber* to measure the beam intensity, (4) a *cooling system* for the window, target and beam absorber to carry off excessive heat.

Solid targets material can be irradiated with internal or external beams. The *internal target systems* make it possible to use higher beam currents compared with *external beam systems* since beam extraction efficients are avoided.

The target material should be distinguished by high stability, high melting point and good heat conductivity at the interface between the target and the backing plate. Fig.7.7 gives one of the solid target designs with sodium chloride for the production of ^{38}K by the ^{35}Cl (α,n) ^{38}K reaction [7.21]. The window is a 50 μm tantalum foil which reduces the alpha-beam energy from 29 to 21 MeV. The sodium iodide layer is placed on an Ag substrate. The Ag plate is mounted on an aluminum water-cooled target holder. The maximum beam current may be as high as 20 μA.

The disadvantage of a solid target is that a new target is needed for each irradiation. If the target material contains an enriched initial isotope, it has to be recovered for re-use. In many cases, however, e.g. in the production of ^{123}I, ^{75}Br and ^{211}At, separation can be carried out by the dry distillation method. In this case, the same target can be used repeatedly.

Gaseous end-products can be removed from solid or molten targets by a *continuous flow technique.* For example, when ^{123}I is produced via the ^{123}Xe precursor by the method of sodium iodide irradiation described above, the end-product is removed by a helium flow over the NaI target. Advances in the technology of solid targets are discussed in [7.25].

Two positron-emitting isotopes, ^{18}F and ^{13}N, commonly produced using *liquid targets* by the $^{18}O(p,n)^{18}F$ and $^{15}O(p,\alpha)^{13}N$ reactions (Sec.7.3) can be mentioned here by way of example. The sketch of the target is shown in Fig. 7.8. A thin layer of $H_2\,^{18}O$ is placed between two 50 m thick titanium foils. The total water volume is 0.9 ml. The input foil is cooled by helium at -4°, whereas the output foil by water at 7°C. Both foils are welded to the target body by an electron beam so that the use of other seals is eliminated.

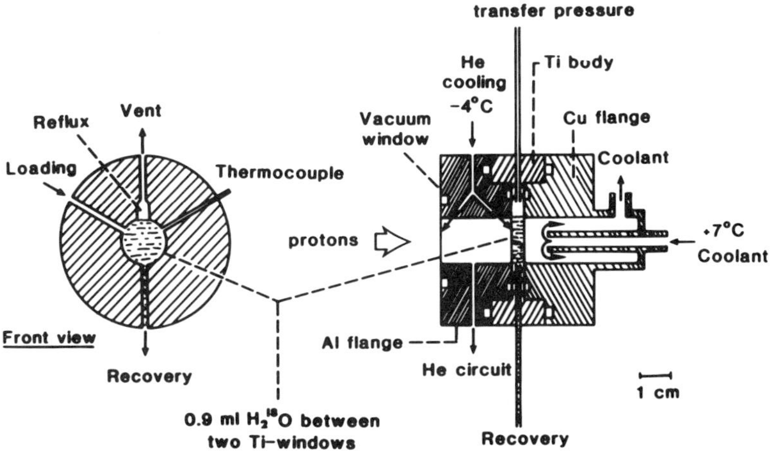

FIGURE 7.8 A diagram of a liquid $H_2\,^{18}O$ target (after S.M.Quaim et al.[7.25])

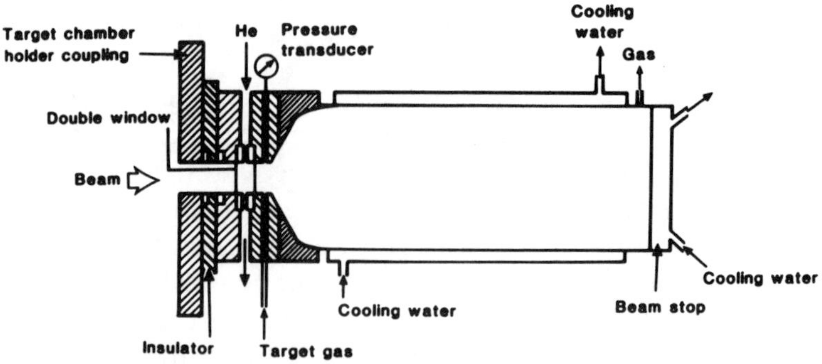

FIGURE 7.9 A gaseous target (after M.Sajjad and L.M.Lambrecht [7.25])

The target like one shown in Fig. 7.8 with a static material may be irradiated with a beam current not higher than 10 μA. Higher production yields can be achieved in flow-type designs.

Gas targets are commonly used for the production of such positron-emitting nuclides as ^{11}C, ^{15}O, ^{15}F, and ^{81}Rb. The construction of a target of this type is given in Fig. 7.9. It consists of a beam collimator and two windows. The first window separates the cyclotron vacuum, and the other separates the irradiation chamber proper. The target containment system is usually cylindrical or conical in shape. The irradiation chamber ends with an electrically insulated beam absorber. Both the target containment and the beam absorber are cooled by circulating water, whereas the windows are usually cooled by helium flow.

In gaseous targets, ^{18}F is produced in three forms, i.e. as $[^{18}F]\ F_2$, $H^{18}F$ and $^{18}F_{aq}$. For the production of $[^{18}F]\ F_2$ a mixture of 0.18% F_2 and high purity neon is used. Details of the technique is presented in [7.22 through 7.26].

7.3 Production of Positron Emitters

7.3.1 Positron Tomography

The main idea behind *positron tomography* is that a chemical substance, being a radionuclide of high biological activity associated with positron emission, is introduced into the patient's body. The emitted *positrons* become totally thermalized within 10^{-10} s over the region about 4 mm in diameter. Then, as a result of quasi-free collisions with one of the free electrons, they undergo *annihilation,* during which two gamma photons are emitted per one event, carrying an energy of about 511 keV. The angle between the two photons is almost 180°, and actual deviations do not exceed 15 mrad. The gamma photons travel through the surrounding tissues and are detected by a *ring-like system of detectors,* which operate in coincidence, and thus decay events which occur along the line connecting two detectors can be determined. Then, for each of these lines of known coordinates a so-called *radial sum* is determined which is a projection of a two-dimensional radionuclide distribution along a given direction. Hence a linear estimation in the integrals of the distribution in a given direction is obtained. With such estimations in all the selected directions the total unknown *radionuclide distribution* can be reconstructed.

Details of positron emission tomography can be found in the textbook [7.27], whereas advances in the construction of computed tomographs are discussed in [7.28 through 7.30]. The spatial resolution of the latest generation of PET tomographs is as good as about 3 mm [7.28] or even better [7.29].

Proper interpretation of images obtained by the PET modality provides a possibility of determining a large number of biochemical processes important in the function of the organ visualized. In addition to being able to precisely localize labeled compounds introduced into the patient's body, positron tomography has an important advantage over other conventional imaging modalities employed in nuclear medicine in that it utilizes radionuclides which are biologically highly essential since biological systems consist mainly of carbon, nitrogen, oxygen and hydrogen compounds. Carbon, nitrogen and oxygen have *short-lived positron-emitting radioisotopes,* such as ^{15}O, ^{13}N and ^{11}C, which are particularly useful in positron tomography. Hydrogen, unfortunately, has no positron-emitting isotopes, but in tomographic studies water can be labeled with ^{15}O. Moreover, several ^{18}F-labeled compounds are being used in PET since the C-F bonds is analogous to the C-H bond in organic molecules. Thus the set of four fundamental radionuclides used in PET consists of ^{11}C, ^{13}N, ^{15}O and ^{18}F (Table 7.7).

However, there is a range of *medium half-life* (a few hours to a few days) *positron-emitting radionuclides* which could be also employed as tracers for use with PET [7.51]. The use of radio-

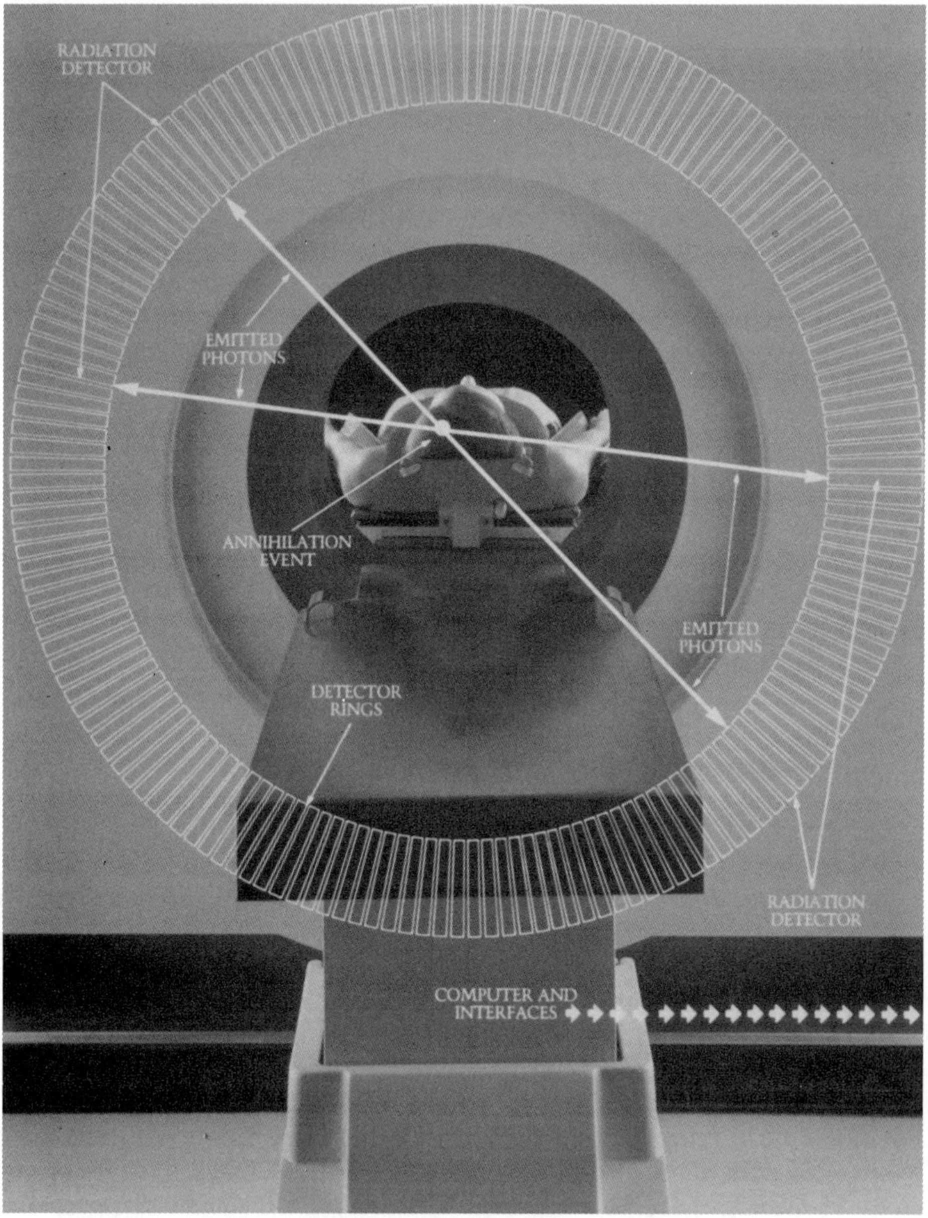

FIGURE 7.10 The operating principle of a positron emission tomograph (Scanditronix Inc.)

TABLE 7.7 Parameters of Conventional Short-Lived Emitters for PET

Radionuclide	Half-life [min]	Labelled precursor	Reaction protons	Reaction deuterons	Decay [%]	Yield[1)] [mCi]
Carbon ^{11}C	20.4	$^{11}CO_2$, ^{11}CO, $H^{11}CN$	$^{14}N(p,\alpha)^{11}C$	$^{10}B(d,n)^{11}C$	99.8	3100
Nitrogen ^{13}N	9.97	$^{13}NH_3$, $^{13}NO_x$, $^{13}NH_4$	$^{16}O(p,\alpha)^{13}N$ $^{13}C(p,n)^{13}N$	$^{12}C(d,n)^{13}N$	100	200
Oxygen ^{15}O	2.07	$^{15}O_2$, $C^{15}O$, $C^{15}O_2$, $H_2{}^{15}O$	$^{15}N(p,n)^{15}O$	$^{14}N(d,n)^{15}O$	99.9	2800
Fluorine ^{18}F	109.8	$^{18}F_{aq}$, $^{18}F_2$	$^{18}O(p,n)^{18}F$	$^{20}Ne(d,\alpha)^{18}F$	96.9	525

1) For EOB, 20-30 μA beam currents, irradiation parameters are given in [7.26]

pharmaceutical labeled with such radionuclides could bring advantages of PET imaging such as better resolution and quantification of the distribution, to the existing nuclear medicine investigations and open new possibilities where quantitative measurements of the pharmaceutical tracer is essential. For example, the use of ^{124}I as a label for MIBG, IAMP, Tomaxofin, Rose Bengal and many other biological compounds will allow their distribution to be followed with a PET scanner and obtain additional information not available by conventional imaging techniques.

The Manchester Group in collaboration with the University of Birmingham has developed procedures for the production of a number of such radionuclides as listed in Table 7.8 [7.51]. This table also contains information on the nuclear reaction employed for the production of each isotope, beam and energy window used, target composition and the yields. Wherever the use of enriched target material has been necessary, procedures have been developed for its recovery and reuse. All irradiations are carried out in air using an external beam from the Nuffield cyclotron (max. energy of alpha particles 25 MeV, deuterons 12.5 MeV, external 150 pμA) of the University of Birmingham. The targets are pressed pellets of elements of 5 mm, 7 mm or 10 mm in diameter.

TABLE 7.8 List of Medium Half-Life Positron Emitting Radionuclides for PET (after H.L. Sharma et al. [7.51]).

Radionuclide (Half-life)	Nuclear Reaction ΔE(MeV)	Target	Yield (Mbq/C)	Contaminants
Iodine-124 (4.1 days)	^{124}Te(d, 2n) 15-8	^{124}Te Enriched	5200	^{125}I, ^{126}I, ^{131}I (0.8% combined)
Cobalt-55 (18 hours)	^{54}Fe(d, n) 12-6	^{54}Fe Enriched	17400	^{57}Co, ^{56}Co
Copper-64 (12.7 hours)	^{64}Ni(d, 2n) 159-13	^{64}Ni Enriched	102,000	^{61}Cu
Gallium-66 (9.4 hours)	^{63}Cu(α, n) 18-8	Cu Natural	7240	^{67}Ga (0.25%)
Arsenic-71 (61 hours)	^{70}Ge(d, n) 12-6	^{70}Ge Enriched	4800	^{74}As (0.05%)
Zirconium-89* (78 hours)	^{89}Y(d, 2n) 16-8	Y Natural	18500	^{88}Zr (0.008%)
Nickel-57* (36 hours)	^{54}Fe(α, n) 30-10	^{54}Fe Enriched		1534 ^{56}Ni (1.1%)
Iron-52 (8.2 hours)	^{50}Cr(α, 2n) 38-20	^{50}Cr Enriched	1667	^{55}Fe (0.008%)

* Preliminary data from on going work

The main clinical applications of positron tomography, apart from tumor diagnosis [7.31], include the diagnosis of epileptic attacks, brain gliomas and mental dementia as well as of the diseases of brain vessels and coronary arteries [7.32].

Although *positron-emitting radioisotope generators* [7.33] can provide an alternative to cyclotrons, still the latter ones serve as the main source of such nuclides. Typically, a PET facility consists of (1) a dedicated cyclotron for the production of short-lived positron emitters (Table 7.7), (2) a laboratory for labeling compounds, and (3) a positron emission tomograph. In what follows we shall describe the instrumentation involved in the production of and labeling with positron emitters.

7.3.2 Positron Emission Tomography Facilities

The most effective reactions are those of the (p,n) type implemented with highly enriched stable nuclides and 11 MeV protons accelerated in small PET cyclotrons, e.g.

$$^{13}C\,(p,n)\,^{13}N\,;\quad ^{15}N\,(p,n)\,^{15}O\,;\quad ^{18}O\,(p,n)\,^{18}F \qquad (7.6)$$

In such cases, the radionuclides obtained are not contaminated with any stable nuclide of the same element, and are referred to as *no carrier added (NCA) substances.* In some cases, positron emitters can be administered in the elemental state, for example, ^{15}O, in lung studies. Some simple compounds can easily be labeled: water with oxygen-15, carbon oxide and dioxide with carbon-11 and oxygen-15, add finally, ammonia with nitrogen-14. In other cases, production of a complex compound may be much more difficult and may require complicated procedures. What is especially troublesome is the problem of *very high specific activities* on the order of a thousand Ci per millimole required for some studies on the nervous cell receptors. This means that most of the molecules, or even all molecules of a compound, must be labeled with a radionuclide.

In general, it is required that all the operations following irradiation in the cyclotron, such as radionuclide extraction, radiotracer synthesis, purification and quality control should not take longer than *three half-lives of the nuclide produced* (Table 7.7). The activity used at a single diagnostic examination should range from a few to one mCi for the initial activities of the radionuclide of about 10 mCi for ^{11}C and ^{13}N, 15-20 mCi for ^{15}O and 5-10 mCi for ^{18}F. For this reason, careful shielding, remote control and, as much as possible, full automation of all the procedures are absolutely essential.

Intensive work has led to the development of *automatic systems,* called *synthesizers,* for the synthesis of compounds labeled with positron-emitting nuclides. Those devices usually contain a microcomputer control of the synthesis. Signals fed to the computer from temperature, pressure and radioactivity detectors help to control loading of reagents, heating, cooling, chromatographic analysis and enrichment. When the synthesis is completed full information is also obtained from the computer.

Fig.7.11 is a schematic of a system *for automatic production of labeled compounds with positron emitters,* integrated with the 370 PET-type cyclotron (Sumitomo + CGR-MeV, [6.33]) which accelerates protons to an energy of 18 MeV and deuterons to an energy of 10 MeV. To prevent contamination in the target gas, a separate target is intended for each of the four positron-emitting nuclides. Target exchange can be carried out by remote control.

Table 7.9 lists various labeled compounds produced by radioisotope synthesizers together with their applications. Each of these synthesizers can be incorporated in the production system at the user's option. Automation of the whole process considerably reduces the radiological hazard to the personnel.

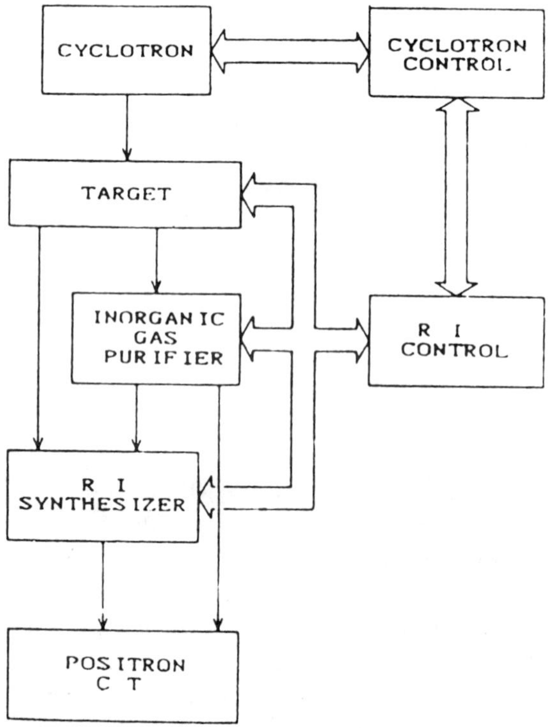

FIGURE 7.11 Schematic diagram of a positron-emitting radioisotope production system (Sumitomo + CGRMeV) (after Y.Adachi et al.[7.34])

TABLE 7.9 Radioisotope Synthesizers in Conjunction with 370 Type (Sumitomo + CRGMeV) cyclotron (after Y. Adachi et al. [7.33])

Labeled compounds		Use
^{11}C	CO gas	Local blood flow
		Pulmonary perfusion
^{11}C	CO_2 gas	Local blood flow
^{13}NC	N_2 gas	Ventilatory function
^{15}O	O_2 gas	Oxygen consumption
^{15}O	CO gas	Local blood flow
		Pulmonary perfusion
^{15}O	CO_2 gas	Local blood flow
^{11}C	hydrogen cyanide	Precursor of amino acid,
		glucose
^{11}C	methyl iodide	Precursor of various drugs
^{11}C	glucose	Brain & myocardium imaging
^{11}C	CO_2 gas	Local blood flow
		Sugar metabolism
^{11}C	methionine	Brain & pancreas imaging
^{11}C	methyl spiperone	Receptor study
^{13}N	ammonia	Brain & heart imaging
		Precursor of amino acid
^{15}O	water	Blood flow
^{18}F	HF aqueous solution	Precursor
^{18}F	fluoro deoxy glucose	Brain & heart imaging
		Cancer diagnosis

Each of the six types of gases, ^{11}C CO, ^{11}C CO_2, ^{13}N N_2, ^{15}O O_2, ^{15}O CO and ^{15}O CO_2, passes through a common inorganic gas purifier unit. Synthesizers are fabricated as standard units (black box system) whose dimensions do not exceed 45 x 48 x 48 cm^2. Control systems of the cyclotron and the synthesizer are coupled by means of a computer link. Systems of this kind are in operation at some Japanese hospitals (Kyoto University Hospital, Chiba University Hospital).

Types of automation. There are three basic types of control systems used in most modern PET facilities [7.52]: (1) programmable logic controllers (PLC), (2) personal computers such as IBM-PC/AT (PC) or Macintosh II (MAC) and (3) minicomputers or workstation (WS) class computers. Each system is primarily employed to run a particular device or set of tasks. At present, vendors of PET equipment do not provide hardware/software solutions that permit operation of a device or task from any location on a network.

The need for automation integration is driven by the high rate at which new PET facilities are being constructed and by the shortage of experimental manpower in the PET field. As the technology improves and the manufacturers implemented enhanced automation in PET, the full throughput potential of the cyclotron and tomograph will be better realized.

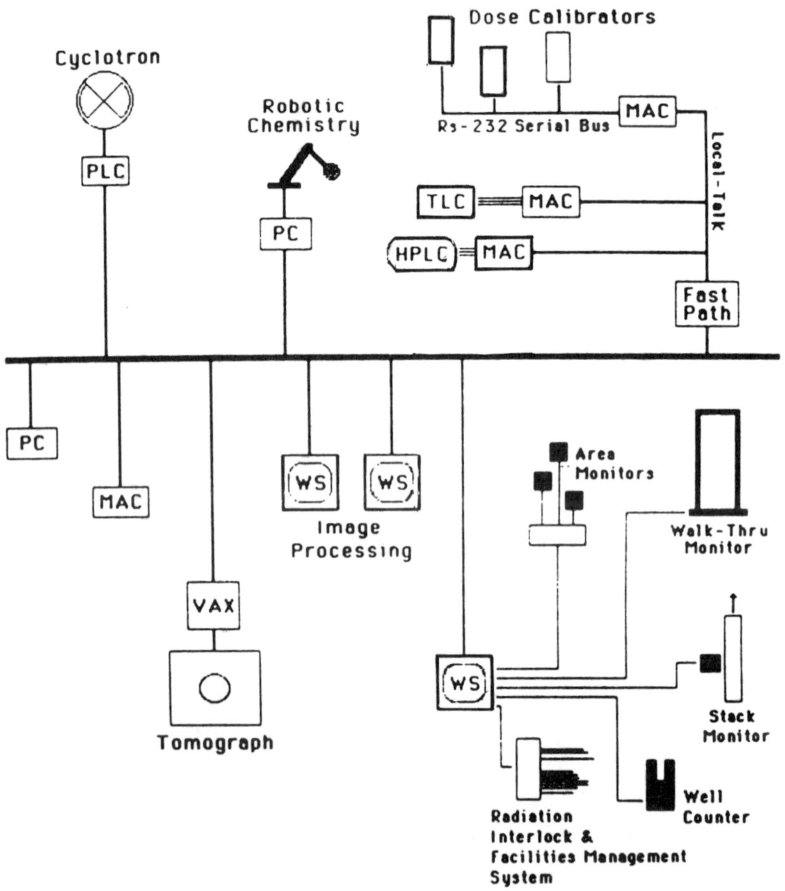

FIGURE 7.12 Schematic diagram of integrated PET computing facilities (University of Iowa PET Imaging Center), (after R.D. Hichwa [7.52])

An example of automation integration is shown in Fig. 7.12 for the newly constructed PET Imaging Center at the University of Iowa. The cyclotron is operated under a Siemens PLC system. The system is capable of automatically starting up the cyclotron, placing the correct target on line for bombardment, filling the target, irradiating the target at the proper beam current and duration to produce the desired amount of radioactivity for radiopharmaceutical production and finally, shutting down to a standby mode. This unit has been modified to provide data logging capabilities for monitoring the number of μAh of beam on target, target pressure during irradiation, beam current stability, vacuum pressure and other operational parameters that provide insight about target failure modes, loss of radioactivity yield and when to perform maintenance on the cyclotron or target systems. Data are automatically transferred to the facility MicroVAX 4000 minicomputer [F33] for storage. Further, an Ethernet interface has been added to the PLC to permit operation of the cyclotron from any X-Window compatible workstation located on the network. The PLC still controls the cyclotron but parameter display and communication with the PLC is now simultaneously possible from different locations on the network.

Syntheses of radiopharmaceuticals are performed with robot or black box systems interfaced to a PC or MAC [F34] computer. All quality assurance and purity determination is achieved with the same type of MAC systems interfaced to high pressure liquid chromatography systems (HPLC), gas-radiochromatography systems (GRC), and thin layer chromatography systems (TLC). Software and data are maintained for all these devices on the VAX. Local-talk and Ethernet networks physically link these systems together while DECnet and DEC LanWORKS for Macintosh provide the software integration.

Several VAXstations 4000 and 3100 [F33] are used to remotely acquire raw PET data and process it into functional images for analysis. A dedicated VAXstation 3200 is used for all realtime data acquisition needs such as (1) arterial blood sampling, (2) automated γ counting, (3) laboratory analysis of blood, (4) radiation monitoring of the entire center, and (5) facilities management. The last item includes status and analog readout of doors ajar, air temperature, air pressure, radiation interlock systems, humidity, water sensors, and power conditions. Faults or out-of-tolerance conditions are logged and if they persist beyond a prescribed period of time, appropriate maintenance personnel are automatically paged through modems attached to the computer system.

Automatic delivery system. Wide spread clinical use of PET requires rapid, efficient, *sterile and pyrogen-free delivery* of these radiotracers from the cyclotron to the scanner. Rapid delivery is required because of the short half-lives and the limited production capabilities of 11 MeV proton only cyclotrons. For example, acceptance tests of the CTI, RDS 11 MeV cyclotron at Vanderbilt University produced the following results: 20 mCi of ^{13}N-ammonia after a 13 min irradiation (30 μA), 400 mCi of ^{15}O-water after a 10 min irradiation (40 μA) and 2 min post-irradiation processing, 300 mCi ^{18}F-FDG after a 75 min irradiation (20 μA) and 60 min post-irradiation processing [7.53].

The extreme difficulty and expense of anticipating all potential problems and incorporating necessary safeguards precludes developing an entirely *automatic delivery and injection system.* For this reason, in the system [7.53] activity is automatically delivered to the scanner but human intervention is required to inject the patient. Controls that operate various valves and a syringe pump are used to remotely operate the infusion system. The technologist has complete control over the injection but her/his radiation exposure is greatly reduced by increasing her/his distance from the radioactive source.

Patient safety concerns are addressed by having the radioactivity automatically delivered into a receiving vial, then requiring active participation by the technologist to complete the injection. The remotely controlled injection is no different in principle than if it were handled manually. The injection can be controlled at a distance, thereby reducing radiation exposure, and one technologist can perform the entire scan.

The stimulus for developing this system is to gain the *capability to perform eight quantitative PET studies per day* using one scanner and three technologists. The following criteria were deemed necessary for a patient injection system: (1) production of the tracer is initiated from the scanner room through a remote terminal connected to the cyclotron, (2) no further interaction with the cyclotron is necessary, (3) coordination of the radioactivity delivery within the whole scan protocol must be simple, (4) the system must be easily testable for pyrogens, (5) the infusion must have sufficient safeguards to insure that the patient is protected from inadvertent injection, (6) the radioactivity must be deliverable to any of several final destinations, and (7) the system must be easy to use by a single imaging technologist whose primary concern is the patient, not the cyclotron or delivery system.

The production of ^{13}N ($t_{1/2}$ = 10 min), ^{15}O ($t_{1/2}$ = 2 min), and ^{18}F ($t_{1/2}$ = 110 min) via proton reactions has been investigated as well as the incorporation of these radionuclides into the PET tracers ^{13}N-ammonia, ^{15}O-water, and ^{18}F-FDG. Activity is transferred in one of two ways depending upon the half-life of the radiopharmaceutical.

Ammonia is directed to either the animal laboratory or the scanner room with valve V2 (Fig. 7.13). Then it is transferred from the target to the switching manifold in aqueous form by an

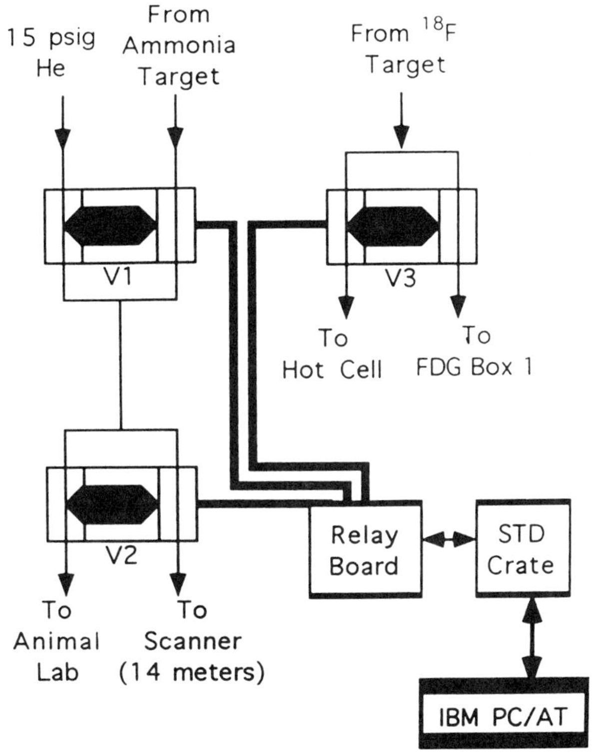

FIGURE 7.13 Target delivery manifold with remote controlled switching valves (Vanderbilt University system), (after J.R. Votaw et al. [7.53])

HPLC (High Performance Liquid Chromatography) pump forces 1.5 ml min^{-1} of water through a carbon slurry target. Once the water has passed the manifold, 15 psig of helium is switched on to complete the 13.1 m transfer to the scanner in ~ 45 s. At its final destination, the ammonia is forced through a vented 0.22 μm filter into the receiving vial of the patient injection system.

^{15}O labelled water is produced with the aid of a platinum catalyst converter at the scanner. Air and hydrogen are mixed with the (^{15}O) O_2 target has which is then transferred to the scanner in a constantly flowing stream of helium. After the gas has been converted to water, it is bubbled through ~ 5 ml of isotonic saline in the receiving vial of the patient injection system.

The patient injection system is used for remote administration of liquid radiopharmaceuticals. It consists of a receiving

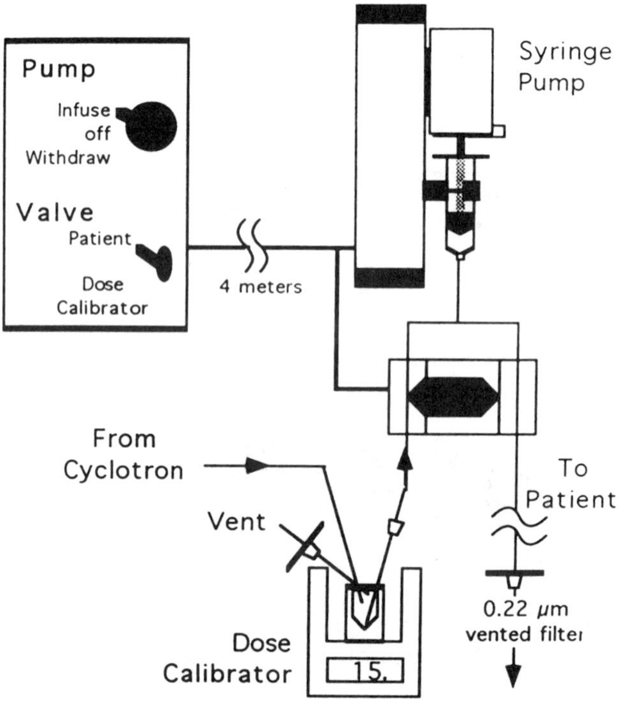

FIGURE 7.14 Patient injection system (Vanderbilt University, Nashville, USA), (after J.R. Votaw et al. [7.53])

vial, dose calibrator, switching valve, and withdrawal/infusion syringe pump as shown in Fig. 7.14. The system is connected with sterile disposable tubing except for silicone tubing at the switching valve. This component is preassembled and gas sterilized in house. The syringe pump has been modified by removing its infusion/withdrawal and on/off switches and placing them in a separate box that is connected to the syringe pump with a 4 m cable. This allows operation of the system at a distance where radiation exposure is greatly reduced. When fully assembled, all of the components are clearly visible to the operator so that if there is a problem, immediate action can be taken.

As activity arrives from the cyclotron, it is collected in the vial in the dose calibrator. The injection sequence is begun by opening the remote controlled three-way valve to the dose calibrator and withdrawing the activity into the syringe. It is infused into the patient using the same pump after the three-way valve is switched to the patient IV line.

To ensure apyrogenicity and sterility of the injected product, the entire system is washed with sterile water before it is used. Since the tracer is delivered in an ~ 8 ml bolus of water, the next delivery through the system is considered safe for injection if pyrogens are not detected at a threshold of 3 *endotoxin units* per ml (EU/ml) in the wash. Time delayed test shows that the system may be left unused for up to 6 h before the wash must be repeated.

Use of the delivery system has made it possible for two PET technologists to perform five quantitative PET studies per day. This is possible because one technologist can execute all phases of the PET study including patient preparation, transmission scanning, radiopharmaceutical production (for ^{13}N-labeled ammonia and ^{15}O-labeled water studies), injection of the radiopharmaceutical, collecting blood samples, emission scanning, analyzing the blood samples, and quantitating and archiving the images.

Currently, patients spend an average of 2 h in the PET facility but only 45 min in the scanner room. The technologist must spend approximately two additional hours after the patient has left to complete the study. Completing five quantitative studies per day using two technologists is possible by interleaving patients as shown in the time line (Fig. 7.15). With two addition-

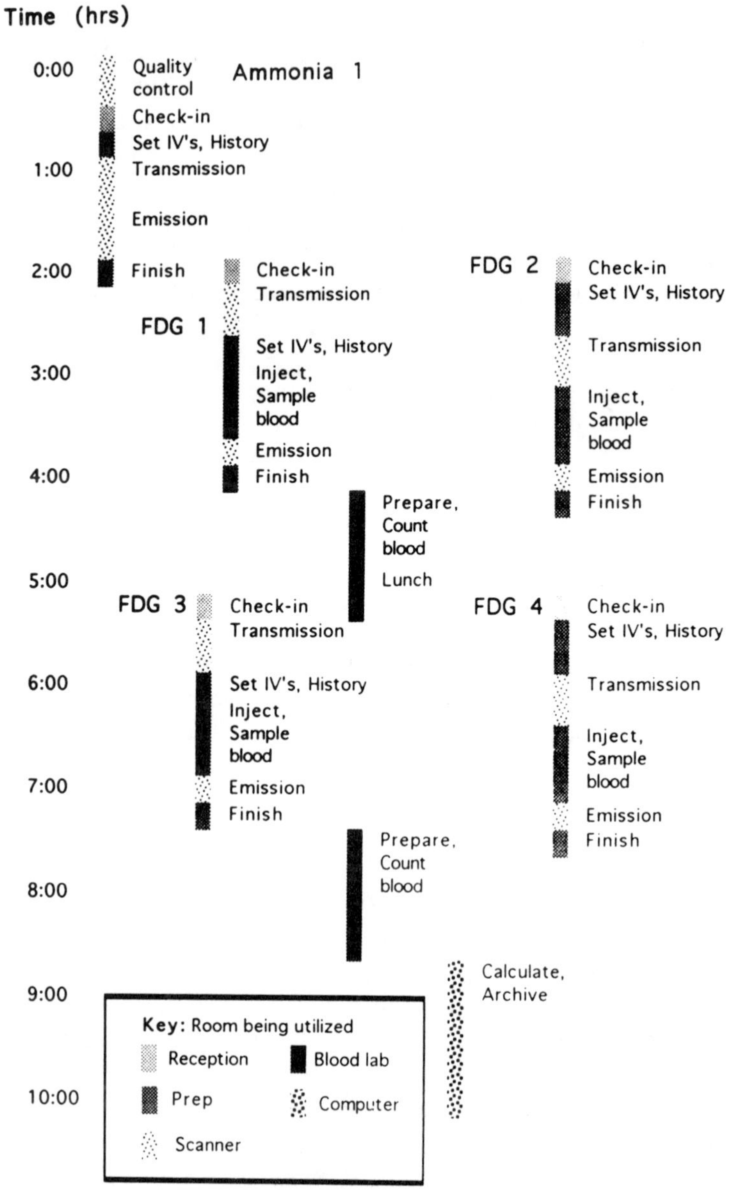

FIGURE 7.15 Timing diagram for a typical day containing 5 quantitative PET scans. The first technologist arrives approximately 2 hours before the second and performs the daily quality control and ammonia scan. Two technologists work together to complete the FDG scans and the second stays late to quantitate and archive the data (after J.R. Votaw et al. [7.53])

al injection systems, one for each patient per room, and three technologists, eight quantitative studies can be performed per day.

Radiation monitoring systems. Production levels of radioactivity in a modern PET center may exceed 10-50 Ci/d [7.52]. Several different precursors are made that will be synthesized into radiopharmaceuticals. These may take the form of volatile liquids or gases. In some instances radiolabeled gases are administered directly to patients, who in turn may exhale radioactivity into the environment. Automatic and continuous monitoring of the production, synthesis, and patient areas is necessary, as well as the *monitoring of environmental releases,* radioactivity in blood samples and transport of radioactivity from the synthesis site to the utilization site. In Iowa PET Imaging Center five to ten radiopharmaceutical syntheses starting from 2 Ci each may be made on a given day. Chances for an unexpected release are high. A typical PET center with a significant number of research and clinical studies performed per day may release up to 400 Ci/yr. Current Nuclear Regulatory Commission regulations require that the exposure to the general public be less than 500 mrem/yr. Pending Environmental Protection Agency legislation may reduce this number to 10 mrem/yr which will require extremely accurate monitoring to demonstrate compliance with regulations.

The types of radiation monitoring systems that are employed in PET are the following [7.52]: (1) hot cell and cyclotron vault exhaust stacks, (2) walk-through monitors, (3) area monitors, (4) auto gamma well counters and (5) dose calibrators. Stack monitoring is more difficult and costly to perform than the other types mentioned. Fig. 7.16 depicts a simplified schematic for an on-line system to measure stack releases that does not require stack penetrations, ancillary pumps or complex calibration procedures. This system was modeled using STELLA [F35] to ascertain efficiency, sensitivity, and in general, the variables that determine the effectiveness of the system to measure radioactivity passing along a stainless steel duct. Design considerations entailed modeling exhaust flow rate, duct size (length and diameter), detector efficiency, sampling volume, detector size, release concentration and release volume. At flow rates of 28.3-254.7 m^3 min^{-1} (1000-9000 cubic feet per minute) a 2 in. x 2 in. NaI detector has sufficient sensitivity to accurately measure

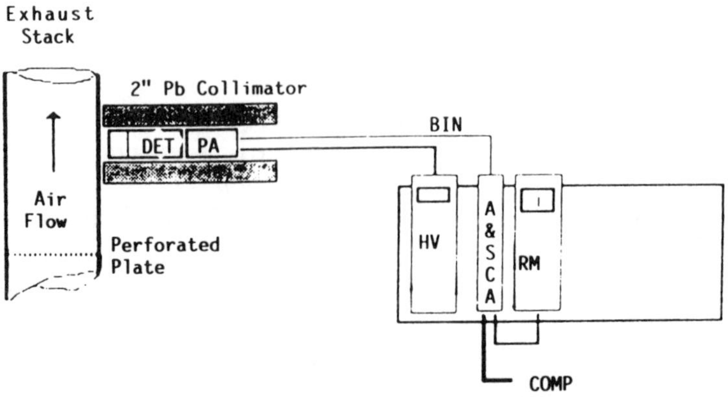

FIGURE 7.16 On-line radiation stack monitoring system using standard NIM modules and detectors (PET Imaging Center, University of Iowa, USA): DET: 2 in. x 2 in. NaI detector; PA: preamp and detector base; HV: high voltage supply; A&SCA: combined amplifier and single channel analyzer; RM: ratemeter; BIN: NIM bin and power supply; COMP: output pulse train to computer (after R.D. Hichwa [7.52])

0.1-10000 mCi releases of ^{11}C - CO_2. The best performance of the system is achieved at approximately 85 m^3 min^{-1} (3000 cfm) to appropriately determine releases in the 10-2000 mCi range which is more typical of general PET operations. A performed plate with 65% open area is used to transform laminar flow to well mixed plug flow. The detector then observes a uniform concentration of radioactivity.

Long-distance transport of PET radionuclides. The total cost of a facility equipped with a cyclotron and a radiochemical laboratory is quite high. Therefore, in large urban areas compounds labeled with PET nuclides can be produced at a *centralized facility* and then shipped do major hospitals by conventional means of transport. It is estimated that for short half-life materials (20 min < $t_{1/2}$ <110 min) such a transport is feasible and cost-effective within a distance of 20-25km [7.35]. This, of course, is not true for compounds labeled with ^{15}O of two-minute half-life.

In the above case, a solution is provided by a *long-distance on-line transport system* of ^{15}O-labeled gases, implemented in Finland by joining Abo Akademi cyclotron with two clinical laboratories at 650 and 970 m distances from the accelerator [7.36]. Fig.7.17 is schematic diagram of the system used. After

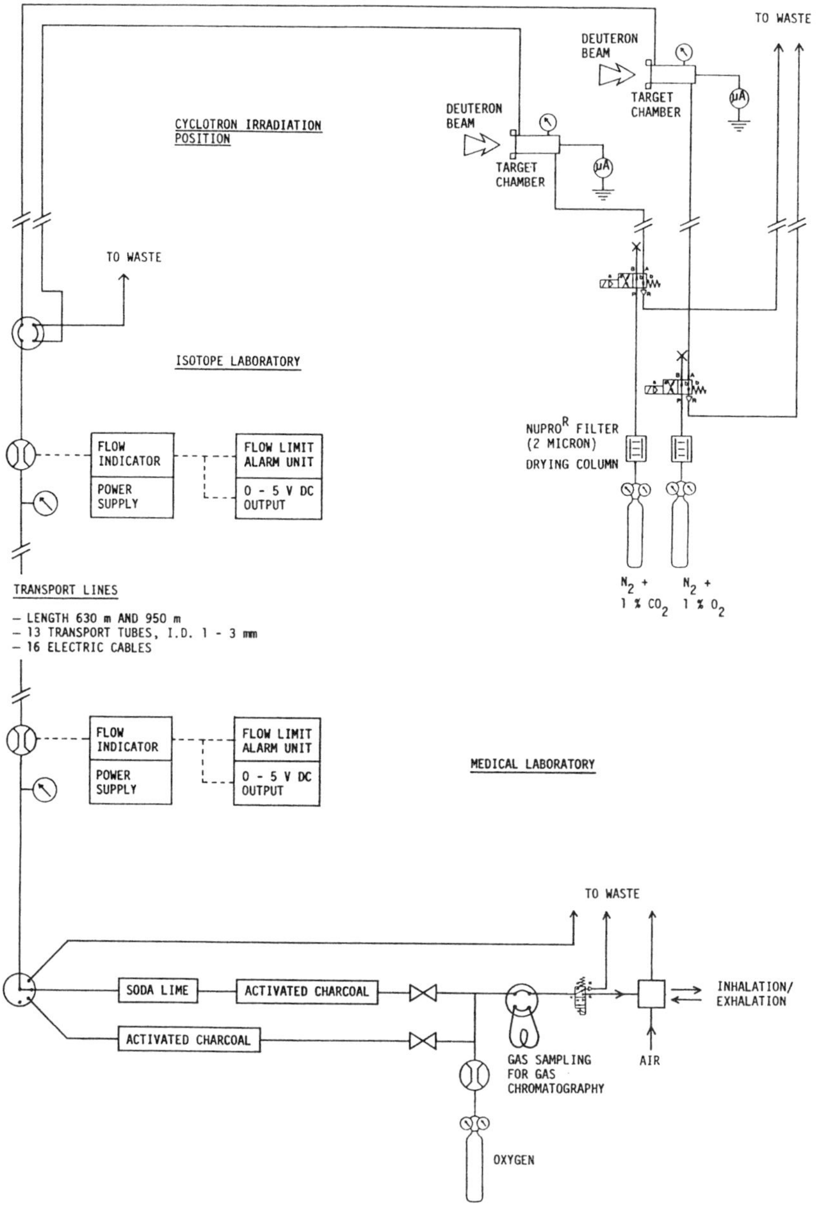

FIGURE 7.17 Schematic diagram of ^{15}O production, long-distance transport and concentration measurements system (Accelerator Laboratory, Abo Akademi) (after S.J.Heselius [7.36])

having been dried and filtered, the mixture of N_2 with 1% admixture of O_2 is bombarded with a beam of 7.4 MeV deuterons. The ^{15}O-labeled molecular oxygen or carbon dioxide is transported through a narrow-bore (1.2 mm internal diameter) polyethylene tube. At the distance of 630 m the drive pressure applied is 550 kPa, the transit time being about 6.5 min and the gas flow rate 470 ml min^{-1}. At 550 kPa, the 6.15 MeV deuteron beam is fully attenuated in the target. For a 5 μA beam current the radionuclide concentration at the tubing outlet is 460 kBq ml^{-1}. For the 950m tubing the transit time increases to 10.3 min with the drive pressure of 750 kPa. For the same beam current, 5 μA, the ^{15}O concentration at the tubing outlet is 97 kBq ml^{-1} [7.36].

At the Canadian TRIUMF facility another design has been adopted, viz that of joining the cyclotron and the hospital with a 2.4 km long *pneumatic pipe-line.* The vials containing radiopharmaceuticals are placed inside capsules that are blown to the receiver through a tube by means of compressed air. Travel times as short as 2 min are achieved. This method is also suitable for the transport of ^{15}O-labeled compounds. The TRIUMF facility is equipped with two cyclotrons for the production of PET-labeled radiopharmaceuticals. The pipe-line is buried 1.20 m underground and covered by a 60 cm thick concrete layer. The radiation level at the surface never exceeds $8 \cdot 10^{-8}$ Sv h^{-1} (8 μrem h^{-1}). The capsules move at a speed of 20 m s^{-1} [7.37].

Cost analysis. The choice of an accelerator for clinical radioisotope production requires careful consideration of the cost of the machine, the desired reaction parameters, and the cost and size of the required shielding. There are many trade-offs to be considered; for example, higher energy beams enable access to greater cross sections but also require greater shielding. The costs include the accelerator itself, accelerator operation and maintenance, shielding, target material, and the time required to produce a sufficient quantity of radionuclide. Each of these costs has constraints, e.g. the accelerator location may impose a limitation on the maximum shield size. The cost also depends on existing resources. Preexisting conditions may allow additional resources to be directed towards the accelerator.

The accelerator will be used primarily to produce the positron emitting isotopes ^{11}C, ^{13}N, ^{15}O, and ^{18}F. Each of these

TABLE 7.10 End of Saturated Bombardment Yields (EOSB) of the Reactions of Interest near 10 MeV Bombardment Energies that Lead to the Production of PET Tracers; the Cost is Approximate and does not Consider Reusing Target Material (after J.K. Votaw and J. Nickles [7.54])

Reaction	Target material	Target cost (US$)	Yield (Mci/μA) (EOSB)	Energy (MeV)	Precursor chemical form
$^{14}N(p, \alpha)^{11}C$	N_2+0.01% O_2	<1	84±10	11	$^{11}CO_2$
$^{16}O(p, \alpha)^{13}N$	H_2O	<1	7±1	10	$^{13}NO_3{}^-$, $^{13}NO_2{}^-$, $^{13}NH_4{}^+$
$^{13}C(p, n)^{13}N$	^{13}C powder in H_2O	30	110	10	$^{13}NH_4$
$^{15}N(p, n)^{15}O$	N_2+0.1% O_2	30	70	9.5	$^{15}O_2$
$^{14}N(d, n)^{15}O$	N_2+0.1% O_2	<1	70	9.5	$^{15}O_2$
$^{18}O(p, n)^{18}F$	$H_2{}^{18}O$	20	120	11	$^{18}F^-$(aq)
$^{20}Ne(d, \alpha)^{18}F$	Ne+1% F_2	13	76	11	$^{18}F_2$
$^{16}O(^3He, p)^{18}F$	H_2O	<1	30	22	$^{18}F^-$(aq)

isotopes can be produced by several different reactions (Table 7.7 and 7.10). The production of ^{15}O illuminates some of the trade-offs that must be considered when choosing an accelerator. With a proton-only machine one must use the $^{15}N(p,n)$ reaction and accept the cost of the separated isotope target material. A machine able to accelerate deuterons allows one to take advantage of cheaper target material with the $^{14}N(d,n)$ reaction. Table 7.10 illustrates that a proton-only accelerator can be used to produce adequate quantities of each of the isotopes of interest. However, such a machine restricts the type of reactions available. The decision to use a single particle accelerator mandates that the target material for each of the reactions (except the $^{14}N(p,\alpha)^{11}C$ reaction) be isotopically enriched. The target material costs are delineated in Table 7.10 [7.54].

Cost Analysis of PET Facilities is specified in detail for the US in [7.32]. The price of a small cyclotron ranges from \$ 1.2 to 1.5 million, and that of a dedicated PET-type cyclotron from \$ 1.2 to 2.0 million. The price of a tomograph is also \$ 1.2 to 2.0 million. The total price of a tomograph, cyclotron as well as a radiochemical laboratory and tomographic image processing laboratory amounts to about $3.55 million.

On assuming 5-year, 10-year and 15-year life times for the tomograph, for either of the laboratories and for the cyclotron, respectively, the annual depreciation costs will be $ 0.36 million. The total annual operation costs, including $ 0.36 million for salaries and other current expenditures, in a PET facility is estimated at $1.2 million. A facility of this type can perform about 2250 examination per year, and including other expenses, the cost of a single examination will thus be about $1000. It has been estimated that, in the USA, there will be some 300 cases of

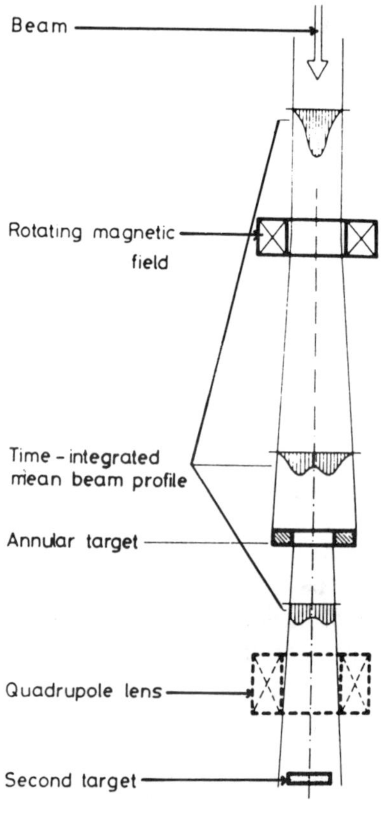

FIGURE 7.18 Two simultaneous irradiations using rotating magnetic field (after I. Gal et al. [7.55])

epilepsia, 200 tumor examinations, 400 cases of mental dementia and 1000 cardiological studies per 1 million inhabitants per year so that one PET facility would be economically cost-effective for a region populated by 1-2 million people [7.32].

The radioisotopes produced with charged particle beams are usually more expensive than the reactor produced radionuclides (Table 1.9) mainly because of the lack of a parallel irradiation possibility [7.55]. There is a renewal interest for simultaneous experiments at multipurpose accelerators where the beam intensity is high enough to allow making two experiments parallel after dividing the primary beam into two parts. The beam splitter based on a rotating electro-magnetic field is presented in [7.55]. The technical arrangement of the splitter is shown in Fig. 7.18.

The beam coming from the accelerator goes through a rotating magnetic field produced by a three-phase electric motor stator mounted on the beam line. As a result of the interaction of the beam and the magnetic field, the beam is translated around a ring. The measure of the radius depends on the strength of the magnetic field. The shapes of the time integrated mean beam profile along the beam line and perpendicular to it can be seen in Fig. 7.18. By applying an annular target after the rotating magnetic field the beam can be divided into two parts. The particles go through the annular target and can be used after focusing for irradiating the second target.

The main advantages of the method are as follows: (1) both targets are free from hot-spots; (2) the intensity ratio of the two parts of the beam could be changed continuously; (3) the time structure of the beam remain unchanged; (4) there is no shadow effect which can lead to intensity loss; (5) low cost and simplicity. The irradiation of the second target is free of charge.

This method can be used at small and medium energy accelerators because of the relatively weak magnetic fields and the space required for the proper deflection of the beam [7.55].

7.4 Commercial Production of Radioisotopes

As already mentioned in Sec. 1.4, each of the world's top producers of medical radioisotopes has several cyclotrons available. Cyclotron systems of this kind have been discussed in Sec. 7.1.4. To provide an example of a commercial radionuclide and labeled compounds production plant, the Amersham International [F29] will be described below. The first cyclotron (Philips) was installed there as early as 1963. In the early 1970 it was computerized and has been operating at a cycle of 150 h weekly ever since. In 1975, its production target exchange was automated. In view of the increased demand for medical radionuclides, and especially for ^{210}Tl, another cyclotron, that of the H^--type, was put into operation. It operates fully automatically and is designed exclusively for ^{201}Tl production. In 1986, a third cyclotron started service.

Using all three cyclotrons the Amersham is currently producing the following radionuclides:

1. the (p,2n) reaction : ^{67}Ga (^{68}Zn target), ^{111}In (^{112}Cd target)
2. the (p,3n) reaction : ^{201}Tl (^{203}Tl target)
3. the (p,2n) reaction: ^{57}Co (^{58}Ni target).

The three most important radionuclides produced are ^{67}Ga, ^{111}In and ^{201}Tl. Table 7.5 lists the products derived from these nuclides and provides their main uses.

CHAPTER 8

Accelerator Radiation Processing

8.1 Fundamentals of Radiation Processing

A *process accelerator,* also referred to as an *accelerator processor,* is a machine for processing various materials with the use of radiation techniques. The material under treatment is usually fed in the form of strips, sheets, plates, blocks and packings. Fig.8.1 shows the *basic system for radiation processing by means of an accelerator.* The beam of accelerated particles is usually circular in cross-section and its diameter ranges from about 1 mm to several centimetres. Since the material under treatment is, as a rule, much larger than the cross-section of the beam, the accelerator must be equipped with a special device, called a *scanner,* which sweeps the beam (Sec.3.1.4.3). In the accelerator processors, the beam is usually swept transversely to the direction of the material moved continuously by means of a suitable *conveyor.* The *swept beam arrangement* is shown on the right-hand side of Fig.8.1.

In order to avoid the troublesome swept beam construction radiation process accelerators have been designed equipped with a *linear or planar cathode* whose length corresponds to the width of the material to be irradiated (left-hand side of Fig. 8.1). In this way, the effective width of the *treatment field,* or dimension w, is considerably increased. However, this system is appropriate only for low beam energies, those not exceeding about 300 keV. Accelerators of this kind are called *linear cathode accelerators or planar cathode accelerators.*

The material under treatment may be irradiated from *one side,* usually from the above, or from *both sides* as shown in Fig. 8.1. Thus we can classify accelerators into *one-side* and *double-side machines.*

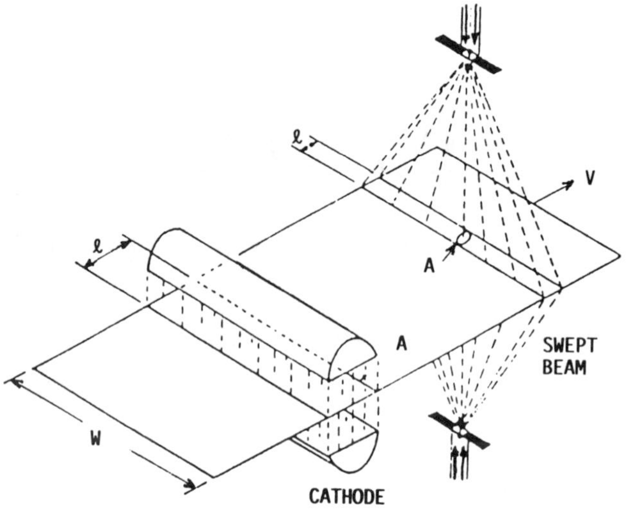

FIGURE 8.1 Radiation processing by means of accelerator with a planar cathode (left hand side) and beam scanning device (right-hand side)

Fig.8.2 illustrates graphically the design of a radiation process accelerator with a separate *hv generator* which is coupled to the accelerating chamber by a flexible cable or rigid transmission line. This facilitates placing the assembly in the assembly line since the accelerator can be located at a distance from the production area. In some cases, one generator can supply power to two *irradiation heads.*

Electron energies in process accelerators range from 0.1 to 10.0 MeV. Energies lower than 0.1 MeV are not used because of the low penetration of the electron beam, whereas those higher than 10 MeV are also not utilized because some radioactivity may be induced in the material under treatment (see Sec.4.3.1 and Fig. 8.12). In respect of the energies employed, radiation process accelerators can be classified into (1) *low-energy (low-voltage) machines,* operating in the 100-300 keV range, (2) *medium-energy machines,* operating in the 300 keV to 3 MeV range, and *high-energy machines,* operating in the 3-10 MeV range.

To make calculations independent of the kind of the material processed, the *useful range of electrons,* which are practically the only kind of particles used in industrial radiation processing, is expressed in terms of the *mass-thickness.* Fig. 8.3 shows a penetration depth curve for electrons as a function of the mass-thickness. A beam of electrons accelerated to their maximum energies first passes through a 10-50 μm thick *accelerator chamber window* made of titanium foil, then crosses an air-gap several centimetres in length, and finally reaches the surface of the material to be irradiated. The absorption in the exit window and in the air-gap cannot be neglected in a low-energy accelerator because it corresponds to a significant portion of the depth-dose curve which exhibits a relatively wide maximum and a slow fall-off (Fig. 8.3). The useful range of the mass-thickness is dependent on the acceptable degree of *inhomogeneity of irradiation.* In the case illustrated in Fig. 8.3, the acceptable inhomogeneity was 80% which is equal to the high-quality processing standard. The end of the useful range thus determines the point at the area-weight scale where the dose drops to 80% of the maximum

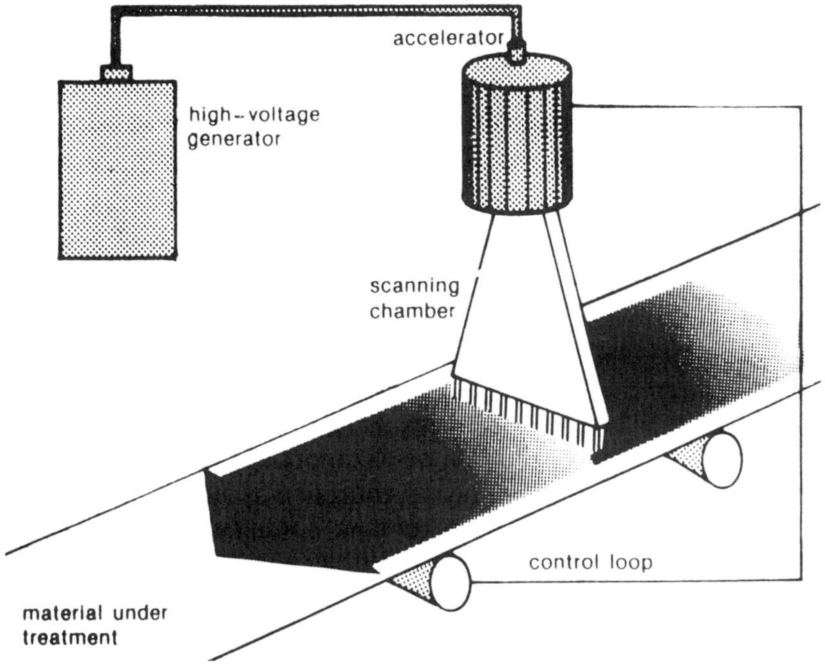

FIGURE 8.2 Radiation process accelerator with a separate hv generator

value. For lower quality requirements, greater inhomogeneity is acceptable, typical values are 60%, which consequently extends the useful range of the mass-thickness, also called *penetration range.*

The *throughput of radiation process accelerators* for bulk materials is usually expressed by the number of kilograms or tons to be irradiated in one hour with a predetermined dose (kg h^{-1} or t h^{-1} for, a dose of 10 kGy or 1 Mrad). Typically, throughputs of this kind range from several tens to several thousand kg h^{-1}. For materials in a packed form (e.g. in medical sterilization), the throughput is expressed in terms of the *volume of packages per unit time* (e.g. m^3 h^{-1} or m^3 year^{-1} for a typical sterilization dose of 25 kGy or 2.5 Mrad).

From the point of view of the *cost analysis,* what is particularly important is the *accelerator energy efficiency.* It denotes the ratio of the beam power to the input accelerator power. The energy efficiency of a radiation process accelerator usually ranges from 50 to 95% and it plays a decisive role in the *radiation processing cost-effectiveness.*

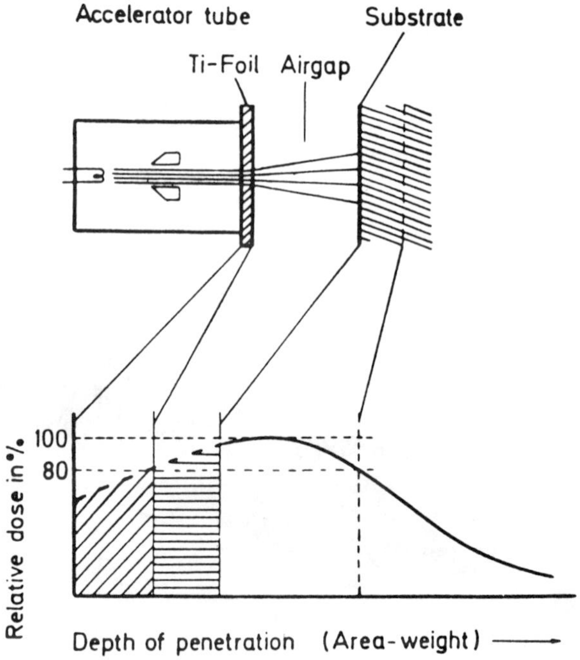

FIGURE 8.3 Penetration depth curve for electrons vs. inhomogeneity of irradiation (relative dose)

TABLE 8.1 Dose Requirements for Various Radiation Effects (after M.R. Cleland)

Radiation effect	Dose requirements
Radiography (film)	1.0 to 10.0 mGy (0.1 - 1.0 rad)
Human lethal dose (LD50)	0.4 to 0.5 Gy (400 to 500 rad)
Sprout inhibition (potatoes, onions)	100 to 200 Gy (10 to 20 krad)
Potable water cleanup	250 to 500 Gy (25 to 50 krad)
Insect control (grains, fruits)	250 to 500 Gy (25 to 50 krad)
Waste water disinfection	0.5 to 1 kGy (50 to 100 krad)
Fungi and mould control	1 to 3 kGy (100 to 300 krad)
Food spoilage bacteria	1 to 3 kGy (100 to 300 krad)
Municipal sludge disinfection	3 to 10 kGy (300 to 100 krad)
Bacterial spore sterilization	10 to 30 kGy (1 to 3 Mrad)
Virus particle sterilization	1 to 30 kGy (1 to 3 Mrad)
Smoke scrubbing (SO_2 and NO_x)	10 to 30 kGy (1 to 3 Mrad)
Aging of rayon pulp	10 to 30 kGy (1 to 3 Mrad)
Polymerization of monomers	10 to 50 kGy (1 to 5 Mrad)
Modification of polymers	50 to 250 kGy (5 to 25 Mrad)
Degradation of cellulosic materials	100 to 500 kGy (10 to 50 Mrad)
Degradation of scrap Teflon®	0.5 to 1.5 MGy (50 to 150 Mrad)

The *range of doses* used in accelerator techniques is very wide, from about 10 mGy (1.0 rad) when taking X-ray radiographs to 1.5 MGy (150 Mrad) when employed in degradation of some polymers. Table 8.1 lists doses used in various irradiation techniques.

In practice, techniques which require higher doses, i.e. those exceeding 10 kGy (1 Mrad), depend in large measure on *dose rate,* which is the basic accelerator parameter that determines the rate at which energy is delivered to the irradiated object. Therefore, this parameter is usually expressed in terms of *gray per second* or by derivative units such as kGy s^{-1} of MGy s^{-1}. In Fig.1.20 the dependence of the irradiation time on the dose rate has already been shown for three main types of processors: ^{60}Co gamma sources, accelerator bremsstrahlung generators and electron accelerators. The curves represent two dose levels: 10 kGy (1 Mrad) and 100 kGy (10 Mrad), thus covering a typical range used in radiation processing. It follows from Fig. 1.20 that the dose rates of the order of 10^2-10^6 Gy s^{-1} (10^4-10^8 rad s^{-1})

can be achieved with electron accelerators. Therefore, the required doses can be delivered to the irradiated product in the period of seconds or even less. In respect to the production throughput rate, electron processors dominate over any other types of radiation processors.

As already mentioned (Table 8.1), the range of accelerator techniques is quite wide. However, in this book we shall only be dealing with medical sterilization (Sec. 8.2) and preservation of food (Sec. 8.3), supplemented with some information on the use of accelerator processors in environmental protection (Sec. 8.4) and microfiltration technique (Sec. 8.5). Detailed discussions on other techniques can be found in [2.16].

Since

$$1\ \text{kGy} = 0.1\ \text{Mrad} = 1\ \text{Ws g}^{-1} = 1\ \text{kWs kg}^{-1} = \frac{1}{3600}\ \text{kWh kg}^{-1}, \tag{8.1}$$

when using a 1.0 kW beam, whose energy is totally absorbed in the irradiated material, an average dose of 10 kGy (1.0 Mrad) can be obtained at the rate of the displaced material of 360 kg h^{-1}. Thus the throughput or processor yield per unit time is given by

$$W_m = \frac{3600\ f}{\overline{D}\ [\text{kGy}]} = \frac{360\ f}{\overline{D}\ [\text{Mrad}]} \quad \left[\frac{\text{kg}}{\text{kWh}}\right] \tag{8.2}$$

where f is the *total radiation energy utilization coefficient,* and $\overline{D}$ is the *mean absorbed dose* in the product. Typically, f values range from 25 to 75%, and are strongly dependent both on the type of radiation employed and on the shape and size of the product processed. For the beam power, P(kW), Eq.(8.2) becomes

$$W_m = = \frac{360\ P\ f}{\overline{D}\ [\text{Mrad}]} \quad \left[\text{kg h}^{-1}\right] \tag{8.3}$$

For example, if the beam power is 100 kW, and coefficient f= 50%, then, using a technique that requires a dose of 10 Mrad, the mass throughput is $W_m = 1800$ kg h^{-1}.

Typically, beam power values of the present-day accelerators range from a few to several tens of kilowatts. For the largest machines commercially available they are as high as 200 kW. Special designs have been developed where the beam powers range from several hundred to several thousand kW.

A typical depth-dose distribution curve in a flat material irradiated with high-energy electrons has already been shown in Fig. 8.3. With increasing material thickness, the dose increases from the entrance value, D_s, to the maximum value, D_{max}, and then falls again (see also Fig. 3.62). The term *useful (effective) penetration or optimum penetration* is most often taken to mean the thickness for which the *surface dose,* D_s, is equal to the *exit dose,* D_e. Generally, $D_s = D_e = 60\%$ of the maximum dose D_{max}. In practice, effective penetrations are expressed in terms of the mass-thickness, F , which for a given material is a product of the layer thickness, d, and the specific weight (density), ρ, i. e.

$$F = d\rho \tag{8.4}$$

where F is the mass-thickness in g cm^{-2}, d is the layer thickness in cm, and ρ is the specific weight in g cm^{-3}. For 5 MeV electrons, effective penetration values, often called *optimum ranges,* do not exceed 2.0 g cm^{-2}, which corresponds to the layer d = 2 cm for unit density $\rho = 1$ g cm^{-3}. The penetration can be raised dramatically by applying two-sided (or double-sided) ir-

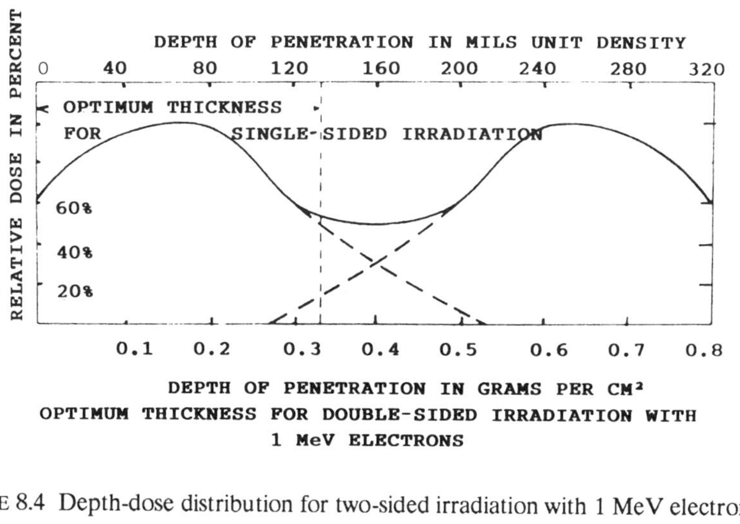

FIGURE 8.4 Depth-dose distribution for two-sided irradiation with 1 MeV electrons

radiation. The product to be irradiated is turned over after exposure on one side, and thus is passed through the processor twice. Fig. 8.4 shows the depth-dose distribution for two-sided irradiation with 1 MeV electrons. In this case the penetration is increased by a factor of 2.4 as compared with one-sided irradiation.

8.2 Medical Sterilization

8.2.1 Present-day Situation and Future Prospects

The chief reasons why radiation sterilization has currently been enjoying increased interest by industries producing medical disposables have been set out in Sec. 1.3. According to [8.1], some 10 million m^3 (361 million ft^3) are being sterilized by radiation the world over, of which 7.1 million m^3 are processed with *ethylene oxide* (ETO). About 2.8 million m^3 are sterilized by radiation with gamma (^{60}Co) processors,whereas only 0.3 million m^3 with high-energy electrons. Fig. 8.5 shows the future growth of all methods of sterilization, up to the year 2000 [8.1].

The main factors responsible for the quantitative development of efficient sterilization techniques are related to the large number of AIDS cases and the increased demand for medical disposables, such as surgical gloves, needles and syringes, protective clothing and a wide range of laboratory equipment. The diminished interest in the *ETO sterilization* in hospitals, which results from practical difficulties described already in Sec. 1.3, has additionally contributed to the ever greater use of disposables. The advances in medical research and the expansion of biotechnological methods employed in industry have also played a part in the increased demand for sterilized laboratory instrumentation. It has been estimated that the yearly rate of increase in the demand for sterilized products in the near future will be 5 to 9% [8.1]. While the share taken by the ETO technique in 1988 was 70% of the world's total production, in 1993 it is expected to fall to about 55%. The rest of the production will be covered by radiation methods, that is, by radioisotopes and accelerators.

8.2.2 Microbiological Inactivation and Dose Ranges

Compared with higher-order organism, *microorganisms* are distinguished by a relatively high *resistance to irradiation.* The

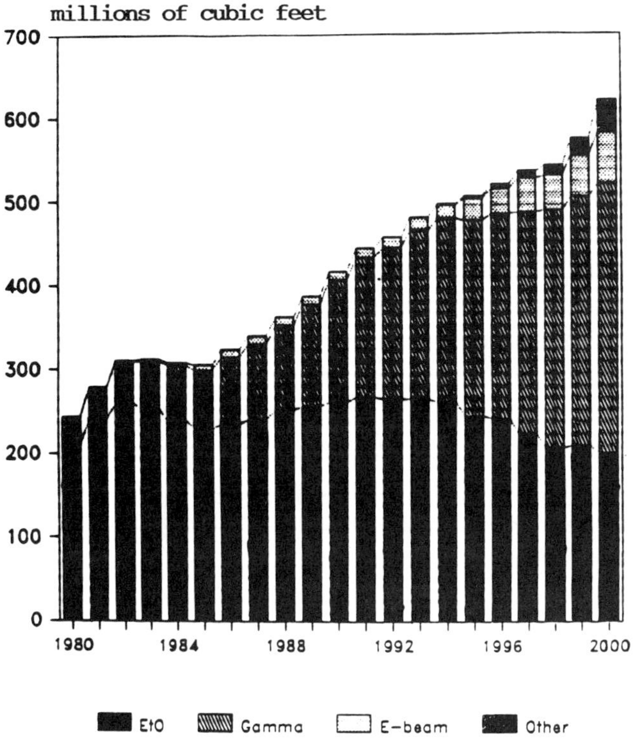

FIGURE 8.5 Development of various methods of sterilization of disposable medical products between 1980-2000 (after R.M.Brinston [8.1])

biological effectiveness of radiation in this case is dependent on the physiological state of microorganisms, their concentration in the culture, as well as on the chemical composition of the environment, presence of water or other chemical compounds, composition of the gaseous atmosphere, temperature, etc. In practice, the *radiation effectiveness* can be expressed by the so-called *inactivation coefficient,* that is the ratio of the number of surviving organisms to the total number of those before irradiation. Thus this ratio denotes the factor by which the population or the reproductive power of the bacteria colony decreases.

Fig. 8.6 shows, by way of example, inactivation curves for Salmonella typhi Ty 2 bacteria. A dose of less than 1 kGy (100 krad) is sufficient to inactive these bacteria in saline at the level

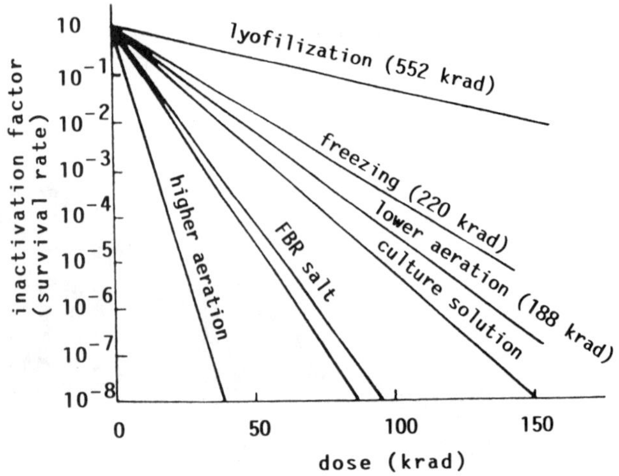

FIGURE 8.6 Inactivation curves for Salmonella typhi Ty 1 as a function of environment: for lower oxygenation 1.8 mM O_2 L^{-1}, and for higher oxygenation 16.2 μM O_2 L^{-1} (in saline)

of 10^{-8}, i.e. of the initial number of 10^8 bacteria only one survives. If, however, the same bacteria are frozen or are present in a lyophilized material, the doses necessary to obtain the same effect are 2.20 kGy and 5.52 kGy, respectively.

In general, the inactivation coefficient is a function of three factors: sterilization method, type of microorganisms and type of environment. When sterilizing medical materials and instruments, the level of about 10^{-6} is regarded at present as a satisfactory value of the inactivation coefficient.

Survival curves for irradiated bacteria are usually shown in the form as in Fig. 8.7. The dose is plotted in a linear scale, while the *survival rate* is given in a logarithmic scale. This is because in radiation sterilization the so-called *10% dose,* denoted by D_{10}, is often employed. It is a dose necessary to reduce the initial number of a given species of microorganisms by 90%, that is to the level of 10%. The curve (1) in Fig. 8.7 referred to as a *sigmoidal curve* represents the radiation-resistant bacteria. In order to obtain a 10-fold reduction in the number of such bacteria, it is necessary to add a *lag dose* to the D_{10} dose. The exponential curve (2) and the *composite curve* (3) represent *radiation-sensitive bacteria.*

The D_{10} doses for various bacteria and substrates have been tabulated and some of them are presented in Table 8.2. It is clear that, in contrast to *oxygen bacteria, gram-negative and gram-positive bacteria* are in general highly sensitive to radiation. Some viruses, however, display a higher degree of radiation resistance. For them, D_{10} is found to be sometimes as high as 8 kGy (0.8 Mrad).

Sterilization doses are, as a rule, determined by special regulations included in national regulations (Farmacopoeias). Except for Scandinavian countries, the basic dose is usually that of 25 kGy (2.5 Mrad). In Nordic countries, however, the dose is dependent on the *initial contamination level,* so-called *initial count.* For example, if the number of microorganisms ranges from 0 to 50 and the sterilization is carried out by means of an electron beam, the required dose is 35 kGy. On the other hand, if the initial count is 500-5000 the dose should be 50 kGy. In

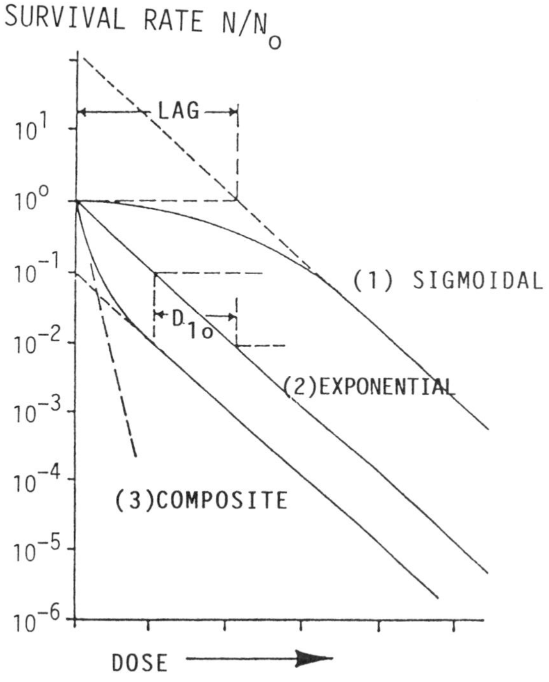

FIGURE 8.7. Hypothetical survival-rate curves for irradiated bacteria

TABLE 8.2 Radiosensitivity of Some Bacteria

Microorganism	Material Irradiated	Dose D_{10} (krad)
GRAM-NEGATIVE BACTERIA		
Pseudomonas aeruginosa	Paper tape	2.9
Escherichia coli	Paper tape	8.5
Aerobacter aerogenes	Paper tape	5.6
Klebsiella pneumoniae	Paper tape	24.0
Salmonella paratyphi B	Phosphate buffer	19.0
GRAM-POSITIVE BACTERIA		
Micrococcus pyogenes var. aureus	Paper tape	18.0
Sarcina lutea	Paper tape	89.0
Diploccus pneumoniae	Paper tape	52.0
Streptococcus phogenes	Paper tape	32.0
Corynebacterium acnes	Paper tape	29.0
SPORING AEROBES		
Bacillus globigii	Paper tape	120.0
Bacillus subtilis	Paper tape	170-250
Bacillus pumilus	Paper tape	170-330
Bacillus stearothermophilus	Paper tape	210.0
SPORING ANAEROBES		
Clostridium novii	Paper tape	220.0
Clostridium Welchii	Paper tape	270.0
Clostridium botulinum	Phosphate buffer	130-340

addition, various types of radiation are distinguished, which relates to the fact that some microorganisms display greater resistance to high-energy electrons than to gamma-rays. That is why in Scandinavian regulations it has been assumed that, in respect to efficiency, the electron dose of 35 kGy corresponds to the gamma-ray dose of 32 kGy.

It should be noted that in some countries there is a tendency to use sterilization doses lower than 25 kGy. At the same time, there is also a tendency to use higher doses. These problems have been discussed in detail in [8.2] with a recommendation that the level of sterilization doses should be standardized before 1993, that is before the planned unification of Europe.

8.2.3 Types of Sterilized Materials and Products

A large assortment of medical materials and devices are now being radiation sterilized. In addition to products for general use, such as disposable needles and syringes, catheters, infusion and transfusion sets, as well as packing materials, surgical threads, devices for implantation and transplantation, cytoscopes, prostheses, artificial arteries, etc. are also sterilized.

What all these products have in common is that they are produced from various *plastic materials* in which high-energy radiation may lead to indirect and undesirable physical and chemical effects, in addition to the desired sterilization effect. Therefore, the full range of polymers used in the manufacture of such products has been carefully investigated with the aim of selecting those that are appropriate for radiation sterilization. This work has been carried out on a large scale both in the USA and Europe, where half of the world's potential of radiation sterilization is located.

As a result of these studies, plastic materials, investigated with respect to such parameters as mechanical strength versus dose, transparency and gas formation, have been classified into four groups. The first group is that of *materials unsuitable for radiation sterilization,* those that undergo strong degradation in the 10 to 50 kGy (1-5 Mrad) dose range. The second group is made up materials that can be *radiation sterilized only in a single procedure,* i.e. cannot be sterilized twice. This group, like the first one, is small and consists of some acrylic resins and polypropylene. The third group are the polymers that *can be sterilized twice.* The last group is comprised of substances that are in theory suitable for *multiple sterilization.*

In practice, the above criteria for radiation resistance of plastic materials can only serve as partial or preliminary indicators. Since plastic materials used in the production of medical appliances should strictly conform to many standards involving a much larger variety of requirements, such as, for example, the *toxicity of the radiation degradation products* since even small amounts of such materials can be highly toxic. For example, in view of its high radiation resistance, polyvinyl chloride could in theory be subjected to multiple sterilization. However, in the light of the requirements imposed by the Pharmacopoeia, only few brands currently manufactured can be sterilized even once, due to the presence of some undesirable degradation products, such as HCl.

A separate problem is that of radiation sterilization of pharmaceuticals and their components. Although, in some countries, rules and regulations permit radiation sterilization of such products (e.g. British Pharmacopoeia allows for sterilization of some powdered pharmaceuticals), this technique has not,as yet, found widespread application in pharmacy. This is because in many cases a sterilizing dose of ionizing radiation leads to changes in the irradiated substance. For details see [8.4], where results have been tabulated of international studies on the radiation sensitivity of 311 pharmaceuticals and their components.

On the other hand, radiation sterilization is absolutely essential in preparing biostatic grafts which, immediately after obtaining them from donors, are frozen to minimize decay processes and then sterilized.

8.2.4 Systems of Accelerator Sterilization

Accelerator sterilization can be implemented either by a beam of accelerated electrons or by X-rays from bremsstrahlung obtained by electron-beam conversion. Due to high dose rates, ranging from 10 to 100 kGy s^{-1} (1-10 Mrad s^{-1}), electron sterilization is cost-effective and enables high throughput rates to be achieved. Typically, a dose of 25 kGy (2.5 Mrad) is a minimum dose, and satisfactory values of the *non-uniformity coefficient,* $D_{max} : D_{min}$, usually lie between 1.25 and 1.5.

Beam energy is an essential problem in electron sterilization. In an *off-line process,* products are separately packaged, e.g. in sealed plastic bags placed in larger shipping cartons. Typically, such packaging materials have densities between 0.08 and 0.25 g cm^{-3}. In two-sided irradiation (Figs. 8.1 and 8.3) for quick estimation of the relation between the maximum mass-thickness, F_{max}, and the electron energy, E_k, the following expression can be given

$$F_{max} = \frac{1.5\ E_k}{(dE/dx)} \tag{8.7}$$

where F_{max} is the maximum mass-thickness in g·cm^{-2}, E_k is the electron energy in MeV, and dE/dx is the stopping power in MeV cm^2·g^{-1}. For most polymers used in the production of medical materials and devices, the stopping power is about 1.75 MeV cm^2 g^{-1}. For example, for polyehtylene and polyvinyl chloride it is 1.93 and 1.67 MeV cm^2 g^{-1}, respectively. Thus, for

3 MeV electrons and two-sided irradiation the maximum mass-thickness is 2.5 g cm^{-2}. If the packaging material has a uniform structure and its density is 0.15 g cm^{-3}, the maximum height of the package cannot be higher than 16 cm. If the packages are higher the electron energy has to be increased. The maximum energy of 10 MeV cannot be, however, exceeded because otherwise some radio-activity could be induced. For example, if plastic materials made up of fluorine and nitrogen are sterilized, induced radioactivity may appear when the threshold energies of 7.9 and 7.5 MeV are exceeded, respectively. The half-lives of induced radioisotopes are, however, so short that the radioactivity levels in products that have been sterilized are quickly non-measurable.

The choice of appropriate instrumentation for accelerator sterilization depends, among other things, on the packaging structure. According to [8.5], over half of all packages used by the US manufacturers of medical devices could be sterilized

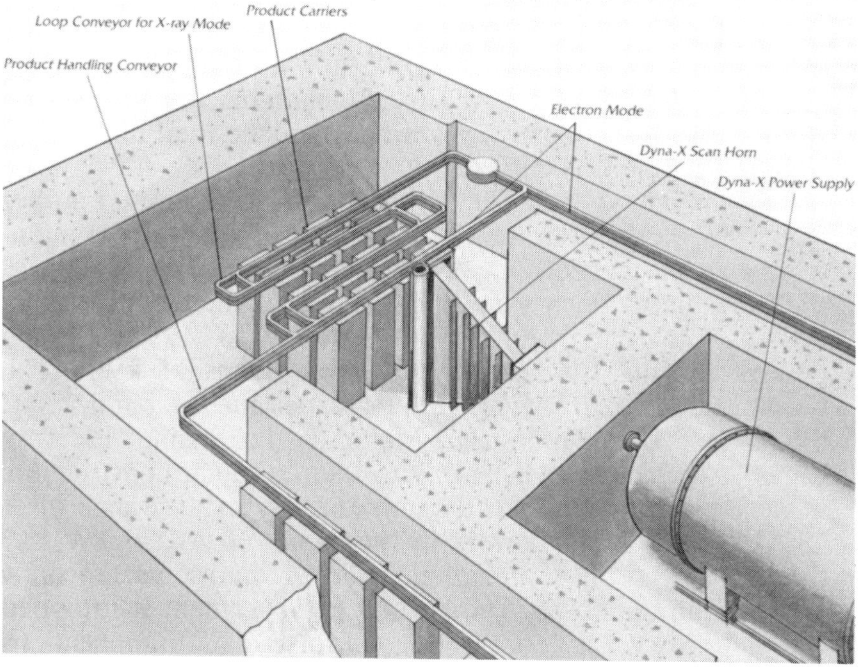

FIGURE 8.8 Dyna-X system for radiation sterilization with 4.5 MeV electrons and bremsstrahlung, mfd. by Radiation Dynamics, Inc. [F 5]

with 4.5 MeV electrons, which is at present the upper energy limit of accelerators equipped with high voltage generators of the dynamitron type (Sec. 2.2.2.3). Fig. 8.8 is a view of the Dyna-type system, equipped with a generator of this kind. The system can operate in an alternate mode, with shipping packages with dimensions of up to 35 cm are sterilized with electrons, whereas larger or the same size packages but with higher density can be sterilized with bremsstrahlung. By using the electron two-sided technique, the throughput is 27 m^3 h^{-1} for the average package density of 0.2 g cm^{-3}. The electron beam accelerator has a continuous use rating of 4.5 MeV and 150 kW [8.6].

The throughput rate of the accelerator system for alternate sterilization shown in Fig. 8.8 will, of course, depend on the structure of sterilized packages. With the use of an industrial 5 MeV, 200 kW electron accelerator employed in a dual electron-bremsstrahlung mode operating in three shifts (6800 h $year^{-1}$) and with identical dose of 25 kGy for both electrons and X-rays, the bulk throughputs will be about 25 m^3 h^{-1} and 4.8 m^3 h^{-1} for electron and X-ray processing, respectively. In a hypothetical case of using only the electron mode,the sterilization volume would be approx. 170,000 m^3, whereas using only the bremsstrahlung it would be 33,000 m^3. In the practical case, however, of 50% of the volume being subjected to electron sterilization and 50% to X-ray processing, the total throughput would be about 55,000 m^3 per year.

Systems employing rf linear accelerators constitute another type of medical sterilization unit. During the last decade intensive work was carried out on developing a new generation of such machines which would be suitable for industrial irradiations over the energy range of at least 5 to 10 MeV and beam powers ranging from several to several tens of kW. The Canadian company AECL [F30] has designed a family of rf linacs IMPELA (Industrial Materials Processing Electron Linear Accelerator) which, due to their modular construction, make it possible to achieve beam energies in the range of 5-18 MeV and beam powers in the range of 20-250 kW [8.7 through 8.9]. A pilot irradiation facility put into service recently at the Whiteshell Nuclear Research Establishment is equipped with a 10 MeV accelerator designed for radiation applications research in the development of new radiation techniques [8.10]. In France, the CGR MeV company [F26] has developed two types of accelerators: Cassitron and CIRCE II [8.11, 8.12]. The Cassitron machine is designed for the 8-10 MeV energy range and 5-10 kW beam power operation, whereas CIRCE II for the 7-10 MeV energies and 10, 20, or 30 kW beam powers. In both cases, the beam is swept by a 50 cm wide scanning system. A new company, MeV Industrie [F31], has also been set up in France with the aim of providing accelerators to industrial medical sterilization and food preservation centers [8.13]. The 10 MeV, 20 kW CIRCE accelerator which operates at the Swedish SCANCARIC facility enables sterilization of 30 to 40 thousand 1-20 ml syringes per hour (12 m^3 h^{-1}). For disposable needles the yield is 600 to 800 thousand per hour (8 m^3 h^{-1}) [8.13]. The Swedish company Scanditronix [F14] has also designed a system for radiation processing, called the EB 10, equipped with a 10 MeV, 30 kW rf linac.

Fig. 8.9 shows a schematic diagram of the Moelnlycke Project which is an example of the use of an rf linac for medical sterilization. It is being put into service by the Belgian subsidiary of the Swedish company Moelnlycke, one of the top manufacturers of medical disposables. The facility will be equipped with two accelerator of the 10 MeV, 20 kW CIRCE type,facing each other and placed vertically. Thus the products will be subjected to two-sided irradiation with two beams. For products with densities ranging from 0.11 to 0.20 g cm^{-3} the annual throughput will

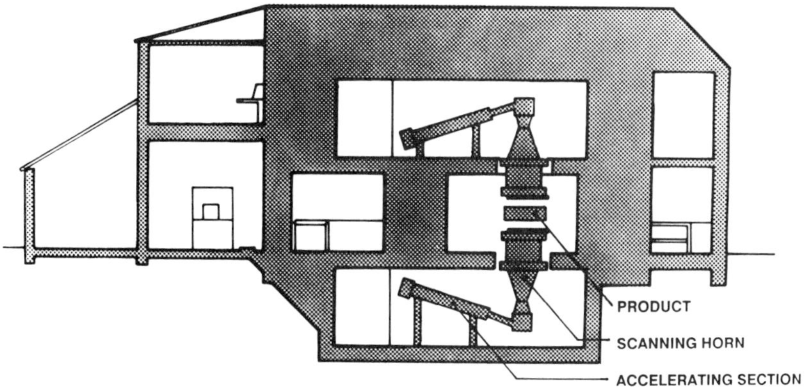

FIGURE 8.9 A system for radiation sterilization equipped with two 10 MeV, 20 kW CIRCE opposite rf linacs (Belgian subsidiary of Moelnlycke company) (after T.Sadat and Ch.Cuillandre [8.13])

be 70,000 m3 [8.13]. The conveyor width is ~1 m and the speed is ~1 m min^{-1}. The repetition rate of the machine is 600 Hz. Each pulse containing 39 J is scanning over the whole width of the product. So, variation of the conveyor speed can be very easily compensated for by adjusting the repetition rate. The installation at Moelnlycke was due to start in April 1991 and commissioning was planned for October 1991 [8.32].

8.2.5 Prospective Applications

Production of aseptic packaging materials. Recently, a good deal of attention on the part of the food industry has been focused on the use of *aseptic packaging materials.* The production of such materials is particularly well developed in Europe where the surface sterilization techniques employ ordinary or superheated steam and chemical methods, of which the action of hydrogen peroxide at elevated temperatures is most popular. These techniques are also being implemented in the USA, where, one company, which is a subsidiary of a European firm, has installed 60 packaging lines which use aseptic materials.

The disadvantage the above techniques have is the presence of moisture and the need to remove chemical residues after processing. Therefore, another approach is now being more often considered, viz that of using energetic electrons which permit the generation of a sterilizing dose of 25 kGy (2.5 Mrad) in a technologically simple and relatively inexpensive way. Since these applications involve foils or containers, sterilization is limited to surfaces. Therefore, low-energy electron accelerators with a linear cathode, operating in the low-energy 150-300 keV range, may be especially well-suited for such purposes (left-hand side of Fig. 8.1). It may be expected that wider application of electron sterilization of aseptic materials for the food industry is only a question of time. This is also strengthened by the fact that, according to the regulations in force in many countries, processing of this type can be performed without special clearance. Such regulations stipulate, in general, that no processes should arise from treating food packaging materials that would lead to undesirable effects in the packaged food. In this respect, electron processing, due to its low energy range, is completely harmless.

Low-energy sterilization should also find application in the sterilization of packages containing radiosensitive biological substances. These may be, for example, pharmaceuticals packaged under aseptic conditions in thin-walled plastic or glass packages. After aseptic filling, the closed product is packaged in an appropriate foil (e.g. Tyvec) and then sterilized by gas. The results of work on replacing ethylene oxide sterilization with a beam of low-energy electrons, which would permit selective treatment of surface and avoid direct treatment of the sensitive product, are presented in [8.16, 8.17].

Cost analysis. It is a fact that recently gas sterilization has been taken over, in large measure, by radiation sterilization (Fig. 8.5). What remains is the question of whether radiation sterilization will be implemented with radioisotope ^{60}Co processors, or whether and to what extent it will become the domain of accelerator techniques. One of the decisive factors in the competition between these two techniques will undoubtedly be the problem of costs. The methods of cost calculation and analysis have been given more attention in [2.16]. Here we shall quote some comparison analysis made in [8.18]. Two sterilization facilities with annual throughputs of 370,000 m^3 and 8,000 h operating time have been analyzed. It is assumed that the cartons have an average bulk density of 0.2 g cm^{-3} and the mini-

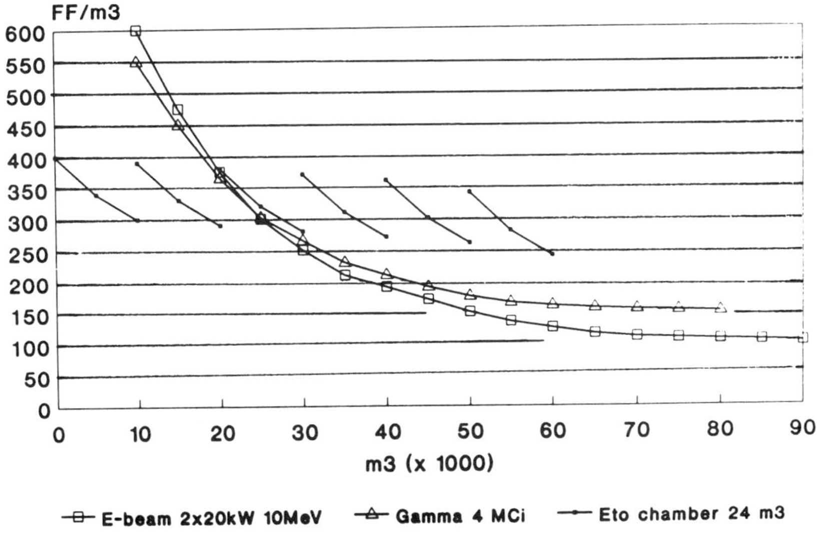

FIGURE 8.10 Cost comparison for Moelnlycke (after T.Sadat [8.32])

mum sterilizing dose is 25 kGy. The first facility would be equipped with a 4 MCi ^{60}Co source, and the other would have an generator for electron and bremsstrahlung sterilization, rated for 500 kW of electron power at an energy of 5 MeV. Total investment cost of the ^{60}Co facility is estimated at $ 7.9 million, whereas that of the accelerator processor at $ 6.2 million. After taking into account all the other costs, the unit cost per cubic meter of product is $ 6.88 with ^{60}Co and $ 5.59 with accelerator. Thus the replacement of the radioisotope processor with an accelerator would save the sterilization unit costs by about 18% [8.18].

The cost comparison for Moelnlycke project (Fig. 8.9) is shown in Fig. 8.10 [8.32].

It is clear from the above figure that both the 2x20 kW, 10 MeV accelerator and the 4 MCi radioisotope processor have become competitive relative to the ETO sterilization (24 m^3 chamber) when the throughput exceeds 30,000 m^3. Economically, the accelerator system is better than the radioisotope processor, especially when the annual throughputs are higher than 50,000 m^3 [8.32].

8.3 Radiation Preservation of Food

8.3.1 Methods of Preservation

Preservation of food with the use of high-energy ionizing radiation has over a 50-year-long history: the first attempt at food irradiation were made in the USA as early as 1943. Since then, radiation preservation of food has been a most controversial subject on which many totally different opinions have been voiced. Nevertheless, radiation preservation techniques have long been the focus of intensive research and interest of the food industry in the USA. One should bear in mind that in the present-day farming industry the losses due to improper storage and preservation of food may be as high as 20% of the total food production. This is especially true of the developing countries in hot regions of the world. For example, in Africa food losses amount to 30-50% of the total production.

Another factor contributing to the prospective use of radiation preservation of food is the problem of energy. Conventional methods of food preservation, i.e. *refrigeration and deep freezing* as well as the *thermal method,* are highly energy-consuming techniques. Fig. 8.11 is a comparison of radiation processing costs for various foods as a function of the average dose required, taking account of various estimates and data. The values expressed in percent relate to the cost of processing compared with the value of the product. The shaded areas represent the equivalent cost range estimates for conventional (non-radiation) methods of preservation. The conclusion that may be drawn from Fig. 8.11 is that radiation processing costs, in spite of considerable differences in their estimation, are quite comparable with those of other conventional techniques. The anticipated increase in the cost of energy will reveal the ad-

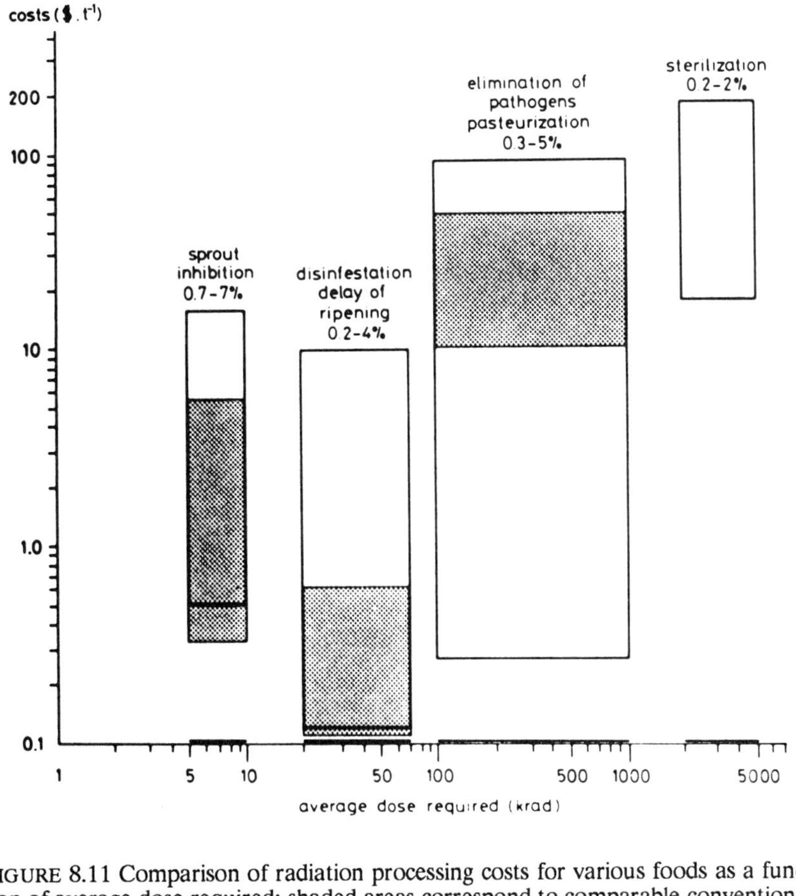

FIGURE 8.11 Comparison of radiation processing costs for various foods as a function of average dose required: shaded areas correspond to comparable conventional (non-radiation) techniques (after J.Farkas [8.19])

vantages of radiation preservation. For example, sprout inhibition in onions and potatoes by radiation together with subsequent storage requires only 67 kJ kg^{-1}, whereas refrigeration over the same period of time requires 1080 kJ kg^{-1}, i.e. 16 times as much energy.

The purpose of radiation preservation of food is most frequently to prevent undesirable changes occurring in the food from the time of its production or gathering. In addition to preservation, radiation processing can also provide a means of destroying pathogens. The main radiation techniques are listed in Table 8.3. By using small doses, i.e. those below 1 kGy (100 krad), sprouting of foods such as potatoes, onions and garlic can be inhibited, or food can be disinfected by eliminating pests and parasites. Processes, such as *radurization,* prolongation of frozen storage time by the reduction of food-spoiling microorganisms, as well as *radicidation or elimination of pathogens,* require slightly higher doses ranging from 1 to 10 kGy (0.1-1.0 Mrad).

Radicidation of animal feed is a separate problem. This technique involves destruction of pathogenic bacteria, fungi and other parasites in animal feed. Studies have demonstrated greater efficacy of radiation processing over conventional methods as concerns the destruction of pathogens and preservation of the nutritious value of the feed with unchanged physical and chemical properties. Radiation processing is primarily used in the feed intended for the so-called *germ-free (GF) animals,* and, to increasingly greater degree, in the so-called *specified pathogen free (SPF) breeders and pedigree animals.*

The total amount of foods processed by irradiation in the world is estimated to be about 500,000 ton per year [8.20]. Mention should be made of sterilization of hospital meals, of the food for US astronauts, of the disinfection of spices, inhibition of potatoe sprouting, etc. The small range and amount of radiation processing in food production are primarily due to the reservations with which administrations in various countries have been issuing licences. In 1958, the US Congress passed a bill which stipulated that radiation processing of food should require special procedures and be treated as a kind of *"additive"* to the preserved foods. In 1964, after having cleared several foods for irradiation, among others, potatoes to inhibit sprouting, the next licence was issued only 19 years later, i.e. in 1983.

TABLE 8.3 Types of Radiation Processing of Foods and Ranges of Applied Doses (after M.A.Welt [8.33])

Technique	Example	Dose	
		kGy	krad
Sprouting inhibition	potatoes, onions, garlic, ginger	0.03 - 0.15	3 - 15
Elimination of insects	wheat, spices, fruit, vegetables, dry fish	0.2 - 0.8	20 - 80
Elimination of parasites	trichinella spiralis ascaris	0.15 - 0.8	15 - 80
Prolongation of storage time-radurization	poultry, fish, shell-fish, strawberries, mushrooms	1 - 3	100 - 300
Eliminations of pathogens	Salmonellae, E.coli, Pseudomonas sp., Staph sp., Campylobacter, etc.	3 - 8	300 - 800
Sterilization	venison, pork, seafood	25 - 46	2300-4600
Elimination of viruses		35 - 60	2500-6000

The situation has now completely changed. The IAEA (International Atomic Energy Agency), FAO (Food and Agricultural Organization) and WHO (World Health Organization) have set up a joint committee of experts, known as JECFI (Joint Expert Committee on Food Irradiation). In 1979, the JECFI stated that *"the irradiation of any food commodity up to an overall dose of 10 kGy would not pose a toxicological hazard and the treatment would not create any special microbiological or nutritional problems"*. Then, in 1983, the co-called Codex Alimentarius Commission (established in 1962 by the FAO and WHO, embracing over 100 countries) approved the above statement and recommended its implementation by the respective member countries.

As for the maximum energies to be used in food radiation preservation, the JECFI has recommended that the electron beam energy should not exceed 10 MeV, and if converted to bremsstrahlung the energy should not be higher than 5 MeV.

These recommendations had been preceded by many years of research on various aspects of the effect of high-energy radiation on food. In Fig. 8.12 results have been tabulated of one of the most controversial issues, viz the activity produced in meat by electrons with energies ranging from 10 to 16 MeV [8.21]. This activity, as well as that from natural background, is expressed here in numbers of atomic disintegrations per second. It follows from Fig. 8.12 that the JECFI level of 10 MeV causes increased activity of meat by 1 decay per week per 1 kg. On the other hand, natural radioactivity per 1 kg of meat is higher than 100 disintegrations per second, whereas natural activity in average human body of 70 kg is about 10,000 disintegrations s^{-1} which corresponds to about 140 disintegrations s^{-1} per 1 kg of human tissue. It is,therefore, clear that, by assuming maximum

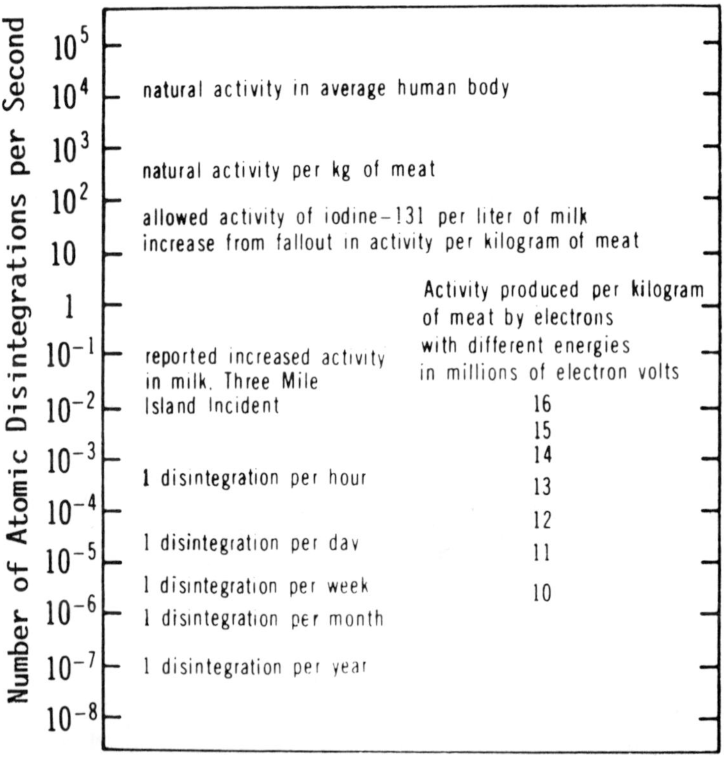

FIGURE 8.12 Natural and induced radiation; the figures are for meat stored for 3 months after treatment with 32 kGy of radiation of different energies (after R.L.Becker [8.21])

electron energy of 10 MeV, the added activity is lower by 7 orders of magnitude than the natural activity.

The practical consequence of the evolution of opinions concerning the lack of adverse effects of radiation processing of food can be seen in the fast growth of radiation clearances issued in the 1980s. At present, in 36 countries health and safety authorities have approved irradiation of more than 40 different foods [8.22]. A detailed list of such clearances is given in [8.23]. Table 8.4 provides, a chronology of clearances issued in France between 1972 and 1990. Farming and food industry have a special place in France's economy since they take a large share in the total number of work places (10%), they are also responsible for 5% of the GNP and, in exports, hold the first place before motor industry. In addition to the clearances listed in Table 8.4, on the waiting list are such products as ham, casein (3 kGy), fresh fish (1.5 kGy), soft cheese (3 kGy) and dried fruit and spices (4 kGy) [8.13].

TABLE 8.4 A list of clearances for radiation processing issued in France between 1972 and 1990 (after T. Sadat and Ch.Cuillandre [8.13]).

Product	Dose [kGy]	Date
Potatoes[1)]	0.15	1972
Feed for laboratory animals	40	1981
Garlic	0.15	1984
Onion	0.15	1984
Shallot	0.15	1984
Deboned poultry	5	1985
Dried legumes	10	1985
Arabic gum	9	1985
Wheat products	10	1985
Spices	11	1986
Blood, plasma	10	1986
Dried fruit[2)]	1	1988
Frog legs	8	1988
Rice flour	5	1988
Strawberries	3	1988
Herbs (for infusion)	10	1990
Chopped poultry	5	1990

1) Clearance revoked
2) Elimination of parasites

Readers interested in radiation preservation of food are referred to the Food Irradiation Newsletter, published jointly by the FAO and IAEA, and to the publications of the Food Research Center in Karlsruhe (Germany),e.g. [8.25, 8.26], which also issues a bibliographical review on radiation processing of food [8.27].

8.3.2 Accelerator Systems

Systems intended for food irradiation, for the same reason as those employed in accelerator sterilization, are currently equipped with rf linacs operating at the maximum permissible energy level of 10 MeV [8.28 through 8.30]. In 1986, the MeV Industrie company was one of the first to install a 7 MeV, 5kW accelerator in one of the British meat factories. Mechanically carved chicken meat was irradiated with a dose of 3 kGy, and the throughput achieved was 3 t h^{-1}. The facility operated in a three-shift system for 230 days per year. In 1990, it was replaced by a 10 MeV, 10 kW CIRCE type accelerator, and, as a result, the throughput doubled [8.13]. Before processing, the meat is mechanically separated and packaged in 10 kg polyethylene bags (53 x 35 x 5.5 cm) and then deep frozen at -18°C [8.29].

In the USA, radiation preservation of food — among others that for astronauts — was carried out with a 10 MeV linear accelerator as early as 1970s [8.30]. Currently, six projects, commissioned by the US Department of Energy, are being implemented in Florida, Oklahoma, Iowa, Hawaii, Washington and Alaska with the aim of determining the optimum methods

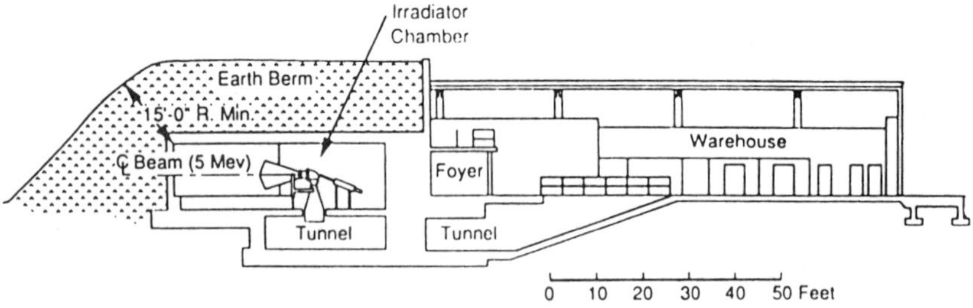

FIGURE 8.13 Longitudinal section of Florida agricultural commodities irradiator equipped with a 10 MeV, 10kW/ 5 MeV, 20 kW CIRCE type rf linac (Gainesville, Florida) (after B. J. Smittle et al. 8.34])

of food preservation. The facility at Gainesville (Florida) is equipped with a 10 MeV, 10 kW - 5 MeV, 20 kW CIRCE type rf linac shown schematically in Fig. 8.13.

The Florida model of the CIRCE III accelerator is equipped with two scanning horns on two levels [8.34]. The lower level has a one meter horn producing 10 MeV electrons at 10 kW to treat shallow layers of commodities. The upper level has a 2.5 meter horn producing 5 MeV X-rays with a 20 kW electron beam to treat large cartons or pallet loads of commodities. A switching magnet is used to continuously control the energy of the accelerated electrons and to control the use of either scanning horn. Automated conveying systems will transport commodities on both levels. The facility is divided for irradiated and unirradiated commodities with freezer and refrigented storage on both sides.

Florida has a $ 6.2 billion dollar agricultural industry with over 300 major commodities produced on 14 million acres (5.68 million hectares) of fruits and vegetables and 18 million acres (7.70 million hectares) of forestland. Interests in Florida range from insect sterilization for biological control of irradiation of citrus to meet regulatory requirements regarding insects to increased seafood safety from Salmonella, Listeria and Vibrio bacteria, to shelf-life extension of food materials, and to non-food items such as potting soil and alligator hides. Food safety is a subject of a paramount interest and concern not only in the US but around the world today. It is estimated that up to 33 million Americans, roughly 14% of population, become ill from microorganisms in their food each year. Microbial foodborne disease is the number one food safety hazard with estimates of $ 4.5 billion impact in medical costs and lost productivity in the US each year.

In addition to food safety, Florida's other main facility objective is non-chemical elimination of pests. With the vast amount of international travel (Florida hosts 28 million visitors each year) and international trade with 27 international ports and airports, pests such as the Mediterranean fruit fly are perennial threats. There have been three outbreaks of mediterranean fruit fly in the past four years with successful quick *eradication* and *reduction of chemical spraying* by 50% with the use of sterile fly release. The state has already completed a $ 1.25 million sterile fly mass rearing facility immediately adjacent to the irradiator site. In a pilot test phase, the Carribbean fruit fly is being mass reared, sterilized by irradiation and released to test control efficiency of the sterile insect technique against this major pest [8.34].

The total federal funding for the Florida project is $ 5.4 million with additional contributions in funds, personnel, management costs, and land through the State of Florida. The facility will be available for utilization and research by any government, university, private or scientific entity.

The facility in Iowa has a similar accelerator equipped with a single electron or X-ray port and is intended for meat preservation studies. In the systems designed for meat radiation processing, a global average dose of 5 kGy ensures a microbiological quality which meets the requirements for cooked meat. The destruction of Salmonellae occurs at the dose of about 3 kGy. Since in the USA the yearly consumption of chicken meat is $5.4\ 10^9$ kg, some 300 accelerators would be required for the expected throughput of 4.5 t h^{-1}. According to [8.30], the cost of accelerator processing in a facility with a yearly throughput greater than 20,000 t should be less than US 7.0 ¢ per kg. In very high throughput facilities this may be lowered to 3.0 ¢ per kg.

The future widespread use of accelerators in food preservation is related to the results of the experiments carried out at the pilot facilities mentioned earlier. However, some less optimistic opinions as to the future of food irradiations are also being voiced [8.31].

8.4 Environmental Protection

For quite a long time now charged particle accelerators have been playing an important role in *environmental protection* since they can be employed as analytical tools (Chapter 9). During the last two decades several radiation techniques have also been developed which may be potentially essential in environmental protection. The feature they all share is the use of beams of accelerated electrons with moderate energies in the range from a few to ten-odd MeV, but with high powers of the order of several hundred or even several thousand kilowatts. One of the reasons why these techniques are still at the experimental stage of development is that accelerators with the above parameters have not been, as yet, commercially manufactured.

Disinfection of drinking water. The *method of chlorination*, which is at present almost exclusively used for this purpose, leads to formation of both chlorine and active chlorides in amounts of about 0.5 mg l^{-1}. The formation of these substances, which are not innocuous to human health, could be prevented by irradiating water with a dose of approx. 1 kGy, which has a bacteria destruction capability higher than that of the maximum amounts of chlorine by a factor of 100-1000. A single 12 MeV accelerator with a beam power of 600 kW could ensure *disinfection of drinking water* for a city with a population of 100 thousand at the daily consumption rate of 380 l per person. With the use of accelerators with 1000 kW beam power the probable unit cost of treatment would be 2.00 ¢ m^{-3}. However, currently the costs of chlorination are lower by a factor of 3.

Disinfection of waste water and sewage sludge. Re-use of *waste water and sewage sludge* along with that of solid wastes constitutes a major problem in municipal economy of the world's industrialised countries. For example, an average Canadian citizen produces 380 l of sewage per day, which results in an annual Canadian production of 10^6 t of sludge. In the USA and Europe the respective figures are 10^7 and $5{\cdot}10^6$ t. These are the amounts of dry (solid) contents of the sewage sludge: the volumes of raw (unprocessed) sewage sludge are several times larger. Therefore, intensive studies have been carried out on the use of *radiation techniques for re-cycling of urban and industrial sewage as well as of excreta from animal farms.* In addition to some substances harmful to the public, such as heavy metals or toxic compounds, sewage sludge also contains numerous pathogens that can effectively be destroyed by irradiation. Doses of 2 to 5 kGy (200 to 500 krad) effectively eliminate most of the pathogenic bacteria and microorganisms.

Typically, raw sludge contains over 10^8 microorganisms per cm^3; the gram-negative bacteria represent 10^4 -10^7 germs per cm^3, viruses 10^2 - 10^3 per cm^3, and Salmonellae 1-10 per cm^3. If conventional disinfection methods are applied, such as fermentation or thermal processes (i.e. heating to 70° C and keeping the material at this temperature for 30 min), a reduction rate of 100 to 1000 can be achieved. However, when the dose of 4 kGy is used, the inactivation coefficient is about 10^{-5}.

Utilization of solid wastes. Cellulose, as a polymer of crystalline structure made up of anhydroglucose units, could become an attractive source of food, animal feed and chemicals. Various kinds of solid wastes constitute an enormous source of cellulose. A US city with 1 million inhabitants produces 1200 t of waste containing about 700 t of cellulosic materials including cellucotton and water: pure cellulose content is estimated at about 300 t per day. Another large source of cellulose are wastes from farms, which in the USA amount to some 550 million tons a year. Assuming that the average cellulose content is 32%, the above waste could yield 28 million tons of ethylene yearly, which represents twice the US consumption of that basic petrochemical compound for 1974.

The crucial step in the production of the above highly useful materials from the cellulose wastes is optimization of the *conversion of cellulose to its monomer glucose.* At present, either *acid or enzymatic hydrolysis* can be used for this purpose. The essential role that the accelerators can play in intensifying these processes consists in pre-treatment to improve the glucose yields. With the assumption that the dose of 100 kGy should be used to irradiate 1000 t of cellulose wastes daily, the total electron beam power should be about 1400 kW. The glucose yield would then increase to about 10% of the dry pure cellulose weight.

Removal of combustion gases. In many highly industrialized countries, increased air pollution by combustion gases from factories, power stations, steel and chemical works and car engines has become a serious problem. In Japan, for example, the operation of some factories is often restricted in summer when the pollution exceeds permissible levels. Sulphur dioxide, SO_2, and nitrogen oxides, NO_x (where x=1 or 2), and especially NO, usually generated in the combustion of fuel oil and coal in electric power stations or other industrial facilities, pose a particularly serious hazard to human health and life.

The electron technique of SO_2 and NO_x removal from combustion gases consists in passing these gases on the way out to the chimney through a reaction chamber, where they are sprayed with ammonia. By irradiation with 1 MeV electron beam of high power, SO_2 and NO_x are precipitated and converted to complex salts of ammonium sulphates and nitrates which constitute useful mineral. The main problem in practical implementation of this technique is the energy efficiency of the system since the accelerator would have to use a few percent of the electrical energy generated by the power unit.

The future conventional use of accelerator methods for the removal of pollutants will depend, in large measure, on the cost-effectiveness of these techniques. A detailed discussion of these problems is presented in [2.16].

Transmutation of nuclear waste. The present plan for disposal of defense and commercial *high-level nuclear waste,* namely *vitrification and long term (10^4-10^5 yr) storage* in geologic repositories, is meeting increasing public scepticism and resistance [8.35]. *Accelerator transmutation,* while not eliminating the need for *short-term storage,* may offer a solution to the overall waste problem. Proton linacs driving high-flux *spallation neutron sources* (about 10^{20} primary neutrons per second) are being considered for transmutation of nuclear waste. After suitable moderation, these neutrons can be used to burn long-lived actinides and fission products from nuclear waste through capture and fission processes. With thermal neutron fluxes in the 10^{16} n cm^{-2} s^{-1} range (100 times greater than in typical thermal reactors) the higher actinides, such as ^{237}Np can be converted by neutron capture to daughter products that are fissioned rapidly by a second neutron interaction before they can decay to nonfissile isotopes. High fluxes of neutrons at thermal energies, where cross sections are large, also permit rapid *conversion of long-lived fission products* such as ^{99}Tc and ^{127}I *to stable or short-lived isotopes.* A single proton linac (1.6 GeV, 250 mA cw current) could continuously burn the waste discharge of about eleven 1000 MW$_e$ light-water reactors while providing enough power to run itself [8.35].

8.5 Microholes and Nuclear Track Filters

Heavy-ion track etching. Nuclear track etching of the permanent *track defects* due to bombarding of a solid by high-energy particles has long been utilized in the so-called *track detectors* in physical measurements. Since the mid-1970s this effect, however, has received an ever greater attention because of its potential use for technological purposes, primarily as a *microhole drilling tool.* Since about 1970 the nuclear track technique has already found wider application in the commercial manufacture of nuclear track filters. In Western Europe, this technology has centered, above all, round the GSI (Gesellschaft für Schwerionenforschung at Darmstadt, Federal Republic of Germany) [8.36]. Intensive work is under way at Dubna (USSR) and GANIL, Caen (France). A detailed review of the production and use of nuclear tracks and a broad literature on this subject are presented in [8.37].

By bombarding an organic foil, mica or many other insulating materials with nuclear fission fragments or with a beam of accelerated ions and then etching it, tiny sharply shaped *microholes or microchannels* distributed statistically on the material surface can be obtained. Among various techniques that can be used, heavy-ion accelerators seem to be most suitable for this purpose, since they can generate a monoenergetic and monoisotopic beam of parallel ions with intensities ranging from 10^0 to 10^{14} ions per second (specific energy 10^{-3} to 10^2 MeV/nucleon).

The effect of the ion passage through insulating materials is a *nuclear track defect* in the bombarded material. It depends on the atomic number of the ion, and attains maximum for the heaviest ions, i.e. those of uranium. Within the specific energy range between 1 and 10 MeV per nucleon the specific energy loss and track damage exhibit a relatively wide and flat maximum depending on the type of ion and material. Typically, in solids the ion ranges correspond to the thicknesses of 10 to 100 μm. With the use of much higher specific energies, for example, those of relativistic beams of heavy ions, the above ranges can be extended up from 100 μm to 10 cm. Such energies are, however, available now only from a few of the largest accelerators. It should be borne in mind that ^{238}U ions accelerated to a specific energy of 1 MeV correspond to a total energy of 238 MeV, i.e. their velocity becomes equal to 5% of the speed of light. As a rule of thumb, the ranges of the most commonly employed heavy ions are very roughly proportional to the specific energy for the 0.1-10 MeV per nucleon range, and are practically independent of the ion mass and the target material and its composition. The corresponding ion ranges are roughly 1-100 μm. Conventional techniques, such as X-ray lithography and laser drilling, permit the drilling of microholes with the minimum diameter of about 0.1 μm. It is only the ion track technique that makes possible microholes with diameters going down to 0.01 μm. The *throughput efficiency* of this new method, as expressed by the number of holes generated per second, is greater than that, for example, of X-ray lithography by more than 4 orders of magnitude. The ion track technique has the outstanding feature of providing holes with a very large relative *aspect ratio* (length divided by width).

The hole diameter can easily be enlarged from the minimum value of ~10 nm to about 1 mm by increasing the time of etching which is linearly proportional to the obtained diameter while the density of holes, which depends on the flow of ions (ions per cm^2), can be controlled with ~10% precision in the range between 1 to 10^{10} holes cm^{-2}.

Apart from ion beams, microholes can also be manufactured with the use of spontaneous fission (^{252}Cf) or uranium fission fragments obtained from nuclear reactors. For this purpose, a ^{235}U foil, which acts as a *neutron-to-fission convertor,* is placed in one of the reactor channels. This method is used for commercial manufacture of so-called *nuclear track filters* in the form of polycarbonate or polyester membranes, known as Nuclepore® filters [F36]. Methods using fission products have, however, some disadvantages, the available ranges are relatively short and, consequently, filter thickness cannot exceed 10 μm. In addition, individual holes are irregular in shape and divergent relative to each other. On the other hand, an ion beam, properly focused and collimated, provides parallel ion tracks within ±0.1° and pore diameters within ~ 2%.

Filtration techniques. Fig. 8.14 lists various types and sizes of particles which are subjected to *microfiltration* currently employed in medicine, biology and industry as well as in environmental protection. Particles liable to filtration can be classified in two groups. The first group is made up of *mechanical pollutants,* such as dusts, aerosols, fibres, small glass and rubber fragments, etc. The second group consists of *microbiological pollutants,* such as bacteria, moulds, fungi and their spores.

Typically, filtration techniques are used for filtering pollutants with diameters in excess of 10 μm. *Microfiltration* includes particles with diameters in the 0.1 to 10.0 μm range, whereas *ultrafiltration* involves macromolecular separations in the 0.001 to 0.1 μm range. Finally, the technique of reverse osmosis is used in the separation of small molecules and ions less than 1 nm (10 Å) in diameter.

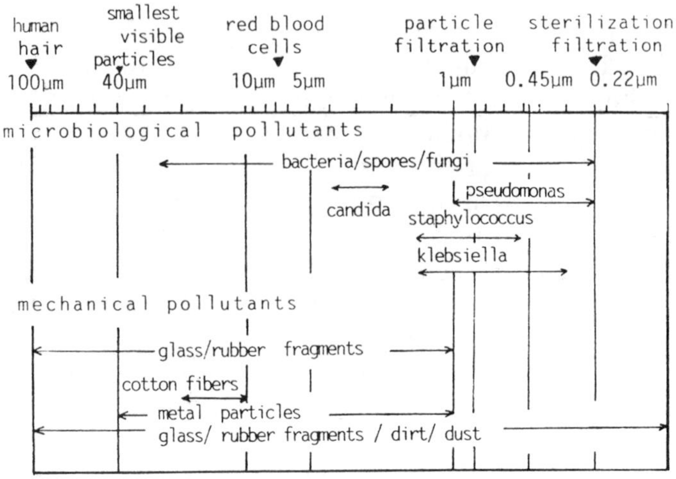

FIGURE 8.14 Range limits of mechanical and microbiological pollutants subjected to micro and ultra-filtration

Thus, nuclear or ion track filters of suitable pore sizes can be employed both in microfiltration and ultrafiltration. Nuclepore® filters with diameters in the 0.015 - 0.10 μm range are used for protein or virus filtration. Those with diameters of about 1.0 μm are used for sterilizing filtration, for beer and wine stabilization, for the production of high purity liquids and for microfiltration of alcohol solutions. The largest-diameter filters (10.0 - 12.0 μm dia.) are used for prefiltration of corrosion-inducing liquids and for identification of large microorganisms.

Systems of Ion Filter Production. For the *on-line production of filters* from polymer films a system has been suggested at Dubna (USSR) as shown in Fig. 8.15. A scanning ion beam is incident on a bent film moving at the speed of several metres per second. The irradiation roll is bent so that tracks become divergent in one plane; thereby the selectivity of the filter at a given filter transmission is improved by multiple pore suppresion. In another version (Fig. 8.15b), after irradiation with an ion beam, the foil is irradiated with UV to increase the etching rate. For example, for a polycarbonate film irradiated with 300-400 nm wavelength, this rate can be increased by a factor of 10. Then the film passes through an etching bath with hot solution of sodium or potassium hydroxide, through a stopping bath and a rinsing water bath, and is finally dried.

At GSI, a facility for the continuous irradiation of 12 cm wide plastic foils has been built as a first step to design a larger one, capable to irradiate foils of 30 cm width at a maximum speed of about 1 m s^{-1}. It is currently (1991) capable of irradiating filter foils of up to 50 μm thickness at a maximum speed of about 30 cm s^{-1} [8.38]

Lately, Cyclopore™ membranes are commercially produced by a company of the same name [F24]. The process of bombarding polymer films with heavy ions accelerated by a cyclotron

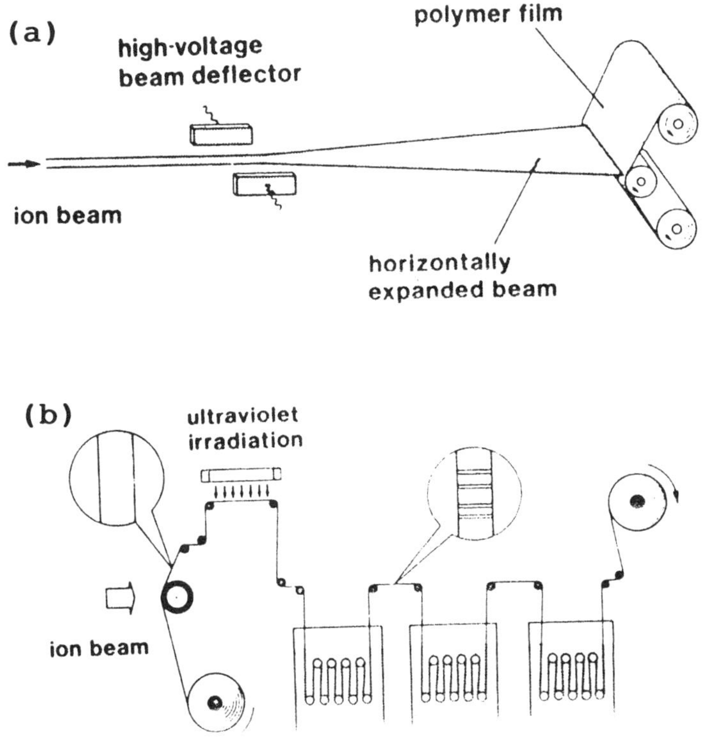

FIGURE 8.15 On-line production of filters using an ion beam: (a) foil irradiation, (b) technological line

was patented by University of Louvain (Belgium) and jointly developed with IBA [F23]. This procedure permits a multitude of applications in the fields of microfiltration, cell culture, separation and purification techniques.

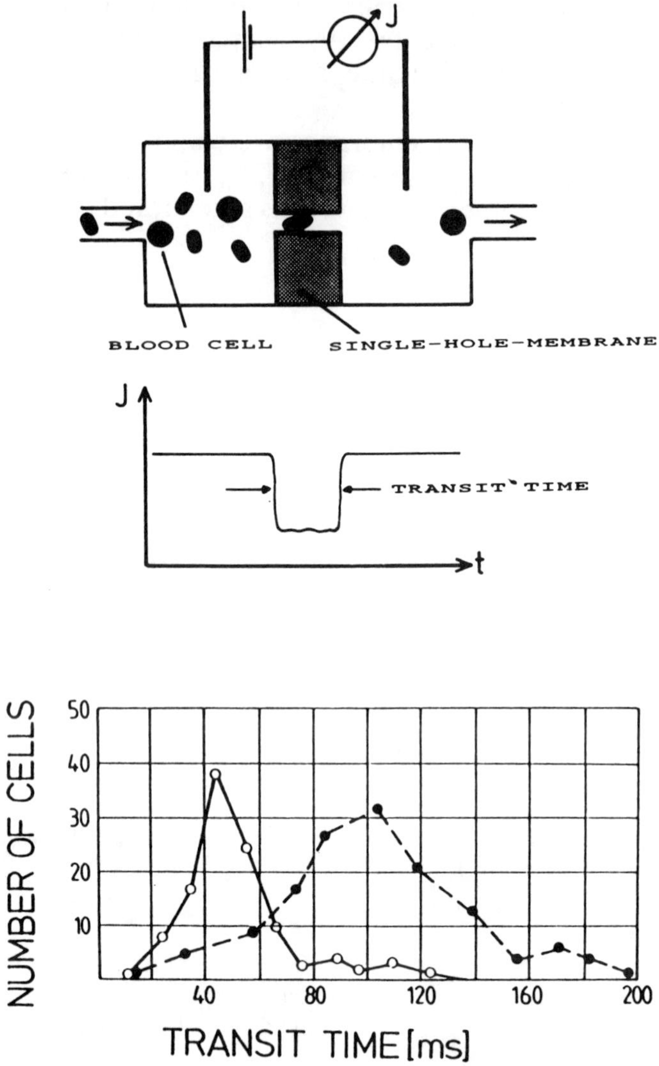

FIGURE 8.16 Top: Measurement of the flexibility of red blood cells by means of a single-pore membrane. Less flexible cells need a longer time to pass the narrow track channel; Bottom: transition time spectra for normal (left peak) and diseased (right peak) red blood cells (after B.E. Fischer [8.39])

In the *off-line treatment,* a foil is placed in a sample changer with a capacity of 20 samples which are consecutively fed into a vacuum chamber and irradiated. These so called *single-pore membranes* have been produced at GSI for many years [8.38]. Membranes of about 50 mm diameter contain exactly one single etched track pore of 50 μm length and 5 μm diameter at their center, and their production represents one of the most economic uses of an ion beam.

Single-particle techniques. Single-pore membranes are used to simulate a short section of *human circulatory system,* where *red blood cell* of about 8 μm diameter have to squeeze themselves through capillaries of about 5 μm diameter. Healthy cells are flexible enough to do that, but there are diseases like diabetes which reduce the flexibility of the red blood cells. Their flexibility can be measured in vitro by using a single pore membrane as a barrier between two small vessels, one of which contains a diluted blood sample under pressure. By measuring the transition time of individual blood cells through that model capillary it is possible to see clear differences between normal and diseased blood (Fig. 8.16).

In the *human microcirculation,* rigid erythrocytes which may be few in number (2 to 3% of the red blood cell, RBC, population) have difficulty in passing through capillaries. Such cells have disproportional effects on flow in vivo and may decisively influence the results in vitro (e.g. filtration through narrow pores). The study of this concept of sub-population is possible using the *Cell Transit Analyser* [8.40], which measures the passing time of individual erythrocytes through cylindrical micropores. The filters have different pore diameters, between 4.5 μm and 5.0 μm, less than the RBC diameter (7.5 μm) and different lengths between 11 μm and 21 μm. The analyser provides knowledge of rheological property distribution of RBCs and makes it possible to study the presence of possible sub-populations.

The superiority of nuclear track microfilters over any other filters is their unitary size of pores in one filter, which enables them to remove all bacteria and particles bigger than their pore size. In the test of ability of Macrofol track microfilters made with heavy ion from UNILAC accelerator at GSI, seven filters were used to filtrate the bacteria, the pore sizes of which were 0.4, 0.6, 0.8, 1.0, 1.5, 2.0 and 3.0 μm respectively; the sizes of

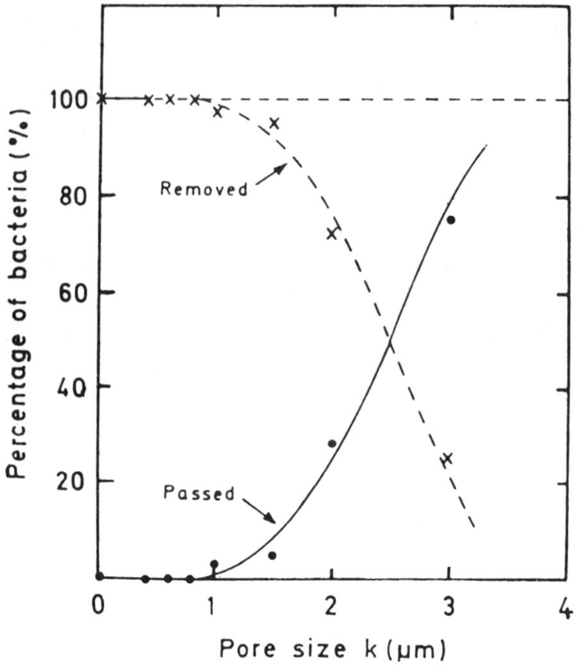

FIGURE 8.17 Percentage of removal and passed bacteria through nuclear track microfilters with pore size from 0.4 to 3.0 μm (after X. Hao et al. [8.41])

most bacteria were from 0.5 to 2.0 μm. The percentages of removed and passed bacteria are shown in Fig. 8.17. From this figure one can see that 100% of bacteria can be removed by Macrofol nuclear track microfilters with pore sizes less than 0.6 μm for a given group of bacteria [8.41].

CHAPTER 9

Accelerator Analytical Methods

9.1 Microbeam Techniques

A concise review of analytical methods implemented with beam of high-energy ions has already been presented in Sec. 1.5 (Fig. 1.23). Initially, the diameters of the accelerated ion beams were of the order of several mm. It was only after new systems generating the so-called *ion microbeams* had been developed that full advantage could be taken of the analytical capabilities, and especially of the high sensitivity to permit analysis of samples with weights even less than 1 μg. *Microbeams* which are now taken to include those whose diameters range between 50 and 1 μm, are now being used in over 50 facilities world wide [9.1], and are becoming ever more popular since they provide a means of investigating surface distributions, i.e. *surface mapping.* If the ion beam is compressed to micrometer dimensions, the analytical methods can be used on the microscale at the target. An instrument where such analyses are performed is normally referred to as a *nuclear microprobe* (NMP). The problems of generation and use of accelerator microbeams have been dealt with in special monographs [9.2 through 9.4] and at periodical conferences [9.5 and 9.6].

The simplest design to obtain microbeams with diameters ranging from a few to several tens of μm consists in using a *collimator* in the form of a solid plate with a hole inside. The smallest diameters of such holes vary from 1 to 2.5 μm (Heidelberg, Chicago). However, by using diameters that are so small, the beam intensity becomes greatly reduced. Another drawback that collimator designs have is the background radiation result-

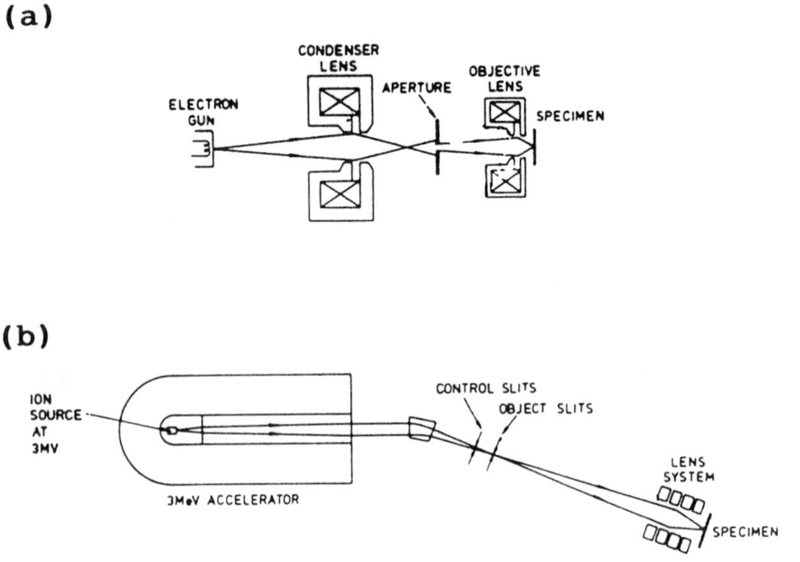

FIGURE 9.1 Focusing microprobes systems for the production of (a) electron microbeams, and (b) ion microbeams (the Harwell system)

ing from bombarding the nozzle edges by an accelerated beam. This disturbs the measurement since the collimator is usually placed close to the measuring unit, and, in addition, collimator nozzle edges become quickly damaged.

Because of the disadvantage of the collimator design, the microbeam units which have recently been put into service have all been equipped with *focusing systems.* Systems of this kind make it possible to increase the beam brightness by several orders of magnitude. They also permit easy electrical or magnetic control in a *fast beam scanning system.* Therefore, work on new designs for the generation of microbeams is mainly centered on various focusing systems, called *microprobes* [9.3].

Typical focusing systems for the production of microbeams are shown in Fig. 9.1. Because of the considerably higher magnetic rigidity of accelerated ions compared with that of electron beams, *electron microprobes* (Fig. 9.1a) are not suitable for ion microbeams. For example, for the same focusing distance, the magnetic rigidity of a 1 MeV proton beam is higher than that of a 30 keV electron beam used typically in electron microprobes by a factor of 230.

Typically, in an *ion microprobe* (Fig. 9.1b) an accelerated beam is delivered to a magnetic mass analyzer. After passing through a control slit reaches an object slit. Then the already divergent beam strikes the focusing system proper made up of four magnetic lenses which demagnify the image of the object slit on the specimen plane. Typically, diameters thus obtained are 10 μm or more, but designs are known, in which they can be as small as 0.5 to 0.1 μm [9.7 through 9.12]. When a new generation of proton sources with brightness elevated by several orders of magnitude have been developed, the diameters of microbeams obtained will feature focal diameters as small as 1 nm [9.1]. However, great difficulties are encountered in the already achievable submicron range of diameters mainly because of the instability of the mechanical subsystems.

With an arbitrary chosen restriction of a beam intensity of 100 pA in a spot less than or equal to 10x10 μm^2 there are roughly twenty nuclear microprobes [9.17] the world over used in biomedicine (Table 9.1). For example, in the cell biology the

TABLE 9.1 Nuclear Microprobes Used in Biomedicine ($\pm 10\ \mu m^2$), (after U. Lindh [9.17])

Location	Size of probe	Applications
SUNY, Albany, USA	1.3x1.3 μm^2	biology, trace elements
Vrije U, Amsterdam, NL	5x5 μm^2	biology 40%
U Birmingham, GB	1.3x1.3 μm^2	biological samples
U Bochum, FRG	10 μm^2	medicine 50%, biology 20%
Bordeaux-Gradignan, F	1.8x1.8 μm^2	biology (1988)
Eindhoven U of Tech, NL	10x10 μm^2	biology 80%
U Oregon, Eugene, Or, USA	8 μm^2	biology 70%
Faure, South Africa	4x4 μm^2	biology
MPI, Heidelberg, FRG	3x2 μm^2	biology 30%
Karslsruhe, FRG	2x3 μm^2	medicine 80%, closed 1987
Queen's U, Kingston, Canada	10 μm^2	biology 30%
Lund U of Tech, S	5x5 μm^2	biomedicine 50%
Manitoba, Canada	5x5 μm^2	medical and biophysics
U Melbourne, Australia	1 μm^2	biomedicine 50%
U Oxford, GB	$\leq$ 1x1 μm^2	biology 50%, medicine 30%
Z Kernforsch, Rossendorf, GDR	6 μm^2	biology 10%
Schonland Res C, South Africa	< 1 μm^2	biology 20%
Waseda U, Tokyo, Japan	?	biology
U Uppsala, S	2.9x2.9 μm^2	biomedicine

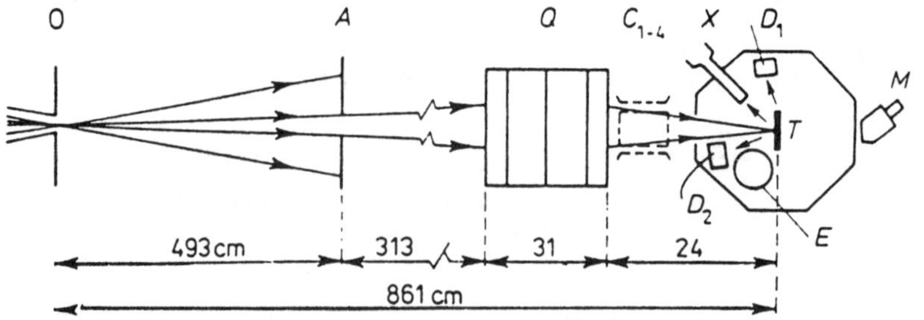

FIGURE 9.2 A system of producing microbeams of protons and helium ions with 1 μm diameter (University of Melbourne Micro Analytical Research Centre) (after G.J.F. Legge et al. [9.13 through 9.16])

average human and animal cells display a diameter of 5-10 μm in the tissue. Some other cells, for example white blood corpuscles, have diameters in the region 10-20 μm. These living cells comprise a very complicated structure with a lot of organelles of varying size. Most of the existing 50 nuclear microprobes will, therefore, be excluded from detailed elemental mapping of cells. Even the best resolution achieved, about 1 μm, is barely adequate to extract detailed spatial information of the smallest organelles [9.17].

Fig. 9.2 is a schematic diagram of a *microprobe beam line* for the production of proton or helium ion microbeams with diameters down to 1 μm and energies ranging from 2 to 3 MeV [9.13]. Ions from a Pelletron-type accelerator are fed to a magnetic mass analyzer, and as a monoenergetic beam are incident on a circular diaphragm with a large diameter (10-300 μm) which acts as an object slit. The beam then reaches slit A (Fig. 9.2) with aperture ranging between 0.25 and 4 mm dia., which selects only the ions whose paths are close to the slit axis. The smallest apertures minimize aberrations, but this occurs at the expense of diminishing the beam current. The resultant beam passes through system Q consisting of four magnetic quadrupole lenses, and finally reaches two pairs of deflecting coils C_{1-4}, supplied from independent triangular wave generators at repetition frequencies of up to 20 kHz, which sweep the beam over an area of 1 x 1 mm^2. Signals are detected by two surface barrier detectors D_1 and D_2 and an X-ray Si (Li) detector. Secondary electrons are detected by detector E. Like the system shown in Fig. 1.23, the above microprobe may be used for the detection of characteristic X-rays, secondary electrons and backscattered ions (RBS), as well as other products of the excited nuclear reactions. The sample investigated is placed on target T and adjusted in the x-y plane within 1 μm. The target can also be rotated with precision about two orthogonal axes in the plane of the target surface. Microscope M permits observation of thin targets and beam focusing by means of thin glass plates mounted in the sample plane.

A useful ion microprobe system requires an adequate distance, called the *working distance,* for detectors and scan coils upstream of the probe target. The current microprobe at Melbourne has a working distance of 200 mm; a new two-stage system is under construction [9.21].

If the beam is collimated and is not scanned, the sample investigated is usually placed on a movable table which afford

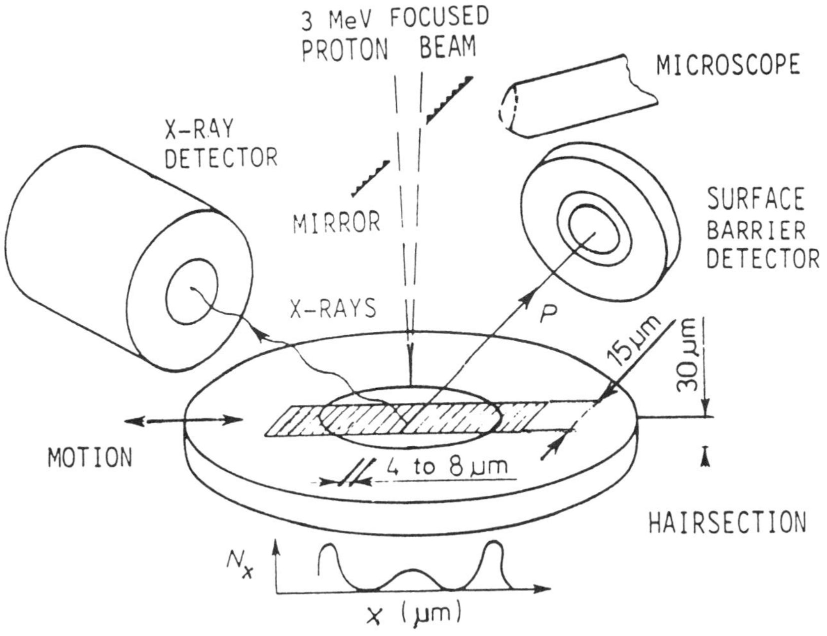

FIGURE 9.3 A schematic arrangement for measuring longitudinal and radial distributions of concentrations of elements in a human hair (Vrije Universiteit, Amsterdam) (after A.J.J.Bos et al.[9.18])

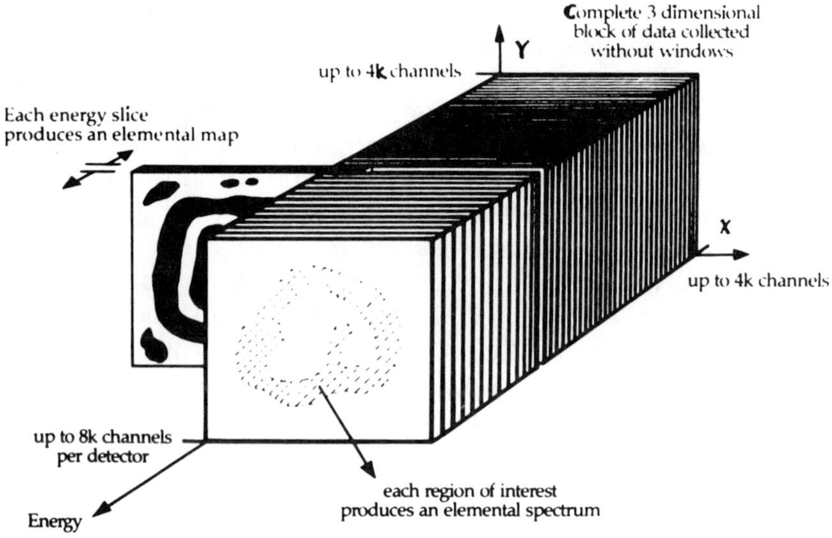

FIGURE 9.4 Schematic diagram of the fast data acquisition system MARC (Micro Analytical Research Centre, Melbourne) (after G.J.F. Legge et al [9.11, 9.12])

micro displacement. Example of such a system is shown in Fig.9.3.

This system is designed for measuring longitudinal and radial distributions of concentrations of elements in a human hair (Vrije Universiteit, Amsterdam, The Netherlands). A 3.0 MeV proton beam is obtained from an isochronous cyclotron, and has transverse dimensions of 8 x 12 or 5 x 12 μm and surface density of 0.5 pA μm^{-2}. The human hair section to be investigated is placed on a circular scanning table which can be moved at 80 $\mu m\ s^{-1}$ by a piezoelectric transducer. The analysis of the concentrations of such elements as S, Cl, Cu and Zn is carried out by measuring the characteristic X-rays (PIXE) with a solid-state Si (Li) detector depicted in the left upper corner of Fig. 9.3. At the same time, the backscattered protons are measured with a surface barrier detector, and thus the thickness of the matrix can be determined.

Analyses performed with microbeams are often denoted by the prefix "micro". For example, the proton induced X-ray emission method in combination with the proton microprobe is called "micro-PIXE", in contrast to "macro-PIXE" implemented with a wide beam.

For details about practical problems with the nuclear microprobes see [9.22 through 9.24].

Systems of data acquisition. When measuring biomedical samples it is necessary to use systems which permit *fast data acquisition.* Since multi-detector systems together with spectrometric methods are usually employed for simultaneous detection of various trace elements. Moreover, the beam is often scanned over a small area of up to 1 mm^2 with frequencies in the 1-10 kHz range. In dedicated systems equipped with four detectors (Fig. 9.2), data from each detector are formatted in the form of 3D blocks depicted in Fig. 9.4. The MARC system can

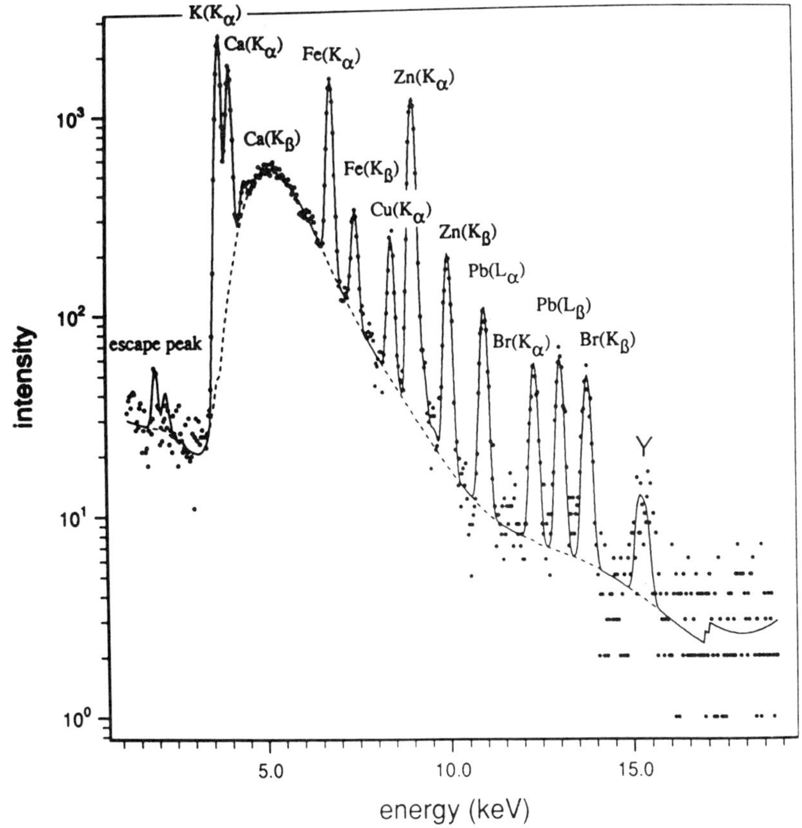

FIGURE 9.5 A PIXE spectrum of glial cells from a rat exposed to lead (PIXE-facility in Turku, Finland), (after R. Brenner et al [9.25])

operate with a frequency of up to 100 thousand words per minute; each event is described with three words, of which the first and the second respectively determine the x and y-coordinates, and the third provides information about energy.

Measurements of this kind are usually carried out in a spectrometric mode, from which the energy spectrum can be obtained. Fig. 9.5 is an example of such a spectrum obtained in the analysis of the glial cells from a rat exposed to lead [9.25]. By selecting an energy region corresponding to the characteristic line of a given element arbitrary spatial distributions of elements can be mapped with the aid of a computer. Some examples of such distributions are shown in Fig. 9.10.

Types of accelerators used. Systems used for generating microbeams are usually operated in conjunction with electrostatic single-stage or tandem Van de Graaff accelerators (Sec. 2.2.2.1). The energies employed are thus in the range of a few MeV (2-8 MeV) [9.26 through 9.28]. Recently, in several microanalytical facilities Pelletron-type machine have been installed (Figs. 2.5 and 2.6). Isochronous cyclotrons are not so often used [9.29 through 9.31]. Since the maximum energies of the accelerated protons in some installations are as high as 30 MeV (Hamburg, Eindhoven) or 40 MeV (Manitoba) microanalysis can be carried out with a wide energy range [9.67].

Typically, currents used in microprobes with focused beams do not exceed a few nanoamperes (1 nA = 10^{-9} A) [9.14]. For the reasons set forth in more detail in Sec. 9.2, systems employed in the analysis of biomedical specimens do not use currents higher than several hundred pA. The value of 100 pA has been adopted internationally as a standard current for comparing spatial resolution of microprobes [9.32].

Details of the design of present-day systems producing ion microbeams can be found in [9.33, 9.34].

9.2 Methods of Accelerator Microanalysis

9.2.1 Preparation of Biomedical Samples

X-ray microanalysis of biological tissue involves the use of thin fragile specimens which can be damaged in various ways by an MeV microbeam [9.43]. Structural damage specimen or element movement, and shrinkage of an area of interest, or mass loss from the tissue matrix will affect concentration measurements, and discoloration and movement of the tissue makes identification of internal structure difficult. The damage incurred will become more severe as microbeam focusing improves towards submicron dimensions, increasing the loading parameter ($C/\mu m^2$) upon which beam damage depends. In order to maintain the integrity of elemental concentrations and distributions within the specimen the damage induced by the incident beam must not cause elemental loss and must be decreased to be below the resolution of microbeam. In electron microscopy the limit in determining biological structure is the damage produced in the specimen due to the action of the irradiating beam, not the resolution of the beam.

Sample preparation procedures for ion microanalysis is carried out using methods similar to those of sample preparation in electron microscopy. However, several additional specific factors resulting from the use of ion microbeams have to be taken into consideration [9.35 through 9.39]. First of all, preparation procedures must ensure that the sample is relatively resistant to the high current of the proton microbeam. Moreover, since the proton ranges are considerably greater than those of the electrons, sample thicknesses used are greater than those generally found in electron microscopy. Another problem encountered in the ion microanalysis is the *destructive effect* of the beam on the sample structure. As the most common analysis is that of the chemical composition of the sample, preparation procedures must maximally limit the effect of factors which cause any changes resulting from, e.g. the evaporation of the trace elements investigated.

Fig.9.6 shows calculated temperature increase as a function of current and microbeam transverse dimension [9.37]. The calculations were made on the assumption that the heat energy dissipation resulted from thermal conduction only, and the thermal conductance was taken to be equal to 2×10^4 W cm^{-1} K^{-1}. The sample ambient temperature was 300 K and the predicted proton energy loss was 0.3 MeV. In order to avoid large temperature increases moderate beam currents have to be used, especially under strong focusing conditions (small radius r). Temperature increase could also be limited by covering the sample with an Al layer. A current of 100 pA or more in a spot of 0.01-0.25 μm^2 implies current densities of at least 400-10000 pAμm^{-1}. It can be questioned if a specimen can withstand such

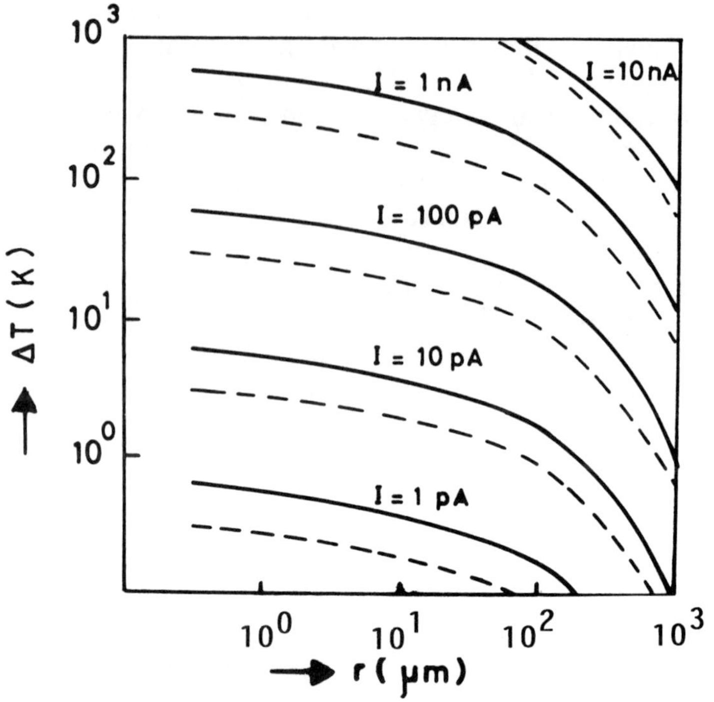

FIGURE 9.6 Increase in sample temperature vs. proton microbeam current and radius (after R.D.Vis et al.[9.73])

an intense radiation even if rapid beam scanning is employed [9.17].

The various techniques for the preparation of biological material prior to analysis are dependent on several factors, e.g. sample type, type irradiation, detection limits required, etc. Since the biological material may be liquid, soft or hard and sometimes the fine structure has to be preserved (semimicro- and microbeam analysis) very different concepts have to be used in sample preparation [9.38].

The typical biomedical sample suitable for PIXE is small or will be so after sample preprocessing. The determination of concentrations from trace to major elements is interesting with or without preparation. Sample processing must be carefully executed using ultrapure chemicals in a clean environment.

Using only a simple calibration PIXE is accurate at the 10-15% level but with an increased use of internal standards accurate and precise results will be obtained at the 5% level, which is the best proof of the reliability of PIXE analysis. The use of biomolecular methods for separation and specific preconcentration techniques should be promoted to a much higher degree than at present.

For wet samples cryocutting combined with freeze-drying is usually employed. Dry samples, such as plant substances, hair, bone tissue,etc. are sealed in araldite or similar materials, and then sliced with a microtome. The sample surface is sometimes covered with a polystyrene layer to protect against radiation damage, or with a carbon or aluminum layer to eliminate the influence of the beam charge.

Details of sample preparation procedures for ion microanalysis can be found in [9.37 through 9.39], whereas the effect of the various procedures on the results of measurements is discussed in [9.36].

9.2.2 Survey of Methods of Ion Microanalysis

For biological systems to function properly appropriate concentrations of a number of trace elements are required. *Trace elements,* which are elements present in tissues and body fluids in concentrations below 0.01%, are indispensable components of the enzyme systems of proteins which are vital for biological organisms. For example, trace elements such as metals are usually embedded deep inside the structure of the protein. The knowledge of both the abundance of trace elements and their distribution is of fundamental importance in the understanding of the vital functions of a biological system.

Trace elements, or *micronutrients,* are intimately involved in physiological function and dysfunction at all levels of organisation. Depending on their action the trace elements have been named either *essential* or *toxic* [9.17]. The feature of being essential, of course, reflects the state of knowledge at a certain point of time. On the other hand, it does not seem logical to establish a group of toxic elements because toxicity is inherent in all elements and depends mainly on the concentration. The essentiality was defined some years ago [9.40]. A trace element is considered to be essential if it meets the following criteria: (1) it is present in all healthy tissues of all living things, (2) its con-

Table 9.2 Analytical Capabilities and Applications of the Atomic and Nuclear Effects Induced by Charged Particles (after Ch. Engelmann [9.55])

Type of effect	Analytical capabilities and specific applications
Atomic reactions	
Channeling of the incident charged partciles	Location of atoms (insertion or substitution) and detection of lattice damage crystals
Emission of secondary electrons	Precise location of beam impact point and secondary electrons microscopy[a)]
Emission of characteristic X-ray photons	Multielement analysis (Z>11)
Slowdown of the incident charged particles	Scanning transmission ion microscopy or radiography of materials transparent to the used projectiles
Elastic interactions[b)]	
Forward-scattered projectiles	Determination of light elements (H, He, B, Li, Be, C, N, O, F, etc,) in ultra-thin samples
Backward-scattered projectiles	Determination of medium and heavy mass elements in the surface region of light substrates
Recoil of bombarded nuclei	Determination of very light isotopes (H, D, T, ^{3}He, ^{4}He, etc.) in ultrathin targets by transmission or in the surface region of thick substrates by reflection
Nuclear reactions[b)]	Selective determination of light elements (Z<15), more particularly their isotopes, in thin targets or in the surface region of thick samples

[a)] While using microbeams.

[b)] These interactions also enable erosion-free concentration profile measurements of detected elements beneath the surface of solids (maximum explorable depth: 5-10 μm according to the chemical composition of the examined material, type and energy of the projectiles employed) and, if microbeams are used, non-destructive 3-D mapping.

centration from one animal to the next is fairly constant, (3) its withdrawal from the body induces reproducibly the same physiological and structural abnormalities, regardless of the species studied, (4) its addition either reverses or prevents these abnormalities, (5) the abnormalities induced by deficiencies are always accompanied by specific biochemical changes and (6) these biological changes can be prevented or cured when deficiency is prevented or cured [9.17].

Most of living matter is made up of the 11 elements: H, C, N, O, Na, Mg, P, S, Cl, K and Ca, whereas essential trace elements for warm-blooded species are F, Si, V, Cr, Mn, Fe, Co, Mi, Cu, Zn, As, Se, Mo, Sn, and I. This list is, of course, not complete, and it will be extended with the development of analytical techniques. The common feature of all essential trace elements is a narrow range of their concentrations suitable for proper functioning of a biological system. If their concentrations are lower or higher than the above range the functioning of the organism is impaired. In excess, they may become toxic. Some elements, such as Cd, Hg, Pb, etc. always have toxic effects even when present in very low concentrations.

Studies of trace elements in biological samples have increased our knowledge of the functioning of biological systems. Some organs and tissues have distributions of concentration patterns which reflect the function or metabolic activity, resulting from the concentration structure of a given element. The knowledge of the distribution of chemical elements in biological systems may be of great use in understanding of the basic phenomena occurring in these systems, such as cell growth or cell division.

Analytical capabilities and applications of the atomic and nuclear effects induced by charged particles are listed in Table 9.2, and fundamental parameters of a microanalysis carried out with ion microbeams are listed in Table 9.3. The most common method is that of the *fluorescence analysis induced with protons* (PIXE). This method permits analysis of elements with $Z > 11$, and is not suitable for the detection of light elements. Among other methods which can be used for the analysis of light elements, such techniques as *nuclear reaction analysis* (NRA), *Rutherford backscattering* (RBS) or *elastic recoil detection analysis* (ERDA), have found wide application [9.41, 9.42].

The PIXE method, prevalent in biomedical measurements [9.17, 9.44 through 9.54], is implemented with inducing beams of protons, and sometimes deuterons or alpha particles. It provides sensitivities of elemental detection of about 1 ppm, whereas in fluorescence analysis with X-ray emission typical sensitivities range from 100 to 1000 ppm [9.45].

One of the main advantages of multielemental analytic methods is that they are unbiased. This means that one can get information on elements that was not primarily sought. Multielemental methods can generate data of 10-15 elements when biological objects are analysed. The effective handling of such large volumes of elemental data calls for multivariate statistical

TABLE 9.3 Parameters of the Microanalysis Implemented with Ion Microbeams

Method	Detected elements	Transverse resolution [μm]	Depth resolution [μm]	Depth range [μm]	Limit detectability [ppm]
Fluorescence analysis (PIXE)	Z > 11	1	very little	5	>1 wt.
Nuclear reaction excitation (NRA)	all light	1 Z=1 to 15	0.005	1-10	0.01
Elastic backscattering (RBS)	all except very low Z	>1	>0.01	1	>0.01 wt.%
Elastic recoil (ERDA)	Z < 15	>1	0.005	1	>500

methods such as principal component analysis, discriminant analysis, cluster analysis and pattern recognition [9.17].

Another feature of the microanalysis using beams of accelerated ions is the complementary character of individual methods (Tables 9.2 and 9.3), often implemented concurrently [9.56]. In this case, each type of radiation is detected with a set of detectors surrounding the sample under investigation (Fig. 1.23). The PIXE analysis may then be supplemented with other methods, such as the RBS or NRA, which considerably adds up to the information on the object investigated. Table 9.4 lists some nuclear reactions used in light element analysis with the NRA method.

The PIXE technique permits elemental detection at the nanogram (1 ng = 10^{-9}) level or lower. The detection sensitivity as low as 10^{-16} g can be achieved for zinc by reducing the beam cross-section to 5 x 5 μm^2 [9.23].

Analysing very minute volumes in biological tissues causes problems in assessing the accuracy of determination. When probing volumes as small as 10^{-9} mm^3, which corresponds to 1 pg or less, one has to bear in mind that trace element analysis can prove meaning-less. With an analytical sensitivity of 1 μg/g, the *absolute mass detection limit* approaches 10^{-18} g. Pushing the limits further may imply practical obstacles. For example, 10^{-18} g of selenium contain some 76000 selenium atoms. Extending the spatial resolution of the nuclear microprobe to

0.1 μm, a goal which is within realistic reach, means that the probed volume can be as small as 10^{-12} mm^3 corresponding to a mass of 1 fg. This would imply, if the detection limit can be pushed down to 100 ng/g, the analysis of 8 selenium atoms. This situation is, of course, absurd. The cross sections will probably not permit such analyses within practical limits [9.17].

The general characteristics of microanalytical methods in biology and medicine is that they use the ever more subtle techniques. The best example is provided by the *accelerator mass spectrometry* (AMS) (Sec. 9.2.6), which helps to attain the sensitivity of 1:10^{17}. This means that one atom of an isotope of the element consisting of several isotopes can be detected in 10^{17} other atoms of the same element (the AMS method), whereas an ordinary mass spectrometer has a limited sensitivity of 1:10^9 [9.57]. By adapting the AMS technique to biomedical studies, interesting possibilities of analyzing the effect of cancerogenic materials have opened up (Sec. 9.2.6).

TABLE 9.4 Nuclear Reactions Used in Detection of Light Elements

ELEMENT	REACTION
Hydrogen	^{2}H (^{3}He, p) ^{4}He
	^{1}H (^{11}B, α) ^{8}Be
Carbon	^{12}C (d, p) ^{13}C
	^{12}C (p, γ) ^{13}N
Nitrogen	^{14}N (d, α) ^{12}C
	^{14}N (d, α) ^{12}C
Oxygen	^{16}O (d, p) ^{17}O
	^{18}O (p, α) ^{15}N
	^{16}O (d, pγ) ^{17}O
Fluorine	^{19}F (p, $\alpha\gamma$) ^{16}O
	^{19}F (p, α) ^{16}O
Magnesium	^{24}Mg (d, p) ^{25}Mg

9.2.3 Methods of Particle-induced Characteristic X-ray Emission

Fundamental principles of the PIXE method. A typical fluorescence analysis is based on identification and quantitative estimation of the chemical composition of the investigated material through registering the characteristic X-ray emission which is produced when an electron is removed from the K, L, M, N... shells out of the atom or to a neighbouring shell as a result of some excitation. The vacancies thus created can be occupied by electrons from the outer orbits (L, M, N,... shells). These events are associated with the emission of soft X-rays whose spectrum consists of a number of peaks denoted by letters $K_{\alpha\beta\gamma}$, $L_{\alpha\beta}$, etc. Initially, characteristic X-ray radiation used to be induced by X-rays from an X-ray tube, radioisotopes, or by a beam of accelerated electrons. In 1970, S.A.E.Johansson and T.B.Johansson were first to use a 1.5 MeV proton beam to induce characteristic X-ray emission, thus creating a foundation for the dynamic development of a novel analytical method referred to as the PIXE *(Particle Induced X-ray Emission)* technique.

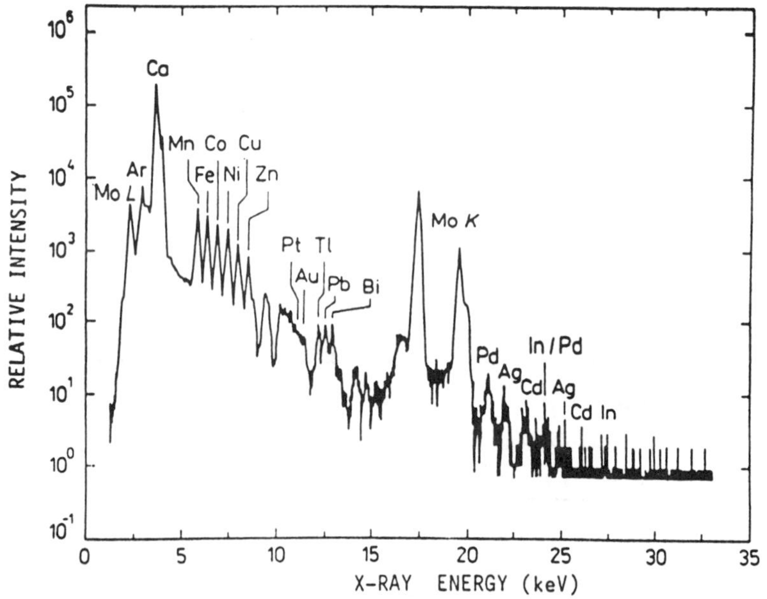

FIGURE 9.7 The PIXE spectrum of an aqueous sample containing 10 μg of each of the investigated elements (after S.Brueggerhoff and E.Jackwerth [9.58])

For the analysis using an internal beam, the sample investigated is introduced into a vacuum chamber to be irradiated with a beam of protons, alpha particles or, less often, heavy ions. The energy of the incident beam is usually in the range of 1 to 5 MeV per nucleon, its intensity varying between 0.1 and 100 nA. X-rays are detected with a solid-state Si(Li) detector featuring good energy resolution, usually 160 eV (FWHM) for $E_\gamma = 5.9$ keV, which makes it possible to resolve K-series lines of the characteristic radiation of the neighbouring nuclides with $Z \geq 12$. Radiation emitted by low Z elements is so soft that it becomes absorbed in the detector beryllium window. Therefore, windowless detectors are mounted directly in the wall of the irradiation chamber or inside it, which permits detection of elements with $Z \geq 5$.

The characteristic X-ray energies analyzed with a window detector usually lie in the range between a few keV and several tens of keV. Fig. 9.7 provides an example of an X-ray emission spectrum of a water sample containing 10 μm of each of the investigated elements.

The sample thicknesses vary between several $\mu g{\cdot}cm^{-2}$ up to the thicknesses at which the incident ions totally lose their energies, the typical values being those with the energy loss of 10-15%. Many samples must be deposited on chemically pure mounts featuring good mechanical strength and electric and thermal conductivity. Continuous bremsstrahlung produced in the mount should be at minimum, and therefore low Z materials must be used. High-purity carbon foils or those made from such plastic materials as Kapton, Mylar, or Nucleopore or Millimpore filters are often employed.

Internal irradiation techniques lead, however, to considerable difficulties in the analysis of samples containing volatile elements which vaporize as a result of the heat produced by the incident beam (Fig. 9.6). In addition, these methods exclude the possibility of analyzing fragile or large-size objects with the PIXE technique. Therefore, a method which is increasingly more often employed is that using an external beam extracted from the accelerator vacuum. The sample is then analyzed either directly in air or in the protective atmosphere of helium. The advantage of using helium is that it permits the detection of light elements with Z less than 14 (Si), whose characteristic X-rays would be otherwise absorbed in air.

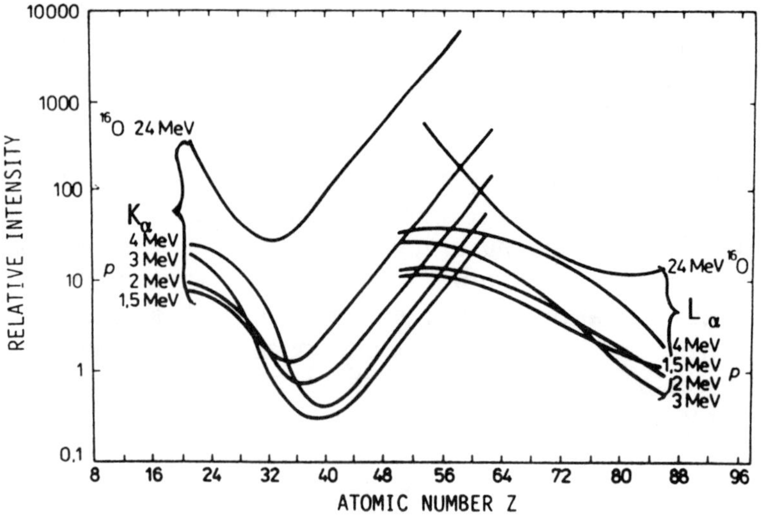

FIGURE 9.8 Dependence of the relative sensitivity of the PIXE analysis on the Z-number of the element investigated and ion energies (after B.Raith et al. [9.59])

The *sensitivity of the PIXE method* is dependent, among other things, on the energy of the inducing radiation, particle species and the Z-number of the element under investigation. Fig. 9.8 shows the relative sensitivity as a function of Z for the irradiation of samples with 1.5 - 4 MeV protons and a 24 MeV ^{16}O beam. It can easily be seen that the best results can be achieved for elements whose Z-numbers vary from 20 to 40 for K_α lines, and from 65 to 85 for L_α lines. Under optimum conditions, it is therefore possible to detect concentrations as low as 0.1-10 $\mu g \cdot g^{-1}$, that is in the relative range of 10^{-7} to 10^{-5}.

A major advantage of the PIXE method is that it permits analysis of very small absolute masses of an element. This is particularly true of the measurements made with the use of proton microbeams (Sec. 9.1). The absolute sensitivity as high as 10^{-18} g can be attained with a microbeam over an area of 10^{-8} cm^2 with a relative sensitivity of 10^{-6}. In this respect, no other analytical method can compete with the PIXE technique which uses ion microbeams.

Investigations of biomedical samples. The PIXE method is especially well suited to trace measurements of elements of medium and high Z-number in a matrix composed of light elements, and therefore it has found wide application in the analysis of biological samples. These are, first of all, medical samples [9.17, 9.44 through 9.53] as well as those used in environmental studies [9.54]. Among medical samples the most frequently analyzed are those of the blood, various tissues of the body and hair. Fig. 9.9 shows a longitudinal distribution of trace elements Zn, Fe and As in a single hair obtained by scanning along the hair's axis. The spacing between successive positions of the proton beam was chosen so that it represented approximately one day's hair growth. The results shown in Fig. 9.9 were obtained from measurements carried out 183 days after accidental As inhalation. The As peak corresponding to the maximum concentration of 200 ppm was localized at the distance of 56 mm from the hair bulb. A rapid increase in As concentration was associated with a drop in Zn concentration which did not return to normal until after 30 days, whereas the Fe levels remained unchanged.

By using a microbeam like one shown in Fig. 9.3, in addition to longitudinal distributions, *radial distributions or concentration*

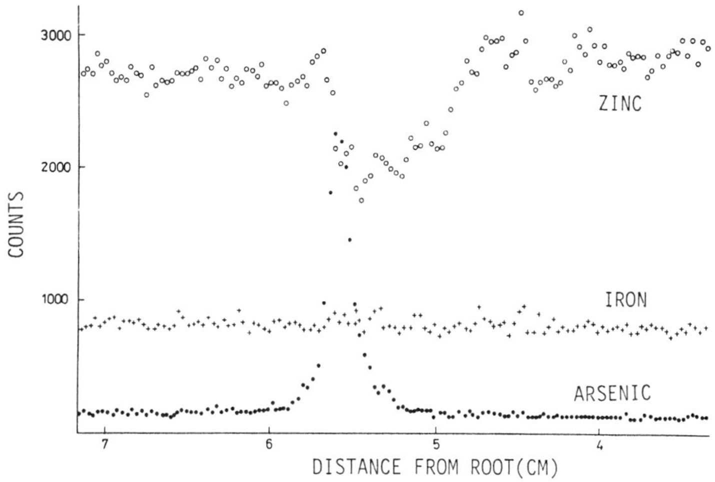

FIGURE 9.9 Longitudinal distributions of Zn, Fe and As in a hair of a patient who accidentally inhaled arsine (after P.Horowitz et al. [9.60])

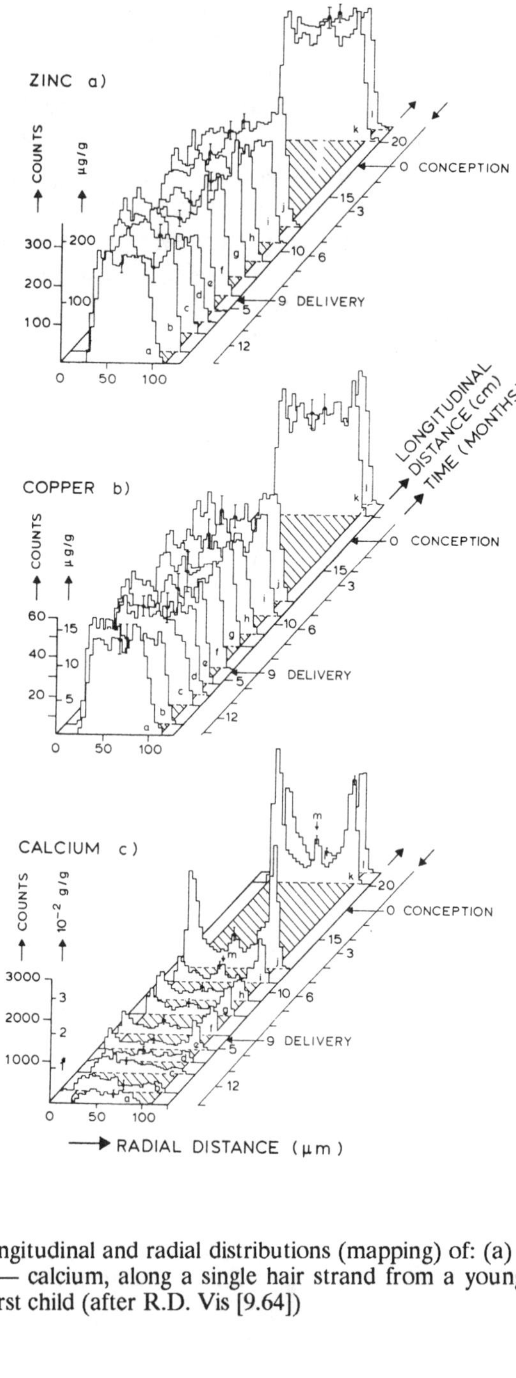

FIGURE 9.10 Longitudinal and radial distributions (mapping) of: (a) — zinc, (b) — copper and (c) — calcium, along a single hair strand from a young woman after delivery of the first child (after R.D. Vis [9.64])

profiles, can also be obtained. Fig. 9.10 shows radial distributions of zinc and copper along a single hair strand from a young woman after delivery of the first child. It can easily be seen that no significant Cu and Zn deviations occurred during gestation. The concentration was measured from the ratio of the number of counts, N_x, detected by an X-ray detector to the number of counts, N_p, detected each time by the proton recoil detector (Fig. 9.3 and also Sec. 9.2.4) since it was found that the elemental concentration is proportional to the ratio $N_x : N_p$. By additionally measuring recoil protons, local inhomogeneities of the sample can be smoothed out and other factors which influence both the X-ray detector and proton detector can be accounted for. Thus the elemental concentration is calculated for each position by dividing the number of counts within the characteristic peak by the total number of protons scattered by recoil, and then by multiplying the result by a calibration factor for a given element. Zn concentrations are of the order of 200 $\mu g\ g^{-1}$, whereas those of copper are by one order of magnitude smaller (Fig. 9.10a, b and c).

The brain, as a highly structured organ, is particularly well suited for trace elemental research [9.53]. Various studies have been dedicated to the elemental distribution in normal and pathological brain. The analysis of different brain regions has revealed a heterogeneous distribution of several elements. It was also shown that neurological disorders may be associated with trace element alterations in the brain. Moreover, these alterations often only occur in certain, sometimes very small brain structures, so that their examination requires the use of sensitive analytical techniques which can handle tiny samples and/or perform a microanalysis. Because of the inherent characteristics of PIXE and micro-PIXE, these two techniques offer significant advantages for such studies.

In the study [9.53] PIXE was applied to investigate the changes in trace element levels in 5 different brain structures, from the cerebral cortex, basal ganglia and cerebellum of patients with atherosclerosis. Samples from both pathological and normal brains from Portuguese individuals were examined, and the results were compared with data for brains obtained from Belgian individuals.

The normal brains originated from persons aged between 28 and 44 years (mean age 34±6 years) who died from traumatic shock or acute cardiac failure. The brains affected by atherosclerosis were taken from persons with ages ranging from 47 to 85 years (mean age 74±10 years). They were removed at autopsy within 48 h after death, superficially washed with double distilled water and immediately dissected. Five structures were selected for trace element determination (K, Ca, Mn, Fe, Cu, Zn, Se and Rb), i.e. the frontal lobe, nucleus caudatus, putamen, substantia nigra, and vermis cerebelli. The samples were weighed to determine their wet weight, freeze-dried for 72 h, and stored in polyethylene or Teflon vials. Typically, four PIXE targets were prepared for each brain sample using a nitric acid digestion procedure. The targets were irradiated in the experimental PIXE setup of the University of Gent (Belgium) with a proton beam of 2.4 MeV. The differences between the data for normal Portuguese and Belgian brains were insignificant or small, except for Rb which showed systematically higher values in the Portuguese samples.

The average concentrations and associated standard deviations for Rb in each of the 5 structures of the Portuguese and Belgian brains are depicted in Fig. 9.11. This figure clearly

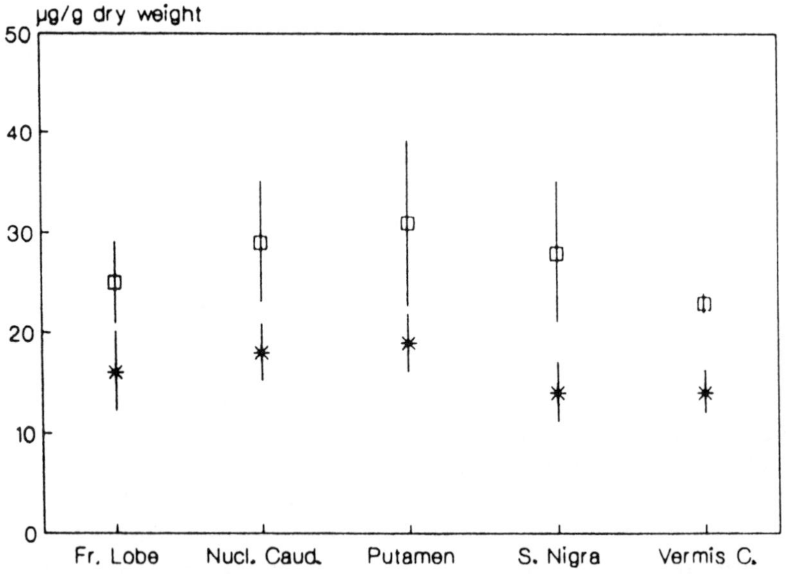

FIGURE 9.11 Rubidium concentration in 5 different regions from normal Portuguese (□) and Belgian brains (*); data points and error bars indicate averages and standard deviations based on up to 5 and 12 brains, respectively (after T.Pinhero [9.53]).

shows that the difference is systematic for the 5 regions studied. The higher Rb level in the brains from Portuguese individuals could indicate a geographical concentration difference [9.53].

PIXE has been used for analysing erythrocyte and plasma samples from humans treated with telecobalt therapy in order to search for possible changes in the concentration of elemental constituents caused by radiation load of bone narrow [9.50].

In partial body irradiation applied in human radiotherapeutic practice (cf Sec. 3.4.2) damages in bone marrow functions may be introduced by large doses received by the bone marrow in the treatment fields. However, it is generally accepted and proved that even a severe damage of a portion of bone marrow may not compromise the overall capacity of the haematopoietic system, since uneffected portions of the marrow will repopulate damaged areas depleted in mature cells. Therefore, the appearance of a possible effect on the elemental composition of peripheral blood depends on how much of the bone marrow volume has been irradiated. A comparison between tumoricidial doses for different cases and the irradiation load of the segments of the bone marrow exposed in the treatment fields led to the conclusion that studies of the elemental composition of blood fractions could be of importance in the case of humans undergoing radiation therapy.

Minimum detection limits of the order of 0.5-5 ppm for essential and trace elements in the PIXE method made it fully possible to search for such effects in low volume separated erythrocyte and plasma samples taken from humans during and after partial body radiotherapeutic treatment.

Measured iron concentrations for erythrocyte and plasma samples are presented in Fig. 9.12 for a patient as a function of total accumulated tumor dose, together with values for haemoglobin, haematocrit and mean corpuscular haemoglobin concentration data. In addition to total tumor dose values, in each step of the irradiation the corresponding actual biological effect CRE, calculated for the irradiation period i is indicated as

$$\mathrm{CRE_i} = (N/T)^{0.11} N^{0.65} d\mathrm{x}100 \tag{9.1a}$$

while for postirradiation times pi

$$\mathrm{CRE_{pi}} = \mathrm{CRE_i}\ \mathrm{e}^{-0.008\,\mathrm{G}} \tag{9.1b}$$

Here, N denotes the number of fractions in telecobalt treatment, T and G denote the lengths of the treatment and postirradiation periods expressed in days, respectively, d stands for

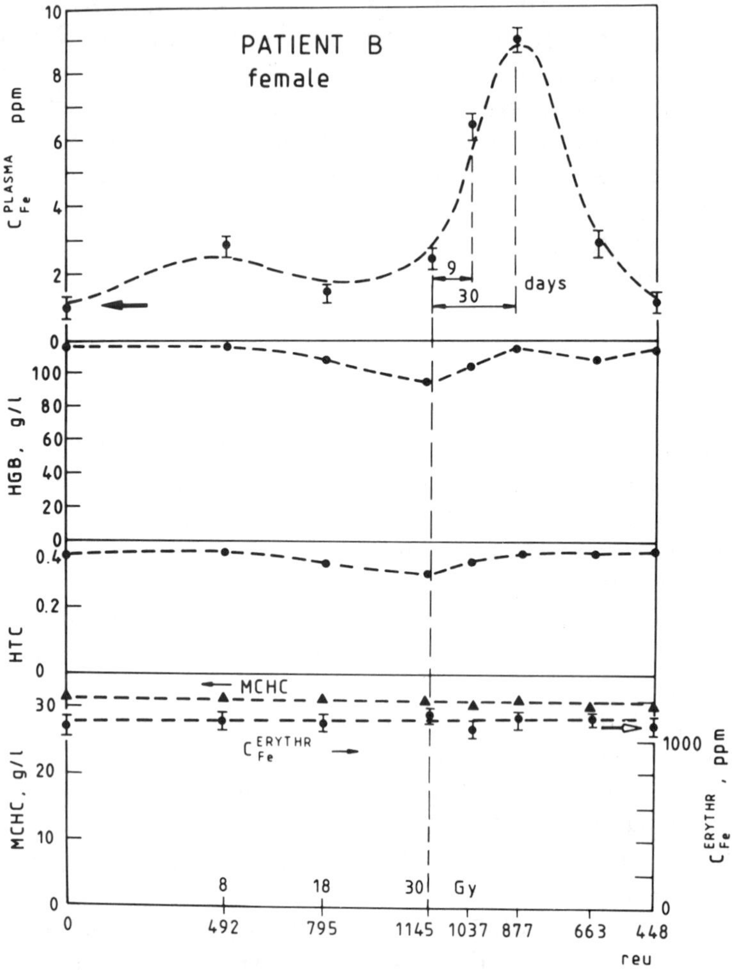

FIGURE 9.12 Measured iron concentrations in plasma and erythrocyte samples together with haemoglobin, haematocrit and mean corpuscular haemoglobin concentration data, for patient, as a function of total tumor dose and actual biological effect (reu) (prompt and prolonged effect); full and empty arrows indicate normal iron concentration values in plasma and erythrocytes, respectively (after I. Borbely-Kiss et al. [9.50]).

the daily dose, given in Gy. The hyperferremia in blood plasma, a strong prolonged effect, was observed after the end of the irradiation (Fig. 9.12) [9.50].

Micro-PIXE quantification of histochemical reactions. Classical *histochemistry* is a branch of science which links morphology (histology, cytology) and biochemistry [9.65]. It is based on biochemical reactions but at the same time preserves tissue structure which enables the localization of reaction products, something that biochemistry lacks. Each histochemical method possesses, of course, its drawbacks. Only such a chemical procedure may be used which gives unsoluble not diffusing reaction products which additionally must be visible in light or electron microscopes. Hence, the number of biochemical methods that may be used for histochemical purposes is limited. Most of the histochemical methods are indirect. It means that using reagents with affinity for certain groups typical for macromolecules, we may, on the basis of the amount of bound reagents, draw conclusions concerning the amount of a detected substance. The routine histochemical procedures are qualitative only. This comes from two main reasons: (a) the deviations from Beer-Lambert law, and (b) the optical density, equal to log(transmittance), proportional to the product of the concentration of the material, its absorptivity and the path length transversed by the light (local section thickness), but not to the concentration alone.

The micro-PIXE eliminates entirely both of the above-mentioned causes of artifacts. Moreover, micro-PIXE enables the determination of concentrations of certain elements typically present in high amounts in the studied macromolecules (P, S and Ca). One example may be proteoglycans (PG) which contain SO_4^{2-} groups. These groups may be detected directly (S concentration) as well as indirectly by positively charged dyes (histochemical staining procedures). The result may be compared and the quantitative power of histochemical staining characterized. However, the quantification of histochemical reactions by the micro-PIXE method is possible only if reaction products contain the element or elements which are not normal constituents of the investigated tissue or, at least, are present in trace amounts. The list of selected histochemical procedures which may be combined with the micro-PIXE method is given in Table 9.5. The comprehensive review is provided in [9.65, 9.66].

TABLE 9.5 The List of Histochemical Reactions Useful for micro-PIXE Investigations (after E. Rokita [9.65]).

Determine organic molecule	Element detectable by micro-PIXE	Name of the method
Proteins (tyrosine)	Hg	Millon reaction
SH groups, SS groups	Fe	ferric ferricyanide
SH groups	Hg	mercury orange
SS groups	Se	hydrogen selenide
SH groups	Zn	hydrazide zinc
Reticulin	Ag	silver impregnation
Nervosecretory substance	Cr	Bargmann's chrome hematoxylin
Nucleic acids	Cr	gallocyaninchromalum
Proteoglycans	Cu	alcian blue
Proteoglycans	Fe	colloidal iron
Mucosubstances	Pb	lead tetracetate-Schiff
Mucosubstances	Bi	sodium bismethate-Schiff
Mucosubstances	Au	colloidal gold
Sialic acid	Cu	Biall reaction
Pectin	Ru	ruthenian red
Ascorbic acid	Ag	silver method for ascorbic acid
Glycogen	I	iodine method for glycogen
Amyloid	Ag	reducing method for amyloid
Phospholipids	Hg	Okamotos mercuric nitrate
Choline-containg lipids	Mo	phosphomolybdic acid
Unsaturated lipids	Ag, Br	bromine-silver
Cholesterol	Bi	bismuth trichloride
Aldehydes	Hg	Plasmal reaction
Alkaline phosphatase	Ca, Co	calcium-cobalt
Acid phosphatase	Pb	lead nitrate
Melanin	Ag	diamine silver stain
Noradrealin	I	iodate method
Lopofusin	Fe	Schmorl procedure
Bilirubin	Cr	trichrome procedure

As an example, the measurements of proteoglycans (PG) content in cartilage (alcian blue method at pH 1.0) are presented in Fig. 9.13. The measurements were performed in the raster scan mode (43x64=2752 points) with the step size equal to 16 μm. The irradiation time was 7200 s and the total charge was equal to 4.78 μC. The thickness was determined by the Rutherford backscattering technique (Sec. 9.2.4). The distribution of concentration of P, S, Cu as well as Cu/S and P/S ratios are presented in Fig. 9.13. The cartilage is stained strongly, which reflects the number of SO_4^{2-} groups and hence the amount of PG. This is in agreement with the highest S content at the cartilage territory. Some sulphur is present also in the bone trabeculae. S in the latter case is incorporated in the collagen molecules. This observation is supported by the fact

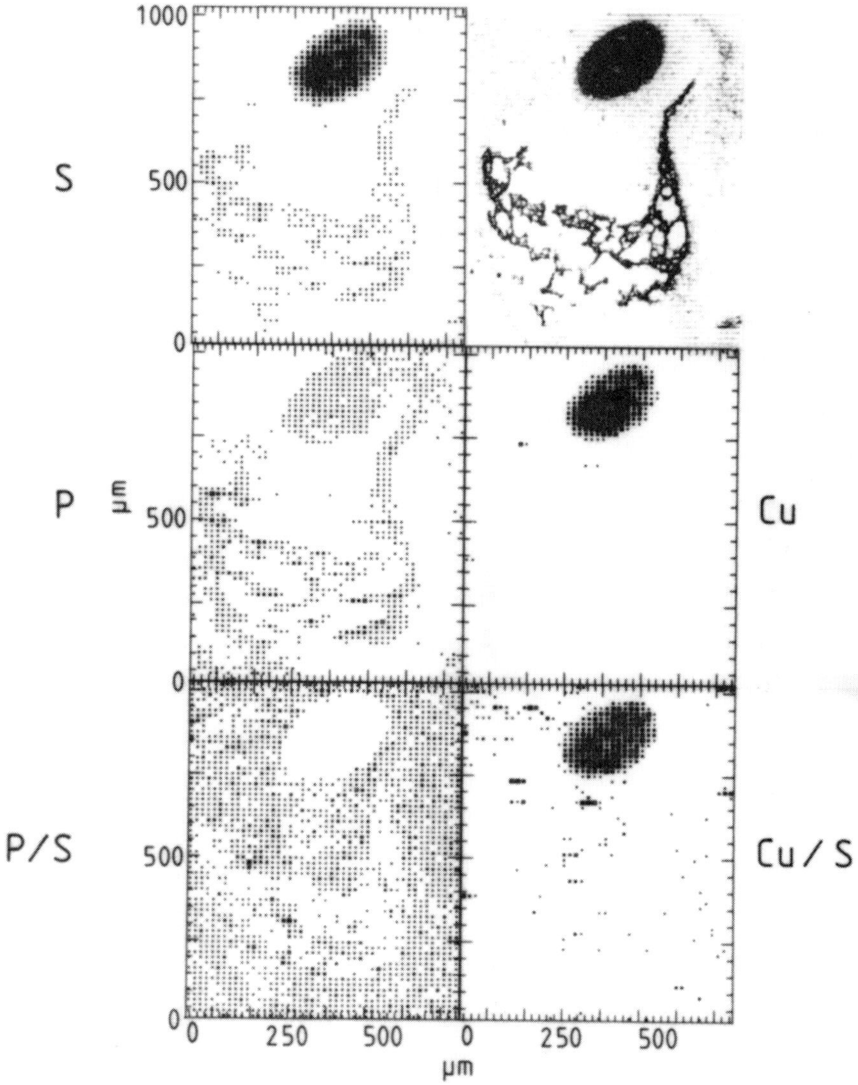

FIGURE 9.13 Skull of a 17 days old mouse embryo; the section stained with the alcian blue method (proton microprobe at the Dynamitron tandem accelerator of the Ruhr University at Bochum, FRG) (after E. Rokita et al. [9.65])

that Cu/S ratio is much higher in the cartilage than in the bone, i.e. S in the bone is not connected with the sulphate groups binding alcian blue. The presence of P in bone trabeculae comes from partial removal of inorganic material. Comparison of the P/S ratio and localization of Cu excludes the possibility that phosphate groups participate in alcian blue staining.

At present, the micro-PIXE method turns out to be supplementary to routine histochemistry [9.65].

Details of other uses of the PIXE method of investigating medical samples are given in [9.17 and 9.36], whereas those of investigating biological samples are discussed in [9.22 and 9.61].

As already mentioned in Sec. 6.7, the fluorescent analysis of biological samples can also be made with the aid of synchrotron radiation. The results obtained so far are very promising although the range of applications is smaller than that with the PIXE technique [9.62 through 9.64].

9.2.4 Methods of Elastic Ion Scattering

Technique of elastic backscattering. If the surface of a solid sample is bombarded with a beam of light ions, such as protons or alpha particles with energies in the 200 keV -3.0 MeV range, a small fraction of the incident beam (about 10^{-6}) will be elastically backscattered and recoiled backwards (Fig. 9.14). This is called *Rutherford backscattering.* The analytic methods based on ion elastic scattering are referred to a PESA *(Proton Elastic Scattering Analysis)* techniques, whereas those using backscattered ions are called RBS *(Rutherford Back-Scattering)* techniques.

In the RBS, the energies of E_{RBS} ions deflected through an angle of 180° are related to the primary energies of the initiating ions, E_0, by the relationship

$$E_{RBS} = E_0 \left(\frac{M - m}{M + m} \right)^2 \tag{9.2}$$

where M and m are masses of the substrate atoms (target) and of the bombarding particles, respectively. Therefore, by analyzing the energies of the deflected ions information can be gained about the composition of the substrate (Fig. 9.15). The depth range of this analysis depends on the primary energy E_0; if

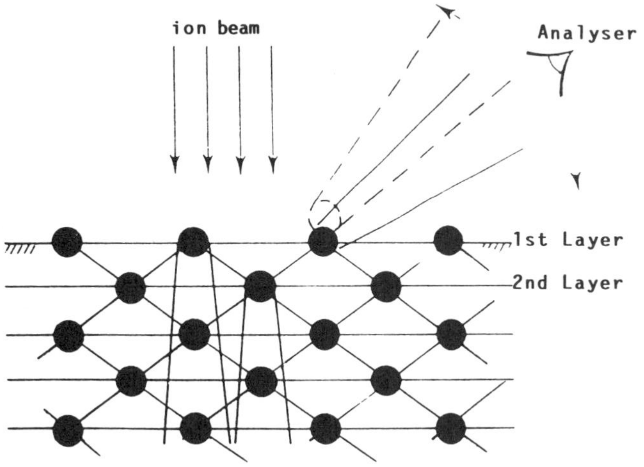

FIGURE 9.14 A diagram showing the elastic-ion-backscattering (Rutherford Scattering) principle

helium ions are used whose energies are controlled over the 1.0-3.0 MeV range, the near-surface layers with respective thicknesses of 0.5 and 3.0 μm can be analyzed. By diminishing the energy E_0, the range can be confined to the *first monoatomic layer* (Fig. 9.14).

Thanks to the simplicity of expressing the energy scale of recoiled ions in terms of a depth scale, the RBS technique enables precision profiling of the composition of individual constituents in multi-constituents structures. The RBS technique is among the methods of medium sensitivity since it permits detection of specific elements in the range between 10^{-1} and 10^{-4} (atomic). The disadvantage of this method is that the detection sensitivity largely depends on the atomic number of the analyzed element. For elements between bismuth and oxygen it falls by a factor of 100.

Rutherford backscattering is widely used for the determination of thickness of surface layers. The thicknesses range from 1 μm, when alpha particles of a few MeV are used as incident projectiles, to 10 μm with 3 MeV protons [9.68]. Computer programs for the simulation and the fitting of experimental data are now available. The program HCNO-MAP specially designed to evaluate data from the analysis of biological samples, evaluates RBS spectra to calculate the partial areal mass densities for carbon, nitrogen and oxygen, and the energy loss of the protons penetrating a sample [9.69]. It produces quantitative maps of the elements ^{1}H, ^{12}C, ^{14}N and ^{16}O from spectra of elastically backscattered and forward-scattered protons. HCNO-MAP will be part of a PIXE spectrum evaluation program for scanning data. The determined areal mass densities will be used both for mass normalization and for the determination of the PIXE background.

Elastic forward scattering and recoil techniques. In order to eliminate the disadvantages of the conventional RBS technique, such as the inability of detecting hydrogen or low sensitivity of the detection of light elements (Fig. 9.15), another method can be employed, viz. that of forward particle scattering denoted by the acronym FAST *(Forward Alpha Scattering Technique)*. By using a beam of 30 MeV alpha particles to bombard aerosol filters, employed in air contamination studies, at an angle of 45°, mass thickness of 0.1 to 2.0 μg cm^{-2} are obtained with good sensitivity for light elements from hydrogen to oxygen inclusive.

In addition to the analysis of the scattered initiating particles, the recoil nuclei of the investigated element can also be directly

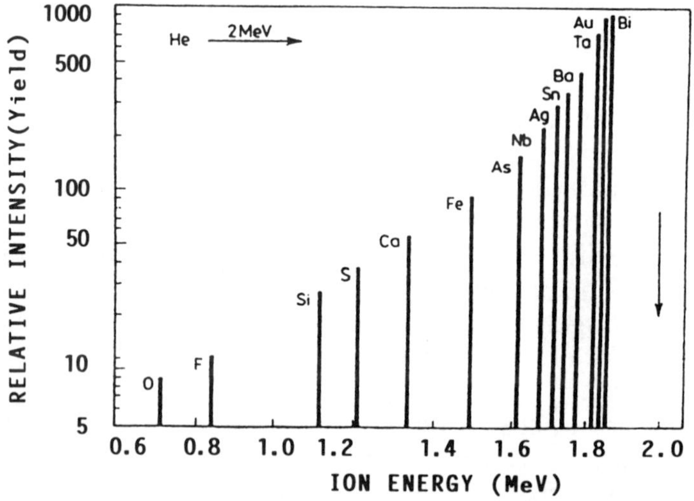

FIGURE 9.15 Dependence of the energy of backscattered 4 He^{+} ions for various elements (primary beam energy E_0 = 2 MeV) (after C.A. Evans Jr. and M.D. Strathman)

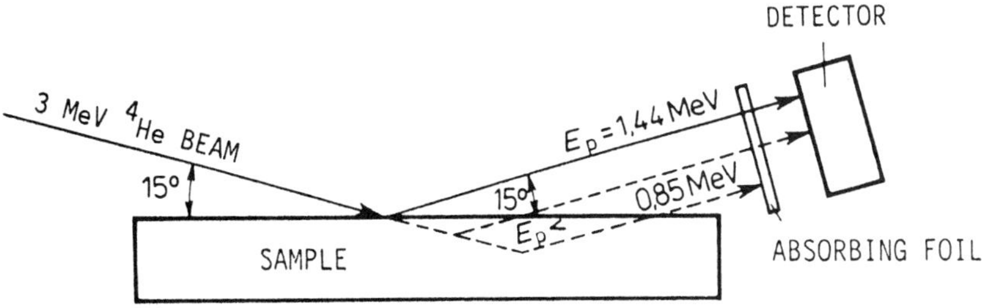

FIGURE 9.16 A diagram of the system for analyzing hydrogen content using recoil protons; E_P is proton energy

analyzed. This is called the ERD *(Elastic Recoil Detection)* method and it is specially suitable for the detection of hydrogen for which, due to kinematic reasons, there is no backscattering. A beam of accelerated alpha-particles strikes the investigated surface at a small angle of 15° (Fig. 9.16). The recoil protons, as a beam emitted at a similar angle forward, are registered by the detector. If the sample is thin compared with the range of the bombarding particle, which is 11 μm for 3 MeV alpha-particles in Al, the incident beam can strike the sample perpendicularly and the recoil protons will be detected from the opposite side. For the geometry as shown in Fig. 9.16, the maximum recoil proton energy is 1.44 MeV, while that of ^{4}He ions scattered at 30° is 2.99 MeV. By placing an absorbing foil, e.g. an 11 μm Al film, in front of the detector, the alpha particle beam is totally eliminated, whereas the energy of the recoil protons is attenuated to the energy of 1.05 MeV. In actual fact, both the initiating particles and recoil protons lose some energy inside the sample depending on the depth of the recoil event. Hence, the above energy of 1.44 MeV corresponds to that of protons which are produced directly on the surface, while the energy spectrum registered by the detector includes a large quantity of protons with lower energies produced at greater depths.

Examples of application. The RBS method is often employed in the analysis of biomedical samples for determining the masses of individual measurement areas or pixels. This is useful because the PIXE method implemented with the aid of a microbeam scanning pixel by pixel requires that the local mass, or the mass of the pixel investigated, be taken into account. In order to obtain true elemental concentrations, the fluorescent signal has to be related to the mass of the area from which the signal is detected. The point mass can be determined with the RBS technique by placing a barrier detector, which registers back recoiled protons, above the sample (Figs. 9.2 and 9.3). Fig. 9.17 is an illustration of a energy spectrum of such recoil protons. For detection, a silicon barrier detector with the useful area of 100 mm^2, placed at about 47 mm from the investigated liver sample was used [9.70]. The sample was analyzed with a 3 MeV proton beam (2.5 μm dia.) at 150-300 pA. The scanned area was 70 x 60 μm and consisted of a 32 x 32 point matrix. The scanning rate was 2 mm s^{-1} to limit local overheating of the sample. The spectrum shown in Fig. 9.17 encompasses the back recoiled protons within the angle range between 173 and 178°.

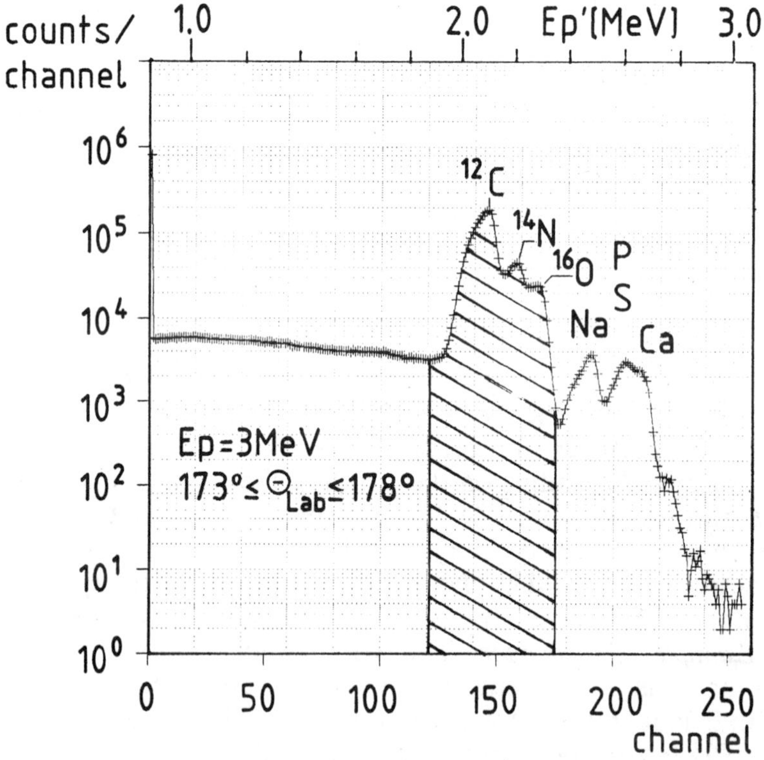

FIGURE 9.17 The RBS spectrum of a liver sample collected during a 30-minute measurement (after D.Heck et al. [9.70])

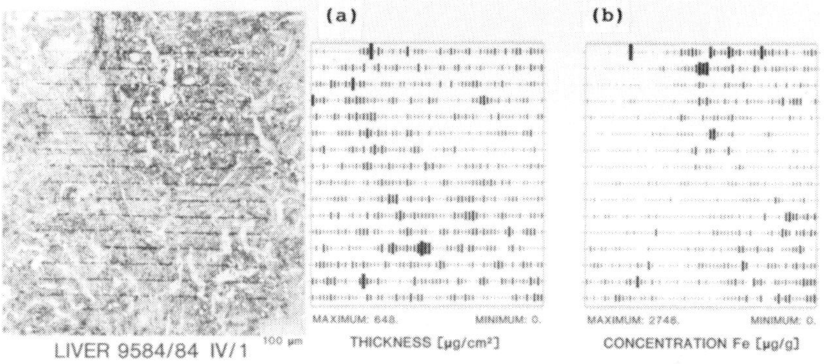

FIGURE 9.18 Microanalysis of a human liver sample (for siderosis): (a) — mass distribution determined with the RBS method, (b) — Fe concentration distribution determined with the PIXE method with allowance made for mass distribution (after D.Heck et al [9.71])

Protons are scattered on the atoms of carbon, nitrogen and oxygen; the contribution from other elements in the sample matrix can be neglected. The local mass of each of the pixels of the organic matrix under investigation can be determined, to a good approximation, from the sum of the pulses from individual spectrometric channels.

The results of the mass measurements obtained with the RBS method are subsequently used in the way shown in Fig. 9.18, which illustrates the microanalysis of a cirrhotic human liver with respect to Fe concentration distribution [9.71]. The sample was analyzed with a 3 MeV proton beam (3-5 μm dia.). The total sample dimensions were 0.6 x 0.6 mm^2 and the mass of one pixel was about 0.1 pg. Fig. 9.18a is a distribution of masses in the investigated area,whereas Fig. 9.18b shows the Fe concentration distribution obtained with the use of the characteristic radiation after making allowance for the masses of individual pixels.

Mapping with microbeams may serve as a basis for dynamic studies of the metabolism of trace elements. Fig. 9.19 provides an example of such studies. It shows a change in copper concentrations in isolated rat hepatocytes and intercellular space [9.70]. The analysis was made with the aid of a proton microbeam with parameters like those in Fig. 9.17. It can easily be seen from Fig. 9.19 that copper concentrations in hepatocytes are still increase 30 min after the beginning of the experiment,

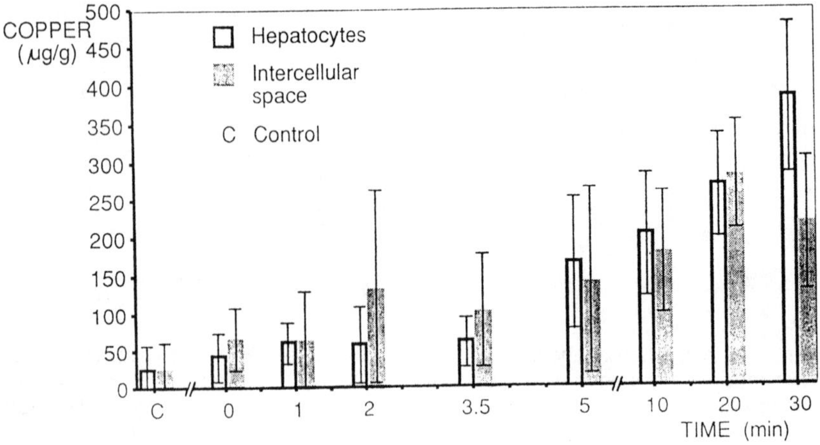

FIGURE 9.19 Time dependence of copper concentration in isolated rat hepatocytes and intercellular space (studies on the metabolism of trace elements in hepatocirrosis) (after D.Hech et al. [9.70])

whereas those in the intercellular space begin to fall 10 minutes earlier.

9.2.5 Nuclear Microscopy

Scanning transmission ion microscopy. Scanning transmission microscopy (STIM) is based on the fact that ions passing through a sample will lose energy and also suffer from geometrical scattering. The ion beam is allowed to pass through the specimen to be collected in a detector at 0° (Fig. 1.23, cf. also Sec 6.6, Fig. 6.18). The energy loss in the specimen is a measure of the specimen areal density (mass per unit area, mass thickness). This technique is almost 100% efficient and has a thickness sensitivity of approximately 3 nm in biological material. Because the beam is stopped down to a very small object and aperture value, the spatial resolution is very small, being limited by mechanical and electrical stability of the microprobe beamline to a present minimum of about 50 nm [9.16].

When using the energy loss technique, the initial sharp energy distribution of the beam will be broadened due to energy straggling. For the energies used in this analysis, the energy loss in an inverse function of the energy: lower energies have higher stopping powers. The main goal with STIM is to produce an *image of the sample based on its structure;* i.e. differences in density or in composition. Obviously, the ultimate lower limit of resolution is set by the energy straggling, which is around 5 keV (FWHM) for a 50 μg/cm^2 sample (corresponding to ~ 0.5 μm thickness) and increases to 30 keV for a 2 mg/cm^2 sample). A good surface barrier detector has an energy resolution around 15 keV, which is considerably more than the straggling contribution for the 50 μg/cm^2 sample. Thus it is of interest to improve the detector energy resolution, if possible [9.72].

The beam transmitted through a real sample will not be strictly monoenergetic as the sample, in addition to the energy straggling, even in the smallest details, exhibits some degree of inhomogeneity. This information is carried over by the beam and will appear as a skewness in the energy distribution. The skewness can be thought of as a mixture of Gaussians each representing different proportions of varying thickness of the irradiated pixel; the median of the energy distribution gives a robust energy selection to be used as a contrast parameter. By also using information on the skewness of the distribution, the degree of inhomogeneity in that spot can serve to increase the contrast.

The STIM technique applied to biomedical specimens is still at an early stage of development. For details see [9.16, 9.72].

Nuclear microscopy. The use of ion microbeams, apart from the application in PIXE methods (Sec. 9.2.3) and the RBS techniques already described (Sec. 9.2.4) provides better analytic possibilities, listed in Table 9.3 and in Fig. 1.23.

MeV ions impinging on surfaces induce large quantities of secondary electrons, which can be detected as in scanning electron microscopy (SEM, Fig. 1.23) to form an image of the sample. Those electrons which escape originate from a region close to the surface and so the electron emission is strongly affected by the surface condition, electrical properties and topography. Although, unlike with SEM, there seems to be very little contrast due to atomic number. Secondary-electron emission (SEE) imaging is used primarily for alignment, locating the sample and mapping surface topography, an important consideration, since surface conditions affect the accuracy of the results. The simultaneous use of particle backscattering analysis and secondary-electron imaging allows the light elements and the surface topography of the sample to be determined. In some cases, *nondestructive three-dimensional elemental analysis* can be carried out [9.73].

The basic trend in the present-day accelerator analytical methods is to make as full use as possible of all the information about a sample, obtained, first of all, from PIXE, RBS and SEM, as well as from STIM or NRA (Sec. 9.2.7) techniques. All these methods, taken together, have often been recently referred to as the technique of *nuclear microscopy* (Oxford group concept).

Nuclear microscopy thus refers to the use of microbeams of nuclear projectiles of MeV energy for qualitative and quantitative elemental mapping of surface and subsurface regions of the sample whatever analytical technique is used.

The main advantage of nuclear microscopy consequently is that elemental mapping of all elements of biomedical interest at the trace concentration level can be performed quantitatively [9.74]. The so called *ultratrace elements* are beyond the determination in nuclear microscopy. The term ultratrace element was recently introduced for elements occurring at very low levels. It has hitherto been used rather arbitrarily. Suggestions to make a distinction between *trace* (0.01-100 μg/g) and *ultratrace levels* (<0.01 μg/g) have not won general acceptance. To be able to detect these ultratrace elements other analytical methods have to be used. Examples of such methods are neutron activation analysis (Sec. 9.2.7), atomic absorption spectrophotometry and inductively coupled plasma mass spectrometry. None of these methods, however, can provide positional information from the sample. Nuclear microscopy and the methods with better sensitivity are not competitive but complementary [9.74].

Comprehensive review of nuclear microscopy is given in [9.73 and 9.74]; the application to diabetes research is described in [9.75].

9.2.6 Accelerator Mass Spectrometry (AMS)

Fundamental principles of the AMS method. The application of an accelerator as an essential component in a mass spectroscopy system in 1977 gave birth to a new technique called at present the *ultrasensitive mass spectrometry* or the AMS *(Accelerator Mass Spectrometry)* technique, which has made it possible to *separate isotopes or isobars,* atoms of elements with very close masses, with the highest possible resolution of 10^{-15} to 10^{-16} atoms per atom. Worldwide, about 30 accelerator facilities are now performing AMS measurements with a total capacity of over 10000 samples per year. By far the largest fraction of facilities is using old tandems originally designed for nuclear physics experiments. The maximum terminal voltages of these facilities range from 6 to 16 MV. They are, therefore, very versatile. For light ions (Be, C) very high stripping efficiency can be obtained and for heavier isotopes up to ^{41}Ca good isobar suppression is possible. New systems designed and built for AMS in general operate around 2 MV. These facilities have mostly been optimized for precise ^{14}C measurements. If isobaric interferences are not a problem, they also allow very sensitive measurements of heavier radioisotopes, e.g. ^{26}Al and ^{129}I. A few large facilities, tandems with boosters or linacs, which are primarily used for nuclear and atomic physics research, make available an energy range above 100 MeV. In the last few years a few experiments have been performed with single-ended Van de Graaffs and cyclotrons on ^{3}H and ^{41}Ca. Due to their small size, minicyclotrons would be very attractive, but first the sensitivity and other problems have to be solved in a satisfactory way [9.76].

Among the first uses of accelerator mass spectrometry was the study in geochronology or dating with radiocarbon-14, which is constantly being generated in the atmosphere by cosmic rays from ^{14}N, and decays with the half-life of 5740 years. In the photosynthesis of CO_2, ^{14}C is taken up by living organisms, and when life processes are stopped no more ^{14}C is absorbed. Therefore, if decaying ^{14}C activity is then measured, the age of the organic material can be determined. This technique, called *radiocarbon dating*, permits determination of sample age back to some 70 000 years.

The counting rate of beta particles in the ^{14}C decay per one gram of the present day carbon of biological origin is about 15 particles per minute. The source of this low counting rate is 6.5×10^{10} atoms of ^{14}C in one gram of biological carbon sample. However, the main bulk of all carbon atoms are the ^{12}C isotope, and the relative concentration of $^{14}C\,/^{12}C$ is very low, i.e. 10^{-12} atoms per atom.

The difficulty of detecting very small activities of ^{14}C is that old samples require large amounts of the initial material, up to several grams, and the investigation damages the sample. Therefore, the use of mass spectrometry and direct nonradioactive measurement of the ^{14}C content have been discussed for quite a long time. It would be possible to reduce the necessary amount of carbon to some micrograms.

The main obstacle in the above method, unavoidable in practice, is the presence of ^{14}N atoms in the analyzed carbon beam. ^{14}N and ^{14}C masses differ only by 10^{-5}, so that no conventional mass spectrometer is capable of separating them. Another factor, equally important, which complicates the measurements is the presence of a large number of $^{12}CH_2$ and ^{13}CH molecules with nearly the same mass as that of ^{14}C.

In 1977 D.E.Nelson and C.L.Bennet, independently of each other, put forward an idea to use a tandem Van de Graaff accelerator in the AMS method. Thanks to the use of a source of negative ions and the known instability of $^{14}N^-$ ions, the contamination effect of nitrogen ions can be reduced. The accelerator also serves as an eliminator of $^{12}CH_2$ and ^{13}CH molecules by causing them to disintegrate after being accelerated to a

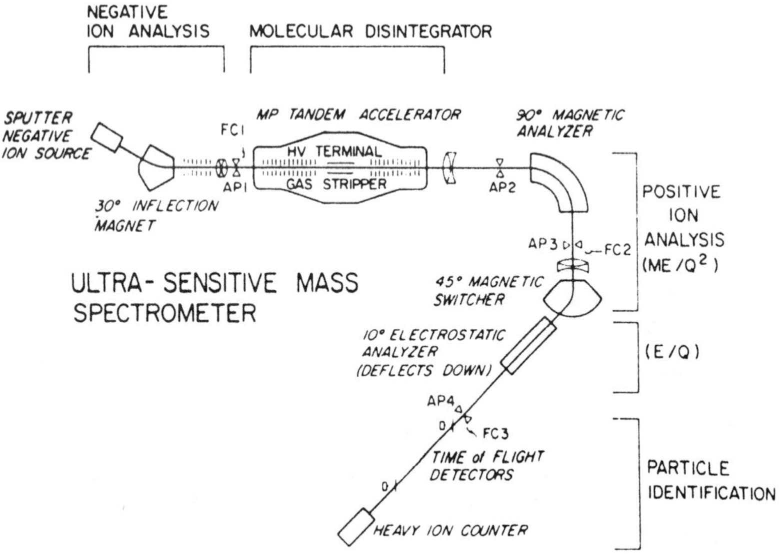

FIGURE 9.20 A schematic diagram of an accelerator system for accelerator mass spectrometry (AMS)(description in the text) (University of Rochester)

suitable velocity or energy. Molecules with high velocities become quickly dissociated in collisions with atoms of gas such as argon. To become completely dissociated, the molecules must attain velocities which would lead to the removal of three electrons. For example, carbon atoms must achieve the energy of 2.6 MeV, which exceeds the energy range of conventional mass spectrometry by a factor of about 100. At the above energy of 2.6 MeV, the $^{12}CH_2$ molecules breaks into C^+ ions, and thus molecular ions which may interfere with $^{14}C^{+3}$ ions are not present any more.

Fig. 9.20 is a schematic diagram of an accelerator system for ultrasensitive mass spectrometry installed at Rochester University. The investigated sample is bombarded with a beam of cesium ions, which, as a result of *ion sputtering,* cause the emission of negative ions from the sample material. The analyzed beam passes through a mass pre-accelerator which selects ions with mass 14 (in ^{14}C analysis), and then travels through a collimating slit AP1 and a movable Faraday cup FC1 for current monitoring. At the first stage of the tandem accelerator the beam is accelerated to the energy of 8 MeV, and then is converted to C^{4+} ions so that molecular ions can be effectively eliminated. After conversion, the beam, which now consists of positive ions, passes the second stage of the accelerator and enters the second mass spectrometer, in which positive ions become separated from molecular fragments. Now the beam reaches the particle identification system which determines the ion energy as well as unit energy loss dE/dx. For heavy ions, an additional time-of-flight detector is employed. Finally, the selected ions are counted.

Fig. 9.21 shows the resolution of the above system by a 3-D representation of the results of measurements of a beam containing ^{14}N ions and a mixture of ^{12}C, ^{13}C and ^{14}C ions. The vertical axis represents the currents of each beam in the logarithmic scale.

Detection limits for AMS measurements can be divided into three categories [9.77]: (1) stable isotope (element) concentrations $\sim 1^{-12}$ (parts per trillion, ppt); (2) radioisotope concentrations (actually radioisotope/stable-isotope ratios) $\sim 10^{-16}$; (3) exotic-particle concentrations $\sim 10^{-19}$. The large differences in the detection limits of the three categories are determined by the level of background. In an accelerator environment it is

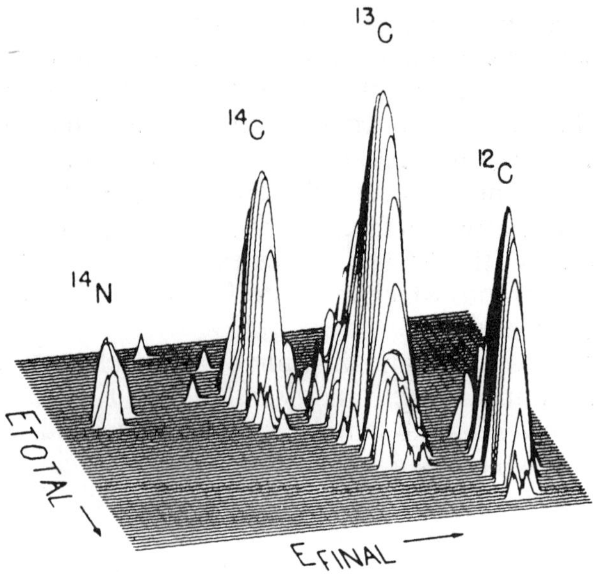

FIGURE 9.21 A 3-D representation of the separation of ^{14}N, ^{12}C, ^{13}C and ^{14}C using an accelerator system shown in Fig. 9.20

extremely difficult to establish the "clean-room" conditions necessary to measure stable isotopes in the ppt range. Below this level the whole periodic table seems to be present in almost any material and/or environment on earth. The conditions are less severe for radioisotope detection, since with proper discrimination radioisotopes can be separated from stable-isobar interferences. However, of increasing concern in measurements of very low natural radioisotope concentrations is laboratory and global contamination with man-made radioisotopes. The lowest detection limits can be reached in searches for exotic particles since a parameter space of mass and charge (e.g. fractionally charged particles) is scanned, where no known species exist [9.77].

Sensitivity of AMS in biomedical research [9.78]. Radioisotope tracers are a useful tool in numerous areas of biomedical research, including metabolism, pharmacokinetics, and the detailed study of biomolecular interactions (cf. Sec. 7.3.2, Table 7.9). Standard techniques exist for the incorporation of radioisotopes of many elements into molecules of biological interest.

At present, the isotopes in common use are most often short-lived. Relatively short counting times can be used to adequately determine the isotopic concentration of reasonably sized samples. However, short decaytimes may limit the usefulness of these isotopes for long-term studies of pharmacokinetics in some situations. Further, the rapid decay promotes radiolysis, limits the "shelf-life" of the compound, and delivers a radiation dose to the material under investigation. Since AMS directly detects the undecayed radioisotopes [9.78], isotopes with long lives are measured as quickly as unstable ones and much smaller amounts of the tracer are needed. More specific chemical separation can be done on small samples to select fractions which may contain very little tracer. The labeled agent can be made with lower specific activity, reducing radiolysis and the efforts presently used to prevent it, minimizing radiation effects which are detrimental to experimental organisms, and decreasing the nonspecific binding effects which plague many uses of radioisotopes in biomedical studies. Moreover, biological variability is easily determined, given the rapid measurement capabilities of AMS. AMS not only improves measurements of traditionally labeled compounds but also allows the study of the biologic behavior of some elements for the first time.

AMD is able to count 10^{-5} to 10^{-1} of the selected atoms (long-lived isotopes) depending on the efficiencies for ionization and transport of that element in the accelerator system. Even after considering the substantial time needed to prepare some elements for measurement, the sensitivity of AMS for detection of long-lived isotopes is many orders of magnitudes larger than that attainable through decay detection for equivalent counting times (Fig. 9.22). There are a number of such isotopes which can now be applied to biomedical research, many of which have already been applied to geoscience research. These isotopes are valuable as elemental tracers and as molecular labels.

Most biomedical applications use short-lived isotopes as tracers, and numerous labeled compounds are available. AMS also offers gains in detecting these isotopes (Fig. 9.22), as long as the sample separation and preparation can be made sufficiently rapid (seconds to hours) to compete with decay counting. Direct conversion and ionization of the product of chromatographs and capillary electrophoresis instruments could greatly reduce the "preparation" time for some of these isotopes. For

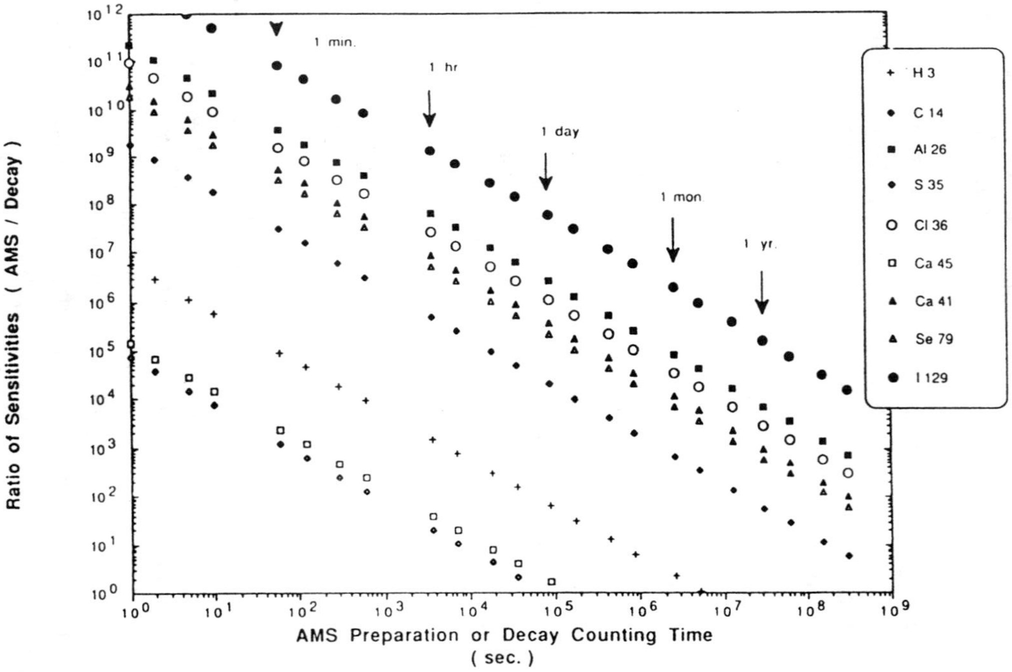

FIGURE 9.22 The ratio of the sensitivity of accelerator mass spectrometry (assumed constant at 1% for this illustration) to the sensitivity of a 100% efficient detection of the decayed isotopes, expressed as a function of the time alloted to counting the decays or the time required for preparation and measurement of the AMS sample. AMS efficiency will be lower for elements such as calcium and aluminium, but higher for the chalcogens and halogens. Even for very short-lived isotopes such as ^{3}H, AMS can provide greater sensitivity if preparation time is reduced to a few hours (after J.S. Vogel et al. [9.78]).

the moment, implementation of the known techniques for sample preparation allow experiments to prove the usefulness of these isotopes while faster preparation schemes are developed. As with the long-lived isotopes, advantages in sample size and lower radioactivity can be gained from the sensitivity of ASM [9.78].

Radiocarbon as the universal biochemical label. The most developed capability in AMS is for radiocarbon [9.77, 9.79 and 9.80]. The application of the accelerator mass spectrometry in dating of organic samples by measuring the ^{14}C content has made it possible to determine the age of 60-thousand-year-old samples to the accuracy of $\pm 1\%$, whereas the age of "younger" samples may be determined within ± 80 years [9.81]. Apart from archeological samples, the life-times of sediments from lakes and bogs can be investigated routinely [9.82]. This method is also used to measure the abundance of fully deuterated methane $^{13}CD_4$ which is a high-sensitivity tracer in studies of atmosphere dynamics [9.83].

Reliable *dating of fossil bone* is of great interest to anthropologists, since bones are the only direct remains of early man. Various new techniques have been developed and some surprisingly old ages of well over 20000 years were reported for some human skeletad samples in North America, using so-called *amino-acid racemization dating.* The method measures the time-dependent deorientation of originally left-oriented amino-acids extracted from bone collagen. Recent AMS radiocarbon dates of these amino acid extracts and of other organic bone fractions gave much younger ages in the 3000-6000 yr range, more consistent with general ideas about the appearance of man in this part of the world [9.77].

The scientific significance of dating a religious relic has been questioned by many scientists. However, the public interest in such objects is undeniable and from the onset of the modern AMS era in 1977 it was clear that ^{14}C dating of milligram samples with a precision of 1% would allow radiocarbon dating of the famous *Shroud of Turin* with an acceptable sacrifice of material. The Shroud is believed by many to be the burial cloth of Christ. After considerable negotiations with the church authorities, three AMS laboratories (Arizona, Oxford, Zurich) were chosen to perform radiocarbon measurements on a small piece of linen from the Shroud, with a total weight of about

50 mg. Excellent agreement for the AMS results of the Shroud and three other samples of known ages were found by the three laboratories. The age of the Shroud was determined to be medieval with a date of 1260-1390 AD (≥95% confidence). This is in agreement with the notion of some art historians that the Shroud could be a medieval icon rather than originating from the time of Christ [9.77].

Systematic studies of great detail can now be performed in the natural storehouses of the earth such as ocean sediments and ice sheets of Greenland and Antarctica. The collection of kilograms rather than tons of sample material make a big difference for these studies [9.80 and 9.84]. Polar ice cores provide perhaps the best way of obtaining samples for study of the earth's atmosphere at times in the past. The most reliable way of dating samples of ice cores older than the limit of annual isotopic stratigraphy (9000-12000 yr P.B.) may be through radiocarbon measurements on the CO_2 contained in the ice.

Radiocarbon can be introduced into nearly every organic molecule, and the site of incorporation can, in most cases, be specified by the production procedure. The ^{14}C does not substantially modify the chemical behavior of the molecule, as do other labels. The ^{14}C can be located at stable sites within the molecule which are not subject to the exchange that make tritium and other (e.g. halogen) labels nonspecific. If the organic compound is modified by metabolism, the portion which contains the ^{14}C will not be lost. Thus, ^{14}C is initially the isotope of choice for many biomedical labeling experiments. With the sensitivity of AMS, single, selective labeling with ^{14}C can be done [9.78].

Some AMS laboratories are exploring molecular events critical to the development of chemically induced cancer (chemical carcinogenesis). Cancer is primarily the result of heritable changes at the molecular level which affect a cell's ability to regulate cell division. The heritable nature of cancer is the result of mutation or changes in the genetic code (deoxynucleotide base sequence of DNA) in the animal. Practically all cancer is initiated through genetic mutation. Chemically induced cancer is dependent on the ability of chemicals to be metabolized to electrophilic species, then attack and covalently bind to DNA. Chemicals capable of such attack (mutagens and carcinogens) are widespread in the environment and human exposure is unavoidable. Since the initiating event in chemical carcinogenesis is believed to be the binding of chemicals to any of the deoxynucleotide bases and since these adducts persist for some time following exposure, DNA adducts are touted as markers of carcinogen exposure [9.85].

An ambitious program of adapting the AMS technique to biomedical and clinical research is being implemented now at the Lawrence Livermore National Laboratory [9.86, 9.87]. Studies on the carcinogenic effect of some substances using the AMS have yielded interesting results. For example, AMS is being used to determine the amount of carcinogen covalently bound to animal DNA (DNA adduct) at levels relevant to human exposure. Detection sensitivities are 1 carcinogen molecule bound in 10^{11} to 10^{12} DNA bases, depending on the specific activity of the radiolabeled carcinogen. Studies have been undertaken utilizing heterocyclic amine food-borne carcinogens and 2,3,7,8-tetrachlorodibenzo-p-dioxin (TCDD), a potent environmental carcinogen, to study the metabolism of carcinogens at low doses. In addition, AMS is being used to detect the presence of rare proteins (mutant forms of protamine) in human sperm. Approximately 1 per 10^6 sperm analyzed contain the rare form of the protamine. Protamine isolated from this small number of cells is being analyzed by AMS, following ^{14}C labeling. Thus, AMS can be used to verify the identity of an extremely small amount of biological material. Furthermore, an additional improvement of 2 orders of magnitude in the sensitivity of biomedical tracer studies is suggested by preliminary work with bacterial hosts depleted in radiocarbon [9.85].

Aluminum metabolism studies. Despite being the third most abundant element in the earth's crust, *aluminum* was until recently believed to have no significant biochemical role in health or disease, nor was it thought to participate in major metabolic processes. It was considered to be very poorly absorbed from the gastrointestinal tract, and that fraction absorbed was assumed to be efficiently eliminated by the kidneys. Research over the past decade, however, presents a very contrary assessment of its effect on the human body — aluminum is now referred to as a toxic metal with sufficient evidence to implicate it in several disease states (e.g. Alzheimer's disease [9.89, 9.100].

Although there is now overwhelming data in support of aluminum toxicity, research into its mechanism has been severely limited by the absence of a suitable isotope with which to perform conventional tracer studies, as well as other methodological difficulties. Since aluminum has only a single stable isotope and no radioactive ones with life-times suitable for traditional radiation detection methods, kinetic studies using isotopic tracers have been precluded. Because of its very low concentration in biological systems (ppm in tissues and ppb in blood) monitoring aluminum is a formidable task. Consequently, the sole method of studying aluminum kinetics prior to the present work had been to administer large dosages of ^{27}Al into the body, followed by measurements of its level in blood, urine and several organs. Since these loads of aluminum were very large compared to the total body burden, experiments in humans are impractical and the information obtained for animals cannot reflect the steady state kinetics [9.89].

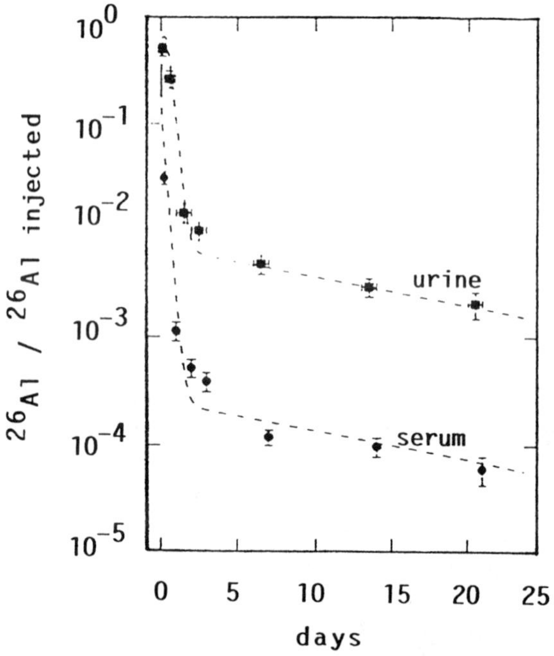

FIGURE 9.23 The data shown is the fraction of the administered ^{26}Al remaining in the blood and excreted in the urine as a function of time after injection. the horizontal bars indicate the duration of the urine collection (after O. Meirav et al. [9.89].

Naturally occurring aluminum is ^{27}Al (100%). Only one long half-life radionuclide exists ^{26}Al, $T_{1/2}$ = 0.716 x 10^6 yr. Use of ^{26}Al coupled with mass spectrometry would be particularly attractive for isotopic tracer work in human toxicology because of the isotope's negligible natural abundance, and the low radiological hazard resulting from the low detection limit. However, use of conventional mass spectrometry is impracticable due to ^{26}Mg interference (^{26}Mg comprises 11% natural Mg, ubiquitous in biological and environmental samples), but high-energy AMS, resulting in fully stripped ions, potentially overcomes this problem.

Previously, most ^{26}Al AMS measurements have been conducted using tandem accelerators at terminal potentials typically under 8 MV, by accelerating the $^{26}Al^-$ ion.

In [9.88], a 20 MV tandem Van de Graaf accelerator was used to conduct $^{26}Al^-$ AMS with fully stripped aluminum ions. The aim was to detect $^{26}Al^{13+}$, which is well discriminated against $^{26}Mg^{12+}$. Discrimination of ^{26}Al from ^{26}Mg was found to be better than 1:10^9, with the estimated detection limit of about 10^{-18} g. Preliminary biomedical applications include identification of Al carrier proteins in the blood, and investigations of Al uptake by cells in culture and Al speciation in natural waters.

AMS is particularly suited to long-lived radioisotopes (Fig. 9.22), thus providing an analytical method by which the radioisotope ^{26}Al can be measured with a sensitivity a million times greater than that of conventional decay counting. Hence, it is now possible to use ^{26}Al as a marker, with amounts of material that are neither prohibitively costly nor hazardous [9.89].

A 385 g rat was injected intravenously with 0.35 ng of ^{26}Al and 40 ng of ^{27}Al. Fig. 9.23 shows the fraction of the administered ^{26}Al remaining in the serum and excreted in the urine as a function of time after injection. There is a rapid expotential decrease of the level of ^{26}Al over the first couple of days. Thereafter, from a concentration of about a tenth of a percent of the injected dosage, it decreases at a much slower rate.

Calcium absorption and metabolism studies. Absorption and metabolism of *calcium* is currently of great interest, because of (1) the widespread incidence of the metabolic bone disease osteoporosis and (2) the effects of space travel on the skeletal system. Osteoporosis manifests itself by a general loss of bone

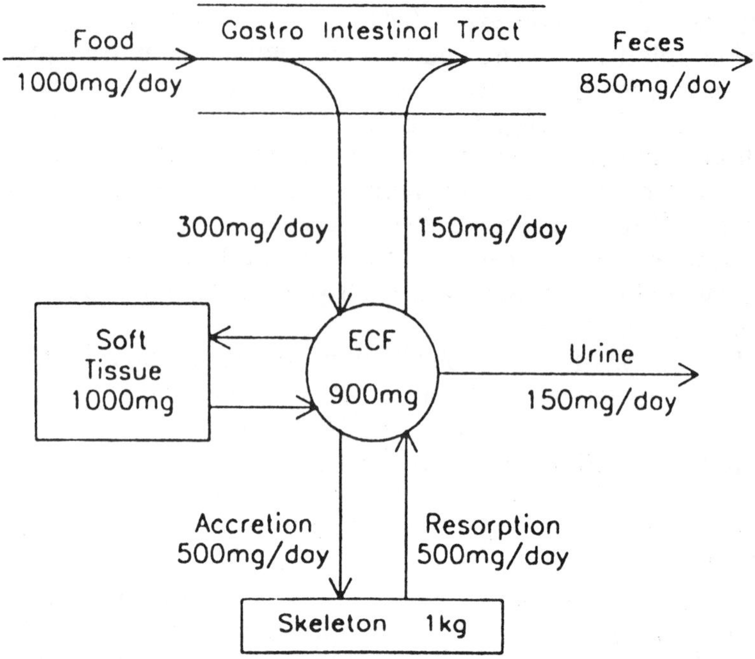

FIGURE 9.24 A scheme for calcium metabolism for a normal adult at zero calcium balance showing approximate pool sizes and transfer rates (EDF stands for extracellular fluid). Older persons often show a negative skeletal calcium balance (bone resorption higher than accretion) (after G.R. Mundy et al.)

mass. It afflicts 20 million in the USA or nearly 10% of the population. It can start early in life and lead to fractures that are disabling or later in life (30% of women over 90 have had hip fractures). Similarly, bone loss has been reported by both the American and Soviet space programs and it thought to be induced by the lack of gravity on the skeletal system, similar to prolonged bed rest [9.90].

A simplified scheme for calcium metabolism showing average transfer rates is shown in Fig. 9.24. Calcium in the diet is partially absorbed with about 30% efficiency through the intestines into the blood. Here it exchanges rapidly with *extracellular fluid* (ECF) and soft tissue. Calcium concentration in ECF is tightly controlled by hormones and has several vital functions including cell growth and differentiation muscle contraction, blood coagulation and mineralization of bone. Calcium is cleared from ECF through the gastro-intestinal tract to faces, through the kidneys to urine and by bone accretion. When dietary calcium is low, resorption of skeletal calcium maintains the required ECF level. During youth, bone accretion exceeds bone resorption and an increase in bone mass occurs. Bone mass reaches a maximum at about 20 years and normally drops by a factor of two between the ages of 40 and 80.

Calcium has six stable isotopes and three radioisotopes with a half-lives longer than a day (Table 9.6), that can be used as in vivo tracers.

Table 9.6 compares these radioisotopes with a view towards a prelabeling study. The *annual limit on intake* (ALI) for occupational exposure is a measure of the radiobiological hazard. Permissable doses can range from several orders of magnitude lower for studies using human subjects taken from the general population to much higher for diseases patients. The relative values for ALI in Table 9.6 given in Bq (decays per second)

TABLE 9.6 A Comparison of Three Calcium Isotopes Used in a Hypothetical Prelabeling Study (after D. Elmore et al. [9.90])

	Isotope		
	^{47}Ca	^{45}Ca	^{41}Ca
Half-life (radiological)	4.5 d	165 d	10^5 yr
ALI[a)]			
[Bq]	$3x10^8$	$6x10^7$	$1x10^8$
[atoms]	$2x10^{14}$	$1x10^{15}$	$5x10^{20}$
Dose			
[atoms]	$2x10^{11}$	$1x10^{12}$	$5x10^{17}$
Activity after 1 yr			
[atoms/10 ml plasma]	<1	$2x10^4$	$5x10^{10}$
[atoms/atom ^{40}Ca]	$1x10^{-15}$	$3x10^{-9}$	
[counts/h/10 ml plasma][b)]	4	23	
AMS counting rate			
[counts/h/10 ml plasma]	$6x10^4$ [c)]		

a) ALI = Annual Limit of Intake (oral)
b) Counting rate for decay counting is based on 100% detector efficiency
c) Counting rate for AMS measurement of ^{41}Ca is conservatively based on a 100 nA CaF_3 negative ion beam with a 1% accelerator efficiency.

reflect competition among several factors including the radiobiological half-life (short for ^{47}Ca making it relatively safe), the biological lifetime (long for skeletal calcium), and the energy per decay (low for ^{41}Ca). The advantage of ^{41}Ca is readily apparent when one looks at the ALI in number of atoms: it is more than five orders of magnitude higher for ^{41}Ca than for ^{45}Ca.

The bottom of Table 9.6 shows the predicted attenuation of a dose of 10^{-3} times the ALI for a hypothetical prelabeling study. The values are based on a one-year study in mice that used a much higher ^{45}Ca dose. The activity after one year is given in three sets of units:atoms/10 ml plasma, atom ratio and decay rate in counts per hour/10 ml plasma. All three radioisotopes are unmeasurable by decay counting methods. After one year the $^{41}Ca/^{40}Ca$ ratio is still six orders of magnitude above the detection limit of AMS indicating multi-year studies should be possible [9.90].

9.2.7 Activation Methods

Neutron methods. The most common *activation analyses* are implemented with fast neutrons with the energy of 14 MeV generated using accelerator methods (Sec. 5.3.2). Neutrons for this purpose are produced by the use of so-called *neutron tubes* [9.91] or *laboratory-type neutron generators* [9.92], which generate neutron fluxes of 10^8 to 10^{11} ns^{-1}, and in top designs even up to 10^{12} ns^{-1} [9.93].

Using the *neutron activation* analysis the elemental mass in the sample to be investigated can be determined from

$$g = \frac{A\,M}{N_A\,\sigma\,\Phi\,(1 - e^{-\lambda t})\,\theta} \tag{9.3}$$

where A is the induced activity after an irradiation time t, M is the atomic mass of the element, N_A is Avogadro's number, σ is the activation cross-section, Φ is the neutron flux, λ the decay constant and θ is the isotopic fraction of the given nuclide in the sample. Because of the difficulties involved in precise determination of such quantities as the neutron flux and the activation cross-section, relative measurements with the aid of a standard are commonly used. The unknown mass of the element investigated may then be calculated from a simple relationship

$$g_X = g_S \frac{A_X}{A_S} \tag{9.4}$$

where g_X, A_X are the mass and activity of the element under investigation, and g_S and A_S are the respective parameters of the standard.

The sensitivity of neutron activation depends, among other things, on the type of element. Fig. 9.25 presents the results of sensitivity calculations for the activation with 14 MeV neutrons with the flux density of 3 x 10^{10} n cm^{-2} s^{-1}. The figure includes 78 elements classified in three groups according to their half-lives: less than 1 min, between 1 min and 1 hour and more than 1 hour. A detailed list of reactions used in the calculations in Fig. 9.25 can be found in [9.93].

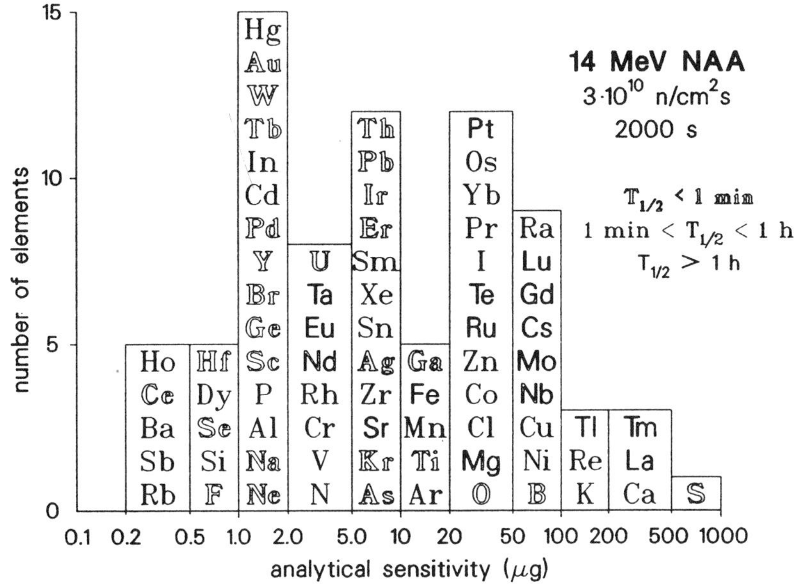

FIGURE 9.25 Calculated detection sensitivities of some elements for the activation with 14 MeV neutrons (cyclical activation with the flux density of 3 x 10^{10} n cm^{-2} s^{-1}, measurement time of 2000 s);(after R.Pepelnik [9.93])

Studies on the simultaneous determination of phosphorus and nitrogen in human tissue serves as an example of the neutron activation analysis [9.94]. The reaction employed in these studies was that of $^{31}P(n, \alpha)^{28}Al$ and $^{14}N(n, 2n)^{13}N$.

Ionic methods. In the analysis of tracer elements in biological samples activation with a beam of high-energy ions can also be used. Reactions suitable for the detection of light elements are listed in Table 9.4 (Sec. 9.2.2).

In the activation analysis with a beam of ions, simultaneous irradiation of the sample and standard is generally impossible, therefore beam monitoring is required, e.g. by means of a Faraday cup, or by placing a thin foil in front of the sample and standard and measuring the induced activity. If thick samples are irradiated the mass of the component investigated can be determined from

$$g_x = g_s \frac{A_x}{A_s} \frac{i_s}{i_x} F(E_1) \tag{9.5}$$

where A_x and A_s are the activities of the sample and standard, and $F(E_1)$ is the correction factor which is a function of the energy, E_1, of the inducing particles. Compared with neutron activation (Eq. 9.4) Eq. 9.5 includes a correction factor $F(E_1,)$ which can be determined analytically from the excitation function expressed as the relationship between the activation cross-section and energy (Sec. 7.2).

The cost of activation analysis induced with high-energy ions is relatively high. The method, however, is useful in the detection of such light elements as Li, B, C, N and O. If concentrations of these elements in the investigated samples are high, i.e. within the range of $\mu g\ g^{-1}$ or more, the accuracy of the results obtained with the ionic method is comparable with that of other chemical techniques. On the other hand, in the low concentration range, ionic analysis is the only available technique, particularly for the detection of elements such as C, N and O which are most abundant in biomedical samples. Details of the potential provided by ionic activation methods are given in [9.95]. Typical application is the fluorine depth profiling in teeth [9.96, 9.97] and the analysis of biological fluids for ^{18}O [9.98]. *Thin-layer activation* as a method of measuring wear has been applied in a tribological study of artificial hip joints [9.99].

9.3 Radiobiological Research Facilities

A early as the 1970s, the wide range of radiobiological research (Sec. 1.2) led to setting up specialized laboratories, of which some became later transformed into dedicated facilities intended for research into the biological effects of ionizing radiation. The Radiological Research Accelerator Facility, (RARAF) located at Nevis Laboratories of Columbia University, is an example of such facilities [9.95, 9.101 and 9.102]. The general plan of the main floor of RARAF II is shown in Fig.9.26. A Van de Graaff accelerator featuring accelerating potential variable over the 0.4-4.2 MV range is a source of high-energy particles, such as hydrogen, deuteron and helium-3 ions, as well as residual ions from the high vacuum. The accelerated particles are subsequently redistributed in four beamlines.

The first beamline delivers a 50° beam to laboratory U (Fig. 9.26) located in an non-shielded area. It is intended for irradiating cell cultures with low intensity molecular ions. Two lines deliver 30°and 45°beams to the shielded laboratory T designed for radiological physics research. The room also includes a 7 Ci

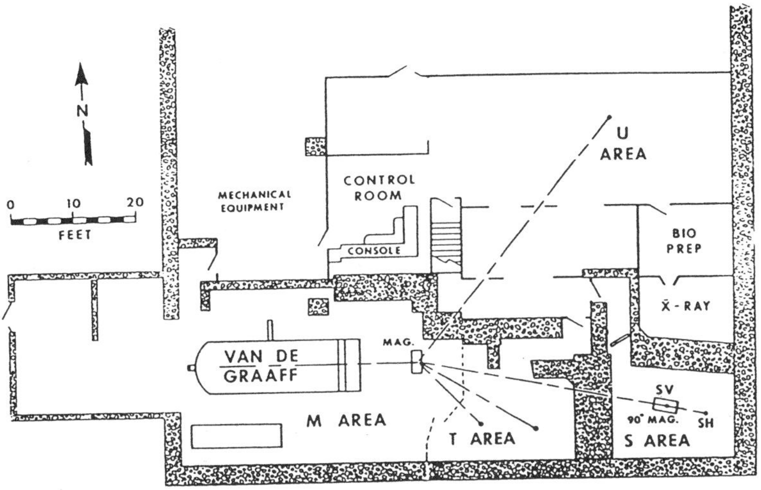

FIGURE 9.26 Layout of the main floor of the RARAF (Radiological Research Accelerator Facility) at Nevis Laboratories, Columbia University,USA [9.101 and 9.102]

^{137}Cs calibration source. The fourth 10.5° beamline has two branches leading to the fully shielded laboratory S. The beam in the main floor may also be used for the production of neutrons, or it may enter the upper floor where single-layer cell cultures are irradiated with particles with strictly defined LET values. A 250 kV_p X-ray unit for photon irradiation, together with another 50 kV_p X-ray unit, is located in a separate lead-shielded room.

Maximum intensities of beams are: 150 μA for H^+, H_3^+, D^+ and D_3^+ ions, 40 μA for H_2^+ and D_2^+ ions, 1 μA for $^3H^{++}$ ions, and 10 nA for C^+ and other residual high vacuum ions. The minimum beam intensities are less than 1 pA. The LET values are variable over the range between 10 keV μm^{-1} (for 4.1 MeV protons) to 170 keV μm^{-1} (for 4.3 MeV 3He ions). Dose rates can be varied between 0.001 and 8 Gy s^{-1}.

Neutrons in the RARAF facility are produced in the $^3H(p,n)^3He$, $^2H(d,n)^3He$ and $^3H(d,n)^4He$ reactions. With a maximum accelerating potential of 4 MV, monoenergetic neutrons can be generated in three bands ranging from 200 keV to 3 MeV, from 3.5 MeV to 7 MeV, and from 13.5 MeV to 20 MeV. These neutrons are subsequently used in radiobiological, dosimetric and microdosimetric research. The results are then utilized in radiation protection and neutron radiotherapy studies.

Radiobiological research at RARAF focuses on the determination of biological activity of radiation, with special emphasis on the effect of energy distribution in the cell. Due to the use of high-LET particle cellular effects induced by single particle can be assayed. An example of such investigations is afforded by the studies on chromosomal alterations induced by small doses of radiation responsible for genetic mutations specifically associated with human cancers, which is of particular interest from the point of view of radiological protection [9.103, 9.106]. A more detailed discussion of research implemented at RARAF is given in [9.102].

Accelerator systems for microdosimetric investigations are expected to meet especially stringent requirement since the behavior of individual cells affected by radiation has to be determined. If the studies are concerned with small doses, e.g. at the permissible levels for the general population and/or professionally occupied personnel, great difficulties are encountered in investigating the random nature of the distribution of individual particle trajectories. Fig. 9.27 is a schematic diagram of a system designed for microdosimetric investigations (Pacific Northwest Laboratory [9.104]). In a 2 MV tandem-type accelerator, ^1H and ^2H ions can be accelerated to energies between 1 and 4 MeV, and ^3He and ^4He ions to energies between 2 and 6 MeV, the beam intensities being about 10 particles per second. Most irradiation experiments carried out at the Pacific Northwest Laboratory make use of single protons or alpha particle with the time-of-flight of the particles through the cell well defined. In some investigations, two or more particle are used, and, in addition to the separation in the irradiation times by individual particles, a proper time sequence relative to the cell division has to be ensured. The feature that all these experiments have in common is that individual cells are irradiated with identical specific energies [9.104]. The aim of the above studies is to determine the cancerogenic effect of radiation.

The accelerator depicted in Fig. 9.26 is situated horizontally and, the beam is bent by 90°. This makes it possible to irradiate cell cultures in the Petri dishes also placed horizontally, which ensures their unimpaired growth. The culture plate is placed

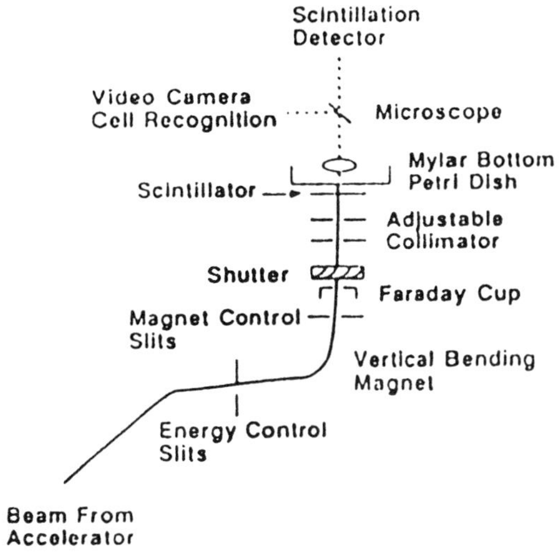

FIGURE 9.27 System for microdosimetric investigations (Pacific Northwest Laboratory, Richland, USA) (after L.A.Braby and W.D.Reece [9.104])

above the beam collimator, whereby the beam diameter can be controlled in the 0.2-10 μm range. The irradiation time is also controlled by a movable 10 ms shutter mechanism. A microscope equipped with a video camera to make observations of cells and of their positioning in relation to the collimator is mounted above the plate. The microscope collects the light emitted from a thin plastic scintillator which covers the exit window of the beamline. The shutters placed in the path of the microscope lamp light and in that of the photomultiplier prevent the strong light necessary to illuminate the specimen from entering the window of the photomultiplier coupled to the thin scintillator.

When the selected cell has been positioned and the photomultiplier sensitized, the shutter which fractionates the beam according to the time necessary to obtain a single trajectory or a desired number of particle trajectories is opened. The cells to be irradiated are cultured on a thin foil built into the Petri plate mounted in a stainless steel ring which permits precise positioning and automatic sample changing. Both the thickness of the foil and that of the scintillator-beam exit window are chosen separately for each experiment in a compromise between the ease of handling thick foils with the increasing energy spread of particle resulting from multiple scattering in foils.

The system shown in Fig. 9.27 makes it possible to study the effect of single ions on the growth of single cells. In this case, stochastic uncertainties in conventional irradiation methods are eliminated.

The last few years have witnessed increased interest in the studies on the properties and reactions caused by highly reactive *short-lived radicals*. This has led to the formation of dedicated facilities, such as the Center for Fast Kinetics Research (CFKR) at the University of Texas, USA [9.105], where spectrophotometric studies on the radiolysis initiated by high-energy pulses with microsecond resolution, as well as investigations on fluorescence and laser stimulation or electronic stimulation with a pair of pulses with controlled delay times have been conducted.

The CFKR facility is equipped with two Van de Graaff machines which accelerate electrons to the nominal energy of 4 MeV adapted for pulse operation, frequency of pulse generation ranges from single pulses to 500 Hz. The CN-type accelerator produces pulses of 30 to 800 ns duration and typical pulse currents of 0.5 to 1 A. Doses of the order of 50 rad ns^{-1} A^{-1} can be generated in sample of unit specific density. A pulse of 100 ns duration produces a dose of about 500 rad. Thus, in a pulse of 0.1 A current for 100 ns, about 500 rads would be absorbed. This would produce about 10^{-6} M of e_{aq}^{-}, OH or O_2 in an appropriate aqueous system. On the other hand, a KN-type accelerator generates pulses of 12-1000 ns duration and pulse beam current of up to 3 A. Thus a dose of about 60 rad ns^{-1} A^{-1} can be produced in a sample of unit specific density. The beam from this accelerator can be delivered to three shielded rooms: two beamlines are situated horizontally and one beamline provides a vertical beam. The accelerated beam can also be converted to bremsstrahlung with peak energy of 4 MV or less.

The CFKR facility is additionally equipped with two lasers of the ND : YAG type: the first one emits 11 s pulses, and the other produces 30 -200 ps pulses.

Bibliography

[1.1] Europe against cancer: A plan of action for the European Community — COM 86-717 (DOC — P 101, Dec. 1986)

[1.2] W.M. Saunders: Radiation oncology: the use of beams of photon or particles for the treatment of tumors, Radiat. Phys. Chem. **24**, no 3/4, 357-64 (1984)

[1.3] M.D. Schutz: The supervoltage story, Am. J. Roentgenology, **124**, 541-59 (1975)

[1.4] W.F. Hanson: The changing role of accelerators in radiation therapy, IEEE Trans. Nucl. Sci. **NS-30**, 1781-83 (1983)

[1.5] D.W. Kerst: The Betatron, Radiology **40**, 115-19 (1943)

[1.6] E. Persilka: Das Betatron — ein Strahlengenerator für die Strahlentherapie, Siemens-Zeitschrift **42**, 355-62 (1968)

[1.7] P. Wideroe: Tumor therapy with high-energy ionizing radiation, Kerntechnik **11**, 312-17 (1969)

[1.8] E.L. Ginzton, C.S. Nunan: History of microwave electron linear accelerators for radiotherapy, Int. J. Radiation Oncology Biol. Phys. **11**, 205-16 (1985)

[1.9] D.K. Bewley: The 8 MeV linear accelerator at the MRC Cyclotron Unit, Brit. J. Radiology **58**, 213-17 (1985)

[1.10] D.W. Fry, W. Walkinshaw: Linear accelerators, Reports on Progress in Physics **12**, 102-32 (1949)

[1.11] C.W. Miller: An 8 MeV linear accelerator for X-ray therapy, Proc. I.E.E. **101**, 207-22 (1954) and **102**, 500 (1955)

[1.12] E.L. Ginzton et al.: The Stanford medical linear accelerator: I. Design and development, Stanford Medical Bull. **15**, 123-40 (Aug. 1957)

[1.13] H.S. Kaplan, M.A. Bagshaw: The Stanford medical linear accelerator: III. Application to clinical problems of radiation therapy, Stanford Medical Bull. **15**, 141-51 (Aug. 1957)

[1.14] L.L. Morgan: Impact of accelerators on technology, IEEE Trans. Nucl. Sci. **NS-20**, 36-39 (1973)

[1.15] L. Rosen: Relevance of particle accelerators to national goals, Science **173**, 490-97 (1971)

[1.16] C.S. Nunan: Design and performance criteria for medical electron accelerators, Nucl. Instrum. Meth. **B10/11**, 881-87 (1985)

[1.17] E. Tanabe, R.W. Hamm: Compact multi-energy electron linear accelerators, Nucl. Instrum. Meth. **B10/11**, 871-76 (1985)

[1.18] G.K.Y. Lam et al.: Cancer radiotherapy using negative pi-mesons, Nucl. Instrum. Meth. **B10/11**, 1096-1101 (1985)

[1.19] Nuclear Safety Bureau: Statistics on the Use of Radiation in Japan 1987, Tokyo (1987)

[1.20] R.W. Hamm: Linacs for medical and industrial applications, Proc. of the 1986 Linear Accelerator Conference (1987)

[1.21] J.R. Stewart: Particle accelerators in cancer therapy, IEEE Trans. Nucl. Sci. **NS-26**, 2992-96 (1979)

[1.22] J.D. Chapman: The cellular basis of radiotherapeutic response, Radiat. Phys. Chem. **24**, 283-91 (1984)

[1.23] J. Purdy: The application of high energy X-rays and electron beams in radiotherapy, IEEE Trans. Nucl. Sci. **NS-26**, 1833-37 (1979)

[1.24] S. Kramer et al.: High-Energy Photons and Electrons: Clinical Application in Cancer Management, John Wiley and Sons, New York (1976)

[1.25] N.M. Bloehan et al. (eds): Radiation Therapy Planning, Marcel Dekker Inc., New York (1983)

[1.26] S. Klevenhagen: Current status of electron therapy — clinical and physical aspects, Brit. J. Radiology Supplement **22**, 34-51 (1988)

[1.27] P. Bey, D. Langlois: Electronthérapie utilisation clinique, J. Eur. Radiother. **7**, 129-132 (1986)

[1.28] L.M. Peacock et al.: Electron arc therapy: Clinical experience with chest wall ir-

radiation, Int. J. Radiat. Oncology **10**, 2149-53 (1984)
[1.29] F.M. Khan: Electron beam therapy, 299-355, in The Physics of Radiation Therapy, Williams and Wilkins, Baltimore-London (1984)
[1.30] S.C. Klevenhagen: The Physics of Electron Beam Therapy, Adam Hilger, Bristol (1985)
[1.31] E. Briot: Clinical application of electrons, in Megavoltage Radiotherapy 1937-87, Brit. J. Radiology Suppl **22**, 51 (1988)
[1.32] R.S. Heusinkveld et al.: Accelerated particles in radiation oncology, IEEE Trans. Nucl. Sci. **NS-22**, 1211-15 (1975)
[1.33] M.R. Raju: Heavy particle therapy — some historical perspectives, LA-UR-86-3502 Report, Working Group Meeting on the Use of Protons in Radiotherapy, Leningrad (Oct. 14-18, 1986)
[1.34] M.R. Raju: Heavy Particle Radiotherapy, Academic Press, New York — London (1980)
[1.35] R.E. Stone: Neutron therapy and specific ionization, Amer. J. Roentg. Radium Therapy **59**, 771-85 (1948)
[1.36] G. Schmitt: V. Probleme der Effektivität der radiologischen Tumortherapie durch schwere Teilchen, in Handbuch der Medizinischen Radiologie, vol. XX, 613-39, Springer-Verlag, Berlin (1985)
[1.37] D.K. Bewley: Physics and Radiobiology of Fast Neutron Beams, Med. Phys. and Biomed. Engn., London (1989)
[1.38] L. Cohen, F.R. Hendrickson: External neutron beam therapy, Illinois Med. J. **171**, 231-36 (1987)
[1.39] G. Schmitt: Der gegenwartige Stand der Therapie mit dicht ionisierenden Strahlenarten, Tumor-Diagnostik und Therapie, **4**, 63-67 (1983)
[1.40] R.R. Wilson: Radiological use of fast protons, Radiology **47**, 487-91 (1946)
[1.41] G. Kraft: On the interpretation of radiobiological experiment performed with heavy charged particles, Preprint GSI-88-33, Darmstadt (July 1988), IAEA Advisory Group Meeting on Atomic and Molecular Data for Radiotherapy, Vienna (June 1988)
[1.42] B. Lentle, H. Singh: Radiation and medicine, Radiat. Phys. Chem. **24**, 267-72 (1984)
[1.43] J. Butler et al.: Chemical mechanism of the effects of high energy radiation on biological systems, Radiat. Phys. Chem. **24**, 273-82 (1984)
[1.44] V.T. Kagiya et al.: Chemical studies on DNA damage by radiation, Radiat. Phys. Chem. **29**, 451-54 (1987)
[1.45] R.A. Floyd: Free Radicals and Cancer, Marcel Dekker, New York (1982)
[1.46] W.A. Pryor: Free Radicals, in Biology, **1** to **5**, Academic Press, New York (1975-1982)
[1.47] M.A.J. Rodgers, E.L. Powers (eds): Oxygen and oxy-Radicals in Chemistry and Biology, Academic Press, New York (1981)
[1.48] P. Wardman: Radiation chemistry in the clinic-hypoxic cell radiosensitizers for radiotherapy, Radiat. Phys. Chem. **24**, 293-305 (1984)
[1.49] M. Quintiliam: Physico-chemical basis of radiosensitization by iodine compound, Radiat. Phys. Chem. **30**, 409-22 (1987)
[1.50] P. Wardman: The mechanism of radiosensitization by electron-affinic compound, Radiat. Phys. Chem. **30**, 423-32 (1987)
[1.51] C.A. Tobias: The repair-misrepair model in radiobiology: comparison to other models, Radiat. Res. **104**, S77-95 (1985)
[1.52] T. Sugahara: PLD repair inhibitors as radiosensitizers, Radiat. Phys. Chem. **24**, 321-28 (1984)
[1.53] C. von Sontag: The Chemical Basis of Radiation Biology, Taylor and Francis, London (1987)
[1.54] G. Kraft et al.: Cellular and subcellular effect of heavy ions: A comparison of the induction of strand breaks and chromosomal aberation, GSI Preprint 88-52 Darmstadt (Oct. 1988), Space Res. Meeting, Helsinki (July 1988)
[1.55] J.S. Sivinski, D.P. Sloan: The role of linear accelerator in industry, Nucl. Instrum. Meth. **B10/11**, 981-86 (1985)
[1.56] M. Brenner et al.: The production and use of cyclotron isotopes in Turku, Radiobiol. Radiother. **24**, 563-68 (1983)
[1.57] G.T. Seaborg: The biomedical cyclotron — Remarks at the dedication of the biomedical cyclotron at UCLA, Report USAEC S-14-71 (1971)
[1.58] G. Wolber: A new cyclotron for biomedical research, Deutches Krebsforschungszentrum, Heildeberg (1988)
[1.59] J.B. Smathers, L.T. Myers: Use of cyclotrons in medical research: past, present, future, Nucl. Instrum. Meth. **B10/11**, 1111-16 (1985)
[1.60] J.C.H. Chu et al.: Patterns of change in the physics and technical support of radiation therapy in the USA 1975-1986, Int. J. Radiation Oncology Biol. Phys. **17**, 437-442 (1989)
[1.61] J.C.H. Chu et al.: Equipment and manpower needs in radiation oncology

dosimetry and treatment delivery, Med. Dosimetry **14**, 265-267 (1989)

[1.62] Report of the Inter-Society Council For Radiation Oncology: "Radiation Oncology in Integrated Cancer Management" (Nov. 1986)

[1.63] B. Sanyal et al.: Therapy with accelerated electrons — a clinical view point, Nucl. Instrum. Meth **B40/41**, 1345-1349 (1989)

[1.64] C. Muir et al. (eds): Cancer Incidence in Five Continents, vol. 5, Int. Agency for Research on Cancer, Sci. Publications No. 88, Lyon (1987)

[1.65] D.M. Parkin et al. (eds): International Incidence of Childhood Cancer, Int. Agency for Research on Cancer, Sci. Publications No. 87, Lyon (1988)

[1.66] J.W.W. Coebergh, L.H. van der Heijden (eds): Cancer Incidence and Survival 1975-1987 in Southeastern Netherlands, IKZ Eindhoven Cancer Registry, Eindhoven (1991); also J.W.W. Coebergh: Incidence and Prognosis of Cancer in the Netherlands — Studies based on Cancer Registries, Thesis, Erasmus University, Rotterdam (1991)

[1.67] H.R. Balvert-Locht et al.: Improved prognosis of ovarian cancer in the Netherlands during the period 1975-85, Gynecologic Onc. **42**, 3-8 (1991)

[1.68] A. Vermorken: European strategy for research on cancer therapy, S-124 to S-126, in Proc. of the 2nd European Particle Accelerator Conf. (EPAC 90), Nice, June 12-16, 1990, Editions Frontiers, Gif-sur-Yvette (1990)

[1.69] B. Palival et al. (eds): Optimization of Cancer Radiotherapy, AAPM Proc. No. 5 (1986)

[1.70] H. Thames, J. Hendry: Fractionation in Radiotherapy, Adam Hilger (1987)

[1.71] S. Shahabi (ed): Blackburn's Introduction to Clinical Radiation Therapy Physics, Medical Physics Publishing, Madison (1988)

[1.72] S. Shahabi: Radiation Therapy Physics Review: Questions and Solutions, Medical Physics Publishing, Madison (1991)

[1.73] P.N. Plowman, A.N. Harnett (eds): Megavoltage Radiotherapy 1937-1987, BJR Supplement **22**, British Institute of Radiology (1988)

[1.74] N.J. McNally (ed): The Scientific Basis of Modern Radiotherapy, BIR Report 19, British Institute of Radiology (1989)

[1.75] G.C. Bentel: Treatment Planning and Dose Calculation in Radiation Oncology, 4th Edition, Pergamon Press, Riverside (1989)

[1.76] I. Bruinvis: Electron Beams in Radiation Therapy: Collimation, Dosimetry and Treatment Planning, Netherlands Cancer Institute (1987)

[1.77] A. Wambersie: Radiobiological and clinical bases of particle therapy (review), S1-S3, in Proc. of the 2nd European Particle Accelerator Conf. (EPAC 90), Nice, June 12-16, 1990, Editions Frontiers, Gif-sur Yvette (1990)

[1.78] Anonym: World wide charged particle patient totals, July 1992, Particles No. 10 (June 1992)

[1.79] M. Zaider et al.: Am. J. Clin. Oncol. **11**, 212-219 (1988)

[1.80] C.G. Orton: Impact of radiobiology on the future of radiation oncology, Post. Fiz. Med. **25**, 87-100 (1990)

[1.81] E. Riklis (ed): Frontiers of Radiation Biology, VCH Publishers, New York (1990)

[1.82] M. Tubiana et al.: Radiobiology, Taylor and Francis, Bristol (1990)

[1.83] R. Granier: Applied Radiobiology and Radiation Protection, Prentice Hall (Ellis Horwood), Old Tappan (1990)

[1.84] E.L. Alpen: Radiation Biophysics, Prentice Hall (Ellis Horwood), Old Tappan (1990)

[1.85] J. Heilmann: Zelluläre und molekulare Wirkungen beschleunigter schwerer Ionen auf kultivierte Säugerzellen, Thesis, Bonn University, GSI Report 91-17, Darmstadt (June 1991)

[1.86] V. Volkovic, G. Moschini: Some advances in medical applications of low energy accelerators, Nucl. Instrum. Meth. **B56/57**, 1274-1278 (1991)

[2.1] J.B. Schroeder et al.: A new 3 MV tandem Pelletron system. Nucl. Instrum. Meth. **B24/25**, 763-66 (1987)

[2.2] Micro Analytical Research Centre: The Scanning Proton Microprobe, p.12, University of Melbourne, School of Physics Victoria, Australia (1987)

[2.3] G.J.F. Legge et al.: Proton microscopy and microanalysis — biological applications, Ultramicroscopy **24**, 283-98 (1988)

[2.4] G. Frick, E. Jaeschke (eds): Proc. of the 5th Int. Conf. on Electrostatic Accelerators and Associated Boosters, Strasbourg, May 24-30 (1989), in Nucl. Instrum. Meth., **A287**, 1-335 (1990)

[2.5] Proc. of the 6th Int. Conf. on Electrostatic Accelerators and Associated

Boosters, Padua, Legnaro (1992) to be published
[2.6] G. Loew: Electron linear accelerators and stretchers, SLAC — Pub-3343, Stanford (May 1984)
[2.7] G. Loew, R. Talman: Elementary Principles of Linear Accelerators, SLAC Publ-3221, Stanford (1983)
[2.8] I.M. Kapchinskyi: Theory of Resonance Linear Accelerators, Harwood Acad. Publ., New York (1987)
[2.9] Proc. of the 1988 Linear Accelerator Conf., CEBAF Rep.90-00, Virginia (1989)
[2.10] Proc. of the 1990 Linear Accelerator Conf., LANL Rep., Los Alamos (1991)
[2.11] M.K. Craddock: Cyclotrons in 1986, Proc. 11th Conf. on Cyclotrons and their Appl., Ionics, Tokyo (1987)
[2.12] W. Scharf: Particle Accelerators and Their Uses, Harwood Academic Publishers, New York (1986), p.178
[2.13] References data on Int. Cyclotron Conferences, 761, in B. Martin, K. Ziegler (eds): Proc. of Twelfth Int. Conf. on Cyclotrons and Their Applications, Berlin, May 8-12, 1989, World Scientific, Singapore — New Jersey (1991)
[2.14] D.J. Clark, R.A. Gough: Synchrotrons in cyclotron territory, in [2.11], p. 748-51
[2.15] J.M. Slater et al.: Development of a hospital — based proton beam treatment center, Int. J. Radiation Oncology Biol. Phys. **14**, 761-75 (1988)
[2.16] W.H. Scharf: Particle Accelerators — Applications in Technology and Research, Research Studies Press, Taunton, John Wiley, New York — Chichester — Toronto (1989)
[2.17] D. Greene: Linear Accelerators for Radiation Therapy, Medical Physics Handbook 17, Adam Hilger (1985)
[2.18] C. Karzmark, R. Morton: A Primer of the Theory and Operation of Linear Accelerators in Radiation Therapy, Medical Physics Publishing Corp., Madison (1989)
[2.19] R.L. Maughan et al.: Clinical charcteristics of a superconducting cyclotron for neutron radiation therapy, 7th Symposium of Neutron Dosimetry, Oct. 14-18, Berlin (1991)
[2.20] H. Blosser et al.: Medical accelerator projects at MSU, 742-746, in Proc. of the 2nd European Particle Acc. Conf. (EPAC 90), Nice, June 12-16, 1990, Editions Frontiers, Gif-sur-Yvette (1990)
[2.21] P. Mandrillon: High energy medical accelerators, S4-S8, in Proc. of the EPAC 90, Edition Frontiers, Gif-sur-Yvette (1990)
[2.22] P. Mandrillon et al.: Advances of the feasibility study of the European Light Ion Medical Accelerator, 767-769, in Proc. of the EPAC 90, Edition Frontiers, Gif-sur-Yvette (1990)
[2.23] B. Larsson: Biomedical program for the converted 200 MeV synchrocyclotron at the Gustav Werner Institute, Radiat. Res. **104**, 310-318 (1985)
[2.24] S. Holm: Operational experience of the reconstructed Uppsala synchrocyclotron, 17-20, in B. Martin, K. Ziegler (eds): Proc. of Twelfth Int. Conf. on Cyclotrons and Their Applications, World Scientific, Singapore — New Jersey (1991)
[2.25] B. Martin, K. Ziegler (eds): Proc. of Twelfth Int. Conf. on Cyclotrons and Their Applications, Berlin, May 8-12, 1989, World Scientific, Singapore — New Jersey (1991)
[2.26] F.T. Cole et al.: Loma Linda medical accelerator project, 737-741, in Proc. of the EPAC 90, Edition Frontiers, Gif-sur-Yvette (1990)

[3.1] J. Purdy, D.A. Goer: Dual energy X-ray beam accelerators in radiation therapy — an overview, Nucl. Instrum. Meth. **B10/11**, 1090-95 (1985)
[3.2] D.F. Herring, D.M.J. Compton: British J. Radiol. Spec. Rept. Series no. 5 (Proc. 3rd Int. Conf. on Computers in Radiotherapy, p.51) (1981)
[3.3] Int. Commision on Radiation Units and Measurements (ICRU) Report 24: Determination of absorbed dose in a patient irradiated by means of X or gamma rays, Washington (1976)
[3.4] IEC Draft Report SC62C (March 1984)
[3.5] AAPM Report: Physical aspects of quality assurance in radiation therapy, AAPM Report 13, New York (1984)
[3.6] B.J. Mijnheer et al.: What degree of accuracy is required and can be achieved in photon and neutron therapy? Radiotherapy and Oncology **8**, 237-52 (1987)
[3.7] IEC Protocol: Medical electron accelerators: Guidelines for functional performance charcteristics, IEC Report 62C (CO) 36, Document No. 85/53705 (1985)
[3.8] ICRU Protocol: Report No. 35 — Radiation Dosimetry: Electron beams with energies between 1 and 50 Mev, Washington (1984)
[3.9] ICRP Report No. 44: Protection of the Patient in Radiation Therapy (1985)

[3.10] HPA: A practical guide to electron dosimetry 5-35 MeV, HPA Report Series No. 4 (1971)
[3.11] HPA: A practical guide to electron dosimetry below 5 MeV for radiotherapy purpose, HPA Report Series No. 13 (1975)
[3.12] Sub-Committee on Radiation Dosimetry of the American Association of Physicists in Medicine (SCRAD): Protocol for the dosimetry of high energy electrons, Phys. Med. Biol. **11**, 505 (1966)
[3.13] SCRAD: Protocol for the dosimetry of X-ray and gamma-ray beams with maximum energies between 0.6 and 50 MeV, Phys. Med. Biol. **16**, 379 (1971)
[3.14] AAPM: A protocol for the determination of absorbed dose from high-energy photon and electron beams, Med. Phys. **10**, 741 (1983)
[3.15] AAPM: Physical aspects of quality assurance in radiation therapy, AAPM Report 13, Task Group 24, Ne York (1984)
[3.16] German Standard Association: (a) DIN 6800/1, Procedures in dosimetry: principles of photon and electron dosimetry with proto-type detectors (1975); German Standard Association: (b) DIN 6800/2 Procedures in dosimetry: ionization dosimetry (1975)
[3.17] DIN Standard 6847/5: Medical electron accelerators; Constancy testing: Suggested cct of periodic test methods and test conditions (1986); also DIN Standard 6847 (1991)
[3.18] NACP: Procedures in external radiation therapy dosimetry with electron and photon beams with maximum energies between 1 and 50 MeV, Acta Radiol. Oncol. **19**, 55-79 (1980), also Electrobeams with mean energies at the phantom surface below 15 MeV, Supplement to the recommendations by NACP, Acta Radiol. Oncol. **20**, 401-415 (1981)
[3.19] W.H. Suthertland: Standard equipment — the linear accelerator, Brit. J. Clinic Equipment 197-209 (1978)
[3.20] V.A. Vaguine: Electron linear accelerator structures and design for radiation therapy machines, IEEE Trans. Nucl. Sci. **NS-28**, 1884-1888 (1981)
[3.21]C. Nunan: Criteria for optimal design of modern multi-mode medical linear accelerator, p.22-28, in [3.22]
[3.22] Proceedings of The Varian Twelfth Users Meeting, Pebble Beach, California, May 1-3 (1988)
[3.23] C.J. Karzmark, N.C. Pering: Electron linear accelerators for radiation therapy, Phys. Med. Biol. **18**, 321-54 (1973)
[3.24] F. Voogt: Computer based equipment enables the development of new radiotherapy techniques, Eight Congress of the Polish Society of Medical Physics, Poznań, Poland (Sept. 1989)
[3.25] T. Taylor et al.: Therac 25: a new medical accelerator concept, IEEE Trans. Nucl. Sci., **NS-30**, 1768-71 (1983)
[3.26] P. O'Brien et al.: Radiation protection aspects of a new high-energy linear accelerator, Med. Phys. **12**, 101-107 (1985)
[3.27] C.J. Karzmark: Advances in linear accelerator design for radiotherapy, Med. Phys. **11**, 105-28 (1984)
[3.28] H. Enge: Achromatic magnetic mirror for ion beams, Rev. Sci. Instrum. **34** no. 4 (1963)
[3.29] J.I.M. Botman: A double focusing magnet system for a medical linear electron accelerator, Nucl. Instrum. Meth. **B10/11**, 796-98 (1985)
[3.30] P.J. Biggs: Review of the improvements in the quality and applications of photon beams from linear accelerators in radiation therapy and a projection of future trends, Nucl. Instrum. Meth. **B24/25**, 1096-99 (1987)
[3.31] F.A. Silber, H. Schmid: Schweiss Techn. **5**, 73-78 (1976)
[3.32] E.B. Podgorsak et al.: X-ray depth doses from linear accelerators in the energy range from 10 to 32 MeV, Am. J. Roentg. **123**, 182-91 (1975)
[3.33] E.B. Podgorsak et sl.: Design of X-ray tergets for high energy linear accelerators in radiotherapy, Am. J. Roentg. **123**, 873-82 (1974)
[3.34] R.E. George, M. Hartson-Eaton: Characteristics of an MM 22 medical microtron 21 MV photon beam, Int. J. Radiation Oncology Biol. Phys. **11**, 1221-24 (1985)
[3.35] P. O'Brien et al.: Characteristics of electron beams from a new 25 MeV linear accelerator, Med. Phys. **12**, 799-805 (1985)
[3.36] H. Forslo: Low energy photon treatment using independent collimators, p.98 in [3.37]
[3.37] Proceedings Varian's Fourth European Clinac Users Meeting, May 25-26, Malta (1984)
[3.38] S. Harly, H.H. Nielsen: Automatic transport system for blocking and compensating filters, 957-58, in XIV ICMB and VII ICMP, Helsinki (1985)
[3.39] J.M. Jensen et al.: A new approach to dose modification in large-field-irradiation. Some physical and technical considerations, 173-75 in [3.40]
[3.40] Proceedings: Fifth Varian European

Clinac Users Meeting, February 12, 14, Flims (1987)

[3.41] G.T.Y. Chen et al.: A computer simulated evaluation of multivane collimator beam delivery systems, 48-50, in [3.22]

[3.42] D.D. Leavitt: Multileaf collimation in electron arc therapy, 63-67, in [3.22]

[3.43] A.R. Smith et al.: Conformal radiation therapy: technical and clinical aspects, 55-62, in [3.22]

[3.44] H. Meertens: Digital processing of high-energy photon-beam images, Med. Phys. **12**, 111-13 (1985)

[3.45] H. Meertens et al.: A liquid ionization detector for digital radiography of therapeutic megavoltage photon beams, Phys. Med. Biol. **30** (1985)

[3.46] P.J. Biggs et al.: A diagnostic X-ray field verification device for a 10 MV linear accelerator, Int. J. Radiation Oncology Biol. Phys. **11**, 635-43 (1985)

[3.47] I.A.D. Bruinvis: Physical aspects of electron therapy up to 20 MeV, Medicamundi, **24**, no. 3, 119-26 (1979)

[3.48] P.L. Petti et al.: Sources of electron contamination for the Clinac-35 25-MV photon beam, Med. Phys. **10**, 856-61 (1983)

[3.49] P.J. Biggs, M.D. Russell: An investigation into the presence of secondary electrons in megavoltage photon beams, Phys. Med. Biol. **28**, 1033-43 (1983)

[3.50] P.J. Biggs: The forward production of high-energy electrons from megavoltage photon beams, Med. Phys. **14**, 767-71 (1987)

[3.51] H.T. Heaton, R. Jacobs (eds): Proc. of a Conf. on Neutrons from Electron Medical Accelerators (Gaithersburg, April 9-10, 1979), National Bureau of Standards Special Publications 554, Washington (1979)

[3.52] M. Sohrabi, K.Z. Morgan: Neutron dosimetry in high energy X-ray beams on medical accelerators, Phys. Med. Biol. **24**, 756-66 (1979)

[3.53] W.P. Swanson, R.C. McCall: Neutrons from medical electron accelerators, p.11, SLAC-Pub-2346, Stanford (1979)

[3.54] A.P. Turner: Regulations standards and guidelines for the use of medical electron linear accelerators, IEEE Trans. Nucl. Sci. **NS-26**, 1608-12 (1979)

[3.55] R.W. Kersey: Estimation of neutron and gamma radiation doses in the entrance mazes of SL 75-20 linear accelerator treatment rooms, Medicamundi **24**, no. 3, 151-55 (1979)

[3.56] W.P. Swanson: An effective radiological safety program, SLAC-Pub-2643, Stanford (1980)

[3.57] W.P. Swanson: Estimate of the risk in radiation therapy due to unwanted neutrons, Med. Phys. **7**, 141-44 (1980)

[3.58] J. Purdy, G.P. Glasgow: Neutron leakage around a 20 MeV (18 Mv X-ray) linac with a modified neutron shield, Health Phys. **46**, 669-74 (1984)

[3.59] P. O'Brien et al.: Radiation protection aspects of a new high-energy linear accelerator, Med. Phys. **12**, 101-107 (1985)

[3.60] H. Schumacher, O. Krauss: Area monitoring of photons and neutrons from medical electron accelerators, Radiat. Protection Dosimetry, **14**, 325-27 (1986)

[3.61] Y. Uwamino et al.: Measurement and calculation of neutron leakage from a medical electron accelerator, Med. Phys. **13**, 374-84 (1986)

[3.62] R.C. McCall: Neutron yield of medical accelerators, SLAC-Pub-4480, Stanford (1987)

[3.63] J.A. Devanney: Radiological safety of accelerators — a comparison of performance standards, IEEE Trans. Nucl. Sci **NS-30**, 1583-87 (1983)

[3.64] J.A. Purdy: Quality assurance of electron linear accelerators, 86-107, in G. Starkschall (ed): Proc. of a Symposium on Quality Assurance of Radiotherapy Equipment, Institute of Physics, New York (1983)

[3.65] P.A. Parrino, J.A. Purdy: The role of maintenance in quality assurance, 157-68 in [3.64]

[3.66] A. Newing et al.: Commissioning measurements on a Clinac-20, 58-63, in [3.40]

[3.67] J.E. Aldrich et al.: Characteristics of the photon beam from a new 25-MV linear accelerator, Med. Phys. **12**, 619-24 (1985)

[3.68] P. O'Brien et al.: Characteristics of electron beams from a new 25 MeV linear accelerator, Med. Phys. **12**, 799-805 (1985)

[3.69] T.J. Godden: the recommended U.K. quality control protocol as applied to the Clinac 2500, 156-59, in [3.40]

[3.70] A. Scheffler: The German quality control protocol applied to Varian Clinac Accelerators, 162-64, in [3.40]

[3.71] O. Mattson: The NACP dosimetry protocol applied to the Varian Clinac 4/100, 151-153, in [3.40]

[3.72] E. Loewinger: The AAPM quality assurance protocol applied to the Varian Clinac 6-x and Clinac 1800 linear accelerators, 154-55, in [3.40]

[3.73] F.M. Khan: Linear accelerator quality assurance, 44-47, in [3.22]

[3.74] P. Ruotsalainen: The NACP quality

control protocol applied to the Varian Clinac 1800, 160-61, in [3.40]
[3.75] A. Brahme, D. Reistad: Microtrons for electron and photon radiotherapy, IEEE Trans. Nucl. Sci. **NS-28**, 1880-83 (1981)
[3.76] R.E. George: Characteristic of an MM 22 medical microtron 6-MV photon beam, Med. Phys. **11**, 862-65 (1984)
[3.77] R.E. George et al.: Characteristics of electron beams from a medical microtron, Med. Phys. **13**, 533-38 (1986)
[3.78] L.H. Lanzl et al.: The physics of electron intraoperative radiotherapy, XIV Int. Conf. on Med. Biol. Engn. and VII Int. Conf. on Med. Phys., Espoo, Aug. 11-16, Finland (1985)
[3.79] L.L. Gunderson et al.: Residual, unresectable, or recurrent colorectal cancer, Int. J. Radiat. Onc. Biol. Phys. **9**, 1597-1606 (1983)
[3.80] A.L. Goldson et al.: Intraoperative radiotherapy, 99-103, in [3.37]
[3.81] M. Abe: Intraoperative radiotherapy for gastrointestinal malignancy, 327-349, in J.J. DeCosse, P. Sherlock (ed): Clinical Management of Gastrointestinal Cancer, Martinus Nijhoff Publishers, Boston (1984)
[3.82] T.A. Rich: Intraoperative radiotherapy, Radiother. Oncology **6**, 207-21 (1986)
[3.83] W.U. Shipley et al.: Intraoperative electron beam irradiations for patients with unresectable pancreatic carcinoma, Annals of Surgery, **200** , 289-96 (1984)
[3.84] P.J. Biggs: The effect of beam angulation on central axis per cent depth dose for 4-29 MeV, Phys. Med. Biol. **29**, 1089-96 (1984)
[3.85] P.J. Biggs: The physics of intraoperative radiotherapy using high energy electron beam, Nucl. Instrum. Meth. **B10/11**, 1102-06 (1985)
[3.86] A.L. Goldson, M.Y. Halyard-Richardson: Intraoperative radiotherapy for intracranial malignancies, 176-80, in [3.40]
[3.87] T.A. Rich, E. Dally: The Siemens Mevatron ME, 102, Proc. of the Second Mevatron User's Conf., Hilton Head Island, South Carolina (1985)
[3.88] T.A. Rich: Dedicated IORT facility within a surgical suite, Intraoperative Radiation Therapy Int. Symposium, Medical College of Ohio, May 15-16 (1986)
[3.89] H. Aget et al.: Utilization of a high energy photon beam for whole body irradiation, Radiology, **123**, 747-51 (1977)
[3.90] J.Van Dyk: Magna-field irradiation: physical consideration, J. Radiation Oncology Biol. Phys. **9**, 1913-18 (1983)
[3.91] J. Van Dyk: Dosimetry for total body irradiation, Radiotherapy and Oncology, **9**, 107-118 (1987)
[3.92] K. Kuphal et al.: Total body irradiation for the conditioning of patients before bone narrow transplantation, 6-7, in [3.37]
[3.93] J.M. Jensen et al.: The "Kiel Model" of total body irradiation, 10-12, in [3.37]
[3.94] O. Soeberg: Single shot and fractionated TBI with Clinac 6 and Clinac 4/100, 8-9, in [3.37]
[3.95] K. Kawashima et al.: Total body irradiation using a moving beam and variable collimator, NIRS Annual Report (1986)
[3.96] W.R. Lutz et al.: Design and characteristics of a facility for total-body and large-field irradiation, Harvard Medical School Rep. (Apr. 1988)
[3.97] E.B. Podgorsak et al.: Radiosurgery with photon beams: physical aspects and adequacy of linear accelerators, Radiotherapy and Oncology, **17**, 349-358 (1990)
[3.98] E.B. Podgorsak et al.: Dynamic stereotactic radiosurgery, Int. J. Radiat. Oncol. Biol. Phys. **14**, 115-26 (1988)
[3.99] W. Lutz et al.: A system for stereotactic radiosurgery with a linear accelerator, Int. J. Radiat. Oncol. Biol. Phys. **14**, 373-81 (1988)
[3.100] J.D. Bradshaw: The Sheffield stereotactic radiosurgery unit, 188-91, in [3.40]
[3.101] C.K. Bomford, L. Walton: The physics of stereotactic radiosurgery, 183-87, in [3.40]
[3.102] T. Greitz et al.: Stereotactic radiation therapy of intracranial lesions, Acta Radiologica **25**, 81-89 (1986)
[3.103] F. Pozza et al.: Stereotactic external radiotherapy for gliomas: technique, dosimetry, results, 139-152, in Karim-Laws (eds): Glioma, Spinger-Verlag, Heidelberg (1991)
[3.104] D. Chariego: Central nervous system external stereotactic irradiation, 18-21, in [3.37]
[3.105] F. Pozza et al.: Stereotactic radiotherapy-results and perspectives, 93-95, in [3.40]
[3.106] O.O. Betti, V.E. Derechinsky: Hyperselective enceploalic irradiation with linear accelerator, Acta Neurochirurgica, Suppl. **33**, 385-90 (1984)
[3.107] F. Colombo et al.: External stereotactic irradiation by linear accelerator, Neurosurgery **16**, 154-160 (1985)
[3.108] G.H. Hartman et al.: Cerebral radiation surgery, Int. J. Radiat. Oncol. Biol. Phys. **11**, 1185-92 (1985)
[3.109] E.D. Bailey: State registration and inspection programs for particle accelerators,

Nucl. Instrum. Meth. **B10/11**, 1013-1016 (1985)

[3.110] J.C.H. Chu et al.: Dose to superficial node for patients with head and neck cancer treated with 6 MV and Co-60 photons, Radiotherapy and Oncology, **21**, 257-260 (1991)

[3.111] F. Brix: Blenden — und Kompensatortechnik in der Strahlentherapie, Georg Thieme Verlag, Stuttgart — New York (1989)

[3.112] A. Brahme, H. Svenson: Electron beam quality parameters and absorbed dose distribution from therapy accelerators, 12-19, in A. Zuppinger et al. (eds): High Energy Electrons in Radiation Therapy, Springer-Verlag Berlin (1980)

[3.113] J. Rassow: Aspects of safety and quality assurance for medical accelerators, Post. Fiz. Med. **26**, 35-54 (19910

[3.114] P.R. Almond, B.E. Bjarngard: Quality assurance of linear accelerators, Med. Phys. World **5**, 6-12 (1989)

[3.115] IEC (Sub-Committee 62C): High-Energy Radiation Equipment and Equipment for Nuclear Medicine: Particular requirements for the safety of medical electron accelerators in the range 1 MeV to 50 MeV. Draft standard in voting procedure (1990), in press

[3.116] ACMP: Report no. 2 Radiation Control and Quality Assurance in Radiation Oncology: A Suggested Protocol, The American College of Medical Physics (1986)

[3.117] J.L. Horton: Handbook of Radiation Therapy Physics, Prentice Hall Inc., New Jersey (1987)

[3.118] J.M. Jensen et al.: General and specific aspects of experimental dose measurement in total body irradiation (TBI), Strahlenth. und Oncologie, **162**, 250-253 (1986)

[3.119] P. Kohr, J.M. Jensen: Some remarks on method and dosimetry of the Kiel compensating system used in TBI before bone marrow transplantation, Strahlenth. und Oncologie, **163**, 228-230 (1987)

[3.120] J.M. Jensen: Planning der Lungendosismodifikation mittels Kompensatoren, in Proc. of a DSMP-Working Group Seminar in Essen, May 14-15, Essen (1990)

[3.121] B. Larsson et al.: The high energy proton beam as a neurosurgical tool, Nature **182**, 1222 (1958)

[3.122] J.H. Lawrence et al.: Heavy-particle irradiation in neoplastic and neurologic disease, J. Neurosurg. **19**, 717 (1962)

[3.123] R.N. Kjellberg et al.: Proton beam therapy in acromegaly, N. Engl. J. Med. **278**, 689 (1968)

[3.124] R.N. Kjellberg et al.: Bragg-peak proton-beam therapy for AVM of the brain, N. Engl. J. Med. **309**, 269 (1983)

[3.125] J.T. Lyman, J. Howard: Dosimetry and instrumentation for helium and heavy ions, Int. J. Radiat. Oncol. Biol. Phys. **3**, 81 (1977)

[3.126] J.T. Lyman et al.: A helium-ion beam for stereotactic radiosurgery of central nervous system disorders, Med. Phys. 13, 695 (1986)

[3.127] B. Pike et al.P: Dose distributions in dynamic stereotactis radiosurgery, Med. Phys. **14**, 780-789 (1987)

[3.128] E.B. Podgorsak: Dynamic stereotactic surgery, 36-40, in [3.22]

[3.129] E.B. Podgorsak: Radiosurgery with high energy photon beams: a comparison among techniques, Int. J. Radiation Oncology Biol. Phys. **16**, 857-865 (1989)

[3.130] L. Souhami et al.: Radiosurgery of cerebral arteriovenous malformations with the dynamic stereotactic irradiation, Int. J. Radiation Oncology Biol. Phys. **19**, 775-782 (1990)

[3.131] B. Pike et al.: Stereotactic external beam calculations for radiosurgical treatment of brain lession, Appl. Neurophysiology **50**, 269-273 (1987)

[3.132] Y. Uchiyama et al.: Integrated system of the computer-controlled conformation radiotherapy at Aichi Centre Hospital, 44-53, in Use of Computers in Radiation Therapy, AMPI Med. Phys. Bull. **16** , no. 2 (1991)

[3.133] T. Ishigaki et al.: Computer-assisted conformation radiotherapy, 54-60, in Use of Computers in Radiotherapy, AMPI Med. Phys. Bull. **16**, no. 2 (1991)

[3.134] E.S. Sternick: A quality management model for radiation oncology physics, 108-125, in AMPI Med. Phys. Bull **16**, no.2 (1991)

[3.135] M.C. Plazas et al.: The influence of dosimetric parameters on the dose distribution for single target, Med. Biol. Engineering and Computing **29** (Suppl.) 1086 (1991)

[3.136] R.S. Cox: Faster treatment planning for the linac knife using a spherical-polar dose grid and non-linear interpolation, Med. Biol. Engineering and Computing **29** (Suppl.) 1089 (1991)

[3.137] E.J. Born: Photon energy spectra of X-ray beams of medical electron accelerators, Phys. Med. Biol. **33** (Suppl. I) 55 (1988)

[3.138] E.J. Born: Simulation of radiation

treatment transport through treatment unit head: photon beam results for three medical accelerators, Med. Biol. Engn Computing **29** (Suppl.) 1152 (1991)

[3.139] J. Gildersleve et al.: Megavoltage imaging systems in clinical practice, Med. Biol. Engn. Computing **29** (Suppl.) 858 (1991)

[3.140] J.W. Wong et al.: Study of treatment variation using a fiber-optic on-line radiotherapy imaging, Med. Biol. Engn. Computing **29** (Suppl.) 858 (1991)

[3.141] S. Shalev et al.: The quantification of radiation treatment accuracy, Med. Biol. Engn. COmputing **29** (Suppl.), 857 (1991)

[3.142] M. van Herk et al.: Portal image processing and interpretation, Med. Biol. Engn. Computing **29**, 857 (1991)

[3.143] L.E. Antonuk: First and second generation system for real-time megavoltage radiotherapy imaging, Med. Biol. Engn. Computing **29** (Suppl.), 856 (1991)

[3.144] H. Meertens et al.: On-line portal imaging: clinical applications, Med. Biol. Engn. Computing **29** (Suppl.), 860 (1991)

[3.145] H. Mota et al.: Monte Carlo study of the Therac 20 electron beam dosimetric parameters, Med. Biol. Engn. Comp. **29** (Suppl.) 448 (1991)

[3.146] H. Neuenschwander et al.: Performace of a macroscopic Monte Carlo method for electron beam dose calculations, Med. Biol. Engn. Comp. **29** (Suppl.) 448 (1991)

[3.147] P.K. Kartha et al.: A new look at the Quality Assurance in the radiation oncology — the six P'S, Med. Biol. Engn. Comp. **29** (Suppl) 239 (1991)

[3.148] W.H.C. Theunissen et al.: Experiences with Orfit, a new moulding material, Catharina Hospital, EindHoven, private communique (1991)

[3.149] M. van Herk: Physical aspects of a liquid-filled ionization chamber with pulsed polarizing voltage, Med. Phys. **18**, 692-702 (1991)

[3.150] E.B. Podgorsak: Physics for radiosurgery with linear accelerators, Neurosurgery Clinics of North America, **3**, 9-34 (1992)

[3.151] P.H. McGinley et al.: A patient rotator for stereotactic radiosurgery, Phys. Med. Biol. **35**, 649-657 (1990)

[3.152] D. Larson, W. Lutz: Radiosurgery (refresher course), Int. J. Radiat. Oncol. Biol. Phys. (Suppl. 1) **17**, 102 (1989)

[4.1] B.J. Mijnheer: Protocol for the determination of absorbed dose in a patient irradiated by means of nuclear particles in radiotherapy procedures, 141-54, in Nuclear and atomic data for radiotherapy and related radiobiology, IAEA, Vienna (1987)

[4.2] ICRU Rep. 21: Radiation Dosimetry: Electrons with initial energies between 1 and 50 MeV, Washington (1972)

[4.3] H. Kubo: Absorbed dose comparison between water and polystyrene-water calorimeter, 568-570 in [4.4]

[4.4] Proc. of the XIV Int. Conf. on Med. Biol. Engn. and VII Int. Conf. on Med. Phys., Espoo, Aug. 11-16, Finland (1985)

[4.5] B.J. Mijnheer et al.: Experimental verification of the air kerma to absorbed dose conversion factor, Radiother. Oncology, **8**, 49-56 (1986)

[4.6] P.F. Brien et al.: Measurements in high-intensity beams from medical linear accelerators, Med. Phys. **14**, 1067-70 (1987)

[4.7] P.V. Houdek et al.: Dosimetry of small radiation fields for 10 MV X-rays, Med. Phys. **10**, 333-36 (1983)

[4.8] G. Chierego et al.: Small field dosimetry, 27-30 in [3.40]

[4.9] B.J. Mijnheer et al.: Consistency and simplicity in the determination of absorbed dose to water in high-energy photon beams, Radiother. Oncology **7**, 371-84 (1986)

[4.10] J.A. Purdy: Basic dosimetry — Radiation therapy machine data, 141-68, in A. Wright, A. Boyer (eds): Advances in Radiation Therapy Treatment Planning Am. Inst. of Physics, New York (1983)

[4.11] Int. Commission on Radiation Units (ICRU) and Measurements: Radiation Dosimetry — X-rays and gamma rays with max. photon energies between 0.6 and 50 MeV, Report 14, Bethesda (1969)

[4.12] Absorbed Dose Determination in Photon and Electron Beams, Technical Report Series No. 277, IAEA, Vienna (1987)

[4.13] A.F. McKinlay: Thermoluminiscence Dosimetry, Hilger, Bristol (1981)

[4.14] G. Rikner: Silicon diodes as detectors in relative dosimetry of photon, electron and proton radiation fields, Ph. D. Thesis, Uppsala University (1983)

[4.15] G. Rikner: The value of solid state semiconductor detectors for dosimetry in high energy beams, 561-62, in [4.4]

[4.16] K. Tahvanainen et al.: In vivo dosimetry with semiconductor detectors, 580-81, in [4.4]

[4.17] K.R. Shortt et al.: Comparison of dosimetric measurements and calculations

for inhomogeneities irradiated by electron beams, 604-605, in [4.4]
[4.18] K.V. Ettinger et al.: Quartz resonators as dose meters, 578-79, in [4.4]]
[4.19] J. Rostkowska et al.: Dozymetria wizek promieniowania X i e- za pomoc techniki filmów (Dosimetry of X-ray and e- beams with films techniques), Post. Fiz. Med., **23**, 243-53 (1988)
[4.20] Ch. S. Reft: Output calibrations in solid water for high energy photon beams, Med. Phys. **16**, 293-301 (1981)
[4.21] D.R. White et al.: Foam epoxy resin based lung substitutes, Brit. J. radiol. **59**, 787-790 (1986)
[4.22] D.M. Galbraith et al.: Dose errors due to charge storage in electron irradiated plastic phantoms, Med. Phys. **11**, 197-203 (1984)
[4.23] J.A. Rawlison et al.: Theoretical and experimental investigation of dose enhancement due to charge storage in electron irradiated phantoms, Med. Phys. **11**, 814-20 (1984)
[4.24] W.G. Pitchford, D.I. Thwaites: Effect on dose of charge storage in electron irradiated insulating phantoms, 610-11, in [4.4]
[4.25] B.J. McParland: Using background substraction for measuring the dose distribution of an electron pencil beam, Med. Phys. **14**, 406-409 (1987)
[4.26] N.E. Ipe et al.: The SLAC pulsed X-ray facility, Report SLAC-Pub-3968, Stanford (1986)
[4.27] R.C. McCall, N.E. Ipe: The response of survey meters to pulsed radiation fields, Report SLAC-Pub-4488, Stanford (1987)
[4.28] K.C. Duvall et al.: The development of a 6 to 7 MeV photon field for instrument calibration, Nucl. Instrum. Meth. **B10/11**, 942-45 (1985)
[4.29] L.H. Lanzl: Magnetic and threshold techniques for energy calibration of high-energy radiations, Ann. of the NY Academy of Sciences, **161**, 101-108 (1969)
[4.30] L. André et al.: Energy distribution of electron beams from clinical accelerators measured with a magnetic spectrometer, 1169-70,in [4.4]
[4.31] R.J. Schultz et al.: Clarification of the AAPM Task Group 21 protocol, Med. Phys. **13**, 5 (1986)
[4.32] J. Iżewska, R. Gajewski: Porównanie jonometrycznych metod obliczania dawki pochłoniętej dla wysokoenergetycznych terapeutycznych wiązek promieniowania X i elektronów (Comparison of ionometric methods of absorbed dose calculation for high-energy X-ray and electron beams), Post. Fiz. Med. **23**, 223-42 (1988)
[4.33] ESTRO (European Society for Therapeutic Radiology and Oncology): Proceedings Symposium of New Protocols for the Dosimetry of High-Energy Photon and Electron Beams, Radiother. Oncol. **4**, 289-328 (1985)
[4.34] O. Mattson: Comparison of different protocols for the dosimetry of high-energy photon and electron beams, Radiother. Oncol. **4**, 313-18 (1985)
[4.35] B.J. Mijnheer, F.W. Wittkämper: Comparison of recent codes of practice for high-energy dosimetry, Phys. Med. Biol. **31**, 407-16 (1986)
[4.36] J.H. Paul et al.: AAPM Group 21 protocol: Dosimetric Ealuation, Med. Phys. **12**, 424-30 (1985)
[4.37] E. Ertan et al.: Phys. Med. Biol. **29**, 789-96 (1984)
[4.38] J. Toivanen, A. Rekonen: Scanning beam electron mapping, 999, in [4.4]
[4.39] E. Martell et al.: A flatness and calibration monitor for accelerator photon and electron beams, Int. J. Radiation Oncology Biol. Phys. **12**, 271-75 (1986)
[4.40] T. Bush, J. Dyet: Proc. of a Workshop on Radiotherapy Simulators: Purchase, Acceptance, Testing and QA, CRC Report 4, 27-42, AAPM, New York (1985)
[4.41] A. Scheffler: Requirements for delicated simulated X-ray equipment, 129-132, in [3.40]
[4.42] C.K. Bomford: Exact simulation-presicion and flexibility, 126-128, in [3.40]
[4.43] Y. Nagata et al.: Clinical eficacy of highly integrated radiotherapy treatment planning system: CT simulator, Med. Biol. Engn. Computing **29**, Suppl. 319 (1991)
[4.44] E.C. McCullough, C.r. Blackwell: Quality assurance of radiotherapy simulators, 42-43, in [3.22]
[4.45] P.W. Horton et al.: Acceptance testing of the Ximatron CX simulator, 49-58, in [3.40]
[4.46] P.W. Horton et al.: Quality control tools for simulators, 133-136, in [3.40]
[4.47]S.H. Crooks: Cross sectional images for computerised treatment planning, 140-41, in [3.40]
[4.48] J. Myles: The future of simulation, 144-45, in [3.40]
[4.49] E.J. Hall: Radiobiology for the Radiologist, Lippincott, Philadelphia, 3rd ed. (1988)
[4.50] G.G. Steel et al. (eds): The Biological Basis of Radiotherapy, Elsevier Science Publishers B.V. Amsterdam (1983)

[4.51] A.Brahme et al.: Application of radiation — Biological data for dose optimization in radiation therapy, AMPI Medical Phys. Bull. **16**, 79-89 (1991)

[4.52] M. Tubiana et al.: Introduction to Radiobiology, Taylor and Francis, London-New York-Philadelphia (1990)

[4.53] B.R. Palival, M.L. Griem (eds): A Categorial Course in Radiation Therapy: Treatment Planning, Radiological Society of North America, Oak Brook IL (1986)

[4.54] I.A.D. Bruinvis et al. (eds): Proc. 9th Int. Conf. on the Use of Computers in Radiation Therapy, Schevemingen, North Holland (1987)

[4.55] I. Bruinvis et al. (eds): Use of Computers in Radiation Therapy, Elsevier Science Publishers (1987)

[4.56] S. Zink (ed): Joint U.S. — Scandinavian Symposium on Future Directions of Computer — Aided Radiotherapy, San Antonio, Aug. 1988, National Cancer Institute Bethesda, MD (1988)

[4.57] J. Eska: Matematyczny model frakcjonowania radioterapii (A mathematical model of radiotherapy fractionation), Post. Fiz. Med. **22**, 119-30 (1987)

[4.58] M. Gotein, A. Niemierko: Biologically based models for scorin treatment plans, 1-2, in [4.56]

[4.59] N. Ulso, A. Brahme: Computer-aided irradiation technique optimization, 1-33, in [4.56]

[4.60] ICRU Report no. 29: Dose specification for reporting external beam therapy with photons and electrons (1978)

[4.61] R.O. Kornelsen, M.E.J. Young: Empirical equations for the representation of depth dose data for computerized treatment planning, Brit. J. Radiol. **48**, 739 (1975)

[4.62] C.K. Bomford: Calculation of percentage depth dose in regions of build-up, Brit. J. Radiol. **42**, 551 (1969)

[4.63] A. Niemierko et al.: Metoda wyznaczania współczynników rozpraszania dla promieniowania X i gamma (A method of determining scattering coefficients for X-rays and gamma radiation), Post. Fiz. Med. **18**, 243-52 (1983)

[4.64] M.R. Sontag, J.R. Cunningham: The equivalent tissue-air ratio method for making absorbed dose calculation in heterogenous medium, Radiology **129**, 787-94 (1978)

[4.65] A. Iwasaki: Uncertainty of the equivalent tissue-air ratio method, Med. Phys. **10**, 716-17 (1983)

[4.66] A. Iwasaki: A new proposal for calculating the equivalent tissue-air ratio correction factor, 598-99, in [4.4]

[4.67] F. Nüsslin: Computerized treatment planning in therapy with fast electrons: a review of procedures for calculating dose distribution, Medicamundi, **24**, 112-18 (1979)

[4.68] N. Suntharalingam, K. Ayyangar: 3 — dimensional dose calculations, 124-35, in F. Bagne (ed): Computerized Treatment Planning Systems, Publication FDA 848223, Washington (1984)

[4.69] J. Cygler et al.: Electron dose distributions in experimental phantoms, Phys. Med. Biol. **32**, 1073-86 (1987)

[4.70] I.A.D. Bruinvis: Electron beams in radiation therapy, Ph. D. Thesis, University of Utrecht, Utrecht, (1987); also I.A.D. Bruinvis, W.A.F. Mathol: Calculation of electron beam depth-dose curves and output factors for arbitrary field shapes, Radiother. Oncol. **11**, 395-404 (1988)

[4.71] Proc. of the 9th Int. Conf. on the Use of Computers in Radiation Therapy, Scheveningen, The Netherlands, North Holland (1987)

[4.72] T.J. Bateman, T. Hobday: Three dimensional radiotherapy planning, 987-88, in [4.4]

[4.73] J.A. Purdy et al.: Three dimensional radiation treatment planning system, 277-79, in [4.4]

[4.74] J.A. Purdy et al.: Treatment planning computers, 495-511, in J.G. Kereiakes et al. (eds): Monograph No. 15 of Amer. Assoc. of Phys. in Med., Amer. Inst. of Phys., New York (1987)

[4.75] H. Weatherburn et al.: The need for three-dimensional radiotherapy planning with electron beams, 1171-72, in [4.4]

[4.76] R.L. Siddon, B.E. Bjarngard: Three-dimensional dose calculations in radiotherapy, 949-50, in [4.4]

[4.77] J.A. Purdy et al.: State-of-the-art of high energy photon treatment planning, in J.M. Vaeth (ed): Frontiers of radiation therapy and oncology, **21** (1986)

[4.78] H.M. Kooy, H. Rashid: A three-dimensional electron pencil-beam algorithm, Phys. Med. Biol. **34**, 229-43 (1989)

[4.79] M. Goitein, M. Abrams: Multi-dimensional treatment planning — II. Beam's eye-view, back projection and projection through CT section, Int. J. Radiat. Oncology Biol. Phys. **9**, 789-97 (1983)

[4.80] S. Peltola: The real time verification, 31-32, in [3.40]

[4.81] A. Scheffler: Practical requirements for treatment verification systems, 33-48, in [3.40]

[4.82] J.A. Purdy: 3D radiation therapy treatment planning, 1-23, in [4.56]
[4.83] G.W. Sheronse et al.: Virtual simulation, 433-36, in [4.55]
[4.84] E.S. Sternick: Developing a knowledge-based information management system, Varian Centerline **10**, no. 2, 6-7 (1988); also 68-71, in [3.22]
[4.85] P. Kolari et al.: CARTES — a prototype decision support system in oncology, 1-10, in J. Talmon, J. Fox (eds): Proc. of the COMAC-BME Workshop Systems Engn in Medicine, Maastricht, March 16-19, Springer Verlag, Berlin (1989)
[4.86] S. Hyödynma et al.: What is a knowledge based system for radiotherapy? A Scandinavian perspective, 1-8, in [4.56]
[4.87] I.L. Lamm et al.: CART — A Nordic challenge in medical computing, 81-86, in Computers in Radiotherapy and Oncology, Session 3, Geneva (March 1984)
[4.88] S. Hyödynma et al.: Decision support for treatment decision making and treatment result evaluation in radiation oncology, in P. Minet (ed): Impact of Personal Computer on Radio-Oncology Departments, European Association of Radiology, 4th Workshop, Liege (1990)
[4.89] S. Hyödynma et al.: Decision support system in oncology, Proc. of EFMI Special Topic Meeting "Expert Systems and Decision Support in Medicine", Hannover (Sept. 1988), 65-68 in Lecture Notes in Medical Informatics no. 36, Springer Verlag, Berlin (1989)
[4.90] I.J. Kalet, W. Paluszyski: Knowledge-based computer systems for radiotherapy planning, 1-14, in [4.56]
[4.91] D. Leavitt: New developments in treatment planning systems, 138-40, in [3.37]
[4.92] S. Zink: The promise of a new technology-knowledge-based systems in radiation oncology, Computerized Medical Imaging and Graphics, **13**, 281-93 (1989)
[4.93] S. Rodenhnis, N.J. Mars: NEO (Netherlands Expertise Center for Oncology) 11-12, in s. Hyödynma et al.: (eds): Report of the Minisymposium on Artificial Intelligence in Oncology Techn. Res. Center of Finland (1988)
[4.94] National Council on Radiation Protection and Measurements (NCRP): structural shielding design and evaluation for medical use of X-rays and gamma rays of energies up to 10 MeV, NCRP Report No. 49, Bethesda (1976)
[4.95] NCRP: Radiation protection design guidelines for 0.1-100 MeV particle accelerators facilities, NCRP Report No. 51, Bethesda (1977)
[4.96] NCRP: Neutron contamination from medical electron accelerators, NCRP Report No. 79, Bethesda (1984)
[4.97] W.P. Swanson: Radiological safety aspects of the operation of electron linear accelerators, Techn. Report Series No. 188, IAEA, Vienna (1979)
[4.98] R.H. Thomas, G.R. Stevenson: Radiological safety aspects of the operation of proton accelerators, Techn. Report Series No. XXX, IAEA, Vienna (1989)
[4.99] R.C. McCall et al.: Health Physics Manual of Good Practices for Accelerator Facilities, SLAC Report No. 327, Stanford (1988)
[4.100] Int. Comm. on Radiat. Protection (ICRP): Cost-benefit analysis in the optimization of radiation protection, ICRP Publ. 37, Pergamon Press, Oxford (1982)
[4.101] NCRP: Recommendations on limits for exposure to ionizing radiation, NCRP Report No. 91, Bethesda (1987)
[4.102] US Department of Energy (DOE): A guide to reducing radiation exposure to as resonably achievable (ALARA) DOE/EV 1830-T 5, National Technical Inf. Service, Springfield (1980)
[4.103] S. Kontroubas et al.: New shielding materials for radiotherapy rooms, 1142-43, in [4.4]
[4.104] R.C. McCall, T.M. Jenkins: Effect of hydrogen shields on the average energy of neutron spectra, SLAC-PUB 4492, Stanford (Nov. 1987)
[4.105] D.G. Bratherton: Planning of radiotherapy departments, Medicamundi **24**, 139-44 (1979)
[4.106] E.F. Hickling: The architect's role in the installation of linear accelerator, 64-66, in [3.40]
[4.107] ICRU Rep. 30: Quantitative Concepts and Dosimetry in Radiobiology, ICRU (1979)
[4.108] ICRU Rep. 35: Radiation Dosimetry — Electron Beams with Energies Between 1-50 Mev, ICRU (1984)
[4.109] ICRU Rep. 44: Tissue Substitutes in Radiation Dosimetry and Measurements, ICRU (1989)
[4.110] IAEA: Dosimetry in Radiotherapy, Unipub, Lanham (1988)
[4.111] J.R. Greening: Fundamentals of Radiation Dosimetry, Med. Phys. Handbook 15, Adam Hilger, Bristol (1985)
[4.112] T. Bruinvis: Electron Beams in Radiation Therapy — Collimation, Dosimetry

and Treatment Planning, Netherlands C.I. (1987)

[4.113] F.H. Attix: Introduction to Radiological Physics and Radiation Dosimetry, J. Wiley, New York (1986)

[4.114] AAPM Rep. 10: Total Skin Electron Therapy, Technique and Dosimetry, AAPM (1988)

[4.115] Chang-ming Ma et al.: Bragg-Gray theory and ion chamber dosimetry for photon beams, Phys. Med. Biol. **36**, 413-428 (1991)

[4.116] H. Saitoh et al.: Buildup region and surface dose measurements for high energy photon beams, Med. Biol. Engn. Computing, **29** (Suppl) 106 (1991)

[4.117] V. Schirrmeister et al.: Corrections for the effective point of measurement in cylindrical ion chamber irradiated, Med. Biol. Engn. Computing **29** (Suppl.) 105 (1991)

[4.118] F.W. Wittkämper et al.: Perturbation correction factors for some ionization chambers commonly applied in electron beams, Phys. Med. Biol. **36**, 1639-1652 (1991)

[4.119] N.A. Attari: High-resolution digital mapping of small radiation fields using an ultrasensitive dosimetry system, Phys. Med. Biol., **33**, 1359-1371 (1988)

[4.120] S.K. Mun et al.: Filmless therapy: feasibility study with a phosphor plate reader, Med. Biol. Engn. Computing **29** (Suppl.), 842 (1991)

[4.121] S. Heukelom et al.: In vivo dosimetry during external photon beam treatment using diodes, Phys. Med. Biol. **33**, Suppl. I, 44 (1988)

[4.122] L. Andre, H. Neuenschwander: Stopping power ratios calculated from measured energy spectra of different electron accelerators, Med. Biol. Engn. Computing **29** (Suppl.) 110 (1991)

[4.123] H.D. Thames, J.F. Fowler: Fractionation in Radiotherapy, Taylor and Francis, London — New York (1987)

[4.124] U.F. Rosennow: Modelling of radiation response, AMPI Medical Phys. Bull. **16**, 67-78 (1991)

[4.125] D. Bakker et al.: Consideration for an integrated dose planning procedure, Proc. of The XIV Int. Conf. on Med. Biol. Engn. and VII Int. Conf. on Med. Phys., Espoo, Aug. 11-16, Finland (1985)

[4.126] J. Bijhold et al.: Fast evaluation of patient set-up during radiotherapy by aligning features in portal and simulator images, Phys. Med. Biol. **36**, 1665-1679 (1991)

[4.127] D. Bakker, M.A. Crommelin: A coherent approach in computed-aided radiotherapy, Proc. of The Ninth Ann. Conf. Engineering in Med. and Biol. Soc., Boston (1987)

[4.128] J.A. Purdy et al.: Radiotherapy treatment planning tools — A collaborative project of the National Cancer Institute, Med. Biol. Engn. Computing **29** (Suppl.) 319 (1991)

[4.129] N. Saranummi et al.: Knowledge-based systems in medicine — a Nordic research and development programme, Comp. Meth. Programs in Biomedicine **34**, 81-89 (1991)

[4.130] K. Inamura: PACS for radiotherapy — technology overview, AMPI Med. Phys. Bull. **16**, 94-103 (1991)

[4.131] G. Read: The clinical interface, AMPI Med. Phys. Bull **16**, 90-993 (1991)

[4.132] K.M. Ayyangar et al.: Three-dimesional radiation treatment planning system based on STARDENT super graphic workstation, AMPI Med. Phys. Bull. **16**, 36-43 (1991)

[4.133] T. Arimoto et al.: PACS for radiotherapy, Med. Biol. Engn. Computing **29** (Suppl.) 843 (1991)

[4.134] Y. Uchiyama et al.: Integrated system of a three-dimensional treatment planning for conformation radiotherapy, Med. Biol. Engn. Computing **29** (Suppl.) 320 (1991)

[4.135] K. Inamura et al.: Application of personal computer communication and ISAC for treatment planning and radiation oncology database, Med. Biol. Engn. Computing **29** (Suppl.) 842 (1991)

[4.136] J.T. Nishidaji et al.: Quality assurance for radiation treatment dose in Japan, Med. Biol. Engn. Computing **29** (Suppl.) 1162 (1991)

[4.137] J. Sobol, F. Pernicka: Photon, neutron and microwave stray radiation from a medical linear accelerator Mevatron-77, Post. Fiz. Med. **25**, 73-80 (1990)

[4.138] J.V. Siebers: Shielding measurements for a 230 MeV proton beam, Wisconsin Medical Physics, Ph. Thesis, Univ. of Wisconsin — Madison, Madison (1990); also J.V. Siebers et al.: Measurement of neutron dose equivalent and penetration in concrete for 230 MeV proton bombardment of Al, Fe and Pb targets, The 7th Symposium on Neutron Dosimetry, Berlin, Oct. 14-18 (1991)

[5.1] J.M. Sisterson: Clinical use of protons and ion beams from a world-wide perspective, Nucl. Instrum. Meth. **B10/11**, 1350-52 (1989)

[5.2] J. Sisterson (ed): World wide charged particle patients totals, in Particles (PTCOG), no 4, 6 (June 1989)

[5.3] B. Larsson: Proton therapy-review of clinical results, EULIMA Workshop on the Potential Value of Light Ion Therapy, Nice (Nov. 3-5, 1988)

[5.4] M.M. Raju: Heavy Particle Radiotherapy, Academic Press, New York (1988)

[5.5] National Accelerator Centre Annual Report (NAC/AR/89-01), Faure, RSA (June 1989)

[5.6] R.A. Gough: The Light Ion Biomedical Research Accelerator (LIBRA), Lawrence Berkeley Laboratory, Report LBL-23413, Berkeley (1987)

[5.7] R.A. Gough (ed), E.L. ALpen: The Heavy Ion Medical Accelerator, Lawrence Berkeley Laboratory, Report PUB-5122, Berkeley (1984)

[5.8] J. Sisterson (ed): Particles — a News-letter for those interested in proton, light ion and heavy charged particle radiotherapy, Harvard Cyclotron Laboratory, 44 Oxford Street, Cambridge MA 02138 (USA)

[5.9] S. Hohm et al.: New accelerators in Uppsala, Physica Scripta, **34**, 513-32 (1986)

[5.10] J.M. Sisterson et al.: Present and future proton treatment facilities at the Harvard Cyclotron Laboratory, Nucl. Instrum. Meth. **B10/11**, 1083-85 (1985)

[5.11] W.G. Davies: The design of ion-optical systems for medical and industrial irradiation, Nucl. Instrum. Meth. **B40/41**, 1178-81 (1989)

[5.12] J.M. Sisterson et al.: Design considerations for a proton beam gantry, MED 86 — MED 88, in [5.13]

[5.13] Proc. of the Int. Symposium on Particle Therapy, March 13-14, 1987, SIN Villigen (1987)

[5.14] M.S.Z. Rabin et al.: Compact designs for comprehensive proton beam clinical facilities, Nucl. Instrum. Meth. **B40/41**, 1335-39 (1989)

[5.15] J.R. ALonso: Magnetically scanned ion beams for radiation therapy, Nucl. Instrum. Meth. **B40/41**, 1340-44 (1989)

[5.16] H. Blattmann et al.: Spot scanning for 250 MeV protons, Paul Scherer Institut, Annual Report 1988, Annex II, 18-20 (1988)

[5.17] H. Tsunemoto, S. Morita: Fast neutrons and protons in cancer treatment with medical cyclotron at NIRS, 602-608, Proc. 11th Conf. on Cyclotrons and Their Applications, Ionics, Tokyo (1987)

[5.18] H.G. Blosser: Compact superconducting synchrocyclotron systems for proton therapy, Nucl. Instrum. Meth. B40/41, 1326-30 (1989)

[5.19] B. Gottschalk: Design of a hospital-based accelerator for proton radiation therapy-scaling rules, Nucl. Instrum. Meth. **B24/25** (1987)

[5.20] R.L. Martin: A proton accelerator for medical applications, Nucl. Instrum. Meth. **B24/25**, 1087-91 (1987)

[5.21] R.L. Martin: Modular design of H^- synchrotrons for radiation therapy, Nucl. Instrum. Meth. **B40/41**, 1331-34 (1989)

[5.22] R.L. Martin: A proton medical accelerator by the SBIR route — an example of technology transfer, Nucl. Instrum. Meth. **B40/41**, 1362-64 (1989)

[5.23] H. Tsunemoto et al.: Proton therapy in Japan, Radiation Research **104**, S-235 to S-243 (1985)

[5.24] K. Endo et al.: Medical synchrotron for proton therapy, KEK Preprint 88-20 (June 1988), also in European Particle Acc. Conf., Rome, June 7-11 (1988)

[5.25] S. Fukumoto et al.: A dedicated proton accelerator for cancer therapy, KEK Preprint 89-91 (Sept. 1989), also in XIV Int. Conf. on High Energy Acc. (Tsukuba, Aug. 22-26, 1989)

[5.26] F. Cole et al.: Design and application of a proton therapy accelerator, 1985-87, in Proc. of IEE Particle Acc. Conf. (1987)

[5.27] J.V. Siebers: Commissioning studies for the horizontal beamline at Loma Linda, Abstracts of the PTCOG XIV Meeting, Cambridge, May 21-23 (1991)

[5.28] J.M. Slater: News from LLUMC, Particles no.9, 4-6 (1992)

[5.29] A. Ito, H. Koyama-Ito: Development of the proton computed tomography, 240-41 in [4.4]

[5.30] A proton computed tomography, 140-41, KEK Annual Report (1986)

[5.31] B.W. Augenstein et al. (eds): Proc. of the RAND Workshop on Antiproton Science and Technology, World Scientific, Teaneck, NJ (1988)

[5.32] T.E. Kalergopulos, R. Muratore: Antiprotons for imaging and therapy, Nucl. Instrum. Meth. **B40/41**, 1322-25 (1989)

[5.33] W. Preston: Charged particle bibliography, 6-7, Particles, no.6 (June 1990)

[5.34] H. Tsunemoto et al.: (eds): Proc. Int. Workshop on the NIRS Heavy Particle Medical Acc, March 6-7, 1986, NIRS Chiba (1986)

[5.35] E.L. Alpen: Charged particle accelerators — The present and future in biological research and medicine, 12-31, in [5.34]
[5.36] G. Kraft: The radiobiological and physical basis for radiotherapy with protons and heavier ions, Strahlentherapie und Oncologie, **166**, 10-13 (1990)
[5.37] W. Schimmerling et al.: The relative biological effectiveness of 670 Mev/A neon as a function of depth in water for a tissue model, Radiat. Res. **112**, 436-48 (1987)
[5.38] W. Saunders et al.: Helium-ion radiation therapy at the LBL, Radiat. Res. **104**, S-227 to S-234 (1985)
[5.39] W.M. Saunders et al.: Precision, high dose radiotherapy: Helium ion treatment of uveal melanoma, Int. J. Radiation Oncology Biol. Phys. **11**, 227-33 (1985)
[5.40] W.M. Saunders et al.: ditto: II Helium ion treatment of tumors adjacent to critical central nervous system structures, Int. J. Radiation Oncology Biol. Phys. **11**, 1339-47 (1985)
[5.41] S.R. Zink et al.: Treatment planning study for carcinoma of the esophagus: helium ions versus photons, Int. J. Radiation Oncology Biol. Phys. **14**, 993-1000 (1988)
[5.42] J.T. Lyman et al.: Protocol for heavy cherged-particle therapy beam dosimetry, AAPM Report no. 16, AIP, New York (1986)
[5.43] A. Chatterje, E. Blakely: Applications of the Bevalac in biophysical research, Nucl. Instrum. Meth. **B40/41**, 1365-71 (1989)
[5.44] P. Mandrillon et al.: The EULIMA project, Centre Antoine Lacassagne, Nice (Oct. 1987)
[5.45] R.A. Gough: Present status of the Bevalac and design outline of proposed medical accelerator, 91-109, in [5.34]
[5.46] J.R. Alonso: Intermediate energy heavy ions — An emerging multidisciplinary research tool, Nucl. Instrum. Meth. **B40/41**, 921-26 (1989)
[5.47] W.T. Chu: Biomedical facility at the Bevalac, 119-140, in [5.34]
[5.48] G. Gademan et al.: The medical heavy ion therapy project at the GSI in Darmstadt, Preprint GSI-88-71, Darmstadt (1988)
[5.49] K. Kawachi et al.: Radiation oncological facilities of the HIMAC, J. Jpn. Soc. Ther. Radiat. Oncol. **1**, 19-29 (1989)
[5.50] Proc. of the EULIMA Workshop on the Potential Value of Light Ion Beam Therapy, EUR 12165 EN of the Commission of the European Communities, CAL Edition, Nice (1988)
[5.51] J.T. Lyman et al.: Charged-particle stereotactic radiosurgery, Nucl. Instrum. Meth. **B10/11**, 1107-10 (1985)
[5.52] B. Larsson: Use of the Uppsala synchrocyclotron in radiotherapy and radiosurgery, Proc. of the PTCOG 2nd Int. Charged Particle Therapy Workshop, Loma Linda (Oct. 2-14, 1987)
[5.53] J.I. Fabrikant et al.: Heavy charged-particle radiosurgery for intracranial vascular disorders, 561-63, in Proc. of the EPAC 90, Edition Frontiers, Gif-sur Yvette (1990)
[5.54] J.I. Fabrikant et al.: Neurosurgical applications of ion beam, Nucl. Instrum. Meth. **B40/41**, 1376-84 (1989)
[5.55] J.T. Lyman et al.: A helium-ion beam for stereotactic radiosurgery of CNS disorders, Med. Phys. **13**, 695-99 (1986)
[5.56] P. Chapman: Stereotaxic alignment for radiosurgery, 5-6, Particles, no. 5 (Jan. 1990)
[5.57] M.R. Raju, A.J. van der Kogel: Radiobiology of pions in comparison with other heavy particles, report LANL LA-UR-87-1251, Los Alamos (1987), also in Radioth. Onc.
[5.58] M. Paciotti et al.: Beam characteristics in the LAMPF biomedical channel, IEEE Trans. Nucl. Sci **NS-28**, 3150-52 (1981)
[5.59] M.A. Paciotti et al.: Muon beam at the LAMPF biomedical channel, IEEE Trans. Nucl. Sci **NS-30**, 2359-61 (1983)
[5.60] C.F. von Essen: Long-term results of pion therapy at Los Alamos, SIN Medical Newsletter no. 4, Villigen (1984)
[5.61] L.D. Hansborough: PIGMI — a design report, report LANL LA-8880, UC-28 and UC-48, Los Alamos (1981)
[5.62] G.K.Y. Lam et al.: Beam scanning system for a clinical pimeson beam, Rev. Sci. Instrum. **57**, 329-35 (1986)
[5.63] G.K.Y. Lam: Technique of range modulation for a clinical pimesonbeam, Rev. Sci. Instrum. **53**, 1067-71 (1982)
[5.64] G.B. Goodman et al.: Pimeson radiotherapy at TRIUMF, J. Japn Soc. Ther. Radiol. Oncol. **2**, 85-99 (1990)
[5.65] G.B. Goodman et al.: Pion therapy at TRIUMF, Int. Symp. on Particle Therapy at SIN, Villigen (March 1987)
[5.66] D.A. Pistenma et al.: Initial performance of the Stanford Medical Pion Generator, Radiology, **122**, 527-529 (1977)
[5.67] C.F. von Essen et al.: The Piotron — initial performance, preparation and experience with pion therapy, Int. J. Radiation Oncology Biol. Phys. **8**, 1499-1509 (1982)
[5.68] H. Schumacher et al.: Proportional counter dosimetry and microdosimetry for

radiotherapy with multiple pion beams, Radiat. Res. **101**, 177-96 (1985)

[5.69] G. Schmitt et al.: Review of the SIN and Los Alamos pion trials, Radiat. Res. **104**, S-272-78 (1985)

[5.70] SIN Int. Symp. on Particle Therapy, March 1987, SIN Medical Newsletter no. 9 (1987)

[5.71] Anonim: World wide charged particle patients totals, Particles no.6, 8 (1990)

[5.72] M. Catterall, D.X. Bewley: Fast Neutron in the Treatment of Cancer, Academic Press, London (1979)

[5.73] L. Cohen, M. Awschalom: Fast neutron radiation therapy, Ann. Rev. Biophys. Bioeng., **11**, 359-90 (1982)

[5.74] H.R. Withers: Neutron radiobiology and clinical consequences, Strahlentherapie, **161**, 739-45 (1985)

[5.75] D.K. Bewley: The Physics and Radiobiology of Fast Neutron Beams, Adam Hilger, Bristol (1989)

[5.76] N. Breteau et al.: Current results of neutron boost irradiation in Orleans for inoperable rectal carcinomas, Med. 48-53, in [5.13]

[5.77] G. Schmitt et al.: Neutron and neutron-boost irradiation of soft tissue sarcomas, Radiother. Oncol. **1**, 23-29 (1983)

[5.78] W. Duncan et al.: Fast neutrons in the treatment of head and neck cancers: The results of a multi-centre randomly controlled trial, Radiother. Oncol. **2**, 293-300 (1984)

[5.79] H.R. Withers: Neutrons and other clinical trials: Impossible dreams? Int. J. Radiat. Oncol. Biol. Phys. **13**, (1987)

[5.80] G.E. Laramore: Fast neutron radiotherapy for inoperable salivary gland tumors: is the treatment of choice, Int. J. Radiat. Oncol. Biol. Phys. **13**, 1421-23 (1987)

[5.81] G.E. Laramore et al.: Fast neutron and mixed beam radiotherapy for inoperable nonsmall cell carcinoma of the lung. Am. J. Clin. Oncol. (CCT), **9**, 233-43 (1986)

[5.82] K.J. Russel et al.: Eight years' experience with neutron radiotherapy in the treatment of stages C and D prostate cancer, Prostate **11**, 183-93 (1987)

[5.83] S. Zink et al.: Fast neutron therapy clinical trials in the United States, Am. J. Clin. Oncol. (CTT) **12**, 277-82 (1989)

[5.84] A. Lennox: Neutron therapy — a progress report, Fermilab Report 89/11, 1-3 (1989)

[5.85] J.A. Stokkes et al.: performance of neutron radiography systems using particle accelerators, IEEE Trans. Nucl. Sci. **NS-30**, 1623-26 (1983)

[5.86] D.K. Bewley et al.: Neutron beams from protons on berylium, Phys. Med. Biol. **25**, 887-92 (1980)

[5.87] D.K. Bewley et al.: New neutron sources for radiotherapy, Phys. Med. Biol. **29**, 341-49 (1984)

[5.88] S. Vynckier et al.: Comparison of neutron therapy beams produced by 50 MeV deutrons and 65 MeV protons on berylium, Acta Radiol. Oncol. **21**, 281-87 (1982)

[5.89] S. Vynckier et al.: Improvement of p (65) + Be neutron beam for radiotherapy at Cyclone, Lovain-la-Neuve, Phys. Med. Biol. **28**, 685-91 (1983)

[5.90] D.T.L. Jones, M. Yudelev: The effects of various filters on a p (66 MeV)/Be neutron beam, NAC Report NAC/88-14, Faure (Sept. 1988)

[5.91] J.J. Battermann, K. Breur: Fast neutrons, and advantage in radiation oncology? J. Eur. Radiother., **2**, 27-36 (1981)

[5.92] D.K. Bewley et al.: Neutron generators for radiotherapy, J. Eur. Radiother. **5**, 187-94 (1984)

[5.93] AAPM Protocol for neutron beam dosimetry, AAPM Report No. 7, Am. Inst. of Physics, New York (1980)

[5.94] J.J. Broerse et al.: European protocol for neutron dosimetry for external beam therapy, Brit. J. Radiol. **54**, 882-98 (1981)

[5.95] B.J. Mijnheer et al.: Uniformity in dosimetry protocols for therapeutic applications of fast neutron beams, Med. Phys. **14**, 1020-26 (1987)

[5.96] B.J. Mijnheer et al.: Standardization of dosimetry for fast neutron therapy, J. Eur. Radiother. **5**, 195-205 (1984)

[5.97] B.J. Mijnheer: Clinical dosimetry of high-energetic: neutron beams-future developments, Six. Symp. on Neutron Dosimetry, Neuherberg (1987), Radiat. Protection Dosimetry

[5.98] J.B. Smathers, L.T. Myers: Neutron therapy and dosimetry at the UCLA Neutron Therapy Facility, Nucl. Instrum. Meth. **B24/25**, 1077-81 (1987)

[5.99] R. Risler et al.: Installation of the cyclotron based clinical neutron therapy system in Seattle, 428-430, Proc. of the 10th Int. Conf. on Cyclotrons and Their Appl., East Lansing (1984)

[5.100] R. Risler et al.: Routine operation of the Seattle clinical cyclotron facility, 160-163, Proc. of the 12 Int. Conf. on Cyclotrons and Their Appl., Berlin, May 8-12, 1989, World Scientific, Singapore (1990)

[5.101] R. Risler et al.: two years of operating experience with the Seattle clinical neutron

therapy facility, 682-84, Proc. 11th Conf. on Cyclotrons and Their Appl., Ionics, Tokyo (1987)
[5.102] D. Reitmann: From promise to proof, Nuclear Active, Int. J. of the Atomic Energy Corp. of South Africa, 13-16, no. 39 (July 1988)
[5.103] D.T.L. Jones: Particles for patients, Nuclear Active, 30-36, no. 40 (Jan. 1989)
[5.104] D. Reitmann (ed): National Accelerator Centre Annual Report, NAC/AR/89-01 CSIR, Faure, RSA (June 1989)
[5.105] A.J. Lennox: Hospital-based proton linear accelerator for particle therapy and radioisotope production, Nucl. Instrum. Meth. **B56/57**, 1197-1200 (1991)
[5.106] D. Raparia, S. Machida: Design study of a medical linac for neutron therapy, Nucl. Instrum. Meth. **B40/41**, 1354-57 (1989)
[5.107] C.K.Ch. Wang et al.: A neutronic study of an accelerator-based neutron irradiation facility for boron neutron capture therapy, Nucl. Technology **84**, 93-107 (1989)
[5.108] C.K.Ch. Wang et al.:: A design study of an accelerator-based epithermal neutron source for boron neutron capture therapy, Strahlether. Onkol. no 2/3, 75-78 (1989)
[5.109] W. Blue et al.: A study of low energy proton accelerators for neutron capture therapy, 147-158, in H. Tanaka (ed): Proc. 2nd Int. Symp. Neutron Capture Therapy, Nishimura Ltd., Tokyo (1985)
[5.110] (a) J.F. Crawford et al.: The potential of internal neutron sources in capture therapy, 4th Int. Symp. on NCT for Cancer, Sydney, Dec. 4-7, 1990, in [5.112b]; (b) J.F. Crawford: The potential of NCT and accelerator neutron sources in brachytherapy, GEC/ESTRO Brachytherapy Meeting, Baden-Baden, May 15-17 (1991)
[5.111] (a) R.G. Fairchild (ed): Clinical Aspects of Neutron Capture Therapy, Plenum Press, New York (1989); (b) O.K. Harling et al. (eds): Neutron Beam Design, Development and Performance for Neutron Capture Therapy, Plenum Press, New York (1990)
[5.112] (a) Proc. of the Third Int. Symp. on NCT, Bremen, May 1988, Strahlenther. Onkol. **165**, no. 2/3 (1989); (b) Proc. of the 4th Int. Symp. on NCT, Sydney, Australia, Dec. 1990, Plenum Press, New York (1991)
[5.113] T.P. Wangler et al.: Conceptual design of an RFQ accelerator-based neutron source for BNCT, 678-680, in Proc. of the 1989 IEEE Particle Accelerator Conference, IEEE 89CH 2669-0 (1989)
[5.114] J. Sisterson: World wide charged particle patient totals, Particles, no. 9, 16 (Jan. 1992)
[5.115] L.J. Peters et al.: Am.. J. Clin. Oncol. **11**, 275-287 (1988)
[5.116] M. Phillips et al.: Charged particle spot scanning method — Effects of motion on dose homogeneity, S70-S72, in Proc. of the EPAC 90, Edition Frontiers, Gif-sur Yvette (1990)
[5.117] S. Fukumoto et al.: Design of an synchrotron for proton therapy, 700-702, in Proc. of 1989 IEEE Particle Acc. Conf., IEEE Pub. 89CH2669-0 (1989)
[5.118] S. Fukumoto et al.: A dedicated proton accelerator for cancer therapy, Particle Accelerators **33**, 153-158 (1990)
[5.119] R.L. Martin: The ACCTEK proton medical accelerator and beam delivery system for cancer therapy and radiography, S103-S105, in Proc. of the EPAC 90, Edition Frontiers, Gif-sur-Yvette (1990)
[5.120] Y. Jongen et al.: Preliminary design of a reduced cost proton therapy facility using a compact, high field isochronous cyclotron, S97-S99, in Proc. of the EPAC 90, Edition Frontiers, Gif-sur-Yvette (1990)
[5.121] W. Beeckman et al.: Preliminary design of a reduced cost proton therapy facility, Nucl. Instrum. Meth. **B56/57**, 1201-04 (1991)
[5.122] Y. Jongen: IBA starts proton therapy prototype project, Particles, no. 7, 7-8 (1991)
[5.123] J. Alonso: Design study for the UC Davis Proton Facility, 6, in Abstracts of the PTCOG XIV Meeting, Cambridge, MA, May 21-23 (1991)
[5.124] J.C. Rosenwald et al.: Recent developments of a high-energy proton therapy project in Orsay (France), S91-S93, in Proc. of the EPAC 90, Editions Frontiers, Gif-sur-Yvetter (1990)
[5.125] J.C. Rosenwald et al.: Transformation of a 200 MeV research cyclotron facility into a medical installation for protontherapy, Med. Biol. Engn. Computing **29**, 1159 (1991)
[5.126] J.L. Habrand, J.F. Desgranges: Development of protontherapy at CPO — medical and administrative aspects, in Proc. of the Proton Workshop, Villigen (March 1991)
[5.127] P. Chauvel et al.: A cooperative group and communication network for proton and neutrontherapy in South East of France, S88-S90, in Proc. of the EPAC 90, Editions Frontiers, Gif-sur-Yvette (1990)
[5.128] Anonym: The National Association

for Proton Therapy, Particles, no. 8, 8-9 (1991)

[5.129] T. Kalogeropoulos et al.: Biomedical potential of antiprotons, 640-678, in [5.31]

[5.130] J.R. Alonso et al.: Relocation of the helium ion radiotherapy program from the 184" Synchrocyclotron to the Bevalac, Proc. of the 1989 IEEE Particle Accelerator Conf., 669-671, IEEE, New York (1989)

[5.131] F.J.M. Farley: Round Table on Eulima, S121-S123, in Proc. of the EPAC 90, Editions Frontiers, Gif-sur-Yvette (1990)

[5.132] W.T. Chu et al.: Three-dimensional conformed therapy using light-ion beams, in Proc. of the Workshop in Chiba, Japan (July 1991), also Med. Biol. Engn. Computing **29**, Suppl. 776 (1991)

[5.133] B. Ludewigt et al.: Scanning system for heavy cherged-particle beams, p.2, in Abstracts of the PTCOG XIV Meeting, Cambridge, MA, May 21-23 (1991)

[5.134] T.R. Renner et al.: Quality Assurance for radiation quality in the clinical use of heavy charged-particle beams, Med. Biol. Engn. Computing, **29**, Suppl. 776 (1991)

[5.135] G. Kraft et al.: The Heavy Ion Therapy Project at GSI, GSI-91-15, Preprint GSI Darmstadt (March 1991)

[5.136] Biophysics Group: Design, construction and first experiments of a magnetic scanning system for therapy, Final Report for the EULIMA Collaboration, GSI-91-18 Report, GSI Darmstadt (June 1991)

[5.137] K. Sato et al.: Heavy Ion Medical Accelerator in Chiba (Japan), Particle Accelerators **33**, 147-152 (1990)

[5.138] G. Cesar et al.: Feasibility study of a synchrotron for the European Light Ion Medical Accelerator, EULIMA Feasibility Study Group, hosted by CERN PS Division, CERN/PS/91-08 (DI) Report, Geneva (1991)

[5.139] G. Berger et al.: Radioactive ion beams — Results and perspectives for light ion therapy and diagnostic purposes, 1793-1795, in Proc. of the EPAC 90, Edition Frontiers, Gif-sur-Yvette (1990)

[5.140] J.R. Beveridge: Accelerator plans at TRIUMF, Workshop on the Future of Muon Physics, Heidelberg, May 7-10, 1991, TRIUMF Report TRI-PP-91-50, Vancouver (1991)

[5.141] J.B. Smathers: Current status and the future of neutron therapy in the United States, Med. Biol. Engn. Computing, **29**, Suppl., 109 (1991)

[5.142] H. Blosser: New from National Superconducting Cyclotron Laboratory, Michigan State University, Particles, no. 9, 7 (Jan. 1992)

[5.143] R.L. Maughan et al.: Preliminary performance characteristics of a 50 MeV deuteron superconducting cyclotron for neutron radiation therapy, Med. Biol. Engn. Computing **29**, Suppl. 1, 109 (1991)

[5.144] F.T. Cole: Accelerator considerations in the design of a proton-therapy facility, Particle Acceleration Corporation Report (1991)

[5.145] H. Conde et al.: Status report on the development of a spallation source for neutron capture therapy, 319-323, in R.G. Fairchild et al. (eds): Clinical Aspects on Neutron Capture Therapy, Plenum Publ. Corp., New York (1989)

[5.146] A. Andersson et al.: Programme for accelerator — based neutron capture therapy, S79-S81, in Proc. of the EPAC 90, Editions Frontiers, Gif-sur-Yvette (1990)

[5.147] S. Vynckier et al.: Code of practice for clinical proton dosimetry, Radiotherapy and Oncology, **20**, 53-63 (1991)

[5.148] NAC Annual Report, Faure NAC/AR/91-01 (March 1991)

[5.149] K.H. Möver et al.: Dose distributional properties obtained by improved treatment techniques for fast neutron therapy, Med. Biol. Engn. Computing **29**, Suppl., 108 (1991)

[5.150] T. Matsumoto, M. Ohya: Depth-dose evaluation for neutron capture therapy with gadolinium, Med. Biol. Engn. Computing **29**, Suppl., 777 (1991)

[5.151] J. F. Crawford et al.: Neutrons for capture therapy produced by 72 MeV protons, Proc. of the 4th Int. Symp. on Neutron Capture Therapy for Cancer, Sydney, Dec. 4-7, 1990, in [5.112b]

[5.152] J.F. Crawford et al.: A scintillation spectrometer for direct comparison of neutron energy spectra at high and low rates, Workshop on Dose Components and Biological Effects in Conventional Radiotherapy and NCT, Essen, Feb. 27-29 (1992)

[5.153] G. Coutrakon et al.: A prototype beam delivery system for the proton medical accelerator at Loma Linda, Med. Phys. **18**, 1093-1099 (1991)

[6.1] S.S. Hasnain (ed): Synchrotron Radiation and Biophysics, Ellis Horwood, Chichester (1990)
[6.2] W. Brefeld, P. Gürtler: Synchrotron radiation sources, in S. Ebashi et al.; (eds) Handbook on Synchrotron Radiation, vol. 4, North Holland, Amsterdam (1990)
[6.3]H. Winick: Overview of synchrotron radiation facilities, 7-11, in F. Bennett and J. Kopta (eds): Proc. of the 1989 IEEE Particle Accelerator Conference, March 20-23, IEEE 89CH2669-0 (1989)
[6.4] E.E. Koch (ed): Handbook on Synchrotron Radiation, Vol. 1a, 1b: Characteristics, Instrumentation and Principles of Research Applications, North-Holland, Amsterdam (1983), Vol. 3 X-ray Research, eds G. S. Brown and D.E. Moncton (ditto, 1991)
[6.5] S. Ebashi et al. (eds): Medical, Biological and Physiological Applications, Handbook on Synchrotron Radiation, vol. **4**, North-Holland-Elsevier Science Publ., Amsterdam (1991)
[6.6] N. Greaves, I. Munro (eds): Synchrotron Radiation Sources and Their Applications, American Inst. of Physics, New York (1989)
[6.7] R.P. Walker: An introduction to the physics of synchrotron radiation sources, Report Daresbury Laboratory Dl/SCI/P513A, Daresbury (1986)
[6.8] M. Cornacchia: Requirements and limitations on beam quality in synchrotron radiation sources, Report SLAC-PUB-5021, Stanford (1989)
[6.9] R. Barbini et al.: Synchrotron radiation from magnetic undulators as a prospective diagnostic tool, Report ENEA-Centro Ricerche Energia Frascati RT/TIB/89/6, Frascati (1989)
[6.10] M.W. Poole: Design and technology of undulators, Report Daresbury Laboratory DL/SCI/P 677A, Daresbury (1990), also in Proc. of CERN Accelerator School, Chester (April 1989)
[6.11] D.M.P. Holland: The Daresbury Synchrotron Radiation Source, Report Daresbury Laboratory DL/SCI/P524E, Daresbury (1986)
[6.12] V.P. Suller et al.: SRS-2: performance and achievements, Report Daresbury Laboratory Dl/SCI/P629A, Daresbury (1989)
[6.13] T. Tomimasu: New radiation sources using high energy electrons, Nucl. Sci. Applications **3**, 29-54 (1987)
[6.14] Anonym: Convention and status signed, ESRF Newsletter no. 1, 1 (Feb. 1989)
[6.15] Anonym: Site work in progress, ESRF Newsletter no. 4, 1 (June 1990)
[6.16] M. Derrick: Argonne accelerator research, CERN Courier **27**, no. 2, 22-24 (1987)
[6.17] A. Bienenstock: Full dedication of SPEAR for synchrotron radiation, SSRL Newsletter, Stanford (June 1989)
[6.18] C. Pellegrini: Free Electron Lasers — Development and Applications, Particle Accelerators **33**, 159-170 (1990)
[6.19] G. Dattoli et al.: An introduction to the theory of FEL, report ENEA-Centro Ricerche Energia Frascati RT/TIB/89/38, Frascati (1989), also in Proc. of the CERN/DL Accelerator School Course on Symchrotron radiation and Free Electron Laser, Chester (U.K.) (Apr. 1989)
[6.20] G. Dattoli et al.: The FEL — a simple quantum picture, report ENEA-Centro Ricerche Energia Frascati RT/TIB/89/52, Frascati (1989)
[6.21] M.W. Poole: Infra-red FELS, Report Daresbury Laboratory DL/SCI/P676A, Daresbury (1990), also in The Proc. of CERN Accelerator School, Chester (Apr. 1989)
[6.22] Proc. of 11th Int. Conf. on Free Electron Lasers, Nucl. Instrum. Meth. (1990)
[6.23] R. Ramponi, O. Svelto: Potential applications of FEL in biomedicine, Nucl. Instrum. Meth. **A239**, 386-89 (1985)
[6.24] E. Rubinstein: Medical imaging with synchrotron radiation, Nucl. Instrum. Meth. **222**, 302-07 (1984)
[6.25] A.C. Thompson et al.: Imaging of coronary arteries using synchrotron radiation, Nucl. Instrum. Meth. **B40/41**, 407-12 (1989)
[6.26] W.R. Dix et al.: Coronary angiography using synchrotron radiation-studies in human subjects with the system NIKOS II, DESY Report SR 91-01, Hamburg (Sept. 1991), to be published in NIM
[6.27] W. Kupper and W. Bleifeld: Koronarangiographie und Ventrikulographie, Internist **31**, 341 (1990)
[6.28] E. Rubenstein: Synchrotron radiation for angiography, Am. Rev. Biophys. Chem. **16**, 161-74 (1987)
[6.29] W.R. Dix et al.: Contribution of the NIKOS group to the 1st Frascati meeting, in [6.36]
[6.30] E. Rubenstein et al.: Synchrotron radiation coronary angiography in a human subject with a dual beam, dual detector imaging system, 87-88, SSRL Activity Report, Stanford (1989)
[6.31] W. Thomlinson et al.: Venous syn-

chrotron coronary angiography, The Lancet **337**, 360 (1991)

[6.32] E.N. Dementyer et al.: First results of experiments with a medical one-coordinate X-ray detector on synchrotron radiation, Nucl. Instrum. Meth. **A246**, 726-30 (1986)

[6.33] W.R. Dix et al.: NIKOS — A system for non-invasive examination of coronary arteries by means of digital substraction angiography, Nucl. Instrum. Meth. **A246**, 702-12 (1986)

[6.34] W.R. Dix et al.: NIKOS II — A system for non-invasive imaging of coronary arteries, report DESY SR-89-03, Hamburg (1989)

[6.35] Takeda T. et al.: High speed iodine K-edge substraction angiography — a preliminary experiment with rats, Med. Imaging Technology **6**, 400-406 (1988)

[6.36] Proc. of the 1st Frascati Meeting on Synchrotron radiation Applications to Digital Substraction Angiography, Societa Italiana di Fisica, Frascati (1987)

[6.37] W. Gudat: Report DESY SR-77/21, Hamburg (1977)

[6.38] K. Shinohara et al.: A new approach to observation of the resist in X-ray contact microscopy, Photochem. Photobiology **44**, 401-03 (1986)

[6.39] D. Sayer et al.: X-ray microscopy II, Springer Series in Optical Sciences, **56**, Springer-Verlag, Berlin (1983)

[6.40] K. Shinohara: Application of X-ray microscopy to mammalian cells, in K. Shinohara et al. (eds): X-ray Microscopy in Biology and Medicine, Japan Sci. Societes Press, Tokyo/Springer-Verlag, Berlin (1990)

[6.41] K. Shinohara et al.: Further analysis of the soft X-ray microimages of unstained human chromosome fibres dried with no fixative, 96, in Photon Factory Activity Report 1989, KEK Tsukuba (1989)

[6.42] B.M. Gordon, K.W. Jones: Design criteria and sensitivity calculations for multielemental trace analysis at the NSLS X-ray microprobe, Nucl. Instrum. Meth. **B10/11**, 293-98 (1985)

[6.43] K.W. Jones et al.: Catalyst analysis using synchrotron X-ray microscope, Nucl. Instrum. Meth. **C56/57**, 427-432 (1991)

[6.44] A.L. Hanson et al.: Trace element measurements using white synchrotron radiation, Nucl. Instrum. Meth. **B24/25**, 400-404 (1987)

[6.45] W.J. Lenglet et al.: Bulk and microprobe analysis for trace elements with synchrotron radiation, Analytica Clinica Acta **195**, 153-62 (1987)

[6.46] C. Jacobsen et al.: Quantitative imaging and microanalysis with a scanning soft X-ray microscopy, Phys. Med. Biol. **32**, 431-37 (1987)

[6.47] A.C. Thompson et al.: Computed tomography using synchrotron radiation, Nucl. Instrum. Meth. **222**, 319-23 (1984)

[6.48] Z.H. Cho et al.: Three-dimensional microtomography with synchrotron radiation, Nucl. Instrum. Meth. **227**, 385-92 (1984)

[6.49] D.K. Bowen et al.: X-ray microtomography with synchrotron radiation, SPIE **691** X-ray Imaging II, 94-98 (1986)

[6.50] P. Spanne, M.L. Rivers: Computerized microtomography using synchrotron radiation from the NSLS, Nucl. Instrum. Meth. **B24/25**, 1063-67 (1987)

[6.51] R. Cesareo (ed): Nuclear Analytical Techniques in Medicine (Techniques and Instrumentation in Analytical Chemistry **8**), North-Holland, Amsterdam (1988)

[6.52] S.S. Hasnain, C.D. Garner: Characterisation of metal centres in biological systems by X-ray absorption spectroscopy, Report Daresbury Laboratory DI/SCI/P554E, Daresbury (1987)

[6.53] S.S. Hasnain: Application of EXAFS to Biochemical Systems, 74-93, in Topics in Current Chemistry, **147**, Springer-Verlag, Berlin-Heidelberg (1988)

[6.54] K.O. Hodgson et al. (eds): EXAFS and XANES near Edge Structure III, Springer-Verlag, Berlin (1984)

[6.55] R. Prins, D.C. Koningsberger (eds): Princip. Application Techniques of EXAFS, SEXAFS and XANES, John Wiley and Sons Publ., New York (1987)

[6.56] S.S. Hasnain (ed): X-ray Absorption Fine Structure, Proc. of 6th Int. Conf. York, Aug. 5-11, 1990, Ellis Harwood, Chichester (1991)

[6.57] A. Bianconi, A. Congiu Castellano (eds): Biophysics and Synchrotron Radiation, Springer-Verlag, Berlin (1987)

[6.58] Proc. of the Six National Conf. on Synchrotron Radiation Instrum., Berkeley, Aug. 1989, Nucl. Instrum. Meth. **A291**, no. 1 and 2, 1-517 (1990); also Proc. of the Ninth USSR National Conf. on Synchrotron Radiation Utilization (SR90), Nucl. Instrum. Meth. **A308**, no. 1 and 2, 1-470 (1991)

[6.59] W.R. Dix et al.: NIKOS II — a system for non-invasive imaging of coronary arteries, in Medical Imaging III: Image Formation, SPIE 1090, 282 (1989)

[6.60] H.D. Zeman et al.: A kinestatic charge detector for intravenous coronary angiography using synchrotron radiation X-rays, IEEE Trans. Nucl. Sci. **30**, 641-47 (1991)

[6.61] M. Lohmann et al.: A fast line scan x-ray detector for medical applications — a Status report of the NIKOS II detecttor, SPIE 1245 (1991)

[6.62] V.P. Barsukov et al.: X-ray monochromator for digital substraction angiography using synchrotron radiation, Nucl. Instrum. Meth. **A308**, 419-422 (1991)

[6.63] H.D. Zeman and D.P. Siddons: Contrast agent choice for intravenous coronary angiography, Nucl. Instrum. Meth. **A291**, 67-73 (1990)

[6.64] B.M. Gordon et al.: The application of synchrotron radiation to microprobe trace — element analysis of biological samples, Nucl. Instrum. Meth. **B45**, 527-31 (1990)

[6.65] S. Hayakawa et al.: X-ray microanalysis with energy tunable synchrotron X-rays, Nucl. Instrum. Meth. **B49**, 555-560 (1990)

[6.66] C. Tuniz et al.: A synchrotron radiation microprobe for X-ray fluorescence and microtomography at ELECTRA, Nucl. instrum. Meth. **B50**, 338-42 (1990)

[6.67] J.R. Chen et al.: The uses of SR sources for elemental and chemical microanalysis, Nucl. Instrum. Meth. **B49**, 533-43 (1990)

[6.68] F.A. Dilmanian et al.: Computed tomography with monochromatic X-rays from the National Synchrotron Light Source, Nucl. Instrum. Meth. **B56-57**, 1208-13 (1991)

[6.69] H.D. Zeman et al.: An X-ray monochromator for dual energy computerized tomography using synchrotron radiation, Nucl. Instrum. Meth. **B56/57**, 1218-22 (1991)

[6.70] Proc. of 4th Int. Conf. on Biophysics and Synchrotron Radiation (BSR92), Aug. 30 to Sept. 5 (1992), to be published

[6.71] Proc. of the 4th Int. Conf. on Synchrotron Radiation Instrum. (SRI91), Chester, July 15-19 (1991), to be published

[6.72] R.D. Vis, F. van Langevelde: On the development of X-ray microprobes using synchrotron radiation, Nucl. Instrum. Meth. **B54**, 417-23 (1991)

[7.1] D. Raparia, S. Machida: Compact linacs for positron emission tomography, 987-989, in Proc. of the 1989 IEEE Particle Acc. Conf., Chicago, March 20-23, 1989, IEEE Pub 88-647453 (1989)

[7.2] R.W. Hamm et al.: A compact proton linac for positron tomography, 141-143, 1986 Linear Acc. Conference, Report SLAC 303 (Sept. 1986)

[7.3] K. Fujii et al.: 20-70 MeV variable energy cyclotron for radio-isotope production, 626-29 in [7.7]

[7.4] J.J. Burgerjon: H^- cyclotrons for radioisotope production, Nucl. Instrum. Meth. **B10/11**, 951-56 (1985)

[7.5] M.F. Finlan et al.: A light, superconducting H^- cyclotron for medical diagnostic and neutron radiography, 689-92, in [7.7]

[7.6] R. Baartman et al.: A 30 Mev H^- cyclotron for isotope production, Report TRI-UMF TRI-PP-89-19, Vancouver (1989)

[7.7] M. Sekiguchi et al. (eds): Proc. of the 11th International Conf. on Cyclotrons and Their Applications, Tokyo, Oct. 13-17, 1986, Ionics Publishers Tokyo (1987)

[7.8] Proc. of the 12th International Conf. and Cyclotrons and Their Applications, Berlin, May 8-12, 1989, World Scientific, Singapore-New Jersey-London,Hong Kong (1991)

[7.9] Svedberg Laboratory: 25th European Cyclotron Progress Meeting, Uppsala (Sept. 21-24, 1988), Program + Abstract, Uppsala (1988)

[7.10] S.J. Heselius at al. (eds): Proc. of the Fifth Symposium on the Medical Application of Cyclotrons, Turku, Finland, May 31 — June 3, 1989, Acta Radiologica, Suppl. 376 (1991)

[7.11] M.A. Chaudri: The yield of cyclotron-produced radioisotopes for medical purposes, 614-21, in [7.7]

[7.12] S.J. Heselius: On the accelerator production of short-lived radionuclides — studies of density reduction in gas targets, Accelerators Laboratory, Abo Akademi, Abo (1986)

[7.13] N.A. Konjachin et al.: On ^{123}I production by irradiation of ^{124}Xe with the protons of mean energy, 542-44, in [7.8]

[7.14] P.V. Kulkarni: I-123 labeled radiotracers in cardiovascular nuclear medicine, IEEE Trans. Nucl. Sci. **NS-30**, 1809-12 (1983)

[7.15] M.C. Lagunas-Solar: Cyclotron production of no-carrier-added medical radionuclides, IEEE Trans. Nucl. Sci. **NS-30**, 1805-08 (1983)

[7.16] H.B. Hupf et al.: The cyclotron radionuclide program at King Faisal Specialist Hospital and Research Centre, Nucl. Instrum. Meth. **B10/11**, 967-68 (1985)

[7.17] P.V. Kulkarni et al.: Recent developments of cyclotron produced radionuclides for nuclear cardiology, Nucl. Instrum. Meth. **B24/25**, 928-31 (1987)

[7.18] P.V. Kulkarni: Single-photon emitting radiotracers produced by cyclotrons for myocardial imaging, Nucl. Instrum. Meth. **B40/41**, 1114-17 (1989)

[7.19] V. Bechtold et al.: Modern gas-target technology for the production of high quality radiopharmaceuticals, 519-22, in [7.8]

[7.20] G. Pinto et al.: Target systems for radioisotope production, IEEE Trans. Nucl. Sci. **NS-30**, 1797-1800 (1983)

[7.21] C. Vandecasteele, K. Strijckmans: Targetry for the cyclotron production of short-lived radionuclides for medical use, Nucl. Instrum. Meth. **A266**, 558-62 (1985)

[7.22] R.D. Hichwa et al.: Gas targetry for the production of ^{15}O, ^{11}C and ^{18}F for PET studies, Nucl. Instrum. Meth. **B24/25**, 932-36 (1987)

[7.23] N.A.M. van Bree et al.: Production of ^{18}F using static and circulating [^{18}O]-water targets, 531-534, in [7.8]

[7.24] R. Finn et al.: Modification and integration of JSW cyclotron gas targets at the National Institutes of Health Cyclotron Facility, Nucl. Instrum. Meth. **B24/25**, 954-56 (1987)

[7.25] M. Sajjad, R.M. Lambrecht: Cyclotron targetry for medical radioisotope production, Nucl. Instrum. Meth. **B40/41**, 1100-04 (1989)

[7.26] R.D. Hichwa et al.: Design of target systems for production of PET nuclides, Nucl. Instrum. Meth. **B40/41**, 1110-13 (1989)

[7.27] Positron Emission Tomography, Alan R. Liss, Inc. (1985)

[7.28] G. Muehllehner, J.S. Karp: Advances in SPECT and PET, IEEE Trans. Nucl. Sci. **35**, 639-43 (1988)

[7.29] S.E. Derenzo et al.: A positron tomograph with 600 BGO crystals and 2,6 mm resolution, IEEE Trans. Nucl. Sci. **35**, 659-64 (1988)

[7.30] T.J. Spinks et al.: Physical performance of the latest generation of commercial positron scanner, IEEE Trans. Nucl. Sci. **35**, 721-25 (1988)

[7.31] M. Tada et al.: Current and future aspects of cancer diagnosis with positron emission tomography, 597-601, in [7.7]

[7.32] R.M. Kessler et al.: Positron Emission Tomography — prospects for clinical utility, Investigate Radiology, **22**, 529-37 (1987)

[7.33] Y. Yano: Radioisotope generators for short-lived positron emitters applicable to positron emission tomography, Nucl. Instrum. Meth. **B40/41**, 1105-09 (1989)

[7.34] Y. Adachi et al.: In-house cyclotron and radioisoytopes production systems, 622-25, in [7.7]

[7.35] R.D. Hichwa et al.: Gas target for the production of ^{15}O, ^{11}C and ^{18}F for PET studies, Nucl. Instrum. Meth. **B24/25**, 932-36 (1987)

[7.36] S.J. Heselius et al.: An on-line for long-distance transport of ^{15}O labeled gases, Nucl. Instrum. Meth. **227**, 576-83 (1984)

[7.37] J.J. Burgerjon et al.: Rapid transfer of short-lived radioisotopes via a 2.4 km "rabbit" system, report TRIUMF TRI-PP-84-93, Vancouver (1984)

[7.38] Amersham Int.: The cyclotrons at Amersham and their use in nuclear medicine, First Reactions no. 8, Amersham (1987)

[7.39] B.F. Milton et al.: A 30 MeV H$^-$ cyclotron for isotope production, 145-48, in Proc. of the Twelfth Int. Conf. on Cyclotrons and Their Applications, Berlin, May 8-12, 1989, World Scientific, Singapore-New Jersey (1991)

[7.40] B.F. Milton et al.: Commissioning and first operation of a 500 A, 30 MeV, H$^-$ cyclotron: The TR30, TRIUMF Report TRI-PP-91-35, Vancouver (May 1991), also in Proc. of the 1991 IEEE Particle Acc. Conf., San Francisco, May 6-9, 1991 (to be published in 1992)

[7.41] M. Abs et al.: A new design of truly selfshielding baby-cyclotrons for positron emitter production, 675-77, in Proc. of the 1989 IEEE Particle Acc. Conf., Chicago, March 20-23, 1989, IEEE Pub 88-647453 (1989)

[7.42] J.L. Bol et al.: High intensity H$^-$ cyclotrons for radioisotope production, 149-51, in Proc. of the Twelfth Int. Conf. on Cyclotrons and Their Applications, Berlin, May 8-12, 1989, World Scientific, Singapore-New Jersey (1991)

[7.43] J.F. Lamb: Commercial use of university accelerators: industrial perspectives, Nucl. Instrum. Meth. **B40/41**, 1118-20 (1989)

[7.44] T.E. Boothe: Utilization of a hospital-based cyclotron for commercial radionuclide production, Nucl. Instrum. Meth. **B56/57**, 1266-69 (1991)

[7.45] G. Hör, R. Standke: Cyclotron produced isotopes for medical applications, 502-06, in [7.8]

[7.46] Sixth Symposium on the Medical Application of Cyclotrons, Turku, Finland, June 1-4 (1992)

[7.47] W. Jacobs et al.: Phys. Rev. **C9**, 2134-36 (1974)

[7.48] F. Tarkanyi et al.: Excitation functions of (p, 2n) and (p, pn) reactions and differential and integral yields of ^{123}I in proton induced nuclear reactions on highly enriched ^{124}Xe, Appl. Radiat. Isotopes **42**, 221-28 (1991)

[7.49] F. Tarkanyi et al.: Nuclear reaction cross sections relevant to the production of the ^{122}Xe ^{122}I generator system using highly enriched ^{124}Xe and a medium-sized cyclotron, Appl. Radiat. Isotopes **42**, 229-33 (1991)

[7.50] R.D. Finn et al.: Consequences of electroplated targets on radiopharmaceutical preparation, Nucl. Instrum. Meth. **B56/57**, 1284-86 (1991)

[7.51] H.L. Sharma et al.: Production of medium half-life positron emitting radionuclides, 16-17, in Abstracts of the Fifth Symp. on the Medical Applications of Cyclotrons, Turku, Finland, May 31 — June 3 (1989)

[7.52] R.D. Hichwa: Automation in PET center operation with special application to radiation monitoring, Nucl. Instrum. Meth. **B56/57**, 1205-07 (1991)

[7.53] J.R. Votaw et al.: Automatic, sterile and apyrogenic delivery of PET radiotracers from the cyclotron to the patient, Nucl. Instrum. Meth. **B56/57**, 1214-17 (1991)

[7.54] J.R. Votaw, R.J. Nickles: Radionuclide production for positron emission tomography: Choosing an appropriate accelerator, Nucl. Instrum. Meth. **B40/41**, 1093-99 (1989)

[7.55] I. Gál et al.: Beam spliting for paralell isotope production, 31-32, in [7.10]

[8.1] R.M. Brinston: Gamma sterilization of disposable medical products (DMP's), Proc. of the 19th Japan Conference on Radiation and Radioisotopes, Nov. 14,16, Tokyo (1989)

[8.2] J.W. Dorpema: Review and state of the art on radiation sterilization of medical devices, Radiat. Phys. Chem. **35**, 357-60 (1990)

[8.3] J. Beenen: Practical problems in radiation sterilization of medical devices made from plasticized PCV, Beta-Gamma, **4**, 12-14 (1989)

[8.4] Ch. Zalewski et al.: Der Einfluss der Strahlenbehandlung auf Arzneimittel und Hilfsstoffe. Eine Literaturstudie, Teil VIII, Institut für Strahlenhygiene, Neuherberg, FRG, Issue no. 121 (1988)

[8.5] J.H. Bly: Electron beam sterilization technology 1985 — an update, J. Indust. Irradiation Techn. **3**, 63-74 (1985)

[8.6] B.S. Johnston, C.C. Thompson: Electron beam sterilization using an integrated systems approach, Nucl. Instrum. Meth. **B24/25**, 982-84 (1987)

[8.7] J. McKeown et al.: An intense radiation source, Nucl. Instrum. Meth. **B10/11**, 846-50 (1985)

[8.8] J.P. Labrie et al.: AECL IMPELA electron beam industrial irradiators, Nucl. Instrum. Meth. **B40/41**, 1153-57 (1989)

[8.9] J. McKeown, S.L. Iverson: Recent development in high energy electron beam technology in Canada, Proc. of the 19th Japan Conference on Radiation and Isotopes, Nov. 14-16, Tokyo (1989)

[8.10] J.W. Barnard, F.W. Stanley: Startup of the Whiteshell irradiator facility, Nucl. Instrum. Meth. **B40/41**, 1158-61 (1989)

[8.11] T. Sadat: Two industrial linacs: Cassitron and CIRCE II, Nucl. Instrum. Meth. **B24/25**, 817-19 (1987)

[8.12] D.T. Tran: Application of electron linacs in medicine, food sterilization and synchrotron light sources, Nucl. Instrum. Meth. **B40/41**, 1350-61 (1989)

[8.13] T.Sadat, Ch. Cuillandre: Various applications of high energy electron beam, Report MeV Industrie, Jouy-en-Josas (1990)

[8.14] I.J. Rangwalla et al.: Electron beam sterilization and its applications to aseptic packaging, Pharmaceutical Technology **11**, 36-47 (1985)

[8.15] J.N. Aaronson, S.V. Nablo: An in-line electron sterilizer for container geometries to 250 ml, Radiat. Phys. Chem. **31**, 711-19 (1988)

[8.16] D.A. Cleghorn, S.V. Nablo: Electron sterilization validation techniques, Radiat. Phys. Chem. **35**, 352-89 (1990)

[8.17] D.A. Cleghorn, S.V. Nablo: Electron sterilization techniques for packaged labile products, Proc. of the PDA/PMA Meeting on Sterilization, Aug. 27-29, Washington (1990)

[8.18] M. Cleland, G.M. Pageau: Comparison of X-ray and gamma-ray sources for industrial irradiation processes, Nucl. Instrum. Meth. **B24/25**, 967-72 (1987)

[8.19] J. Farkas: Economic feasibility of food irradiation, Food Irr. Newsletter, **8**, 34-38 (1984)

[8.20] Anonym: Status of practical application of food irradiation, Food Irr. Newsletter, **12**, 42-43 (1988)

[8.21] Council for Agricultural Science and Technology: Ionizing energy in food processing and pest control, I. Wholesomeness

of food treated with ionizing energy, Report no. 109, Ames (1986)
[8.22] IAEA: Food processing by irradiation — World facts and trends, no. 5, Vienna (1989)
[8.23] List of clearances, Food Irr. Newsletter, **14**, Suppl. to no.1 (1990)
[8.24] FAO-IAEA, Vienna: Food Irradiation Newsletter (P.O. Box 100,, A-1400 Vienna, Austria)
[8.25] K.W. Bögl et al. (eds): Health impact, identification and dosimetry of irradiated food, Institut für Strahlenhygiene, ISH-Heft 125, Neuherberg (1988)
[8.26] L. Heide et al.: Thermolumineszenz — und Chemilumineszenzmessungen zur Identifizierung strahlenbehandelter Gewürze — Ein europäischer Ringversuch, ditto, ISH-Heft 130, Neuherberg (1989)
[8.27] Federal Research Centre for Nutrition: Bibliography for Food Irradiation (Engesserstr. 20, D-7500 Karlsruhe, FRG)
[8.28] J. McKeown, R.T. Jones: Important aspects of linac beams for food irradiation, Nucl. Instrum. Meth. **B24/25**, 976-81 (1987)
[8.29] T. Sadat, Ch. Cuillandre: A linear accelerator in a chicken factory, Food Irr. Newsletter, **12**, 61-62 (1988)
[8.30] A. Brynjolfsson: Future radiation sources and identification of irradiated foods, Food Technology, 84-90 (July 1989)
[8.31] M. Karel: The future of irradiation applications on earth and in space, Food Technology, 95-97 (1989)
[8.32] T. Sadat: Dual linear accelerator system for use in sterilization of medical supplies, Nucl. Instrum. Meth. **B56/57**, 1226-28 (1991)
[8.33] M.A. Welt: Food irradiation in the United States, Trichina-Safe Pork Task Force Meeting, National Pork Producers Council, Des Moines, Iowa (Nov. 1983)
[8.34] B.J. Smittle et al.: A new linear facility for the treatment of agricultural commodities, Nucl. Instrum. Meth. **B56/57**, 1229-31 (1991)
[8.35] G.P. Lawrence: High power linear accelerators for tritum production and transmutation of nuclear waste, Nucl. Instrum. Meth. **B56/57**, 1000-04 (1991)
[8.36] B.E. Fischer, R. Spohr: Teilchenspuren in der Mikrotechnik, Naturwissenschaften **75**, 57-66 (1988)
[8.37] R. Spohr: Ion Tracks and Microtechnology: Principles and Applications, Vieweg Verlag, Wiesbaden (1990)
[8.38] B.E. Fischer et al.: Facility for continuous irradiation of filter foils, 259, GSI Scientific Report 1990, GSI 01-1, Darmstadt (1990)
[8.39] B.E. Fischer: Single-particle techniques, Nucl. Instrum. Meth. **B51**, 401-06 (1991)
[8.40] D. Koutsouris et al.: Determination of red blood cell transit time through micropores, 256, in GSI Scientific Report 1989, GSI Darmstadt (1990)
[8.41] X. Hao et al.: Filtration of bacteria with Makrofol track microfilters, 257, in GSI Scientific Report 1989, GSI Darmstadt (1990)

[9.1] J.S.C. McKee et al.: The Manitoba automated high-energy proton microprobe and its application to research in the geological and biological sciences, Nucl. Instrum. Meth. **B40/41**, 680-84 (1989)
[9.2] F. Watt, G.W. Grime: Principles and Applications of High-Energy Ion Microbeams, A. Hilger, Bristol (1987)
[9.3] R.D. Vis: The Proton Microprobe: Applications in Biomedical Field, CRC Press, Boca Raton, Florida (1986)
[9.4] E. Fuchs et al.: Particle beam microanalysis — Fundamentals, methods and applications, VCH, Weinheim (1991)
[9.5] G.W. Grime, F. Watt (eds): Proc. 1st Int. Conf. on Nuclear Microprobe technology and Appl., Nucl. Instrum. Meth. **B30** (1988)
[9.6] G.J.G. Legge, D.N. Jamieson (eds): Proc. of the Second Int. Conf. on Nuclear Microprobe Technology and Appl., Melbourne, Feb. 6-9, 1990, Nucl. Instrum. Meth. **B54**, 1-441 (1991)
[9.7] G.W. Grime et al.: Submicron elemental mapping with the Oxford scanning proton microprobe, Nucl. Instrum. Meth. **B22**, 109-14 (1987)
[9.8] F.W. Martin: Optical parameters of MeV microprobes, Nucl. Instrum. Meth. **B54**, 17-23 (1991)
[9.9] F.W. Martin, R. Goloskie: An achromatic proton microbeam of 1.06 MeV energy and 0.1 m linewidth, Nucl. Instrum. Meth. **B54**, 64-67 (1991)
[9.10] U.A.S. Tapper: Achromatic focusing for probeforming of high resolution microbeams, Nucl. Instrum. Meth. **B56/57**, 712-16 (1991)
[9.11] G.W. Grime et al.: The Oxford submicron microscopy facility, Nucl. Instrum. Meth. **B54**, 52-63 (1991)
[9.12] A. Saint et al.: High resolution techniques for the scanning proton microprobe, Nucl. Instrum. Meth. **B40/41**, 685-90 (1989)
[9.13] P.M. O'Brien et al.: The scanning pro-

ton microprobe and the elemental microanalysis of haemopoietic cells, Australian Phys. Engn. Sci. in Medicine **5**, 30-34 (1982)
[9.14] G.J.F. Legge et al.: The design of a versatile scanning proton microprobe of high resolution and efficiency, Nucl. Instrum. meth. **B30**, 252-59 (1988)
[9.15] G.J.F. Legge: The proton (nuclear) microprobe, Nucl. Instrum. Meth. **B40/41**, 675-79 (1989)
[9.16] G.J.F. Legge, A. Saint: Microprobe techniques and instrumentation, Nucl. Instrum. Meth. **B49**, 418-24 (1990)
[9.17] U. Lindh: Micron and submicron nuclear probes in biomedicine, Nucl. Instrum. Meth. **B49**, 451-64 (1990)
[9.18] A.J.J. Boss et al.: On the incorporation of trace elements into human hair measured with micro-PIXE, Nucl. Instrum. Meth. **B3**, 654-59 (1984)
[9.19] G.J.F. Legge et al.: Microbeam imaging at micron and submicron resolution, Nucl. Instrum. Meth. **B15**, 669-74 (1986)
[9.20] P.M. O'Brien, G.J.F. Legge: High speed acquisition and handling of scanning proton microprobe data, Nucl. Instrum. Meth. **B30**, 312-16 (1988)
[9.21] G.R. Moloney et al.: Ion-optical review of the new Melbourne two-stage microprobe, Nucl. Instrum. Meth. **B54**, 68-70 (1991)
[9.22] Y. Llabador et al.: Advantages of high speed scanning for microprobe analysis of biological samples, Nucl. Iinstrum. Meth. **B49**, 435-40 (1990)
[9.23] U.A.S. Tapper et al.: The Lund nuclear microprobe: Instrumentation data collection and data evaluation, Nucl. Instrum. Meth. **B49**, 425-29 (1990)
[9.24] J.L. Campbell et al.: Practical problems with a proton probe, Nucl. Instrum. Meth. **B56/57**, 694-98 (1991)
[9.25] R. Brenner et al.: The PIXE-facility in Turku, 61-65, in M. Brenner (ed): The Abo Akademi Accelerator Laboratory Triennial Report 1987-1989, Abo (1990)
[9.26] U. Lindh: Biomedical application of the nuclear microprobe, Nucl. Instrum. Meth. **B24/25**, 638-42 (1987)
[9.27] P.M. Read et al.: Beam developments for the Harwell microprobe system, Nucl. Instrum. Meth. **B24/25**, 627-42 (1987)
[9.28] R.P. Chaturvedi, H. Bakhru: Application of microbeam analysis of biomedical samples, Nucl. Instrum. Meth. **B40/41**, 702-04 (1989)
[9.29] H. Brückmann et al.: The microprobe at the Hamburg isochronous cyclotron, 719-21, in Proc. of 9th Int. Conf. on Cyclotron and Their Application, Ionics, Tokyo (1987)
[9.30] P.H.A. Mutsaers et al.: Improvements of the Eindhoven proton microbeam, Nucl. Instrum. Meth. **B24/25**, 648-51 (1987)
[9.31] Y.H. Yeo et al.: Commissioning and characterization of the Manitoba automated high energy proton microprobe, Nucl. Instrum. Meth. **B49**, 430-34 (1990)
[9.32] U. Lindh: A new standard for nuclear microscopy in Uppsala, Uppsala Accelerator News, no. 8, p.4 (1990)
[9.33] G.W. Grime, F. Watt: Focusing high-energy ion beams, 79-112, in [9.2]
[9.34] G.W. Grime, F. Watt: Practical aspects of heavy-ion microbeams, 113-53, in [9.2]
[9.35] D.J.T. Vaux et al.: Microbeam applications in medicine, 203-37, in [9.2]
[9.36] W. Maenhaut et al.: Comparison of different sample and target preparation procedures for PIXE analysis of biological materials, Nucl. Instrum. Meth. **B3**, 135-40 (1984)
[9.37] R.D. Vis et al.: The proton microprobe used for trace element mapping of biological systems, Nucl. Instrum. Meth. **B10/11**, 683-86 (1985)
[9.38] K.G. Malmqvist: Quantitative PIXE analysis of biomedical material — sample preparation, irradiation and quality control, Nucl. Instrum. Meth. **B49**, 183-90 (1990)
[9.39] M. Pakarinen: Preparation of biomedical samples for PIXE, Master thesis, Turku, Finnland (1986)
[9.40] W. Mertz (ed): Trace elements in Human and Animal Nutrition, 5th ed., vol. 1, Academic Press, New York (1986)
[9.41] J.A. Cookson: Analytical techniques, 21-78, in [9.2]
[9.42] J.A. Cookson: The nuclear microprobe and competing techniques, Nucl. Instrum. Meth. **B54**, 433-40 (1991)
[9.43] B.J. Kirby, G.J.G. Legge: Ion beam induced damage and element loss during a microanalysis of biological tissue, Nucl. Instrum. Meth. **B54**, 98-100 (1991)
[9.44] V. Valkovic: Application of nuclear analytical techniques in the study of trace element — role in biology and medicine, Nucl. Instrum. Meth. **B40/41**, 848-52 (1989)
[9.45] G.J.F. Legge: Proton microscopy and microanalysis — biological applications, Ultramicroscopy **24**, 283-98 (1988)
[9.46] B. Gonsior et al.: Application of a proton microprobe to biomedical research — limits and potentialities, Biol. Trace Element Res. **12**, 33-44 (1987)
[9.47] T. Cichocki et al.: Measurements of mineralization process in the femur growth

plate and rib cartilage of the mouse using PIXE in combination with a proton microprobe, Histochemistry, **89**, 99-104 (1988)

[9.48] T. Cichocki et al.: Intramembranaceous ossification analyses by a proton microprobe, Histochemistry 94, 171-77 (1990)

[9.49] E. Rokita et al.: Proton microprobe studies of the mineralization process in selected organic matrices, Nucl. Instrum. Meth. **B50**, 217-20 (1990)

[9.50] I. Borbely-Kiss et al.: The effect of radiotherapeutic on the elemental concentration of iron in human erythrocytes and blood plasma, Nucl. Instrum. Meth. **B49**, 254-60 (1990)

[9.51] G. Weber et al.: Trace elements in human alveolar macrophages studied by PIXE, Nucl. Instrum. Meth. **B49**, 250-53 (1990)

[9.52] T. Tadic et al.: Diffusion of metals from crown into teeth, Nucl. Instrum. Meth. **B49**, 211-15 (1990)

[9.53] T. Pinhero et al.: Brain trace element alterations in atherosclerosis, Nucl. Instrum. Meth. **B49**, 191-94 (1990)

[9.54] W. Maenhant et al.: Study of the respirable immission levels for a cyclist in Brussels trafic using PIXE as analytical technique, Nucl. Instrum. Meth. **B49**, 406-13 (1990)

[9.55] Ch. Engelmann: A complementary technique to PIXE: The nuclear interaction analysis method — advantages possibilities and typical application examples, Nucl. Instrum. Meth. **B49**, 33-38 (1990)

[9.56] J.R. Bird: Total analysis by IBA, Nucl. Instrum. Meth. **B45**, 516-18 (1990)

[9.57] A. Riedel: Accelerator mass spectrometry ^{14}C dating and its application to small shell samples, Report TLU 136/86, Tandem Acc. Lab., Uppsala (1986)

[9.58] S. Brüggerhoff, E. Jackwerth: Beitrag zur Präparation von Spurenkonzentrat — Targets für die PIXE-Multielementalanalyse, Fresenius Z. Anal. Chem. **311**, 252-58 (1982)

[9.59] B. Raith et al.: Forschungsbericht des Landes Nordrhein-Westfalen no. 2982, Westdeutscher Verlag (1980)

[9.60] P. Horowitz et al.: Elemental analysis of biological specimens in air with a proton microprobe, Science **194**, 1162-65 (1976)

[9.61] F. Watt, G.W. Grime: Microbeam application in biology, 154-202, in [9.2]

[9.62] W. Lenglet: On distribution of trace elements in biological material — studies based on the use of an accelerator, Ph. D. Thesis, Vrije Universiteit, Amsterdam (1988)

[9.63] R.M. Wheeler et al.: Trace elements in human nasal cavity bones, Nucl. Instrum. Meth. **B24/25**, 658-61 (1987)

[9.64] R.D. Vis, M. Prins: Trace element analysis in biological specimens, Chem. Br. **22**, 851-52 (1986)

[9.65] E. Rokita et al.: Micro-PIXE quantification of histochemical reactions, Nucl. Instrum. Meth. **B49**, 441-45 (1990)

[9.66] M. Pakarinen et al.: Elemental analysis of long bones of infants from Ficana excavations, Nucl. Instrum. Meth. **B49**, 241-44 (1990)

[9.67] J.S.C. McKee et al.: Micro- and milli-PIXE analysis at 40 MeV — why, how and when?, Nucl. Instrum. Meth. **B45**, 513-515 (1990)

[9.68] G. Demortier et al.: Complementarity of RBS, PIGE and PIXE for the determination of surface layers of thicknesses up to 30 microns, Nucl. Instrum. Meth., **B49**, 46-51 (1990)

[9.69] M. Hult, K. Themner: Quantitative microanalysis of matrix elements in biological samples by MeV proton scattering, Nucl. Instrum. Meth. **B50**, 154-58 (1990)

[9.70] D. Heck et al.: Trace element metabolism in liver cells after copper sulphate uptake, Nucl. Instrum. Meth. **B30**, 420-23 (1988)

[9.71] D. Heck et al.: Trace element distribution in human liver lobules determined with micro-PIXE, 367-74, in P. Brätter, P. Schramed (eds): Trace Element — Analytical Chemistry in Med. Biol. **4**, Walter de Gruyter Co., Berlin (1987)

[9.72] J. Pallon, U.A.S. Tapper: Imaging of thin biological sections using STIM and related techniques — a comparative study, Nucl. Instrum. Meth. **B56/57**, 1270-73 (1990)

[9.73] G.W. Grime, F. Watt: Nuclear microscopy — elemental mapping using high-energy ion beam technique, Nucl. Instrum. Meth. **B50**, 197-207 (1990)

[9.74] U. Lindh: Nuclear microscopy. Its role and future in medicne and trace-element biology, Nucl. Instrum. Meth. **B54**, 160-70 (1991)

[9.75] F. Watt et al.: Nuclear microscopy of biological specimens, Nucl. Instrum. Meth. **B54**, 123-43 (1991)

[9.76] M. Suter: Accelerator mass spectrometry: state of the art in 1990, Nucl. Instrum. Meth. **B52**, 211-23 (1990)

[9.77] W. Kutschera: Accelerator mass spectrometry — a versatile tool for research, Nucl. Instrum. Meth. **B50**, 252-61 (1990)

[9.78] J.S. Vogel et al.: Application of AMS to

the biomedical sciences, Nucl. Instrum. Meth. **B52**, 524-30 (1990)

[9.79] S.P.H.T. Freeman et al.: The design of a radiocarbon microprobe for tracer mapping in biological specimens, Nucl. Instrum. Meth. **B52**, 405-09 (1990)

[9.80] A.T. Wilson, D.J. Donahue: AMS carbon dating of ice: progress and future prospects, Nucl. Instrum. Meth. **B52**, 473-76 (1990)

[9.81] W.E. Kieser et al.: Accelerator mass spectrometry at Toronto, Nucl. Instrum. meth. **B24/25**, 667-71 (1987)

[9.82] G. Possnert: Accelerator mass spectrometry (AMS) at the Uppsala tandem laboratory, Uppsala Accelerator News no. 6, 7 (1988)

[9.83] M.M. Fowler et al.: The Accelerator-based mass spectrometry of methane, Nucl. Instrum. Meth. **B24/25**, 672-75 (1987)

[9.84] C. Tumiz et al.: Probing the environment with accelerator-based techniques, Nucl. Instrum. Meth. **B56/57**, 877-81 (1991)

[9.85] J.S. Felton: Accelerator mass spectrometry in the biomedical sciences: applications in low-exposure biomedical and environmental dosimetry

[9.86] I.D. Proctor: The AMS program at LLNL, Nucl. Instrum. Meth. **B40/41**, 727-30 (1989)

[9.87] J.C. Davis et al.: LLNL/UC AMS facility and research program, Nucl. Instrum. Meth. **B52**, 269-72 (1990)

[9.88] J. Barker et al.: Development of ^{26}Al accelerator mass spectrometry for biological and toxicological applications, Nucl. Instrum. Meth. **B52**, 540-43 (1990)

[9.89] A. Meirav et al.: Application of accelerator mass spectrometry in aluminium metabolism studies, Nucl. Instrum. Meth. **B52**, 536-39 (1990)

[9.90] D. Elmore et al.: Calcium 41 as a long-term biological tracer for bone resorption, Nucl. Instrum. Meth. **B52**, 531-35 (1990)

[9.91] J.H. Weinlein et al.: A pulsed neutron generator for in vivo body composition studies, Nucl. Instrum. Meth. **B56/57**, 904-06 (1991)

[9.92] J.J. Kahayias et al.: Use of a high repetition rate neutron generator for in vivo body composition measurements, Nucl. Instrum. Meth. **B24/25**, 1006-09 (1987)

[9.93] R. Pepelnik: Sensitivities of high-flux 14 MeV neutron activation analysis, Nucl. Instrum. Meth. **B40/41**, 1205-07 (1989)

[9.94] Y. Mao et al.: Simultaneous determination of P and N in human tissue by 14 MeV neutron activation analysis, Nucl. Instrum. Meth. **B24/25**, 1003-05 (1987)

[9.95] H.H. Rossi: The Radiological Research Accelerator Facility at Columbia University, Nucl. Instrum. Meth. **B10/11**, 1086-89 (1985)

[9.96] H.E. Zschau et al.: Fluorine profiling after application of various anti-caries gels, Nucl. Instrum. Meth. **B50**, 74-76 (1990)

[9.97] F.T. Cua: PIXE-PIGE analysis of teeth from children with and without cystic fibrosis, Nucl. Instrum. Meth. **B49**, 205-09 (1990)

[9.98] D.D. Cohen et al.: Determination of ^{18}O concentrations in microsamples of biological fluids, Nucl. Instrum. Meth. **B50**, 43-47 (1990)

[9.99] W. Neumann et al.: Thin-layer activation of hip-joint prosthese for tribological tests, Nucl. Instrum. Meth. **B50**, 57-61 (1990)

[9.100] J. Barker et al.: Aluminium absorption in humans using ^{26}Al isotopic tracer, 98-99, in K.A. Connell, M.A. Bentley (eds): Nuclear Physics, Appendix to the Daresbury Annual Rep. 1990/91, SERC Daresbury Laboratory, Warrington (1991)

[9.101] RARAF User's Guide, Radiological Research Accelerator Facility, Columbia University, Irvington (1985)

[9.102] E.J. Hall, P. Goldhagen: The Radiological Research Accelerator Facility, Progress Report DOE/ER/60439-2, Nevis Laboratories, Columbia University, Irvington (1987)

[9.103] Ch. R. Geard: Chromosamal aberrations per charged particle per cell nucleus, Nucl. Instrum. Meth. **B40/41**, 1372-75 (1989)

[9.104] L.A. Braby, W.D. Reece: Using charged-particle microbeams to study mechanisms of carcinogenesis, Nucl. Instrum. Meth. **B40/41**, 1284-86 (1989)

[9.105] CFKR — Center for fast Kinetics Research The University of Texas at Austin, Austin (1990)

[9.106] C.R. Geard et al.: Single-particle irradiation of mammalian cells at the RARAF: induction of chromosomal changes, Nucl. Instrum. Meth. **B54**, 411-16 (1991)

List of Firms

[F1] High Voltage Engineering Corp., F.C.Box 416, South Bedford Street, Burlington, Mass. 01803, USA

[F2] National Electrostatics Corp. (Electrostatics International Inc.) Graber Road,Box 310, Middleton, Wisconsin 53562, USA

[F3] Emil Haefely Co, Ltd., Lehenmattstr. 353, CH-4028 Basel, Switzerland

[F4] Nissin,High Voltage Corp., Ukyo-ku, Kyoto 615, Japan

[F5] Radiation Dynamics Inc., 316 So.Service Road, Melville, Long Island, N.Y. 11747, USA

[F6] High Voltage Engineering Europe B.V., Amsterdamseweg 61, Amersfoort, The Netherlands

[F7] Varian Radiation Division, 611 Hansen Way, Palo Alto, California, USA

[F8] Siemens Medical Laboratories, 2404 N.Main Street, Walnut Creek, CA 94596, USA

[F9] Atomic Energy of Canada Ltd. — Medical Products, P.O.Box 13140, 413 March Road, Kanata, Ontario, Canada K2K 2B7

[F10] Philips Medical Systems, Eidhoven, The Netherlands

[F11] ABB — Asea Brown Boveri Ltd., CH — 5401, Baden, Switzerland

[F12] Nuson A/S, Stamholmen 153, DK-2650 Hvidovre, Danemark

[F13] Gammex, Inc.Patient Positioning Systems, 6685 N.Sidney Place, Milwaukee, Wisconsin 53209, USA

[F14] Scanditronix AB, Husbyborg, S-75229 Uppsala, Sweden

[F15] Instrument AB Therados, P.O.Box 15027, S-75015 Uppsala, Sweden

[F16] Radiation Measurements Inc., Middleton, WI 53562, USA

[F17] PTW-Freiburg, Loerracherstr.7, D-7800 Freiburg, Germany

[F18] ACCTEK Associates, Inc., 901 S.Kengiston, Lagrange, IL 60525, USA

[F19] Product Genesis Inc, 300 Bent Street, Suite 200, Cambridge, MA 02141, USA

[F20] AccSys Technology Inc. 1177 Quarry Lane, Pleasanton, CA 94566, USA

[F21] Mitsubishi Electric Corp., Mitsubishi Denki Bldg., Marunouchi, Tokyo 100, Japan

[F22] D.V. Efremov Scientific Research Institute of Electrophysical Apparatus, 188631 Sankt Petersburg (Leningrad), USSR

[F23] Ion Beam Applications SA, Chemin du Cyclotron 2, B-1348 Louvain-la-Neuve, Belgium

[F24] Cyclopore, Louvain-la-Neuve, Scientific Park, Louvain-la-Neuve, Belgium

[F25] Sumitomo Heavy Industries, Ltd., Niihama-shi, Ehime 792, Japan

[F26] CGR-MeV, 78530 Buc, France

[F27] Computer Technology and Imaging, 810 Innovation Dr., Knoxville, TN 37922, USA

[F28] Oxford Instruments Ltd, Osney Mead, Oxford OX2 ODX, England

[F29] Amersham International PLC, Amersham Laboratories, Amersham, Buckinghamshire, HP7 9Il, England

[F30] Atomic Energy of Canada Ltd. — Accelerator Business Unit, 436B Hazeldean Road, Kanata, Canada K2L 1T9

[F31] MeV Industrie, 21 rue Albert Calmette, 78353 Jony-en-Josas, France

[F32] EBCO Technologies Ltd., 7851 Alderbridge Way, Richmond, B.C., Canada V6X 2A4

[F33] Digital Equipment Corporation, P.O. Box. 9501, Merrimack, NH 03054, USA

[F34] Apple Computer, Inc., 20525 Mariani Ave., Cupertino, CA 95014, USA

[F35] STELLA, High Performance Systems, 13 Dartmouth College Highway, Lyme, NH 03768, USA

[F36] Nuclepore Corp., 7035 Commerce Circle, Pleasanton, CA 84566, USA, also Nuclepore GmbH, Falkenweg 47, 74 Tübingen, Germany

Index

A

B

C

D

E

F

H

I

K

L

M

N

O

P

Q

R

S

T

U

V

W

X

Z